Computational Differential Equations Using MATLAB®

Second Edition

Textbooks in Mathematics

Series editors:
Al Boggess and Ken Rosen

CRYPTOGRAPHY: THEORY AND PRACTICE, FOURTH EDITION

Douglas R. Stinson and Maura B. Paterson

GRAPH THEORY AND ITS APPLICATIONS, THIRD EDITION

Jonathan L. Gross, Jay Yellen and Mark Anderson

A FIRST COURSE IN FUZZY LOGIC, FOURTH EDITION

Hung T. Nguyen, Carol L. Walker, and Elbert A. Walker

EXPLORING LINEAR ALGEBRA

Crista Arangala

A TRANSITION TO PROOF: AN INTRODUCTION TO ADVANCED MATHEMATICS

Neil R. Nicholson

COMPLEX VARIABLES: A PHYSICAL APPROACH WITH APPLICATIONS, SECOND EDITION

Steven G. Krantz

GAME THEORY: A MODELING APPROACH

Richard Alan Gillman and David Housman

FORMAL METHODS IN COMPUTER SCIENCE

Jiacun Wang and William Tepfenhart

AN ELEMENTARY TRANSITION TO ABSTRACT MATHEMATICS

Gove Effinger and Gary L. Mullen

ORDINARY DIFFERENTIAL EQUATIONS: AN INTRODUCTION TO THE FUNDAMENTALS, SECOND EDITION

Kenneth B. Howell

SPHERICAL GEOMETRY AND ITS APPLICATIONS

Marshall A. Whittlesey

COMPUTATIONAL PARTIAL DIFFERENTIAL EQUATIONS USING MATLAB®, SECOND EDITION

Jichun Li and Yi-Tung Chen

https://www.crcpress.com/Textbooks-in-Mathematics/book-series/CANDHTEXBOOMTH

Computational Partial Differential Equations Using MATLAB®

Second Edition

Jichun Li
Yi-Tung Chen

CRC Press
Taylor & Francis Group
Boca Raton London New York

CRC Press is an imprint of the
Taylor & Francis Group, an **informa** business

A CHAPMAN & HALL BOOK

MATLAB® is a trademark of The MathWorks, Inc. and is used with permission. The MathWorks does not warrant the accuracy of the text or exercises in this book. This book's use or discussion of MATLAB® software or related products does not constitute endorsement or sponsorship by The MathWorks of a particular pedagogical approach or particular use of the MATLAB® software.

CRC Press
Taylor & Francis Group
6000 Broken Sound Parkway NW, Suite 300
Boca Raton, FL 33487-2742

First issued in paperback 2022

ISBN 13: 978-1-03-247519-6 (pbk)
ISBN 13: 978-0-367-21774-7 (hbk)

DOI: 10.1201/9780429266027

Visit the Taylor & Francis Web site at
http://www.taylorandfrancis.com

and the CRC Press Web site at
http://www.crcpress.com

Contents

Preface

The purpose of this book is to provide a quick but solid introduction to advanced numerical methods for solving various partial differential equations (PDEs) in science and engineering. The numerical methods covered in this book include not only the classic finite difference and finite element methods, but also some recently developed meshless methods, high-order compact difference methods, and finite element methods for Maxwell's equations in complex media.

This book is based on the material that we have taught in our numerical analysis courses, MAT 665/666 and MAT 765/766, at the University of Nevada Las Vegas since 2003. The emphasis of the text is on both mathematical theory and practical implementation of the numerical methods. We have tried to keep the mathematics accessible for a broad audience, while still presenting the results as rigorously as possible.

This book covers three types of numerical methods for PDEs: the finite difference method, the finite element method, and the meshless method. In Chapter 1, we provide a brief overview of some interesting PDEs coming from different areas and a short review of numerical methods for PDEs. Then we introduce the finite difference methods for solving parabolic, hyperbolic, and elliptic equations in Chapters 2, 3, and 4, respectively. Chapter 5 presents the high-order compact difference method, which is quite popular for solving time-dependent wave propagation problems. Chapters 6 through 9 cover the finite element method. In Chapter 6, fundamental finite element theory is introduced, while in Chapter 7, basic finite element programming techniques are presented. Then in Chapter 8, we extend the discussion to the mixed finite element method. Here both theoretical analysis and programming implementation are introduced. In Chapter 9, we focus on some special finite element methods for solving Maxwell's equations, where some newly developed algorithms and Maxwell's equations in dispersive media are presented. Chapter 10 is devoted to the radial basis function meshless methods developed in recent years. Some Galerkin-type meshless methods are introduced in Chapter 11.

The book is intended to be an advanced textbook on numerical methods applied to diverse PDEs such as elliptic, parabolic, and hyperbolic equations. Each chapter includes about 10 exercises for readers to practice and enhance their understanding of the materials. This book supplies many MATLAB® source codes, which hopefully will help readers better understand the presented numerical methods. We want to emphasize that our goal is to provide readers with simple and clear implementations instead of sophisticated usages of MATLAB functions. The skilled reader should be able to easily modify or improve the codes to solve similar problems of his or her interest.

This book can be used for a two-semester graduate course that provides an introduction to numerical methods for partial differential equations. The first semester can cover the elementary chapters such as Chapters 1 through 5. This part can also be used at the undergraduate level as a one-semester introductory course on numerical methods or scientific computing. The rest of the chapters are more advanced and can be used for the second semester or a stand-alone advanced numerical analysis course.

MATLAB is a registered trademark of The MathWorks, Inc. For product information, please contact:

The MathWorks, Inc.
1 Apple Hill Drive
Natick, MA 01760-2098 USA
Tel: 508-647-7000
Fax: 508-647-7001
Email: info@mathworks.com
Web: www.mathworks.com

New to this edition:

In this edition, we first corrected those typos in the first edition. Then we provided updates to recent textbooks on related subjects, added more exercises, and expanded some sections, such as a detailed finite element analysis of parabolic equations, and a new section on the popular time-domain finite difference method for Maxwell's equations. Of course, we apologize for any unavoidable omissions and potential typos.

Acknowledgments

We would like to thank Professor Goong Chen for his kind support in accepting our book into this book series. We also want to thank Shashi Kumar, Karen Simon, and Bob Stern for their kind help during the first edition production process, and Bob Ross for his kind help during the second edition production process.

University of Nevada Las Vegas	Jichun Li
University of Nevada Las Vegas	Yi-Tung Chen

I am very grateful for the mathematics department's kind support of my numerical PDE courses, and to those students suffering during my lectures over the years, which made this book possible. This work has benefited greatly from many collaborations with my colleagues over the years. Very special thanks go to Professor Qun Lin who introduced me to the finite element superconvergence world; to Professor Mary Wheeler who led me to the wonderful parallel computing and multidisciplinary research world; to Professors C.S. Chen and Benny Hon who enlightened me on the meshless method; and to Dr. Miguel Visbal who directed me to the compact difference method.

Last, but not least, I want to express my gratitude to my wife, Tao, and our children, Jessica and David, for their love, help, and unwavering support over the years. Finally, I dedicate this book to the memory of my parents, Shu-Zhen Zhao and Hao-Sheng Li.

Jichun Li

This book is dedicated to my wife, Yulien, and our children, Jessica and Jonathan, for their endless love and support.

Yi-Tung Chen

1

Brief Overview of Partial Differential Equations

Generally speaking, partial differential equations (PDEs) are equations containing some partial derivatives of an unknown function u with respect to spatial variables such as $\boldsymbol{x} = (x_1, \cdots, x_d)' \in R^d, d \geq 1$, and sometimes with respect to time variable t.

It turns out that many problems can be modeled by PDEs from various science and engineering applications. Furthermore, most problems cannot be solved analytically, hence finding good approximate solutions using numerical methods will be very helpful. Learning various numerical methods for solving PDEs becomes more and more important for students majoring in science and engineering. These days we are immersed with many "computational" words such as "computational mathematics", "computational physics", "computational chemistry", "computational biology", "computational geosciences", "computational fluid mechanics", "computational electromagnetics", etc. In this chapter, we first review some interesting PDEs to motivate our readers. Then we provide a very quick review of several popular numerical methods for solving differential equations.

1.1 The parabolic equations

The one-dimensional (1-D) differential equation

$$\frac{\partial u}{\partial t} = \alpha^2 \frac{\partial^2 u}{\partial x^2} \quad \text{or} \quad u_t = \alpha^2 u_{xx}, \quad 0 < x < L, \tag{1.1}$$

is a standard 1-D parabolic equation. This equation can be used to model many physical phenomena such as heat distribution in a rod, in which case $u(x, t)$ represents the temperature at point x and time t, and α^2 is the thermal diffusivity of the material. Hence (1.1) is also called heat/diffusion equation. Furthermore, if there is an additional source/sink along the rod, given by $f(x, t)$, then we end up with an inhomogeneous heat equation

$$u_t = \alpha^2 u_{xx} + f(x, t). \tag{1.2}$$

If the material property is not uniform along the rod (e.g., density or specific heat depends on the location x), then the coefficient α will be a function of x. In the case that the thermal conductivity K depends on x, then the heat equation becomes

$$u_t = \frac{\partial}{\partial x}\left(K(x)\frac{\partial u}{\partial x}\right). \tag{1.3}$$

Of course, to make the heat equation well posed, we must supply an initial condition and proper boundary conditions at both ends of the rod. Similarly, 2-D and 3-D heat equations can be derived, which are often written as

$$u_t = \alpha^2(u_{xx} + u_{yy} + u_{zz}) + f, \tag{1.4}$$

and

$$u_t = (au_x)_x + (bu_y)_y + (cu_z)_z + f. \tag{1.5}$$

1.2   The wave equations

The second-order hyperbolic differential equation

$$u_{tt} = c^2 u_{xx}, \quad 0 < x < L, \tag{1.6}$$

is often called the wave equation. Equation (1.6) can be used to describe the vibration of a perfectly flexible string when it is nailed down at both ends, pulled, and let go. In that case, u denotes the displacement of the string.

If some vertical force f is imposed at point x, then we have the inhomogeneous wave equation

$$u_{tt} = c^2 u_{xx} + f(x, t). \tag{1.7}$$

If the string has non-uniform density, then the coefficient c^2 becomes a function of x. To make the wave equation well posed, we need boundary conditions at each end of the string and initial conditions for both u and u_t at each point of the string.

The two-dimensional wave equation

$$u_{tt} = c^2(u_{xx} + u_{yy}) + f(x, y, t) \tag{1.8}$$

can be used to describe the vibration of a membrane, where $u(x, y, t)$ is the height of the membrane at point (x, y) at time t, and $f(x, y, t)$ represents the load effect.

The three-dimensional wave equation

$$u_{tt} = c^2(u_{xx} + u_{yy} + u_{zz}) + f(x, y, z, t) \tag{1.9}$$

can be obtained when we model the propagation of waves in 3-D media. Examples include the propagation of elastic waves in solids and electromagnetic waves.

1.3 The elliptic equations

The second-order elliptic equations are obtained as the steady state solutions (as $t \to \infty$) of the parabolic and wave equations. For example, a general Poisson's equation results in electrostatics

$$(\epsilon u_x)_x + (\epsilon u_y)_y = -\rho, \tag{1.10}$$

where the permittivity ϵ may vary with position and the given charge distribution ρ can vary with position too. When ϵ is a constant, we get the so-called Poisson's equation

$$u_{xx} + u_{yy} = -f(x, y). \tag{1.11}$$

A further simplified form of Poisson's equation with homogeneous right-hand side (i.e., $f = 0$) is the so-called Laplace equation

$$u_{xx} + u_{yy} = 0. \tag{1.12}$$

Solutions of (1.10) are important in many fields of science, such as electromagnetics, astronomy, heat transfer, and fluid mechanics, because they may represent a temperature, electric or magnetic potential, and displacement for an elastic membrane. In three dimensions, we have the corresponding Poisson and Laplace equations.

1.4 Differential equations in broader areas

In this section, we show readers a few more interesting PDEs from different disciplines.

1.4.1 Electromagnetics

With the advancement of modern technology, we live in a world surrounded by electromagnetic waves. For example, the TV programs we watch daily are generated and broadcast by waves propagating in the air; we warm our food using a microwave oven, which warms the food by generating microwaves; ultrasound wave propagation plays a very important role in medical applications, such as hypothermia therapy, detection of tumors and bond density; we talk to our friends by mobile phones, whose signals are transported as ultra-short wave pulses through optical fibers; we build "invisible planes" by absorbing the detecting waves sent by enemies. Hence, the study of electromagnetic waves interacting with different materials is an unavoidable important issue in our daily life and in national defense.

The fundamental equations in describing the electromagnetic phenomena are the Maxwell equations:

$$\text{Faraday's law: } \nabla \times \boldsymbol{E} = -\frac{\partial \boldsymbol{B}}{\partial t},$$

$$\text{Maxwell-Ampere's law: } \nabla \times \boldsymbol{H} = \frac{\partial \boldsymbol{D}}{\partial t},$$

$$\text{Gauss's law: } \nabla \cdot \boldsymbol{D} = \rho,$$

$$\text{Gauss's law - magnetic: } \nabla \cdot \boldsymbol{B} = 0,$$

where $\boldsymbol{E}(\boldsymbol{x}, t)$ and $\boldsymbol{H}(\boldsymbol{x}, t)$ are the electric and magnetic fields, $\boldsymbol{D}(\boldsymbol{x}, t)$ and $\boldsymbol{B}(\boldsymbol{x}, t)$ are the corresponding electric and magnetic flux densities, ρ is the electric charge density. Furthermore, $\boldsymbol{D}$ and $\boldsymbol{B}$ are related to $\boldsymbol{E}$ and $\boldsymbol{H}$ through the constitutive relations

$$\boldsymbol{D} = \epsilon \boldsymbol{E}, \quad \boldsymbol{B} = \mu \boldsymbol{H},$$

where ϵ and μ denote the permittivity and permeability of the medium, respectively. Here we use the conventional notation for the curl operator $\nabla \times$ and the divergence operator $\nabla \cdot$, i.e., for any vector $\boldsymbol{\phi} = (\phi_1, \phi_2, \phi_3)'$, we have

$$\nabla \times \boldsymbol{\phi} = i\left(\frac{\partial \phi_3}{\partial y} - \frac{\partial \phi_2}{\partial z}\right) - j\left(\frac{\partial \phi_3}{\partial x} - \frac{\partial \phi_1}{\partial z}\right) + k\left(\frac{\partial \phi_2}{\partial x} - \frac{\partial \phi_1}{\partial y}\right),$$

and $\nabla \cdot \boldsymbol{\phi} = \frac{\partial \phi_1}{\partial x} + \frac{\partial \phi_2}{\partial y} + \frac{\partial \phi_3}{\partial z}$.

For more detailed discussions on Maxwell's equations in electromagnetics, readers can consult books by Chew [7] and Jackson [17], and references cited therein.

1.4.2   Fluid mechanics

In modeling fluid flow, we will come across the Navier-Stokes (N-S) equations. Consider laminar flows of viscous, incompressible fluids; the two-dimensional N-S equations in dimensionless form are given by:

Momentum equations:

$$\frac{\partial u}{\partial t} + \frac{\partial p}{\partial x} = \frac{1}{Re}\left(\frac{\partial^2 u}{\partial x^2} + \frac{\partial^2 u}{\partial y^2}\right) - \frac{\partial(u^2)}{\partial x} - \frac{\partial(uv)}{\partial y} + g_x,$$

$$\frac{\partial v}{\partial t} + \frac{\partial p}{\partial y} = \frac{1}{Re}\left(\frac{\partial^2 v}{\partial x^2} + \frac{\partial^2 v}{\partial y^2}\right) - \frac{\partial(uv)}{\partial x} - \frac{\partial(v^2)}{\partial y} + g_y,$$

Continuity equation:

$$\frac{\partial u}{\partial x} + \frac{\partial v}{\partial y} = 0,$$

where Re is the dimensionless Reynolds number, g_x and g_y denote the body forces, u and v are the velocity fields in the x- and y-directions, respectively, and p is the pressure.

In the three-dimensional case, the N-S equations are given as follows. Momentum equations:

$$\frac{\partial u}{\partial t} + \frac{\partial p}{\partial x} = \frac{1}{Re}\left(\frac{\partial^2 u}{\partial x^2} + \frac{\partial^2 u}{\partial y^2} + \frac{\partial^2 u}{\partial z^2}\right) - \frac{\partial(u^2)}{\partial x} - \frac{\partial(uv)}{\partial y} - \frac{\partial(uw)}{\partial z} + g_x,$$

$$\frac{\partial v}{\partial t} + \frac{\partial p}{\partial y} = \frac{1}{Re}\left(\frac{\partial^2 v}{\partial x^2} + \frac{\partial^2 v}{\partial y^2} + \frac{\partial^2 v}{\partial z^2}\right) - \frac{\partial(uv)}{\partial x} - \frac{\partial(v^2)}{\partial y} - \frac{\partial(vw)}{\partial z} + g_y,$$

$$\frac{\partial w}{\partial t} + \frac{\partial p}{\partial z} = \frac{1}{Re}\left(\frac{\partial^2 w}{\partial x^2} + \frac{\partial^2 w}{\partial y^2} + \frac{\partial^2 w}{\partial z^2}\right) - \frac{\partial(uw)}{\partial x} - \frac{\partial(vw)}{\partial y} - \frac{\partial(w^2)}{\partial z} + g_z.$$

Continuity equation:

$$\frac{\partial u}{\partial x} + \frac{\partial v}{\partial y} + \frac{\partial w}{\partial z} = 0.$$

Here w denotes the velocity component in the z-direction, and g_z the z-component of the body forces.

Using Einstein summation convention, the above momentum equations can be simply written as:

$$\left(\frac{\partial}{\partial t} + u_j\frac{\partial}{\partial x_j} - \frac{1}{Re}\cdot\frac{\partial^2}{\partial x_j x_j}\right)u_i = -\frac{\partial p}{\partial x_i} + g_i, \quad i,j = 1,2,3,$$

where we denote $(x_1, x_2, x_3) := (x, y, z)$, $(u_1, u_2, u_3) := (u, v, w)$ and $(g_1, g_2, g_3) = (g_x, g_y, g_z)$.

In a more compact vector form, the momentum equations are often written as follows:

$$\frac{\partial \mathbf{u}}{\partial t} + (\mathbf{u}\cdot\nabla)\mathbf{u} - \frac{1}{Re}\nabla^2\mathbf{u} = -\nabla p + \mathbf{g},$$

where we denote $\mathbf{u} := (u, v, w)$ and $\mathbf{g} := (g_x, g_y, g_z)$.

1.4.3 Groundwater contamination

Groundwater contamination and soil pollution have been recognized as important environmental problems over the past three decades. Very often, ground water is an important pathway for wastes found in the land's subsurface. Contaminants leak from the waste, move forward through the unsaturated zone to the water table, and then migrate in the saturated groundwater system. The contaminants may discharge ultimately to well water or surface water, which may threaten our living environments. Hence better understanding of groundwater flow and contaminant transport is essential to the national and global economies, people's health and safety, and billions of dollars for future environmental remediation and crisis management.

The governing equations for groundwater flow and contaminant transport can be classified into two big categories: saturated and unsaturated. Generally speaking, the saturated zone is the region below the groundwater table, and the unsaturated zone is the part of the subsurface between the land surface and the groundwater table. Unsaturated means that the pore spaces are partially filled with water, and partially with air. The unsaturated zone varies from meters to hundreds of meters deep.

The saturated groundwater flow in three dimensions can be modeled by the equation

$$S_s \frac{\partial h}{\partial t} = \frac{\partial}{\partial x}(K_x \frac{\partial h}{\partial x}) + \frac{\partial}{\partial y}(K_y \frac{\partial h}{\partial y}) + \frac{\partial}{\partial z}(K_z \frac{\partial h}{\partial z}) + Q,$$

where h is the water head, K_x, K_y and K_z denote the hydraulic conductivity in the x-, y-, and z-directions, respectively. S_s is the specific storage, and Q is a source or sink term.

In an unsaturated zone, the groundwater flow is governed by the equation

$$c(\psi)\frac{\partial \psi}{\partial t} = \frac{\partial}{\partial x}(K_x(\psi)\frac{\partial \psi}{\partial x}) + \frac{\partial}{\partial y}(K_y(\psi)\frac{\partial \psi}{\partial y}) + \frac{\partial}{\partial z}(K_z(\psi)\frac{\partial \psi}{\partial z}) + Q, \quad (1.13)$$

where we denote ψ the pressure head, $K_x(\psi), K_y(\psi)$ and $K_z(\psi)$ the unsaturated hydraulic conductivity in the x-, y-, and z-directions, respectively, $c(\psi)$ is the specific moisture capacity, and Q is a source or sink term. Equation (1.13) is the multidimensional form of the well-known Richards equation.

The contaminant transport equation in an unsaturated zone is given by

$$\frac{\partial}{\partial t}(\theta_w RC) = \sum_{i,j=1}^{3} \frac{\partial}{\partial x_i}(\theta_w D_{ij}\frac{\partial C}{\partial x_j}) - \sum_{i=1}^{3} \frac{\partial}{\partial x_i}(v_i C) + q_s, \quad (1.14)$$

where both the moisture content θ_w and the retardation factor R are functions of time, and D_{ij} is the dispersion coefficient. We use q_s to describe the addition or removal of the contaminant due to chemical and/or biological reactions. The velocity v_i is related to the pressure head by Darcy's law:

$$v_x = -K_x(\psi)\frac{\partial \psi}{\partial x}, \quad v_y = -K_y(\psi)\frac{\partial \psi}{\partial y}, \quad v_z = -K_z(\psi)\frac{\partial \psi}{\partial z}.$$

While the transport in a saturated zone is the same as (1.14) except that θ_w is replaced by the porosity θ, which is the fraction of a representative elementary volume for the fluid. In this case the porous medium reaches its full saturation, which represents the fraction of the void volume filled by this phase in a porous medium.

More details on groundwater flow and contaminant transport modeling can be found in books such as [25, 26].

1.4.4   Petroleum reservoir simulation

A petroleum reservoir is a porous medium which contains hydrocarbons. Simulation of petroleum reservoirs helps us in predicting the performance of a reservoir and optimizing the oil recovery. The so-called two-phase immiscible flow is used to model the simultaneous flow of two fluid phases such as water and oil in a porous medium. We assume that the two phases are immiscible and there is no mass transfer between them. Hence we have the mass conservation for each phase, which is given by

$$\frac{\partial(\phi\rho_\alpha S_\alpha)}{\partial t} + \nabla \cdot (\rho_\alpha \boldsymbol{u}_\alpha) = q_\alpha, \quad \alpha = w, \, o,$$

where w and o denote the water phase and oil phase, respectively. Here we denote ϕ the porosity of the porous medium, ρ_α the density of each phase, q_α the mass flow rate, and S_α the saturation. The Darcy velocity $\boldsymbol{u}_\alpha$ is described by Darcy's law

$$\boldsymbol{u}_\alpha = -\frac{1}{\mu_\alpha}\boldsymbol{k}_\alpha(\nabla p_\alpha - \rho_\alpha g\nabla z), \quad \alpha = w, \, o,$$

where $\boldsymbol{k}_\alpha$ is the effective permeability, p_α is the pressure, μ_α is the viscosity for each phase α, g is the magnitude of the gravitation acceleration, z is the depth of the reservoir, and $\nabla z = (0,0,1)'$. Note that both phases occupy the whole void space, hence we have $S_w + S_o = 1$.

Thorough discussions on reservoir simulation can be found in books such as [5, 9].

1.4.5 Finance modeling

The celebrated Black-Scholes differential equation is

$$\frac{\partial u}{\partial t} + \frac{1}{2}\sigma^2 S^2 \frac{\partial^2 u}{\partial S^2} + rS\frac{\partial u}{\partial S} - ru = 0. \tag{1.15}$$

Here $u(S(t),t)$ denotes the price of an option on a stock, $S(t)$ is the price of the stock at time t, σ is the volatility of the stock, and r is the risk-free interest rate.

Note that (1.15) is a backward parabolic equation, which needs a final condition at time T. For a vanilla European call, we have the condition

$$u(S,T) = \max\{S - X, 0\},$$

while for a put option, we have

$$u(S,T) = \max\{X - S, 0\},$$

where X is the exercise price at the expiration date T. We want to remark that the call option means the right to buy the stock, while the put option gives the right to sell the stock.

More details on mathematical finance can be found in books such as [11, 16].

1.4.6   Image processing

In recent years, PDEs have become popular and useful tools for image denoising, image segmentation, and image restoration. Such PDE-based methodology has been successfully applied to many problems in image processing and computer vision. Examples include image denoising, segmentation of textures and remotely sensed data, object detection, optical flow, stereo, and enhancing textures such as fingerprints, image sharpening, and target tracking.

For example, the homogeneous linear diffusion equation

$$u_t = \triangle u \equiv \frac{\partial^2 u}{\partial x^2} + \frac{\partial^2 u}{\partial y^2} \tag{1.16}$$

can be used to reduce noise, where u denotes the image value. But this simple model can blur important features such as edges. Hence the inhomogeneous linear diffusion

$$u_t = \text{div}(g(|\nabla u_0|^2)\nabla u) \tag{1.17}$$

was proposed, where u_0 is a given image, $g : R^+ \rightarrow R^+$ is a decreasing function with $g(0) = 1, \lim_{s\rightarrow\infty} g(s) = 0$ and $g(s)$ is smooth, e.g., $g(s^2) = (1 + \frac{s^2}{\lambda^2})^{-1}, \lambda > 0$ is a constant. Using the model (1.17), edges remain better localized and their blurring is reduced. But for large t the filtered image reveals some artifacts which reflect the differential structure of the initial image. To reduce the artifacts of inhomogeneous linear diffusion filtering, a feedback can be introduced in the process by adapting the diffusivity g to the gradient of the actual image $u(\mathbf{x}, t)$, which leads to the nonlinear isotropic diffusion model

$$u_t = \text{div}(g(|\nabla u|^2)\nabla u). \tag{1.18}$$

Model (1.18) can increase the edge localization and reduce the blurring at edges, but the absolute contrast at edges becomes smaller. To overcome this shortcoming, the orthonormal system of eigenvectors v_1 and v_2 of the diffusion tensor D can be constructed such that v_1 is parallel to ∇u_σ, and v_2 is perpendicular to ∇u_σ, where $u_\sigma = G_\sigma \star u$ represents the standard convolution of G_σ with u. Here G_σ can be any smoothing kernel such as the often used Gaussian filter $G_\sigma(\mathbf{x}) = \frac{1}{4\pi\sigma} \exp(-|\mathbf{x}|^2/(4\sigma))$. If we want to smooth the image along the edge instead of across the edge, we can choose the eigenvalues λ_1 and λ_2 as

$$\lambda_1 = g(|\nabla u_\sigma|^2), \quad \lambda_2 = 1,$$

which creates fairly realistic segments.

To circumvent the stopping time problem caused by the diffusion filters with a constant steady state, an additional reaction term can be added to keep the steady-state solution close to the original image, which gives us another model

$$u_t = \text{div}(g(|\nabla u|^2)\nabla u) + \beta(u_0 - u), \quad \beta > 0.$$

For more discussions on PDEs in image processing, readers can consult books such as [1, 4, 19].

1.5 A quick review of numerical methods for PDEs

Due to the ubiquitousness of partial differential equations, there are many different numerical methods for solving them. Generally speaking, we can classify the numerical methods into six big categories: finite difference, spectral method, finite element, finite volume, boundary element, and meshless or mesh-free methods.

The finite difference method seems to be the easiest understandable technique to solve a differential equation. The basic idea is to use finite differences to approximate those differentials in the PDEs. Due to its simplicity, the difference method is often the first choice for those who are interested in numerical solutions of PDEs. One disadvantage of this method is that it becomes quite complex when solving PDEs on irregular domains. Another disadvantage is that it is not easy to carry out the mathematical analysis (such as solvability, stability, and convergence) for the difference methods especially for PDEs with variable coefficients, and nonlinear PDEs. Considering that the finite difference method serves as a basis for other numerical methods, we will cover it in more detail in future chapters.

Spectral methods are powerful technologies for solving PDEs if the physical domain is simple and the solution is smooth. Early literature on spectral methods are the short book by Gottlieb and Orszag [13] and the monograph by Canuto, Hussaini, Quarteroni and Zang [3]. More recent advances can be found in books by Fornberg [10], Trefethen [23], Shen, Tang and Wang [21], and Hesthaven et al. [14]. Due to our lack of training and experience in this area, we will not cover the spectral methods in this book. Interested readers can consult the above books for more details.

The finite element method is arguably the most popular method for solving various PDEs. Compared to other methods, it has well-established mathematical theory for various PDEs and is a good choice for solving PDEs over complex domains (like cars and airplanes). The finite element method works by rewriting the governing PDE into an equivalent variational problem, meshing the modeled domain into smaller elements and looking for approximate solutions at the mesh nodes when using a linear basis function over each element. But the implementation of the finite element method may be the most complicated compared to other numerical methods. Due to its popularity, there are many commercial finite element packages and some free packages too. We will discuss both the mathematical theory and implementation of the finite element method in later chapters.

The boundary element method is used to solve those PDEs which can be formulated as integral equations. The boundary element method attempts to use the given boundary conditions to fit boundary values into the integral equation, rather than values throughout the space defined by the PDE. Conceptually, the boundary element method can be thought of as a finite element method over the modeled surface (i.e., meshing over the surface) instead of over the modeled physical domain [15]. Hence, the boundary element method is often more efficient than other methods in terms of computational resources for problems when the surface-to-volume ratio is small. However, boundary element formulations typically yield fully populated matrices, which makes the storage requirements and computational time increase in the square order of the problem size. While matrices from finite element methods are often sparse and banded, the storage requirements only grow linearly with the problem size. Therefore, for many problems boundary element methods are significantly less efficient than those volume-based methods such as finite element methods. Another problem for the boundary element method is that not many problems, such as nonlinear problems, can be written as integral equations, which restricts the applicability of the boundary element method. For detailed mathematical theory on the boundary element method, the reader can consult books such as [6, 8] and references cited therein. For engineering applications and implementation, details can be found in books [2, 20] and references cited therein.

The finite volume method is another method for solving PDEs. This method is very popular in computational fluid dynamics. The basic idea of the finite volume method is to integrate the differential equation over a finite-sized control volume surrounding each nodal point on a mesh, then changing the volume integrals (those involving the divergence term) to surface integrals which can be evaluated as fluxes at the surfaces of each finite volume. Hence the finite volume method is conservative locally on each volume. Another advantage of the finite volume method is that it can be easily used for irregularly shaped domains. Detailed discussions about the finite volume method can be found in books such as [18, 24] and references cited therein.

The meshless or mesh-free method is a more recently developed technique for solving PDEs. The mesh-based methods (the finite element, boundary element and finite volume methods) share the drawbacks such as the tedious meshing and re-meshing in crack propagation problem, the melting of a solid or the freezing process, large deformations, etc. The meshless method aims to overcome those drawbacks by getting rid of meshing or re-meshing the entire modeled domain and only adding or deleting nodes, instead. The initial idea dates back to the smooth partical hydrodynamics method developed by Gingold and Monaghan in 1977 [12] for modeling astrophysical phenomena. Since the 1990s, the research into meshless methods has become very active and many mesh-free methods have been developed, e.g., Element Free Galerkin Method; Reproducing Kernel Particle Method; Partition of Unity Method; hp-Cloud Method; Radial Basis Function Method; Meshless Local Petrov-Galerkin method, etc. We will introduce them in more detail in later chapters.

References

[1] G. Aubert and P. Kornprobst. *Mathematical Problems in Image Processing: Partial Differential Equations and the Calculus of Variations.* Springer, New York, NY, 2nd edition, 2006.

[2] C. A. Brebbia and L. C. Wrobel. *Boundary Element Techniques: Theory and Applications in Engineering.* WIT Press, Southampton, UK, 2nd edition, 1996.

[3] C. Canuto, M.Y. Hussaini, A. Quarteroni and T.A. Zang. *Spectral Methods in Fluid Dynamics.* Springer-Verlag, Berlin, 1988.

[4] T. Chan and J. Shen. *Image Processing and Analysis: Variational, PDE, Wavelet, and Stochastic Methods.* SIAM, Philadelphia, PA, 2005.

[5] Z. Chen, G. Huan and Y. Ma. *Computational Methods for Multiphase Flows in Porous Media.* SIAM, Philadelphia, PA, 2006.

[6] G. Chen and J. Zhou. *Boundary Element Methods.* Academic Press, New York, NY, 1992.

[7] W.C. Chew. *Waves and Fields in Inhomogenous Media.* Wiley-IEEE Press, New York, NY, 1999.

[8] D. Colton and R. Kress. *Integral Equation Methods in Scattering Theory.* John Wiley & Sons, New York, NY, 1983.

[9] R.E. Ewing (ed.). *The Mathematics of Reservoir Simulation.* SIAM, Philadelphia, PA, 1987.

[10] B. Fornberg. *A Practical Guide to Pseudospectral Methods.* Cambridge University Press, Cambridge, UK, 1996.

[11] J.-P. Fouque, G. Papanicolaou and K.R. Sircar. *Derivatives in Financial Markets with Stochastic Volatility.* Cambridge University Press, Cambridge, UK, 2000.

[12] R.A. Gingold and J.J. Monaghan. Smoothed particle hydrodynamics: theory and application to non-spherical stars. *Mon. Not. Roy. Astron. Soc.*, 181:375–389, 1977.

[13] D. Gottlieb and S.A. Orszag. *Numerical Analysis of Spectral Methods: Theory and Applications.* SIAM, Philadelphia, PA, 1977.

[14] J.S. Hesthaven, S. Gottlieb and D. Gottlieb. *Spectral Methods for Time-Dependent Problems.* Cambridge University Press, Cambridge, UK, 2007.

[15] C. Johnson. *Numerical Solution of Partial Differential Equations by the Finite Element Method.* Cambridge University Press, Cambridge, UK, 1987.

[16] I. Karatzas and S.E. Shreve. *Methods of Mathematical Finance.* Springer, New York, NY, 1998.

[17] J.D. Jackson. *Classical Electrodynamics.* 3rd edn. John Wiley & Sons, Hoboken, NJ, 2001.

[18] R.J. LeVeque. *Finite Volume Methods for Hyperbolic Problems.* Cambridge University Press, Cambridge, UK, 2002.

[19] S.J. Osher and R.P. Fedkiw. *Level Set Methods and Dynamic Implicit Surfaces.* Springer, New York, NY, 2002.

[20] C. Pozrikidis. *A Practical Guide to Boundary Element Methods with the Software Library BEMLIB.* CRC Press, Boca Raton, FL, 2002.

[21] J. Shen, T. Tang and L.-L. Wang. *Spectral Methods: Algorithms, Analysis and Applications.* Springer-Verlag, Berlin, 2011.

[22] J.C. Tannehill, D.A. Anderson and R.H. Pletcher. *Computational Fluid Mechanics and Heat Transfer.* Taylor & Francis, Philadelphia, PA, 2nd edition, 1997.

[23] L.N. Trefethen. *Spectral Methods in MATLAB.* SIAM, Philadelphia, PA, 2000.

[24] H. Versteeg and W. Malalasekra. *An Introduction to Computational Fluid Dynamics: The Finite Volume Method.* Prentice Hall, Upper Saddle River, NJ, 2nd edition, 2007.

[25] D. Zhang. *Stochastic Methods for Flow in Porous Media: Coping with Uncertainties.* Academic Press, New York, NY, 2001.

[26] C. Zheng. *Applied Contaminant Transport Modeling.* Wiley-Interscience, New York, NY, 2nd edition, 2002.

2

Finite Difference Methods for Parabolic Equations

The finite difference method (FDM) seems to be the simplest approach for the numerical solution of PDEs. It proceeds by replacing those derivatives in the governing equations by finite differences. In this chapter, we will introduce various difference methods for parabolic equations. In Sec. 2.1, we present both the explicit and implicit schemes for a simple heat equation. Then we introduce some important concepts (such as stability, consistence, and convergence) used in analyzing finite difference methods in Sec. 2.2. Then in Sec. 2.3, we demonstrate a few examples for using those concepts. In Sec. 2.4, we extend the discussion to two-dimensional and three-dimensional parabolic equations. Here we cover the standard difference methods and the alternate direction implicit (ADI) method. Finally, in Sec. 2.5, we present a MATLAB code to show readers how to solve a parabolic equation.

2.1 Introduction

We start with a simple heat equation model: find $u(x,t)$ such that

$$u_t = u_{xx}, \quad \forall \, (x,t) \in (0,1) \times (0, t_F), \tag{2.1}$$

$$u(0,t) = u(1,t) = 0, \quad \forall \, t \in (0, t_F) \tag{2.2}$$

$$u(x,0) = u_0(x), \quad \forall \, x \in [0,1], \tag{2.3}$$

where t_F denotes the terminal time for the model. Here without loss of generality, we assume that the spatial domain is $[0,1]$.

To solve the problem (2.1)–(2.3) by FDM, we first divide the physical domain $(0, t_F) \times (0,1)$ by $N \times J$ uniform grid points

$$t_n = n\triangle t, \quad \triangle t = \frac{t_F}{N}, \quad n = 0, 1, \cdots, N,$$

$$x_j = j\triangle x, \quad \triangle x = \frac{1}{J}, \quad j = 0, 1, \cdots, J.$$

We denote the approximate solution u_j^n to the exact solution u at an arbitrary point (x_j, t_n), i.e., $u_j^n \approx u(x_j, t_n)$. To obtain a finite difference scheme, we need to approximate those derivatives in (2.1) by some finite differences. Below we demonstrate two simple finite difference schemes for solving (2.1)-(2.3).

Example 2.1
(Explicit scheme: Forward Euler) Substituting

$$u_t(x_j, t_n) \approx (u_j^{n+1} - u_j^n)/\Delta t,$$
$$u_{xx}(x_j, t_n) \approx (u_{j+1}^n - 2u_j^n + u_{j-1}^n)/(\Delta x)^2,$$

into (2.1), we obtain an explicit scheme for (2.1):

$$u_j^{n+1} = u_j^n + \mu(u_{j+1}^n - 2u_j^n + u_{j-1}^n), \quad 1 \le j \le J-1, 0 \le n \le N-1, \quad (2.4)$$

where we denote

$$\mu = \frac{\Delta t}{(\Delta x)^2}. \quad (2.5)$$

The boundary condition (2.2) can be approximated directly as

$$u_0^n = u_J^n = 0, \quad 0 \le n \le N-1,$$

and the initial condition (2.3) can be approximated as

$$u_j^0 = u_0(j\Delta x), \quad 0 \le j \le J. \quad (2.6)$$

Note that with the scheme (2.4), the approximate solution u_j^{n+1} at any interior points can be obtained by a simple marching in time.   □

Example 2.2
(Implicit scheme: Backward Euler) Similarly, by substituting

$$u_t(x_j, t_n) \approx (u_j^n - u_j^{n-1})/\Delta t,$$
$$u_{xx}(x_j, t_n) \approx (u_{j+1}^n - 2u_j^n + u_{j-1}^n)/(\Delta x)^2,$$

into (2.1), another difference scheme for (2.1) can be constructed as:

$$\frac{u_j^n - u_j^{n-1}}{\Delta t} = \frac{u_{j+1}^n - 2u_j^n + u_{j-1}^n}{(\Delta x)^2}, \quad 1 \le j \le J-1, 1 \le n \le N. \quad (2.7)$$

In which case, we obtain an implicit scheme

$$-\mu u_{j-1}^n + (1+2\mu)u_j^n - \mu u_{j+1}^n = u_j^{n-1}, \quad 1 \le j \le J-1. \quad (2.8)$$

Note that in this case, we have to solve a linear system at each time step in order to obtain the approximate solutions u_j^n at all interior points. That is why the scheme (2.8) is called implicit in order to distinguish it from the explicit scheme (2.4).

More specifically, the scheme (2.8) is equivalent to a linear system $A\mathbf{u}^n = \mathbf{u}^{n-1}$ with vectors $\mathbf{u}^n = (u_1^n, \cdots, u_{J-1}^n)'$, $\mathbf{u}^{n-1} = (u_1^{n-1}, \cdots, u_{J-1}^{n-1})'$, and the $(J-1) \times (J-1)$ tridiagonal matrix A:

$$
A = \begin{bmatrix}
1+2\mu & -\mu & 0 & & & 0 \\
-\mu & 1+2\mu & -\mu & & & \\
& & \ddots & \ddots & \ddots & \\
& & & \ddots & \ddots & \ddots \\
& & & -\mu & 1+2\mu & -\mu \\
0 & & & 0 & -\mu & 1+2\mu
\end{bmatrix},
$$

which can be solved by the efficient Thomas algorithm (e.g., [11]). □

Example 2.3
(Implicit scheme: Crank-Nicolson (CN)) Similarly, by substituting

$$
u_t(x_j, t_n) \approx (u_j^n - u_j^{n-1})/\triangle t,
$$

$$
u_{xx}(x_j, t_n) \approx \frac{1}{2} \left(\frac{u_{j+1}^n - 2u_j^n + u_{j-1}^n}{(\triangle x)^2} + \frac{u_{j+1}^{n-1} - 2u_j^{n-1} + u_{j-1}^{n-1}}{(\triangle x)^2} \right),
$$

into (2.1), we can obtain the CN scheme for (2.1): For $1 \leq j \leq J-1$, $1 \leq n \leq N$,

$$
\frac{u_j^n - u_j^{n-1}}{\triangle t} = \frac{1}{2} \left(\frac{u_{j+1}^n - 2u_j^n + u_{j-1}^n}{(\triangle x)^2} + \frac{u_{j+1}^{n-1} - 2u_j^{n-1} + u_{j-1}^{n-1}}{(\triangle x)^2} \right), \quad (2.9)
$$

which can be written as:

$$
-\mu u_{j-1}^n + (2+2\mu)u_j^n - \mu u_{j+1}^n = \mu u_{j-1}^{n-1} + (2-2\mu)u_j^{n-1} + \mu u_{j+1}^{n-1}. \quad (2.10)
$$

For this CN scheme, we also have to solve a linear system at each time step in order to obtain the approximate solutions u_j^n at all interior points. Using the same notation as the Backward-Euler scheme, the scheme (2.10) can be written as a linear system $B\mathbf{u}^n = C\mathbf{u}^{n-1}$, where both B and C are $(J-1) \times (J-1)$ tridiagonal matrices given as follows:

$$
B = \begin{bmatrix}
2+2\mu & -\mu & 0 & & & 0 \\
-\mu & 2+2\mu & -\mu & & & \\
& & \ddots & \ddots & \ddots & \\
& & & \ddots & \ddots & \ddots \\
& & & -\mu & 2+2\mu & -\mu \\
0 & & & 0 & -\mu & 2+2\mu
\end{bmatrix},
$$

and

$$C = \begin{bmatrix} 2-2\mu & \mu & 0 & & & 0 \\ \mu & 2-2\mu & \mu & & & \\ & & \cdot & \cdot & \cdot & \\ & & \cdot & \cdot & \cdot & \\ & & & \cdot & \cdot & \\ & & & \mu & 2-2\mu & \mu \\ 0 & & & 0 & \mu & 2-2\mu \end{bmatrix}.$$

∎

2.2   Theoretical issues: stability, consistency, and convergence

From the previous section, we see that constructing a finite difference scheme seems quite simple, but how do we know which scheme is better and how do we compare different schemes? To answer these questions, we need to introduce some important concepts in FDM analysis.

DEFINITION 2.1   *(Truncation error)*
   The **truncation error** *of the scheme (2.4) is defined as*

$$TE(x,t) = \frac{D_+^t u(x,t)}{\triangle t} - \frac{D_+^x D_-^x u(x,t)}{(\triangle x)^2},$$

where we denote the backward and forward difference operators D_-^x and D_+^x in variable x as follows:

$$D_-^x v(x,t) = v(x,t) - v(x - \triangle x, t),$$
$$D_+^x v(x,t) = v(x + \triangle x, t) - v(x,t).$$

Similarly, we can define the forward difference operator D_+^t in variable t as

$$D_+^t v(x,t) = v(x, t + \triangle t) - v(x,t).$$

DEFINITION 2.2   *(Consistence)*
   If $TE(x,t) \to 0$ as $\triangle t, \triangle x \to 0$ at any point (t,x) in the physical domain (for our example $(x,t) \in (0,1) \times (0, t_F)$), then we say that the scheme is **consistent** *with the differential equation (2.1).*

DEFINITION 2.3 *(Convergence)*
 If for any point $(x,t) \in (0,1) \times (0,t_F)$,

$$x_j \to x, t_n \to t \quad implies \quad u_j^n \to u(x,t),$$

i.e., the numerical solution at node (x_j, t_n) approximates the exact solution $u(x,t)$ as (x_j, t_n) gets close to the point (x,t), then we say that the scheme is **convergent**.

DEFINITION 2.4 *(Order of accuracy)*
 If for a sufficiently smooth solution u, there exists a constant $C > 0$ such that

$$TE(x,t) \le C[(\triangle t)^p + (\triangle x)^q], \quad as \ \triangle t, \triangle x \to 0,$$

where p and q are the largest possible integers, then we say that the scheme has pth **order of accuracy** *in $\triangle t$ and qth* **order of accuracy** *in $\triangle x$.*

DEFINITION 2.5 *(Well-posedness)*
 A partial differential equation (PDE) is **well-posed** *if a solution of the PDE exists, the solution is unique, and the solution depends continuously on the data (such as initial conditions, boundary conditions, right-hand side).*

DEFINITION 2.6 *(Numerical stability)*
 For a time-dependent PDE, the corresponding difference scheme is **stable** *in a norm $\| \cdot \|$ if there exists a constant $M > 0$ such that*

$$\|u^n\| \le M\|u^0\|, \quad \forall \ n\triangle t \le t_F,$$

where M is independent of $\triangle t, \triangle x$ and the initial condition u^0.

The following Lax-Richtmyer equivalence theorem connects the stability, consistency, and convergence concepts altogether, which proof can be found in many classic finite difference books (e.g.,[1, 5, 9, 13]).

THEOREM 2.1
(Lax-Richtmyer equivalence theorem) For a consistent difference approximation to a well-posed linear time-dependent problem, the stability of the scheme is necessary and sufficient for convergence.

The classic Lax-Richtmyer equivalence theorem implies that convergence is equivalent to consistency and stability. The consistency is usually easy to check, hence the analysis of stability plays a very important role for numerical methods.
 In general, there are two fundamental approaches for proving stability: the Fourier analysis (also called von Neumann stability analysis), and the energy method. Generally speaking, the Fourier analysis applies only to linear

constant coefficient problems, while the energy method can be used for more general problems with variable coefficients and nonlinear terms. However, the energy method can become quite complicated and the proof is problem dependent. We will demonstrate both methods in the following sections.

2.3   1-D parabolic equations

In this section, we present several popular finite difference methods and their stability analysis for the 1-D parabolic equations.

2.3.1   The θ-method and its analysis

For the model problem (2.1), we can construct the so-called θ-scheme:

$$\frac{u_j^{n+1} - u_j^n}{\Delta t} = \frac{\theta D_+^x D_-^x u_j^{n+1} + (1 - \theta)D_+^x D_-^x u_j^n}{(\Delta x)^2}, \quad 0 \le \theta \le 1, \ 1 \le j \le J - 1,$$

$$(2.11)$$

or

$$-\theta\mu u_{j-1}^{n+1} + (1 + 2\theta\mu)u_j^{n+1} - \theta\mu u_{j+1}^{n+1} = [1 + (1 - \theta)\mu D_+^x D_-^x]u_j^n. \quad (2.12)$$

Note that the special case $\theta = 0$ is just the Forward Euler scheme (2.4). $\theta = 1$ corresponds to the Backward Euler scheme (2.7), while $\theta = \frac{1}{2}$ corresponds to the CN scheme (2.9).

2.3.1.1   Stability analysis with the Fourier analysis technique

To study the stability of (2.11), we can use the Fourier analysis technique. This technique is motivated by the fact that the exact solution of a parabolic equation can be expressed as a Fourier series. For example, by the method of separation of variables, we can prove that the exact solution of the problem (2.1)-(2.3) can be written as

$$u(x,t) = \sum_{n=1}^{\infty} a_n e^{-(n\pi)^2 t} \sin n\pi x, \quad a_n = 2 \int_0^1 u_0(x) \sin n\pi x dx,$$

i.e., the exact solution is a linear combination of all Fourier modes. Hence, we can assume that a similar Fourier mode should be an exact numerical solution of the difference scheme. Substituting

$$u_j^n = \lambda^n e^{ik(j\Delta x)}, \quad (2.13)$$

into (2.11), and dividing by u_j^n, we obtain

$$\lambda - 1 = \mu[\theta\lambda + (1 - \theta)](e^{ik\Delta x} - 2 + e^{-ik\Delta x})$$

$$= \mu[\theta\lambda + (1 - \theta)](-4\sin^2\frac{k\Delta x}{2}). \quad (2.14)$$

Here $\lambda = \lambda(k)$ is called the amplification factor corresponding to the wavenumber k.

Solving (2.14) for λ gives

$$\lambda = \frac{1 - 4(1 - \theta)\mu \sin^2 \frac{k\triangle x}{2}}{1 + 4\theta\mu \sin^2 \frac{k\triangle x}{2}}. \tag{2.15}$$

When $\theta = 0$, we see that $0 < 2\mu \leq 1$ implies $1 \geq 1 - 4\mu \sin^2 \frac{k\triangle x}{2} \geq -1$, i.e., $|\lambda(k)| \leq 1$ for all k, so that the scheme (2.11) in this case is stable.

When $\theta = 1$, then $\lambda = \frac{1}{1 + 4\mu \sin^2 \frac{k\triangle x}{2}} \leq 1$ for all k, in which case, the scheme is said to be unconditionally stable, i.e., there is no constraint on the time step size $\triangle t$ and the mesh size $\triangle x$.

When $\theta = \frac{1}{2}$, then $\lambda = \frac{1 - 2\mu \sin^2 \frac{k\triangle x}{2}}{1 + 2\mu \sin^2 \frac{k\triangle x}{2}}$. When $0 < 2\mu \leq 1$, we have $|\lambda| \leq \frac{1}{1 + 2\mu \sin^2 \frac{k\triangle x}{2}} \leq 1$. When $2\mu \geq 1$, we have $|\lambda| \leq \frac{2\mu \sin^2 \frac{k\triangle x}{2}}{1 + 2\mu \sin^2 \frac{k\triangle x}{2}} < 1$. Hence, for any μ, the scheme (2.11) with $\theta = \frac{1}{2}$ is also unconditionally stable.

2.3.1.2 Stability analysis with the energy method

Example 2.4

To see how the energy method can be used for stability analysis, let us first consider the Forward Euler scheme (2.4). Denote the discrete maximum norm

$$||u^n||_\infty = \max_{0 \leq i \leq J} |u_i^n|.$$

Under the condition $0 < 2\mu \leq 1$, we see that

$$|u_j^{n+1}| = |(1 - 2\mu)u_j^n + \mu u_{j+1}^n + \mu u_{j-1}^n|$$
$$\leq (1 - 2\mu)|u_j^n| + \mu|u_{j+1}^n| + \mu|u_{j-1}^n| \leq ||u^n||_\infty.$$

Taking the maximum of both sides with respect to j, and using the induction method, we obtain

$$||u^{n+1}||_\infty \leq ||u^n||_\infty \leq \cdots \leq ||u^0||_\infty,$$

i.e., the explicit scheme is stable under the condition $0 < 2\mu \leq 1$.   ◻

Example 2.5

Now let us consider the Backward Euler scheme (2.8). Using inequality

$$|a - b| \geq |a| - |b|,$$

we have

$$|u_j^{n-1}| \geq (1+2\mu)|u_j^n| - \mu|u_{j-1}^n + u_{j+1}^n| \geq (1+2\mu)|u_j^n| - 2\mu||u^n||_\infty. \quad (2.16)$$

Taking the maximum of (2.16) with respect to all j, we obtain

$$||u^{n-1}||_\infty \geq (1+2\mu)||u^n||_\infty - 2\mu||u^n||_\infty = ||u^n||_\infty,$$

which shows that the scheme (2.8) is unconditionally stable.   ☐

Example 2.6

Finally, let us consider the CN scheme (2.9). We will use a different energy method for the stability analysis. Let us denote the inner product

$$(v,w)_h \equiv \sum_{j=1}^{J-1}(\triangle x)v_j w_j,$$

and its associated norm

$$||v||_h \equiv (v,v)_h^{1/2}.$$

Furthermore, to simplify the analysis, we introduce the following difference operators:

$$\delta_x v_{j+\frac{1}{2}} \equiv v_{j+1} - v_j, \quad \delta_x^2 v_j \equiv \delta_x \cdot \delta_x v_j = \delta_x v_{j+\frac{1}{2}} - \delta_x v_{j-\frac{1}{2}} = v_{j+1} - 2v_j + v_{j-1}.$$

Using operator δ_x^2, we can rewrite the CN scheme (2.9) simply as: For any $j \in [1, J-1], n \in [1, N]$,

$$\frac{u_j^n - u_j^{n-1}}{\triangle t} = \frac{1}{(\triangle x)^2}\delta_x^2\left(\frac{u_j^{n-1} + u_j^n}{2}\right). \quad (2.17)$$

Using the definition of δ_x^2 and condition $v_0 = v_J = 0$, we easily see that

$$\sum_{j=1}^{J-1} v_j \delta_x^2 v_j = \sum_{j=1}^{J-1} v_j(\delta_x v_{j+\frac{1}{2}} - \delta_x v_{j-\frac{1}{2}}) = \sum_{j=1}^{J-1} v_j \delta_x v_{j+\frac{1}{2}} - \sum_{j=0}^{J-2} v_{j+1}\delta_x v_{j+\frac{1}{2}}$$

$$= \sum_{j=0}^{J-1}(v_j - v_{j+1})\delta_x v_{j+\frac{1}{2}} = -\sum_{j=0}^{J-1}\delta_x v_{j+\frac{1}{2}} \cdot \delta_x v_{j+\frac{1}{2}},$$

which leads to the following useful lemma.

LEMMA 2.1

Under the assumptions $v_0 = v_J = 0$, we have

$$\sum_{j=1}^{J-1} v_j \delta_x^2 v_j = -\sum_{j=0}^{J-1}\delta_x v_{j+\frac{1}{2}} \cdot \delta_x v_{j+\frac{1}{2}}. \quad (2.18)$$

Multiplying (2.17) by $\triangle t \triangle x (u_j^n + u_j^{n-1})$, then summing up the result over $j = 1$ to $j = J - 1$, and using Lemma 2.1, we have

$$||u^n||_h^2 - ||u^{n-1}||_h^2 = 2\mu \sum_{j=1}^{J-1} \triangle x \left(\delta_x^2 \left(\frac{u_j^{n-1} + u_j^n}{2} \right), \frac{u_j^{n-1} + u_j^n}{2} \right)$$

$$= -2\mu \sum_{j=0}^{J-1} \triangle x \mid \delta_x \left(\frac{u_{j+\frac{1}{2}}^{n-1} + u_{j+\frac{1}{2}}^n}{2} \right) \mid^2 \le 0,$$

which leads to

$$||u^n||_h^2 \le ||u^{n-1}||_h^2 \le \cdots \le ||u^0||_h^2.$$

This proves that the CN scheme (2.9) is unconditionally stable. ☐

2.3.1.3 Truncation error analysis

Below, we want to study the truncation error of (2.11). Since the scheme (2.11) is symmetric about the point $(x_j, t_{n+\frac{1}{2}})$, we can consider the truncation error

$$TE_j^{n+\frac{1}{2}} \equiv \frac{u(x_j, t_{n+1}) - u(x_j, t_n)}{\triangle t}$$

$$- \frac{\theta D_+^x D_-^x u(x_j, t_{n+1}) + (1 - \theta) D_+^x D_-^x u(x_j, t_n)}{(\triangle x)^2}. \quad (2.19)$$

By the Taylor expansion, we obtain

$$u(x_j, t_{n+1}) - u(x_j, t_n) = [\triangle t u_t + \frac{(\triangle t)^3}{24} u_{t^3} + \cdots](x_j, t_{n+\frac{1}{2}}),$$

and

$$\theta D_+^x D_-^x u(x_j, t_{n+1}) + (1 - \theta) D_+^x D_-^x u(x_j, t_n)$$

$$= [(\triangle x)^2 u_{x^2} + \frac{(\triangle x)^4}{12} u_{x^4} + \cdots]|_j^{n+\frac{1}{2}}$$

$$+ (\theta - \frac{1}{2}) \triangle t [(\triangle x)^2 u_{x^2 t} + \frac{(\triangle x)^4}{12} u_{x^4 t} + \cdots]|_j^{n+\frac{1}{2}}$$

$$+ [\frac{1}{8}(\triangle t)^2 (\triangle x)^2 u_{x^2 t^2} + \cdots]|_j^{n+\frac{1}{2}}.$$

Substituting the above expansions into (2.19), we obtain

$$TE_j^{n+\frac{1}{2}} = -[(\theta - \frac{1}{2}) \triangle t u_{x^2 t} + \frac{(\triangle x)^2}{12} u_{x^4}]|_j^{n+\frac{1}{2}} + [\frac{(\triangle t)^2}{24} u_{t^3} - \frac{(\triangle t)^2}{8} u_{x^2 t^2}]|_j^{n+\frac{1}{2}}$$

$$+ [\frac{1}{12}(\frac{1}{2} - \theta) \triangle t (\triangle x)^2 u_{x^4 t}]|_j^{n+\frac{1}{2}} + \text{higher-order-terms}, \quad (2.20)$$

which shows that $TE_j^{n+\frac{1}{2}}$ goes to zero as $\triangle t$ and $\triangle x$ go to zero. Hence the scheme (2.11) is consistent for any $\theta, \triangle t$ and $\triangle x$. Furthermore, for any $\theta \neq \frac{1}{2}$, the scheme has 1st-order accuracy in $\triangle t$ and 2nd-order accuracy in $\triangle x$, since $TE_j^{n+\frac{1}{2}}$ can be simply written as

$$TE_j^{n+\frac{1}{2}} = O(\triangle t + (\triangle x)^2) \quad \text{when } \theta \neq \frac{1}{2}.$$

When $\theta = \frac{1}{2}$, (2.20) can be simplified to

$$TE_j^{n+\frac{1}{2}} = -[\frac{(\triangle x)^2}{12}u_{x^4} + \frac{(\triangle t)^2}{12}u_{t^3}]_j^{n+\frac{1}{2}} + \text{higher-order-terms}$$
$$= O((\triangle t)^2 + (\triangle x)^2).$$

2.3.1.4 An unconditionally unstable scheme

Here we would like to mention an interesting scheme for solving (2.1):

$$\frac{u_j^{n+1} - u_j^{n-1}}{2\triangle t} = \frac{u_{j+1}^n - 2u_j^n + u_{j-1}^n}{(\triangle x)^2}, \quad 1 \leq n \leq N-1,\ 1 \leq j \leq J-1, \quad (2.21)$$

which is obtained by approximations

$$u_t(x_j, t_n) \approx (u_j^{n+1} - u_j^{n-1})/(2\triangle t),$$
$$u_{xx}(x_j, t_n) \approx (u_{j+1}^n - 2u_j^n + u_{j-1}^n)/(\triangle x)^2.$$

First, the truncation error of this scheme at (x_j, t_n) is

$$TE(x_j, t_n) \equiv \frac{u(x_j, t_{n+1}) - u(x_j, t_{n-1})}{2\triangle t}$$
$$- \frac{u(x_{j+1}, t_n) - 2u(x_j, t_n) + u(x_{j-1}, t_n)}{(\triangle x)^2}$$
$$= [u_t(x_j, t_n) + \frac{(\triangle t)^2}{6}u_{t^3}(x_j, t_n) + \cdots]$$
$$- [u_{x^2}(x_j, t_n) + \frac{(\triangle x)^2}{12}u_{x^4}(x_j, t_n) + \cdots] = O((\triangle t)^2 + (\triangle x)^2) \to 0,$$

when $\triangle x, \triangle t \to 0$. This shows that the scheme is consistent.

To study its stability, we substitute $u_j^n = \lambda^n e^{ik(j\triangle x)}$ into (2.21), and we obtain

$$\frac{\lambda - \frac{1}{\lambda}}{2\triangle t} = \frac{e^{ik\triangle x} - 2 + e^{-ik\triangle x}}{(\triangle x)^2} = \frac{-4\sin^2 \frac{1}{2}k\triangle x}{(\triangle x)^2}$$

or

$$\lambda^2 + (8\mu \sin^2 \frac{1}{2}k\triangle x)\lambda - 1 = 0,$$

from which we see that the two real roots λ_1 and λ_2 should satisfy

$$\lambda_1 \cdot \lambda_2 = -1.$$

Hence the magnitude of one root must be greater than one no matter how you pick $\triangle t$ and $\triangle x$, in which case we say that the scheme is unconditionally unstable! This warns us that we need to be careful when we develop a finite difference scheme. It is not just simply putting any combinations of difference approximations together.

2.3.2 Some extensions

In this section, we continue developing the finite difference schemes and stability analysis for more general 1-D parabolic equations. More specifically, we demonstrate how the previously introduced techniques be extended to equations with lower-order terms, variable coefficients, more general boundary conditions, and nonlinear terms.

2.3.2.1 Influence of lower-order terms

First, let us consider the governing equation (2.1) augmented by some lower-order terms:

$$u_t = u_{xx} + au_x + bu, \tag{2.22}$$

where a and b are constants.

We can construct an explicit scheme

$$\frac{u_j^{n+1} - u_j^n}{\triangle t} = \frac{u_{j+1}^n - 2u_j^n + u_{j-1}^n}{(\triangle x)^2} + a\frac{u_{j+1}^n - u_{j-1}^n}{2\triangle x} + bu_j^n. \tag{2.23}$$

Using the same stability analysis as in previous examples, we can obtain the amplification factor

$$\lambda = 1 - 4\mu \sin^2 \frac{1}{2}k\triangle x + \frac{a\triangle t}{\triangle x}i \sin k\triangle x + b\triangle t,$$

which gives

$$|\lambda|^2 = (1 - 4\mu \sin^2 \frac{1}{2}k\triangle x + b\triangle t)^2 + (\frac{a\triangle t}{\triangle x})^2 \sin^2 k\triangle x.$$

For model problems with lower-order terms such as (2.22), we need to relax the stability condition $|\lambda| \leq 1$ to

$$|\lambda(k)| \leq 1 + M\triangle t, \quad \text{for any wavenumber } k, \tag{2.24}$$

where the positive constant M is independent of $\triangle t$ and k. Equation (2.24) is often called the von Neumann condition [11, p. 144]. Hence, under the condition $0 < 2\mu \leq 1$, we have

$$|\lambda|^2 = (1 - 4\mu \sin^2 \frac{1}{2}k\triangle x)^2 + 2(1 - 4\mu \sin^2 \frac{1}{2}k\triangle x)b\triangle t$$
$$+ b^2(\triangle t)^2 + a^2\mu\triangle t \sin^2 k\triangle x$$
$$\leq 1 + 2|b|\triangle t + b^2(\triangle t)^2 + \frac{1}{2}a^2\triangle t \leq \left(1 + (\frac{1}{4}a^2 + |b|)\triangle t\right)^2,$$

which yields (2.24). Hence, the added lower-order terms do not change the stability condition imposed on the explicit scheme for (2.1).

It is easy to check that the truncation error of (2.23) is

$$TE_j^n = \frac{D_+^t u(x_j, t_n)}{\triangle t} - \frac{D_+^x D_-^x u(x_j, t_n)}{(\triangle x)^2} - a\frac{u(x_{j+1}, t_n) - u(x_{j-1}, t_n)}{2\triangle x} - bu(x_j, t_n)$$

$$= O(\triangle t + (\triangle x)^2).$$

2.3.2.2   General boundary conditions

Next let us see how to deal with a general boundary condition such as

$$u_x = a(t)u + b(t), \quad \text{at } x = 0. \tag{2.25}$$

One simple scheme for discretizing (2.25) is:

$$\frac{u_1^n - u_0^n}{\triangle x} = a^n u_0^n + b^n, \quad a^n \equiv a(t_n), b^n \equiv b(t_n), \tag{2.26}$$

which coupling with a difference scheme such as (2.4) or (2.8) yields all the numerical solutions u_i^n.

Another popular way to discretize (2.25) is the so-called ghost point technique:

$$\frac{u_1^n - u_{-1}^n}{2\triangle x} = a^n u_0^n + b^n, \tag{2.27}$$

where u_{-1}^n denotes the approximate solution at the grid point to the left of $x = 0$, which is outside of the physical domain. In this case, we need to assume that the scheme (2.4) or (2.8) holds true for $j = 0$, then solve it together with (2.27) for the numerical solution. Note that the scheme (2.27) is more accurate than (2.26).

2.3.2.3   Variable coefficients

In this subsection, we show how to construct and analyze difference schemes for the variable coefficient parabolic equations through two examples.

Example 2.7
A general parabolic equation in self-adjoint form is

$$u_t = (a(x, t)u_x)_x, \quad (x, t) \in (0, 1) \times (0, t_F), \tag{2.28}$$

where the function $a(x, t) > 0$. A nice way to discretize (2.28) is to use central difference twice to approximate the derivative with respect to x, i.e.,

$$(au_x)_x|_j^n \approx \frac{(au_x)_{j+\frac{1}{2}}^n - (au_x)_{j-\frac{1}{2}}^n}{\triangle x} \approx \frac{a_{j+\frac{1}{2}}^n \cdot \frac{u_{j+1}^n - u_j^n}{\triangle x} - a_{j-\frac{1}{2}}^n \cdot \frac{u_j^n - u_{j-1}^n}{\triangle x}}{\triangle x},$$

from which we obtain the explicit scheme

$$\frac{u_j^{n+1} - u_j^n}{\Delta t} = \frac{1}{(\Delta x)^2}[a_{j+\frac{1}{2}}^n(u_{j+1}^n - u_j^n) - a_{j-\frac{1}{2}}^n(u_j^n - u_{j-1}^n)], \quad 1 \le j \le J-1,$$

$$(2.29)$$

where we denote $a_{j\pm\frac{1}{2}}^n = a((j \pm \frac{1}{2})\Delta x, n\Delta t)$. In practical implementation, it is better to rewrite this scheme as follows:

$$u_j^{n+1} = \mu[a_{j+\frac{1}{2}}^n u_{j+1}^n + a_{j-\frac{1}{2}}^n u_{j-1}^n] + [1 - \mu(a_{j+\frac{1}{2}}^n + a_{j-\frac{1}{2}}^n)]u_j^n, \quad (2.30)$$

where we recall the notation $\mu = \frac{\Delta t}{(\Delta x)^2}$.

Note that under the condition

$$a_{max} \cdot \frac{\Delta t}{(\Delta x)^2} \le \frac{1}{2}, \quad a_{max} = \max_{(x,t) \in [0,1] \times [0,t_F]} a(x,t), \quad (2.31)$$

all the coefficients on the right-hand side of (2.30) are non-negative. Hence from (2.30) we have

$$|u_j^{n+1}| \le \mu[a_{j+\frac{1}{2}}^n + a_{j-\frac{1}{2}}^n]\max(|u_{j+1}^n|, |u_{j-1}^n|) + [1 - \mu(a_{j+\frac{1}{2}}^n + a_{j-\frac{1}{2}}^n)]|u_j^n|$$

$$\le \left(\mu[a_{j+\frac{1}{2}}^n + a_{j-\frac{1}{2}}^n] + [1 - \mu(a_{j+\frac{1}{2}}^n + a_{j-\frac{1}{2}}^n)]\right)||u^n||_\infty = ||u^n||_\infty, \quad (2.32)$$

where we used the notation $||u^n||_\infty = \max_{0 \le i \le J}|u_i^n|$. Taking the maximum of (2.32), we immediately have

$$||u^{n+1}||_\infty \le ||u^n||_\infty \le \cdots \le ||u^0||_\infty,$$

which shows that the scheme (2.29) is stable under the constraint (2.31). □

Example 2.8
A more general linear parabolic equation is represented as

$$u_t = a(x,t)u_{xx} + b(x,t)u_x + c(x,t)u + d(x,t), \quad (2.33)$$

where we assume that $a > 0$ and $b > 0$. For the convection-dominated problem (i.e., $b \gg a$), to avoid numerical oscillation, the upwind differencing is preferred: when $b > 0$, use forward difference for the u_x term; otherwise, use backward difference for u_x. For (2.33), the upwind difference scheme becomes:

$$\frac{u_j^{n+1} - u_j^n}{\Delta t} = a_j^n \frac{u_{j+1}^n - 2u_j^n + u_{j-1}^n}{(\Delta x)^2} + b_j^n \frac{u_{j+1}^n - u_j^n}{\Delta x} + c_j^n u_j^n + d_j^n, \quad (2.34)$$

where we denote $a_j^n \equiv a(x_j, t_n)$. Similar notation is used for other functions.

It is easy to check that the truncation error of scheme (2.34) is

$$TE_j^n \equiv \frac{D_+^t u(x_j, t_n)}{\Delta t} - a_j^n \frac{D_+^x D_-^x u(x_j, t_n)}{(\Delta x)^2} - b_j^n \frac{u(x_{j+1}, t_n) - u(x_j, t_n)}{\Delta x}$$
$$-c_j^n u(x_j, t_n) - d_j^n$$
$$= O(\Delta t + \Delta x). \tag{2.35}$$

Finally, let us perform a convergence analysis for the scheme (2.34) with simple boundary condition (2.2) and the initial condition (2.3). Denote the error at point (x_j, t_n) as $e_j^n = u_j^n - u(x_j, t_n)$. Using (2.34) and (2.35), we obtain

$$e_j^{n+1} = e_j^n + \mu a_j^n (e_{j+1}^n - 2e_j^n + e_{j-1}^n) + b_j^n \frac{\Delta t}{\Delta x}(e_{j+1}^n - e_j^n) + \Delta t c_j^n e_j^n - \Delta t \cdot TE_j^n$$

$$= (\mu a_j^n + b_j^n \frac{\Delta t}{\Delta x}) e_{j+1}^n$$

$$+ (1 - 2\mu a_j^n - b_j^n \frac{\Delta t}{\Delta x} + \Delta t c_j^n) e_j^n + \mu a_j^n e_{j-1}^n - \Delta t \cdot TE_j^n. \tag{2.36}$$

Under the constraint

$$2a_j^n \frac{\Delta t}{(\Delta x)^2} + b_j^n \frac{\Delta t}{\Delta x} - c_j^n \Delta t \leq 1, \tag{2.37}$$

all the coefficients on the right-hand side of (2.36) are non-negative. Hence we have

$$|e_j^{n+1}| \leq (1 + \Delta t |c_j^n|) \max\{|e_{j+1}^n|, |e_j^n|, |e_{j-1}^n|\} + \Delta t \cdot M(\Delta t + \Delta x), \tag{2.38}$$

where we used the estimate (2.35).

Let $e^n = \max_{0 \leq j \leq J} |e_j^n|$ and $c = \max_{0 \leq j \leq J} |c_j^n|$. Choosing the maximum of both sides of (2.38) with respect to j, we obtain

$$e^{n+1} \leq (1 + c\Delta t)e^n + M\Delta t(\Delta t + \Delta x),$$

which yields

$$e^n \leq (1 + c\Delta t)^2 e^{n-2} + [(1 + c\Delta t) + 1]M\Delta t(\Delta t + \Delta x) \leq \cdots$$
$$\leq (1 + c\Delta t)^n e^0 + [(1 + c\Delta t)^{n-1} + \cdots (1 + c\Delta t) + 1]M\Delta t(\Delta t + \Delta x),$$

from which, along with the fact $e^0 = 0$ and the identity

$$1 + x + \cdots + x^{n-1} = \frac{x^n - 1}{x - 1},$$

we obtain

$$e^n \leq \frac{(1 + c\Delta t)^n - 1}{c\Delta t} M\Delta t(\Delta t + \Delta x) \leq e^{nc\Delta t} \cdot \tilde{M}(\Delta t + \Delta x) \leq c^{ct_F} \cdot \tilde{M}(\Delta t + \Delta x),$$

i.e., the maximum pointwise error of the scheme (2.34) is $O(\Delta t + \Delta x)$. ∎

2.3.2.4 Nonlinear parabolic PDEs

For nonlinear PDEs, we can construct the finite difference schemes by simply freezing those nonlinear terms. But it is difficult to prove stability and convergence. Here we just provide one example. More discussions of nonlinear PDEs can be found in other books (e.g., [1, 7]).

Let us consider the problem

$$u_t = (u^3)_{xx} = (3u^2 u_x)_x. \tag{2.39}$$

A simple explicit scheme can be constructed as

$$\frac{u_j^{n+1} - u_j^n}{\Delta t} = 3(u^2)_j^n \frac{u_{j+1}^n - 2u_j^n + u_{j-1}^n}{(\Delta x)^2}. \tag{2.40}$$

An implicit scheme can be developed as

$$\frac{u_j^{n+1} - u_j^n}{\Delta t} \approx \frac{(3u^2 u_x)|_{j+\frac{1}{2}}^{n+1} - (3u^2 u_x)|_{j-\frac{1}{2}}^{n+1}}{\Delta x}$$

$$\approx [(3u^2)|_{j+\frac{1}{2}}^n \frac{u_{j+1}^{n+1} - u_j^{n+1}}{\Delta x} - (3u^2)|_{j-\frac{1}{2}}^n \frac{u_j^{n+1} - u_{j-1}^{n+1}}{\Delta x}]/\Delta x$$

$$= \frac{(3u^2)_{j+\frac{1}{2}}^n (u_{j+1}^{n+1} - u_j^{n+1}) - (3u^2)_{j-\frac{1}{2}}^n (u_j^{n+1} - u_{j-1}^{n+1})}{(\Delta x)^2}.$$

2.4   2-D and 3-D parabolic equations

In this section, we discuss how to construct various finite difference schemes and carry out the corresponding stability analysis for both 2-D and 3-D parabolic equations. The ideas and techniques are illustrated through many examples.

2.4.1   Standard explicit and implicit methods

Let us consider a 2-D parabolic differential equation

$$u_t = u_{xx} + u_{yy}, \quad (x, y, t) \in (0, 1)^2 \times (0, t_F), \tag{2.41}$$

with proper boundary conditions and initial condition.

To construct a difference scheme, we assume that the domain $(0, 1)^2$ is partitioned into a uniform rectangular grid, with a spacing Δx in the x-direction and Δy in the y-direction, i.e.,

$$0 = x_0 < x_1 \cdots < x_{J_x} = 1, \quad \Delta x = \frac{1}{J_x},$$

$$0 = y_0 < y_1 \cdots < y_{J_y} = 1, \quad \Delta y = \frac{1}{J_y}.$$

For the time domain $(0, t_F)$, we can use a uniform grid $t_n = n\Delta t, 0 \leq n \leq N$, where $\Delta t = \frac{t_F}{N}$. We denote the approximate solution

$$u_{r,s}^n \approx u(x_r, y_s, t_n), \quad 0 \leq r \leq J_x, \; 0 \leq s \leq J_y, 0 \leq n \leq N.$$

Below we will use quite often the second-order difference operator δ_x^2, which is defined as

$$\delta_x^2 u_{r,s}^n = u_{r+1,s}^n - 2u_{r,s}^n + u_{r-1,s}^n.$$

Example 2.9

Using the operator δ_x^2, we can obtain an explicit scheme for (2.41):

$$\frac{u_{r,s}^{n+1} - u_{r,s}^n}{\Delta t} = \frac{\delta_x^2 u_{r,s}^n}{(\Delta x)^2} + \frac{\delta_y^2 u_{r,s}^n}{(\Delta y)^2}. \tag{2.42}$$

By the Taylor expansion, it is easy to see that the truncation error is

$$TE_{r,s}^n = [\frac{D_+^t u}{\Delta t} - (\frac{\delta_x^2 u}{(\Delta x)^2} + \frac{\delta_y^2 u}{(\Delta y)^2})](x_r, y_s, t_n) = O(\Delta t + (\Delta x)^2 + (\Delta y)^2).$$

The stability of the scheme (2.42) can be obtained by the von Neumann technique. Substituting

$$u_{r,s}^n = \lambda^n e^{i(k_x r \Delta x + k_y s \Delta y)}$$

into (2.42) and simplifying the result, we obtain the amplification factor

$$\lambda = 1 - 4\frac{\Delta t}{(\Delta x)^2}\sin^2\frac{k_x \Delta x}{2} - 4\frac{\Delta t}{(\Delta y)^2}\sin^2\frac{k_y \Delta y}{2}.$$

It is not difficult to see that the condition

$$\frac{\Delta t}{(\Delta x)^2} + \frac{\Delta t}{(\Delta y)^2} \leq \frac{1}{2}, \tag{2.43}$$

guarantees that $|\lambda| \leq 1$ for all wavenumbers k_x and k_y. Hence (2.43) guarantees that the scheme (2.42) is stable. ∎

Example 2.10

Similarly, we can obtain the Crank-Nicolson (CN) scheme for (2.41):

$$\frac{u_{r,s}^{n+1} - u_{r,s}^n}{\Delta t} = \frac{1}{2}[\frac{\delta_x^2(u_{r,s}^n + u_{r,s}^{n+1})}{(\Delta x)^2} + \frac{\delta_y^2(u_{r,s}^n + u_{r,s}^{n+1})}{(\Delta y)^2}]. \tag{2.44}$$

Denote $\mu_x = \frac{\Delta t}{(\Delta x)^2}, \mu_y = \frac{\Delta t}{(\Delta y)^2}$. The CN scheme can be rewritten as

$$(1 - \frac{1}{2}\mu_x \delta_x^2 - \frac{1}{2}\mu_y \delta_y^2)u_{r,s}^{n+1} = (1 + \frac{1}{2}\mu_x \delta_x^2 + \frac{1}{2}\mu_y \delta_y^2)u_{r,s}^n. \tag{2.45}$$

The truncation error of the scheme (2.44) can be easily obtained as

$$TE^n_{r,s} = O((\triangle t)^2 + (\triangle x)^2 + (\triangle y)^2).$$

As for the stability, we can easily obtain the amplification factor

$$\lambda(k) = \frac{1 - 2\mu_x \sin^2 \frac{1}{2}(k_x \triangle x) - 2\mu_y \sin^2 \frac{1}{2}(k_y \triangle y)}{1 + 2\mu_x \sin^2 \frac{1}{2}(k_x \triangle x) + 2\mu_y \sin^2 \frac{1}{2}(k_y \triangle y)},$$

whose amplitude is always less than or equal to one for any mesh sizes $\triangle x$ and $\triangle y$. Hence the CN scheme is unconditionally stable.

Note that at each time step, the CN scheme requires the solution of a $(J_x - 1) \times (J_y - 1)$ pentadiagonal matrix, which is not a feasible approach unless the mesh points are very small. ▯

Example 2.11
Now let us consider the 3-D parabolic equation

$$u_t = u_{xx} + u_{yy} + u_{zz}, \tag{2.46}$$

with proper initial and boundary conditions.

The explicit difference scheme can be obtained by simple extensions of 2-D problems:

$$\frac{u^{n+1} - u^n}{\triangle t} = \frac{\delta_x^2 u^n}{(\triangle x)^2} + \frac{\delta_y^2 u^n}{(\triangle y)^2} + \frac{\delta_z^2 u^n}{(\triangle z)^2}, \tag{2.47}$$

where for simplicity we skipped the explicit dependence on nodal indices. To study its stability condition, we substitute

$$u^n = \lambda^n e^{i(k_x r \triangle x + k_y s \triangle y + k_z j \triangle z)}$$

into (2.47) and obtain the amplification factor

$$\lambda = 1 - 4\frac{\triangle t}{(\triangle x)^2} \sin^2 \frac{1}{2}(k_x \triangle x) - 4\frac{\triangle t}{(\triangle y)^2} \sin^2 \frac{1}{2}(k_y \triangle y) - 4\frac{\triangle t}{(\triangle z)^2} \sin^2 \frac{1}{2}(k_z \triangle z). \tag{2.48}$$

To satisfy the stability condition $|\lambda| \leq 1$ for all wavenumbers, $\triangle t$ must obey the condition

$$\frac{\triangle t}{(\triangle x)^2} + \frac{\triangle t}{(\triangle y)^2} + \frac{\triangle t}{(\triangle z)^2} \leq \frac{1}{2}. \tag{2.49}$$

We want to remark that the condition (2.49) is quite stringent by comparing it to the corresponding 1-D problem. Let $\triangle x = \triangle y = \triangle z = h$, then (2.49) is equivalent to

$$\frac{\triangle t}{h^2} \leq \frac{1}{6},$$

which implies that in 3-D $\triangle t$ must be three times smaller than in the 1-D problem. This is one of the major reasons that the explicit method is rarely used for 3-D parabolic problems. ▯

2.4.2   The ADI methods for 2-D problems

For the Crank-Nicolson scheme (2.45), we see that at each time step, we have
to solve a system of $(J_x - 1) \times (J_y - 1)$ linear equations for unknowns $u_{r,s}^{n+1}$.
Solving such a system is quite laborious if we consider that first it is not easy
to form the coefficient matrix, and second it is quite expensive to store and
solve the matrix directly. More details will be discussed when we come to the
elliptic problems in a later chapter.

A simple and efficient method, the so-called alternate direction implicit
(ADI) method, for solving 2-D parabolic problems was first proposed by
Peaceman and Rachford in 1955 [12]. The basic idea is to break a 2-D problem into two 1-D problems solved by implicit schemes without sacrificing the
stability constraint.

Example 2.12
Let us start with the Peaceman-Rachford (PR) scheme [12]:

$$\frac{u_{r,s}^* - u_{r,s}^n}{\frac{1}{2}\Delta t} = \frac{\delta_x^2 u_{r,s}^*}{(\Delta x)^2} + \frac{\delta_y^2 u_{r,s}^n}{(\Delta y)^2}, \tag{2.50}$$

$$\frac{u_{r,s}^{n+1} - u_{r,s}^*}{\frac{1}{2}\Delta t} = \frac{\delta_x^2 u_{r,s}^*}{(\Delta x)^2} + \frac{\delta_y^2 u_{r,s}^{n+1}}{(\Delta y)^2}. \tag{2.51}$$

We can rewrite this scheme as

$$(1 - \frac{1}{2}\mu_x \delta_x^2)u_{r,s}^* = (1 + \frac{1}{2}\mu_y \delta_y^2)u_{r,s}^n, \tag{2.52}$$

$$(1 - \frac{1}{2}\mu_y \delta_y^2)u_{r,s}^{n+1} = (1 + \frac{1}{2}\mu_x \delta_x^2)u_{r,s}^*. \tag{2.53}$$

The stability of the PR scheme (2.50)-(2.51) can be obtained using the von
Neumann technique. The amplification factor for (2.52) is

$$\lambda_1 = (1 - 2\mu_y \sin^2 \frac{1}{2}k_y \Delta y)/(1 + 2\mu_x \sin^2 \frac{1}{2}k_x \Delta x),$$

while for (2.53) the amplification factor is

$$\lambda_2 = (1 - 2\mu_x \sin^2 \frac{1}{2}k_x \Delta x)/(1 + 2\mu_y \sin^2 \frac{1}{2}k_y \Delta y).$$

Hence, the combined two-step scheme has an amplification factor

$$\lambda = \lambda_1 \cdot \lambda_2 = \frac{(1 - 2\mu_y \sin^2 \frac{1}{2}k_y \Delta y)(1 - 2\mu_x \sin^2 \frac{1}{2}k_x \Delta x)}{(1 + 2\mu_x \sin^2 \frac{1}{2}k_x \Delta x)(1 + 2\mu_y \sin^2 \frac{1}{2}k_y \Delta y)},$$

whose magnitude is always less than or equal to one. Hence the PR scheme
is unconditionally stable.

We can eliminate u^* from (2.52)-(2.53) to obtain a single step algorithm

$$(1 - \frac{1}{2}\mu_x\delta_x^2)(1 - \frac{1}{2}\mu_y\delta_y^2)u_{r,s}^{n+1} = (1 + \frac{1}{2}\mu_x\delta_x^2)(1 + \frac{1}{2}\mu_y\delta_y^2)u_{r,s}^n, \qquad (2.54)$$

which is a perturbation of the CN scheme (2.45). Hence the truncation error is $O((\triangle t)^2 + (\triangle x)^2 + (\triangle y)^2)$. Furthermore, the PR scheme (2.52)-(2.53) is easy to implement, since at each step we only need to solve two tridiagonal equations.

To implement the PR scheme, we need boundary conditions for u^*. For example, let us assume that the original problem imposes the simple Dirichlet boundary condition

$$u(x, y, t) = g(x, y, t) \quad \text{on} \quad x = 0, \ 1.$$

Subtracting (2.51) from (2.50), we obtain

$$u_{r,s}^* = \frac{1}{2}(u_{r,s}^n + u_{r,s}^{n+1}) + \frac{\triangle t}{4(\triangle y)^2}\delta_y^2(u_{r,s}^n - u_{r,s}^{n+1}),$$

which yields the boundary condition for u^* as follows:

$$u^* = \frac{1}{2}(1 + \frac{1}{2}\mu_y\delta_y^2)g^n + \frac{1}{2}(1 - \frac{1}{2}\mu_y\delta_y^2)g^{n+1} \quad \text{on} \quad x = 0, \ 1.$$

☐

Example 2.13
Douglas and Rachford [3] proposed another ADI scheme

$$(1 - \mu_x\delta_x^2)u_{r,s}^* = (1 + \mu_y\delta_y^2)u_{r,s}^n, \qquad (2.55)$$

$$(1 - \mu_y\delta_y^2)u_{r,s}^{n+1} = u_{r,s}^* + \mu_y\delta_y^2 u_{r,s}^n. \qquad (2.56)$$

Eliminating the intermediate variable u^*, the Douglas-Rachford (DR) method leads to the formula

$$(1 - \mu_x\delta_x^2)(1 - \mu_y\delta_y^2)u_{r,s}^{n+1} = (1 + \mu_x\delta_x^2\mu_y\delta_y^2)u_{r,s}^n,$$

from which it is easy to prove that the DR method is unconditionally stable. Furthermore, the truncation error can be proved to be $O((\triangle t)^2 + \triangle t \cdot (\triangle x)^2)$. The boundary condition for u^* can be obtained from (2.56), i.e.,

$$u^* = (1 - \mu_y\delta_y^2)g^{n+1} - \mu_y\delta_y^2 g^n \quad \text{on} \quad x = 0, \ 1.$$

☐

Example 2.14
Mitchell and Fairweather [10] proposed a high-order ADI scheme:

$$[1 - \frac{1}{2}(\mu_x - \frac{1}{6})\delta_x^2]u_{r,s}^* = [1 + \frac{1}{2}(\mu_y + \frac{1}{6})\delta_y^2]u_{r,s}^n, \qquad (2.57)$$

$$[1 - \frac{1}{2}(\mu_y - \frac{1}{6})\delta_y^2]u_{r,s}^{n+1} = [1 + \frac{1}{2}(\mu_x + \frac{1}{6})\delta_x^2]u_{r,s}^*. \tag{2.58}$$

Eliminating the intermediate variable u^*, we obtain

$$[1 - \frac{1}{2}(\mu_x - \frac{1}{6})\delta_x^2][1 - \frac{1}{2}(\mu_y - \frac{1}{6})\delta_y^2]u_{r,s}^{n+1}$$
$$= [1 + \frac{1}{2}(\mu_x + \frac{1}{6})\delta_x^2][1 + \frac{1}{2}(\mu_y + \frac{1}{6})\delta_y^2]u_{r,s}^n,$$

which gives the amplification factor

$$\lambda = \frac{[1 - 2(\mu_x + \frac{1}{6})\sin^2 \frac{1}{2}k_x\triangle x][1 - 2(\mu_y + \frac{1}{6})\sin^2 \frac{1}{2}k_y\triangle y]}{[1 + 2(\mu_x - \frac{1}{6})\sin^2 \frac{1}{2}k_x\triangle x][1 + 2(\mu_y - \frac{1}{6})\sin^2 \frac{1}{2}k_y\triangle y]}$$
$$= \frac{(1 - \frac{1}{3}\sin^2 \frac{1}{2}k_x\triangle x) - 2\mu_x \sin^2 \frac{1}{2}k_x\triangle x}{(1 - \frac{1}{3}\sin^2 \frac{1}{2}k_x\triangle x) + 2\mu_x \sin^2 \frac{1}{2}k_x\triangle x} \cdot \frac{(1 - \frac{1}{3}\sin^2 \frac{1}{2}k_y\triangle y) - 2\mu_y \sin^2 \frac{1}{2}k_y\triangle y}{(1 - \frac{1}{3}\sin^2 \frac{1}{2}k_y\triangle y) + 2\mu_y \sin^2 \frac{1}{2}k_y\triangle y},$$

which guarantees that $|\lambda| \leq 1$ for any μ_x and μ_y. Thus this ADI scheme is also unconditionally stable. It is not difficult to check that the truncation error is $O((\triangle t)^2 + (\triangle x)^4 + (\triangle y)^4)$. □

2.4.3   The ADI methods for 3-D problems

Similar ADI ideas developed for 2-D problems can be extended to 3-D parabolic equations as demonstrated below by three examples.

Example 2.15
The Douglas-Rachford (DR) ADI method for 3-D problems makes one dimension implicit each time while leaving the other two dimensions explicit:

$$\frac{u^{n+\frac{1}{3}} - u^n}{\triangle t} = \frac{\delta_x^2 u^{n+\frac{1}{3}}}{(\triangle x)^2} + \frac{\delta_y^2 u^n}{(\triangle y)^2} + \frac{\delta_z^2 u^n}{(\triangle z)^2}, \tag{2.59}$$

$$\frac{u^{n+\frac{2}{3}} - u^{n+\frac{1}{3}}}{\triangle t} = \frac{\delta_y^2(u^{n+\frac{2}{3}} - u^n)}{(\triangle y)^2}, \tag{2.60}$$

$$\frac{u^{n+1} - u^{n+\frac{2}{3}}}{\triangle t} = \frac{\delta_z^2(u^{n+1} - u^n)}{(\triangle z)^2}, \tag{2.61}$$

where we skipped the explicit dependence on mesh points. We can rewrite (2.59)-(2.61) as

$$(1 - \mu_x\delta_x^2)u^{n+\frac{1}{3}} = (1 + \mu_y\delta_y^2 + \mu_z\delta_z^2)u^n,$$
$$(1 - \mu_y\delta_y^2)u^{n+\frac{2}{3}} = u^{n+\frac{1}{3}} - \mu_y\delta_y^2 u^n,$$
$$(1 - \mu_z\delta_z^2)u^{n+1} = u^{n+\frac{2}{3}} - \mu_z\delta_z^2 u^n.$$

Multiplying the above second equation by $(1-\mu_x\delta_x^2)$ and the third equation by $(1-\mu_x\delta_x^2)(1-\mu_y\delta_y^2)$, respectively, then adding up the results and using the first equation, we have

$$(1-\mu_x\delta_x^2)(1-\mu_y\delta_y^2)(1-\mu_z\delta_z^2)u^{n+1}$$
$$= (1+\mu_y\delta_y^2+\mu_z\delta_z^2)u^n - (1-\mu_x\delta_x^2)\mu_y\delta_y^2 u^n - (1-\mu_x\delta_x^2)(1-\mu_y\delta_y^2)\mu_z\delta_z^2 u^n$$
$$= (1+\mu_x\delta_x^2\mu_y\delta_y^2+\mu_x\delta_x^2\mu_z\delta_z^2+\mu_y\delta_y^2\mu_z\delta_z^2-\mu_x\delta_x^2\mu_y\delta_y^2\mu_z\delta_z^2)u^n. \qquad (2.62)$$

Note that equation (2.62) can also be written in the form

$$u^{n+1} = u^n + (\mu_x\delta_x^2+\mu_y\delta_y^2+\mu_z\delta_z^2)u^{n+1}$$
$$- [(\mu_x\delta_x^2\mu_y\delta_y^2+\mu_x\delta_x^2\mu_z\delta_z^2+\mu_y\delta_y^2\mu_z\delta_z^2) - \mu_x\delta_x^2\mu_y\delta_y^2\mu_z\delta_z^2](u^{n+1}-u^n),$$

which is a perturbation of backward implicit approximation to (2.46). Hence the truncation error is $O(\triangle t + (\triangle x)^2 + (\triangle y)^2 + (\triangle z)^2)$.

The standard von Neumann stability analysis leads to the amplification factor

$$\lambda = \frac{1+a_xa_y+a_xa_z+a_ya_z+a_xa_ya_z}{(1+a_x)(1+a_y)(1+a_z)}, \qquad (2.63)$$

where $a_x = 4\mu_x\sin^2\frac{1}{2}k_x\triangle x$. Similar notations are used for a_y and a_z. From (2.63), we see that $|\lambda| \le 1$ without any constraints, i.e., the algorithm is unconditionally stable. $\square$

Example 2.16
Douglas [2] developed a more accurate ADI scheme, which is also a perturbation of the Crank-Nicolson scheme:

$$\frac{u^*-u^n}{\triangle t} = \frac{1}{h^2}[\frac{1}{2}\delta_x^2(u^*+u^n)+\delta_y^2 u^n+\delta_z^2 u^n], \qquad (2.64)$$

$$\frac{u^{**}-u^n}{\triangle t} = \frac{1}{h^2}[\frac{1}{2}\delta_x^2(u^*+u^n)+\frac{1}{2}\delta_y^2(u^{**}+u^n)+\delta_z^2 u^n], \qquad (2.65)$$

$$\frac{u^{n+1}-u^n}{\triangle t} = \frac{1}{2h^2}[\delta_x^2(u^*+u^n)+\delta_y^2(u^{**}+u^n)+\delta_z^2(u^{n+1}+u^n)],(2.66)$$

where for simplicity we assume that $\triangle x = \triangle y = \triangle z = h$. When the mesh sizes are different, similar schemes can be built up easily.

Let $\mu = \frac{\triangle t}{h^2}$. We can rewrite (2.64)-(2.66) as

$$(1-\frac{1}{2}\mu\delta_x^2)u^* = (1+\frac{1}{2}\mu\delta_x^2+\mu\delta_y^2+\mu\delta_z^2)u^n, \qquad (2.67)$$

$$(1-\frac{1}{2}\mu\delta_y^2)u^{**} = u^* - \frac{1}{2}\mu\delta_y^2 u^n, \qquad (2.68)$$

$$(1-\frac{1}{2}\mu\delta_z^2)u^{n+1} = u^{**} - \frac{1}{2}\mu\delta_z^2 u^n. \qquad (2.69)$$

Eliminating the intermediate variables u^* and u^{**}, (2.67)-(2.69) can be simplified to

$$(1 - \frac{1}{2}\mu\delta_x^2)(1 - \frac{1}{2}\mu\delta_y^2)(1 - \frac{1}{2}\mu\delta_z^2)u^{n+1}$$

$$= [1 + \frac{\mu}{2}(\delta_x^2 + \delta_y^2 + \delta_z^2) + \frac{\mu^2}{4}(\delta_x^2\delta_y^2 + \delta_x^2\delta_z^2 + \delta_y^2\delta_z^2) - \frac{\mu^3}{8}\delta_x^2\delta_y^2\delta_z^2]u^n, \quad (2.70)$$

or

$$u^{n+1} - u^n = \frac{\mu}{2}(\delta_x^2 + \delta_y^2 + \delta_z^2)(u^{n+1} + u^n)$$

$$+ [\frac{\mu^2}{4}(\delta_x^2\delta_y^2 + \delta_x^2\delta_z^2 + \delta_y^2\delta_z^2) - \frac{\mu^3}{8}\delta_x^2\delta_y^2\delta_z^2](u^{n+1} - u^n),$$

which is a perturbation of the Crank-Nicolson scheme to (2.46). Thus the truncation error of the scheme (2.64)-(2.66) is $O((\triangle t)^2 + h^2)$.

From (2.70), the amplification factor of the Douglas ADI scheme can be found as

$$\lambda = \frac{1 - (a_x + a_y + a_z) + (a_x a_y + a_x a_z + a_y a_z) + a_x a_y a_z}{(1 + a_x)(1 + a_y)(1 + a_z)}$$

from which we see that $|\lambda| \leq 1$ always holds true, i.e., the scheme is unconditionally stable. Here we denote

$$a_x = 2\sin^2\frac{1}{2}k_x\triangle x, \quad a_y = 2\sin^2\frac{1}{2}k_y\triangle y, \quad a_z = 2\sin^2\frac{1}{2}k_z\triangle z.$$

□

Example 2.17

Fairweather and Mitchell [4] constructed a number of high-order ADI schemes which are unconditionally stable. For example, a direct extension of (2.57)-(2.58) gives the following ADI scheme:

$$[1 - \frac{1}{2}(\mu_x - \frac{1}{6})\delta_x^2]u^* = [1 + \frac{1}{2}(\mu_y + \frac{1}{6})\delta_y^2]u^n, \quad (2.71)$$

$$[1 - \frac{1}{2}(\mu_y - \frac{1}{6})\delta_y^2]u^{**} = [1 + \frac{1}{2}(\mu_z + \frac{1}{6})\delta_z^2]u^*, \quad (2.72)$$

$$[1 - \frac{1}{2}(\mu_z - \frac{1}{6})\delta_z^2]u^{n+1} = [1 + \frac{1}{2}(\mu_x + \frac{1}{6})\delta_x^2]u^{**}. \quad (2.73)$$

We can eliminate the intermediate variables u^* and u^{**} of (2.71)-(2.73) to obtain

$$[1 - \frac{1}{2}(\mu_x - \frac{1}{6})\delta_x^2][1 - \frac{1}{2}(\mu_y - \frac{1}{6})\delta_y^2][1 - \frac{1}{2}(\mu_z - \frac{1}{6})\delta_z^2]u^{n+1}$$

$$= [1 + \frac{1}{2}(\mu_x + \frac{1}{6})\delta_x^2][1 + \frac{1}{2}(\mu_y + \frac{1}{6})\delta_y^2][1 + \frac{1}{2}(\mu_z + \frac{1}{6})\delta_z^2]u^n,$$

from which the unconditional stability can be proved easily. □

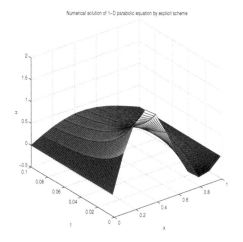

FIGURE 2.1
The numerical solution for the 1-D parabolic problem.

2.5 Numerical examples with MATLAB codes

Here we provide a MATLAB code for solving a general parabolic equation

$$u_t(x,t) = u_{xx}(x,t), \quad x_l < x < x_r, 0 < t < t_F$$
$$u(x,t)|_{t=0} = f(x), \quad x_l \leq x \leq x_r$$
$$u(x,t)|_{x=0} = g_l(t), \quad u(x,t)|_{x=1} = g_r(t) \quad 0 \leq t \leq t_F.$$

The code solves the above problem with $x_l = 0, x_r = 1, t_F = 0.1$ by the explicit scheme. The boundary and initial conditions are chosen properly such that our problem has the analytic solution

$$u(x,t) = \sin(\pi x)e^{-\pi^2 t} + \sin(2\pi x)e^{-4\pi^2 t}.$$

With a 11×51 mesh grid for the domain $(x,t) \equiv (0,1) \times (0,0.1)$, the numerical solution (cf., Fig. 2.1) looks no different from the analytic solution. Of course detailed pointwise errors can be calculated by modifying the code.

```
%-----------------------------------------------------------------
% para1d.m:
%      use the explicit scheme to solve the parabolic equation
%      u_t(x,t) = u_{xx}(x,t),              xl < x < xr, 0 < t < tf
%      u(x,0) = f(x),                       xl < x < xr
%      u(0,t) = gl(t), u(1,t) = gr(t),      0  < t < tf
```

```
%
% A special case is choosing f and g properly such that the
% analytic solution is:
%   u(x,t)= sin(pi*x)*e^(-pi^2*t) + sin(2*pi*x)*e^(-4*pi^2*t)
%
% we solve this program by the explicit scheme:
%      u(j,n+1) = u(j,n) + v*(u(j+1,n) - 2*u(j,n) + u(j-1,n))
%-----------------------------------------------------------------
clear all;                      % clear all variables in memory

xl=0; xr=1;                     % x domain [xl,xr]
J = 10;                         % J: number of division for x
dx = (xr-xl) / J;               % dx: mesh size
tf = 0.1;                       % final simulation time
Nt = 50;                        % Nt: number of time steps
dt = tf/Nt;

mu = dt/(dx)^2;

if mu > 0.5          % make sure dt satisfy stability condition
      error('mu should < 0.5!')
end

% Evaluate the initial conditions
x = xl : dx : xr;                   % generate the grid point
% f(1:J+1) since array index starts from 1
f = sin(pi*x) + sin(2*pi*x);

% store the solution at all grid points for all time steps
u = zeros(J+1,Nt);

% Find the approximate solution at each time step
for n = 1:Nt
      t = n*dt;           % current time
      % boundary condition at left side
      gl = sin(pi*xl)*exp(-pi*pi*t)+sin(2*pi*xl)*exp(-4*pi*pi*t);
      % boundary condition at right side
      gr = sin(pi*xr)*exp(-pi*pi*t)+sin(2*pi*xr)*exp(-4*pi*pi*t);
      if n==1    % first time step
         for j=2:J    % interior nodes
         u(j,n) = f(j) + mu*(f(j+1)-2*f(j)+f(j-1));
         end
         u(1,n) = gl;    % the left-end point
         u(J+1,n) = gr; % the right-end point
      else
```

```
        for j=2:J    % interior nodes
           u(j,n)=u(j,n-1)+mu*(u(j+1,n-1)-2*u(j,n-1)+u(j-1,n-1));
        end
        u(1,n) = gl;   % the left-end point
        u(J+1,n) = gr; % the right-end point
    end

    % calculate the analytic solution
    for j=1:J+1
        xj = xl + (j-1)*dx;
        u_ex(j,n)=sin(pi*xj)*exp(-pi*pi*t) ...
                 +sin(2*pi*xj)*exp(-4*pi*pi*t);
    end
end

% Plot the results
tt = dt : dt : Nt*dt;
figure(1)
colormap(gray);    % draw gray figure
surf(x,tt, u');    % 3-D surface plot
xlabel('x')
ylabel('t')
zlabel('u')
title('Numerical solution of 1-D parabolic equation')

figure(2)
surf(x,tt, u_ex');    % 3-D surface plot
xlabel('x')
ylabel('t')
zlabel('u')
title('Analytic solution of 1-D parabolic equation')
```

2.6 Bibliographical remarks

In this chapter, we only describe some classic difference schemes for linear parabolic equations on rectangular domains. For discussions on irregular domains, nonlinear problems, and more advanced schemes, interested readers are encouraged to consult other books [1, 5, 6, 8, 9, 13, 14].

2.7   Exercises

1. For the θ-scheme (2.11), prove that: when $0 \le \theta < \frac{1}{2}$, it is stable if $\mu \le \frac{1}{2(1-2\theta)}$; when $\frac{1}{2} \le \theta \le 1$, it is stable for all μ.

2. For the θ-scheme (2.12), the coefficient matrix is a tridiagonal matrix, which can be solved very efficiently by the Thomas algorithm (e.g., [11]). For a nonsingular tridiagonal matrix

$$A = \begin{bmatrix} a_1 & c_1 & & & & 0 \\ b_2 & a_2 & c_2 & & & \\ & & \cdot & \cdot & \cdot & \\ & & & \cdot & \cdot & \cdot \\ & & & & \cdot & \cdot & \cdot \\ & & & b_{n-1} & a_{n-1} & c_{n-1} \\ 0 & & & & b_n & a_n \end{bmatrix}$$

we have its LU factorization, i.e., $A = LU$, where

$$L = \begin{bmatrix} 1 & 0 & & & & 0 \\ \beta_2 & 1 & 0 & & & \\ & & \cdot & \cdot & \cdot & \\ & & & \cdot & \cdot & \cdot \\ & & & \beta_{n-1} & 1 & 0 \\ 0 & & & & \beta_n & 1 \end{bmatrix}, \quad U = \begin{bmatrix} \alpha_1 & c_1 & & & & 0 \\ 0 & \alpha_2 & c_2 & & & \\ & & \cdot & \cdot & \cdot & \\ & & & \cdot & \cdot & \cdot \\ & & & 0 & \alpha_{n-1} & c_{n-1} \\ 0 & & & & 0 & \alpha_n \end{bmatrix}.$$

Prove that the α_i and β_i can be computed by the relations

$$\alpha_1 = a_1, \quad \beta_i = b_i/\alpha_{i-1}, \quad \alpha_i = a_i - \beta_i c_{i-1}, \quad i = 2, \cdots, n.$$

Hence solving a tridiagonal system $A\vec{x} = \vec{f}$ is equivalent to solving $L\vec{y} = \vec{f}$ and $U\vec{x} = \vec{y}$, i.e.,

$$y_1 = f_1, \quad y_i = f_i - \beta_i y_{i-1}, \quad i = 2, \cdots, n,$$
$$x_n = y_n/\alpha_n, \quad x_i = (y_i - c_i x_{i+1})/\alpha_i, \quad i = n-1, \cdots, 1.$$

Prove that the total number of multiplication/division is $5n - 4$.

3. Use the method (2.26) and modify the code *para1d.m* to solve the problem

$$u_t = u_{xx}, \quad 0 < x < 1, \ 0 < t < 0.1,$$
$$u(x,0) = x, \quad 0 < x < 1,$$
$$u_x(0,t) = u_x(1,t) = 0, \quad 0 < t < 0.1.$$

Plot the numerical solution at $t = 0.1$ solved with a 21×101 grid on the (x, t) domain. Compare the numerical solution to its analytical solution:

$$u(x,t) = \frac{1}{2} - \sum_{k=1}^{\infty} \frac{4}{(2k-1)^2 \pi^2} e^{-(2k-1)^2 \pi^2 t} \cos(2k-1)\pi x.$$

Try to solve the problem using various grids to see the convergence rate.

4. Consider the nonlinear heat equation

$$u_t = (a(u)u_x)_x, \quad x \in (0,1), \quad t \in (0,T), \tag{2.74}$$
$$u(0,t) = u(1,t) = 0, \tag{2.75}$$
$$u(x,0) = f(x), \tag{2.76}$$

where we assume that there exist constants a_* and a^* such that

$$0 < a_* \leq a(u) \leq a^*$$

for all u. Derive the following explicit scheme for this problem

$$\frac{u_j^{n+1} - u_j^n}{\Delta t} = \frac{a_{j+\frac{1}{2}}^n (u_{j+1}^n - u_j^n) - a_{j-\frac{1}{2}}^n (u_j^n - u_{j-1}^n)}{(\Delta x)^2}, \tag{2.77}$$

where $a_{j+\frac{1}{2}}^n = \frac{1}{2}(a(u_{j+1}^n) + a(u_j^n))$. The stability condition can be found by freezing the nonlinear coefficient and using the von Neumann method. Recall that when $a(u) = a_0$ is a constant, we require that $a_0 \frac{\Delta t}{(\Delta x)^2} \leq \frac{1}{2}$. Hence for our nonlinear problem, we need to require that the time step

$$\Delta t \leq \frac{(\Delta x)^2}{2a^*}. \tag{2.78}$$

Program the scheme (2.77) and solve the problem with

$$a(u) = \frac{1 + 2u^2}{1 + u^2}, \quad f(x) = \sin(2\pi x), \quad \Delta x = 0.02, \quad T = 0.1.$$

Try different time steps Δt to see what happens if the condition (2.78) is not satisfied.

5. Prove that under the condition (2.78), the scheme (2.77) satisfies the estimate

$$\min_{0 \leq j \leq J} u_j^n \leq u_j^{n+1} \leq \max_{0 \leq j \leq J} u_j^n,$$

where we assume that the spatial grid points are $0 = x_0 \leq x_1 \leq \cdots \leq x_J = 1$.

6. Douglas and Rachford [3] designed another ADI scheme for solving 2-D parabolic equation $u_t = u_{xx} + u_{yy}$ as follows:

$$(1 - \frac{1}{2}\mu_x \delta_x^2)u_{r,s}^* = (1 + \frac{1}{2}\mu_x \delta_x^2 + \mu_y \delta_y^2)u_{r,s}^n, \tag{2.79}$$

$$(1 - \frac{1}{2}\mu_y\delta_y^2)u_{r,s}^{n+1} = u_{r,s}^* - \frac{1}{2}\mu_y\delta_y^2 u_{r,s}^n. \tag{2.80}$$

Find the leading term of its truncation error. Prove further that this scheme is unconditionally stable.

7. Let $\Omega = (0,1)^2$. Implement the Douglas method (2.79)-(2.80) with $\triangle x = \triangle y = h$ to solve the 2-D parabolic problem

$$u_t = u_{xx} + u_{yy} \quad \forall\ (x,y,t) \in \Omega \times (0,t_F), \quad t_F = 0.1,$$

subject to the initial condition

$$u|_{t=0} = \sin \pi x \sin \pi y, \quad \forall\ (x,y) \in \Omega$$

and proper Dirichlet boundary condition such that the analytic solution to the problem is given by

$$u(x,y,t) = e^{-2\pi^2 t}\sin \pi x \sin \pi y.$$

Solve the problem using different grids to confirm that the convergence rate is $O(h^2 + (\triangle t)^2)$.

8. Consider the convection-diffusion equation

$$u_t + au_x - u_{xx} = 0, \tag{2.81}$$

with Dirichlet boundary conditions and $a \geq 0$ is a constant. Show that

$$u(x,t) = e^{-(ik\pi a + (k\pi)^2)t}e^{ik\pi x} \tag{2.82}$$

is a set of particular solutions of the problem, i.e., (2.82) satisfies the PDE (2.81). Use the von Neumann technique to derive the stability condition for the upwind scheme

$$\frac{u_j^{n+1} - u_j^n}{\triangle t} + a\frac{u_j^{n+1} - u_{j-1}^{n+1}}{\triangle x} = \frac{u_{j+1}^{n+1} - 2u_j^{n+1} + u_{j-1}^{n+1}}{(\triangle x)^2}.$$

9. Consider the following problem

$$u_t = u_{xx} + u(1-u), \quad (x,t) \in (0,1) \times (0,t_F), \tag{2.83}$$
$$u_x(0,t) = u_x(1,t) = 0, \quad t \in [0,t_F], \tag{2.84}$$
$$u(x,0) = f(x), \quad x \in [0,1]. \tag{2.85}$$

Note that this problem comes from the population growth model and (2.83) is called Fisher's equation. Develop an explicit scheme to solve this problem with

$$f(x) = \sin^2(\pi x), \quad \triangle x = 0.02, \quad \triangle t = 0.0001, \quad t_F = 5.$$

What do you observe for $t_F \to \infty$?

References

[1] W.F. Ames. *Numerical Methods for Partial Differential Equations.* Academic Press, New York, NY, 3rd edition, 1992.

[2] J. Douglas. Alternating direction methods for three space variables. *Numer. Math.*, 4:41–63, 1962.

[3] J. Douglas and H.H. Rachford. On the numerical solution of heat conduction problems in two and three space variables. *Trans. Amer. Math. Soc.*, 82:421–439, 1956.

[4] G. Fairweather and A.R. Mitchell. A new computational procedure for A.D.I. methods. *SIAM J. Numer. Anal.*, 4:163–170, 1967.

[5] B. Gustafsson, H.-O. Kreiss and J. Oliger. *Time Dependent Problems and Difference Methods.* Wiley-Interscience, New York, NY, 1996.

[6] B.S. Jovanović and E. Süli. *Analysis of Finite Difference Schemes.* Springer-Verlag, London, UK, 2014.

[7] L. Lapidus and G.F. Pinder. *Numerical Solution of Partial Differential Equations in Science and Engineering.* Wiley-Interscience, New York, NY, 1999.

[8] R.J. LeVeque. *Finite Difference Methods for Ordinary and Partial Differential Equations: Steady-State and Time-Dependent Problems.* SIAM, Philadelphia, PA, 2007.

[9] A.R. Mitchell and D.F. Griffiths. *The Finite Difference Method in Partial Differential Equations.* John Wiley & Sons, Chichester, 1980.

[10] A.R. Mitchell and G. Fairweather. Improved forms of the alternating direction methods of Douglas, Peaceman and Rachford for solving parabolic and elliptic equations. *Numer. Math.*, 6:285–292, 1964.

[11] K.W. Morton and D.F. Mayers. *Numerical Solution of Partial Differential Equations: An Introduction.* Cambridge University Press, Cambridge, UK, 1995.

[12] D.W. Peaceman and H.H. Rachford. The numerical solution of parabolic and elliptic differential equations. *J. Soc. Indust. App. Math.*, 3:28–41, 1955.

[13] J. Strikwerda. *Finite Difference Schemes and Partial Differential Equations.* SIAM, Philadelphia, PA, 2nd edition, 2004.

[14] A. Tveito and R. Winther. *Introduction to Partial Differential Equations: A Computational Approach.* Springer, New York, NY, 1998.

3

Finite Difference Methods for Hyperbolic Equations

In this chapter, we will continue our discussion on finite difference methods. In Sec. 3.1, we briefly introduce the hyperbolic problems. Then in Sec. 3.2, we present some basic finite difference schemes for the convection problem and review the von Neumann stability analysis technique. We introduce the important concepts on dissipation and dispersion errors in Sec. 3.3. We extend the development of various finite difference schemes to conservation laws and the second-order hyperbolic problems in Sec. 3.4 and Sec. 3.5, respectively. In Sec. 3.6, we present a MATLAB code to solve the convection problem using the Lax-Wendroff scheme.

3.1 Introduction

Consider the initial value problem

$$\boldsymbol{u}_t + A\boldsymbol{u}_x = 0 \tag{3.1}$$

$$\boldsymbol{u}|_{t=0} = \boldsymbol{u}_0(x) \tag{3.2}$$

where $A = (a_{ij})$ is an $m \times m$ matrix and $\boldsymbol{u}$ is a vector function with m components, respectively. The problem (3.1)-(3.2) is well-posed if and only if all eigenvalues of A are real and there is a complete set of eigenvectors [3, §4.3]. Such a system is called strongly hyperbolic [3, §4.3]. Here we will restrict our discussions to such hyperbolic problems.

Assume that the complete set of eigenvectors is $\phi_1, \cdots, \phi_m$. Let matrix $S = [\phi_1, \cdots, \phi_m]$. From linear algebra theory, we know

$$S^{-1}AS = \Lambda = \operatorname{diag}(\lambda_1, \cdots, \lambda_m),$$

using which we can change the problem (3.1)-(3.2) to m scalar equations

$$\tilde{u}_t^{(i)} + \lambda_i \tilde{u}^{(i)} = 0, \quad i = 1, 2, \cdots, m, \tag{3.3}$$

$$\tilde{u}^{(i)}|_{t=0} = (S^{-1}\boldsymbol{u}_0)^{(i)} = \tilde{u}_0(x), \tag{3.4}$$

where $\tilde{u}^{(i)}$ is the ith component of the vector $S^{-1}\boldsymbol{u}$. Hence we will concentrate on the discussion of the scalar equation:

$$u_t + au_x = 0, \quad \forall\, (x,t) \in (0,1) \times (0,t_F), \tag{3.5}$$
$$u|_{t=0} = u_0(x). \tag{3.6}$$

For the hyperbolic equation (3.5), the solutions of

$$\frac{dx}{dt} = a(x,t)$$

are called the **characteristics**. Along a characteristic curve, the solution is a constant since

$$\frac{du}{dt} = \frac{\partial u}{\partial t} + \frac{\partial u}{\partial x}\frac{dx}{dt} = 0.$$

It is easy to check that the solution to (3.5)-(3.6) with constant a can be expressed as $u(x,t) = u_0(x - at)$. If a is a function of u only, then the solution to (3.5)-(3.6) can still be presented as $u(x,t) = u_0(x - a(u(x,t))t)$ until the time when the characteristics cross each other [9, p.85].

3.2   Some basic difference schemes

In this section, we shall develop some simple schemes for (3.5)-(3.6). We assume that uniform grid points are used, i.e.,

$$x_j = j\triangle x, 0 \le j \le J, \triangle x = 1/J,$$
$$t_n = n\triangle t, 0 \le n \le N, \triangle x = t_F/N.$$

First let us consider a central difference scheme:

$$\frac{u_j^{n+1} - u_j^n}{\triangle t} + a\frac{u_{j+1}^n - u_{j-1}^n}{2\triangle x} = 0. \tag{3.7}$$

Applying von Neumann technique to (3.7), we obtain the amplification factor

$$\lambda \equiv \lambda(k) = 1 - a\frac{\triangle t}{2\triangle x}(e^{ik\triangle x} - e^{-ik\triangle x}) = 1 - a\frac{\triangle t}{\triangle x}i\sin k\triangle x,$$

which gives

$$|\lambda|^2 = 1 + (a\frac{\triangle t}{\triangle x})^2 > 1,$$

i.e., the scheme (3.7) is unconditionally unstable. Hence we have to be very careful in choosing proper difference approximations. Below we show three stable schemes.

Example 3.1

A popular scheme for (3.5)-(3.6) is the so-called upwind scheme:

$$u_j^{n+1} = u_j^n - a\frac{\Delta t}{\Delta x}(u_j^n - u_{j-1}^n), \quad \text{if } a > 0, \tag{3.8}$$

i.e., when $a > 0$, we approximate $u_x|_j^n$ by backward difference. When $a < 0$, we approximate $u_x|_j^n$ by $\frac{u_{j+1}^n - u_j^n}{\Delta x}$, i.e., in this case by forward difference.

To check its stability, let us denote $\mu = a\frac{\Delta t}{\Delta x}$. Substituting

$$u_j^n = \lambda^n e^{ik(j\Delta x)} \tag{3.9}$$

into (3.8), we see that the amplification factor

$$\lambda(k) = 1 - \mu(1 - e^{-ik\Delta x}) = 1 - \mu + \mu\cos k\Delta x - i\mu\sin k\Delta x. \tag{3.10}$$

Hence we have

$$\begin{aligned}
|\lambda(k)|^2 &= (1-\mu)^2 + 2\mu(1-\mu)\cos k\Delta x + \mu^2\cos^2 k\Delta x + \mu^2\sin^2 k\Delta x \\
&= (1-\mu)^2 + 2\mu(1-\mu)\cos k\Delta x + \mu^2. \tag{3.11}
\end{aligned}$$

If $0 \le \mu \le 1$, then

$$|\lambda(k)|^2 \le (1-\mu)^2 + 2\mu(1-\mu) + \mu^2 = (1-\mu+\mu)^2 = 1,$$

i.e., the scheme is stable provided that $0 \le \mu \le 1$. By Taylor expansion, it is easy to see that the truncation error of (3.8) is $O(\Delta t + \Delta x)$.   □

Example 3.2

Now we want to derive a more accurate scheme. Using (3.5) in the Taylor expansion, we obtain

$$u(x, t + \Delta t) = u(x, t) + \Delta t\, u_t(x, t) + \frac{(\Delta t)^2}{2} u_{tt}(x, t) + O(\Delta t)^3 \tag{3.12}$$

$$= u(x, t) - a\Delta t\, u_x(x, t) + \frac{(\Delta t)^2}{2} a^2 u_{xx}(x, t) + O(\Delta t)^3.$$

Then approximating those x-derivatives at (x_j, t_n) by central difference, we obtain the so-called Lax-Wendroff scheme:

$$u_j^{n+1} = u_j^n - \frac{\mu}{2}(u_{j+1}^n - u_{j-1}^n) + \frac{\mu^2}{2}(u_{j+1}^n - 2u_j^n + u_{j-1}^n). \tag{3.13}$$

Using von Neumann stability analysis, we obtain the amplification factor

$$\lambda(k) = 1 - \frac{\mu}{2}(e^{ik\Delta x} - e^{-ik\Delta x}) + \frac{\mu^2}{2}(e^{ik\Delta x} - 2 + e^{-ik\Delta x})$$

$$= 1 - i\mu \sin k\triangle x - 2\mu^2 \sin^2 \frac{1}{2} k\triangle x, \tag{3.14}$$

from which we have

$$|\lambda(k)|^2 = 1 - 4\mu^2(1 - \mu^2)\sin^4 \frac{1}{2} k\triangle x. \tag{3.15}$$

Hence the Lax-Wendroff scheme is stable whenever $|\mu| \leq 1$. It is easy to see that the truncation error is $O((\triangle t)^2 + (\triangle x)^2)$. □

Example 3.3

Another classic method for (3.5) is the leap-frog scheme:

$$\frac{u_j^{n+1} - u_j^{n-1}}{2\triangle t} + a\frac{u_{j+1}^n - u_{j-1}^n}{2\triangle x} = 0, \tag{3.16}$$

whose truncation error is $O((\triangle t)^2 + (\triangle x)^2)$. Substituting

$$u_j^n = \lambda^n e^{ik(j\triangle x)}$$

into (3.16), we obtain

$$\lambda^2 + (i2\mu \sin k\triangle x)\lambda - 1 = 0,$$

solving which gives

$$\lambda(k) = -i\mu \sin k\triangle x \pm (1 - \mu^2 \sin^2 k\triangle x)^{1/2}. \tag{3.17}$$

Hence the roots $\lambda(k)$ are complex for all k if and only if $|\mu| \leq 1$, in which case $|\lambda(k)|^2 = 1$, i.e., the scheme is stable provided that $|\mu| \leq 1$.

From the above several examples, we see that a necessary condition for the above difference schemes to be stable is $|\mu| = |a\frac{\triangle t}{\triangle x}| \leq 1$, which is the so-called Courant-Friedrichs-Lewy (CFL) condition: the domain of dependence of the difference scheme must contain the domain of dependence of the differential equation so that any change of initial conditions can be captured by the numerical scheme. Recall that for the Lax-Wendroff scheme (3.13), the value of u_j^{n+1} depends on u_{j-1}^n, u_j^n, and u_{j+1}^n, which then depend on some values at time level $n-1$, i.e., $u_i^{n-1}, j-2 \leq i \leq j+2$, and so on. Hence the domain of dependence of u_j^{n+1} is formed by the function values inside the triangle, whose bottom side includes those values $u_i^0, j-(n+1) \leq i \leq j+(n+1)$. Note that for the continuous problem (3.5)-(3.6), its exact solution is $u(x,t) = u_0(x - at)$. Hence the analytic solution at point (x_j, t_{n+1}) is obtained by tracing this point back along the characteristic to the point $(x_j - at_{n+1}, 0)$, where it meets the initial time line. To satisfy the CFL condition, we need $x_j - (n+1)\triangle x \leq x_j - at_{n+1}$, or $a\frac{\triangle t}{\triangle x} \leq 1$. □

3.3 Dissipation and dispersion errors

Considering that the solutions of hyperbolic equations are waves, we have to study some properties associated with waves, such as the amplitude and the phase of the waves. The corresponding errors introduced by the numerical schemes are called the dissipation (or amplitude) error, and the dispersion (or phase) error. Furthermore, if the amplitude factor λ satisfies

$$|\lambda(k)|^2 \le 1 - C \cdot (k\triangle x)^r, \quad \forall\, |k\triangle x| \le \pi,$$

for some positive constant C and positive integer r, then the corresponding difference method is said to be *dissipative of order r*. Below we present three examples to demonstrate how to carry out the dissipation and dispersion error analysis.

Example 3.4
Let us first look at the upwind scheme for $a > 0$. Consider a special solution of (3.5)

$$u(x,t) = e^{i(kx+\omega t)}, \quad \omega = -ka, \tag{3.18}$$

at $x = j\triangle x, t = n\triangle t$. From (3.10) and representing $\lambda = |\lambda|e^{iarg\lambda}$, the phase of the numerical mode is given by

$$arg\lambda = -\tan^{-1}\frac{\mu \sin k\triangle x}{1 - \mu + \mu \cos k\triangle x}. \tag{3.19}$$

Let $\xi = k\triangle x$. Expanding (3.19) and using the following expansions:

$$\frac{1}{1-z} = 1 + z + z^2 + \cdots, \quad \forall\, |z| < 1, \tag{3.20}$$

$$\sin z = z - \frac{z^3}{3!} + \frac{z^5}{5!} - + \cdots, \quad \forall\, |z| < 1, \tag{3.21}$$

$$\cos z = 1 - \frac{z^2}{2!} + \frac{z^4}{4!} - + \cdots, \quad \forall\, |z| < 1, \tag{3.22}$$

$$\tan^{-1} z = z - \frac{z^3}{3} + \frac{z^5}{5} - + \cdots, \quad \forall\, |z| < 1, \tag{3.23}$$

we obtain: for $\xi \ll 1$,

$$arg\lambda = -\tan^{-1}\frac{\mu(\xi - \frac{\xi^3}{6} + \cdots)}{1 - \mu + \mu(1 - \frac{\xi^2}{2} + \frac{\xi^4}{24} - \cdots)}$$

$$= -\tan^{-1}\{\mu(\xi - \frac{\xi^3}{6} + \cdots)[1 + \mu(\frac{\xi^2}{2} - \frac{\xi^4}{24} + \cdots) + \mu^2(\frac{\xi^4}{4} - \cdots)]\}$$

$$= -\tan^{-1}[\mu(\xi - \frac{\xi^3}{6} + \cdots) + \mu^2(\frac{\xi^3}{2} + \cdots)]$$

$$= -[\mu(\xi - \frac{\xi^3}{6}) + \mu^2(\frac{\xi^3}{2} + \cdots) - \frac{1}{3}\mu^3\xi^3 + \cdots]$$

$$= -\mu\xi[1 + (-\frac{1}{6} + \frac{\mu}{2} - \frac{1}{3}\mu^2)\xi^2 + \cdots]$$

$$= -\mu\xi[1 - \frac{1}{6}(1 - \mu)(1 - 2\mu)\xi^2 + \cdots]. \tag{3.24}$$

From (3.18), at each time step, the phase increases by

$$\omega\Delta t = -ka\Delta t = -k \cdot \Delta x \cdot a\frac{\Delta t}{\Delta x} = -\xi\mu,$$

from which, along with (3.24), we see that the relative phase error is

$$\frac{arg\lambda - (-\mu\xi)}{-\mu\xi} = -\frac{1}{6}(1 - \mu)(1 - 2\mu)\xi^2 + \cdots,$$

i.e., the relative phase error is $O(\xi^2)$.

From (3.11) and (3.22), we have

$$|\lambda(k)|^2 = 1 - \mu(1 - \mu)[\xi^2 - \frac{\xi^4}{12} + \cdots],$$

which implies that the amplitude error of the upwind scheme is $O(\xi^2)$ in one time step. □

Example 3.5
Now let us study the behavior of the Lax-Wendroff scheme. From (3.15), we see that the amplitude error in one time step is $O(\xi^4)$, which means that the solution from the Lax-Wendroff scheme shall have much less amplitude damping compared to the upwind scheme.

From (3.14) and expansions (3.21) and (3.23), we have

$$arg\lambda = -\tan^{-1}\frac{\mu\sin k\Delta x}{1 - 2\mu^2\sin^2\frac{1}{2}k\Delta x}$$

$$= -\mu\xi[1 - \frac{1}{6}(1 - \mu^2)\xi^2 + \cdots],$$

which gives the relative phase error $-\frac{1}{6}(1 - \mu^2)\xi^2$. Note that this corresponds to a phase lag, since the sign is always negative. □

Example 3.6
Finally, let us look at the leap-frog scheme. For all modes k, we have $|\lambda(k)| = 1$, which means that the leap-frog scheme is non-dissipative (i.e., no damping). But there is a problem with this scheme: it has a parasitic solution (i.e., non-physical solution), which oscillates rapidly and travels in the opposite direction

to the true solution. The positive root of (3.17) corresponds to the true mode, in which case,

$$arg\lambda = -\sin^{-1}(\mu \sin k\triangle x)$$
$$= -\mu\xi[1 - \frac{1}{6}(1-\mu^2)\xi^2 + \cdots],$$

i.e., the relative phase error is $-\frac{1}{6}(1-\mu^2)\xi^2$, same as the Lax-Wendroff scheme. Here we used the expansion

$$\sin^{-1} z = z + \frac{1}{2}\frac{z^3}{3} + (\frac{1\cdot 3}{2\cdot 4})\frac{z^5}{5} + (\frac{1\cdot 3\cdot 5}{2\cdot 4\cdot 6})\frac{z^7}{7} + \cdots, \quad \forall\, |z| \le 1.$$

☐

3.4 Extensions to conservation laws

In many practical applications, the hyperbolic equation often appears in the conservation form

$$\frac{\partial u}{\partial t} + \frac{\partial f(u)}{\partial x} = 0. \tag{3.25}$$

We can derive the Lax-Wendroff scheme directly for (3.25), in which case

$$u_{tt} = -(f_x)_t = -(\frac{\partial f}{\partial u}\frac{\partial u}{\partial t})_x = (a(u)f_x)_x, \quad a(u) \equiv \frac{\partial f}{\partial u},$$

substituting this into the Taylor expansion (3.12) yields

$$u(x,t+\triangle t) = u(x,t) - \triangle t\frac{\partial f}{\partial x} + \frac{1}{2}(\triangle t)^2(a(u)\frac{\partial f}{\partial x})_x + O((\triangle t)^3).$$

Approximating those x-derivatives by central differences, we obtain the Lax-Wendroff scheme

$$u_j^{n+1} = u_j^n - \frac{\triangle t}{2\triangle x}[f(u_{j+1}^n) - f(u_{j-1}^n)] + \frac{1}{2}(\frac{\triangle t}{\triangle x})^2$$
$$[a(u_{j+\frac{1}{2}}^n)(f(u_{j+1}^n) - f(u_j^n)) - a(u_{j-\frac{1}{2}}^n)(f(u_j^n) - f(u_{j-1}^n))], \tag{3.26}$$

where we can set $a(u_{j\pm\frac{1}{2}}^n) = a(\frac{1}{2}(u_{j\pm 1}^n + u_j^n))$.

Note that the special case $f(u) = au$ (where a is a constant) reduces to the scheme (3.13) we obtained earlier. The vector form of scheme (3.26) can be extended directly to systems of equations

$$\frac{\partial \boldsymbol{u}}{\partial t} + \frac{\partial \boldsymbol{f}(\boldsymbol{u})}{\partial x} = 0. \tag{3.27}$$

A more convenient variant of the Lax-Wendroff scheme (3.26) is implemented in a two-step procedure (e.g., [9, p. 108] or [3, p. 237]):

$$u_{j+\frac{1}{2}}^{n+\frac{1}{2}} = \frac{1}{2}(u_j^n + u_{j+1}^n) - \frac{\Delta t}{2\Delta x}[f(u_{j+1}^n) - f(u_j^n)], \tag{3.28}$$

$$u_j^{n+1} = u_j^n - \frac{\Delta t}{\Delta x}[f(u_{j+\frac{1}{2}}^{n+\frac{1}{2}}) - f(u_{j-\frac{1}{2}}^{n+\frac{1}{2}})]. \tag{3.29}$$

For the special case $f = au$ with constant a, it can be seen that elimination of $u_{j+\frac{1}{2}}^{n+\frac{1}{2}}$ leads to scheme (3.13). While for nonlinear problems, the two schemes are not equivalent.

The leap-frog scheme for (3.25) can be easily constructed:

$$\frac{u_j^{n+1} - u_j^{n-1}}{2\Delta t} + \frac{f(u_{j+1}^n) - f(u_{j-1}^n)}{2\Delta x} = 0. \tag{3.30}$$

Finally, we want to mention that similar schemes can be developed for conservation laws in two and three space dimensions (cf. [9] or [3]).

3.5 The second-order hyperbolic equations

In this section, we demonstrate how to develop and analyze the schemes for solving the second-order hyperbolic equations. Discussions are subdivided into the 1-D and 2-D subsections.

3.5.1 The 1-D case

Let us consider the second-order hyperbolic equation in one space dimension

$$u_{tt} = a^2 u_{xx}, \quad \forall\, (x,t) \in (0,1) \times (0,t_F), \tag{3.31}$$
$$u(x,0) = f(x), \ u_t(x,0) = g(x), \quad \forall\, 0 \le x \le 1, \tag{3.32}$$
$$u(0,t) = u_L(t), \ u(1,t) = u_R(t), \quad \forall\, t \in (0,t_F), \tag{3.33}$$

where the constant $a > 0$.

Let $\xi = x + at, \eta = x - at$, and $\phi(\xi,\eta) = u(x,t)$. By simple calculus, we have

$$\partial_t u = \partial_\xi \phi \cdot (a) + \partial_\eta \phi \cdot (-a), \quad \partial_{tt} u = \partial_{\xi\xi}\phi \cdot a^2 + 2\partial_{\xi\eta}\phi \cdot (-a^2) + \partial_{\eta\eta}\phi \cdot a^2,$$
$$\partial_x u = \partial_\xi \phi + \partial_\eta \phi, \qquad \partial_{xx} u = \partial_{\xi\xi}\phi + 2\partial_{\xi\eta}\phi + \partial_{\eta\eta}\phi.$$

Hence, (3.31) can be reduced to

$$\phi_{\xi\eta} = 0. \tag{3.34}$$

Integrating (3.34) twice gives

$$u(x,t) = \psi_1(\xi) + \psi_2(\eta) = \psi_1(x + at) + \psi_2(x - at),$$

where ψ_1 and ψ_2 are arbitrary twice differentiable functions. The lines

$$x \pm at = \text{constant}$$

are the characteristics of (3.31).

For an arbitrary point $p(x_*, t_*)$, the two characteristics passing through it are

$$t - t_* = \pm\frac{1}{a}(x - x_*),$$

which intersect the x-axis (i.e., the $t = 0$ line) at points $x = x_* \mp at_*$. These two points along with the point p form a triangle, which is called the domain of dependence for the solution $u(x_*, t_*)$. The domain of dependence of numerical difference solution can be defined as we did previously for the first-order hyperbolic equation. We want to emphasize that the domain of dependence of the difference scheme must include the domain of dependence of the corresponding PDE.

If we define $v = u_t$ and $w = au_x$, we can rewrite (3.31) as

$$\frac{\partial}{\partial t}\begin{bmatrix} v \\ w \end{bmatrix} + \begin{bmatrix} 0 & -a \\ -a & 0 \end{bmatrix}\frac{\partial}{\partial x}\begin{bmatrix} v \\ w \end{bmatrix} = 0,$$

i.e, (3.31) can be reduced to a system of first-order hyperbolic equations we discussed previously. But we shall discuss numerical schemes for solving (3.31) directly.

Example 3.7

First, we consider the simple explicit scheme: for any interior nodal points (i.e., $1 \leq j \leq J - 1, 1 \leq n \leq N - 1$),

$$\frac{u_j^{n+1} - 2u_j^n + u_j^{n-1}}{(\Delta t)^2} = a^2\frac{u_{j+1}^n - 2u_j^n + u_{j-1}^n}{(\Delta x)^2}, \tag{3.35}$$

which needs two level initial conditions.

Approximating the first equation of (3.32), we obtain

$$u_j^0 = f(x_j), \quad \forall 0 \leq j \leq J. \tag{3.36}$$

The second initial condition of (3.32) can be approximated as

$$\frac{u_j^1 - \frac{1}{2}(u_{j+1}^0 + u_{j-1}^0)}{\Delta t} = g(x_j), \quad \forall 1 \leq j \leq J - 1,$$

or

$$u_j^1 = \frac{1}{2}(u_{j+1}^0 + u_{j-1}^0) + \Delta tg(x_j), \quad \forall 1 \leq j \leq J - 1. \tag{3.37}$$

The truncation error of the explicit scheme (3.35)-(3.37) is $O((\triangle t)^2 + (\triangle x)^2)$. As for the stability of (3.35), denoting $\mu = a\frac{\triangle t}{\triangle x}$ and using the von Neumann technique, we obtain

$$\lambda - 2 + \frac{1}{\lambda} = \mu^2(e^{-ik\triangle x} - 2 + e^{ik\triangle x})$$

or

$$\lambda^2 + (-2 + 4\mu^2 \sin^2 \frac{1}{2}k\triangle x)\lambda + 1 = 0, \tag{3.38}$$

where λ is the amplification factor. Solving (3.38), we obtain

$$\lambda = (1 - 2\mu^2 \sin^2 \frac{1}{2}k\triangle x) \pm \sqrt{(1 - 2\mu^2 \sin^2 \frac{1}{2}k\triangle x)^2 - 1}.$$

In order for $|\lambda| \leq 1$, it is necessary that

$$-1 \leq 1 - 2\mu^2 \sin^2 \frac{1}{2}k\triangle x \leq 1,$$

or $|\mu| \leq 1$, under which condition the scheme is stable.   ⧠

Example 3.8
For (3.31), the Crank-Nicolson scheme can be written as

$$\frac{u_j^{n+1} - 2u_j^n + u_j^{n-1}}{(\triangle t)^2} = \frac{1}{2(\triangle x)^2}[(u_{j+1}^{n+1} - 2u_j^{n+1} + u_{j-1}^{n+1}) + (u_{j+1}^{n-1} - 2u_j^{n-1} + u_{j-1}^{n-1})],$$
$$\tag{3.39}$$

whose truncation error is $O((\triangle t)^2 + (\triangle x)^2)$. As for stability, using the von Neumann technique, we obtain

$$\lambda - 2 + \frac{1}{\lambda} = \frac{1}{2}\mu^2[-4\lambda \sin^2 \frac{1}{2}k\triangle x - \frac{4}{\lambda} \sin^2 \frac{1}{2}k\triangle x]$$

or

$$(1 + 2\mu^2 \sin^2 \frac{1}{2}k\triangle x)\lambda^2 - 2\lambda + (1 + 2\mu^2 \sin^2 \frac{1}{2}k\triangle x) = 0. \tag{3.40}$$

Solving (3.40), we have

$$\lambda = (1 \pm i\sqrt{(1 + 2\mu^2 \sin^2 \frac{1}{2}k\triangle x)^2 - 1})/(1 + 2\mu^2 \sin^2 \frac{1}{2}k\triangle x)$$

from which we obtain

$$|\lambda|^2 = 1$$

for any k and μ, i.e., the scheme (3.39) is unconditionally stable.   ⧠

3.5.2 The 2-D case

Now let us consider a 2-D hyperbolic equation

$$u_{tt} = u_{xx} + u_{yy}, \tag{3.41}$$
$$u(x, y, 0) = f(x, y), \quad u_t(x, y, 0) = g(x, y), \tag{3.42}$$
$$u(0, y, t) = u_L(y, t), \quad u(1, y, t) = u_R(y, t), \tag{3.43}$$
$$u(x, 0, t) = u_B(x, t), \quad u(x, 1, t) = u_T(x, t), \tag{3.44}$$

which holds true for any $(x, y, t) \in (0, 1)^2 \times (0, t_F)$.

The explicit scheme and Crank-Nicolson scheme developed for 1-D hyperbolic problems can be extended straightforward to (3.41)-(3.44).

Example 3.9

First, let us consider the following forward Euler explicit scheme

$$\frac{u_{i,j}^{n+1} - 2u_{i,j}^n + u_{i,j}^{n-1}}{(\Delta t)^2} = \frac{u_{i+1,j}^n - 2u_{i,j}^n + u_{i-1,j}^n}{(\Delta x)^2} + \frac{u_{i,j+1}^n - 2u_{i,j}^n + u_{i,j-1}^n}{(\Delta y)^2}. \tag{3.45}$$

Denote $\mu_x = \frac{\Delta t}{\Delta x}$ and $\mu_y = \frac{\Delta t}{\Delta y}$. By von Neumann stability analysis, we substitute the solution $u_{r,s}^n = \lambda^n e^{i(k_x \cdot r\Delta x + k_y \cdot s\Delta y)}$ into (3.45) and easily see that the amplification factor λ satisfies the equation

$$\lambda - 2 + \lambda^{-1} = \mu_x^2 (e^{ik_x \Delta x} - 2 + e^{-ik_x \Delta x}) + \mu_y^2 (e^{ik_y \Delta y} - 2 + e^{-ik_y \Delta y}),$$

or

$$\lambda^2 - 2[1 - 2(\mu_x^2 \sin^2 \frac{k_x \Delta x}{2} + \mu_y^2 \sin^2 \frac{k_y \Delta y}{2})]\lambda + 1 = 0.$$

Under the conditions $\mu_x \leq \frac{1}{\sqrt{2}}$ and $\mu_y \leq \frac{1}{\sqrt{2}}$, we have

$$-1 \leq c \equiv 1 - 2(\mu_x^2 \sin^2 \frac{k_x \Delta x}{2} + \mu_y^2 \sin^2 \frac{k_y \Delta y}{2}) \leq 1,$$

which leads to

$$|\lambda| = |(2c \pm \sqrt{4c^2 - 4})/2| = |c \pm i\sqrt{1 - c^2}| = 1,$$

i.e., the scheme is stable for $\max(\frac{\Delta t}{\Delta x}, \frac{\Delta t}{\Delta y}) \leq \frac{1}{\sqrt{2}}$. □

Example 3.10

Now let us consider the following backward Euler implicit scheme

$$\frac{u_{i,j}^{n+1} - 2u_{i,j}^n + u_{i,j}^{n-1}}{(\Delta t)^2} = \frac{u_{i+1,j}^{n+1} - 2u_{i,j}^{n+1} + u_{i-1,j}^{n+1}}{(\Delta x)^2} + \frac{u_{i,j+1}^{n+1} - 2u_{i,j}^{n+1} + u_{i,j-1}^{n+1}}{(\Delta y)^2}. \tag{3.46}$$

By substituting the solution $u_{r,s}^n = \lambda^n e^{i(k_x \cdot r \triangle x + k_y \cdot s \triangle y)}$ into (3.46), we easily see that the amplification factor λ satisfies the equation

$$\lambda - 2 + \lambda^{-1} = \lambda \left[\mu_x^2 \cdot (-4 \sin^2 \frac{k_x \triangle x}{2}) + \mu_y^2 \cdot (-4 \sin^2 \frac{k_y \triangle y}{2}) \right],$$

or

$$(1 + 4\mu_x^2 \sin^2 \frac{k_x \triangle x}{2} + 4\mu_y^2 \sin^2 \frac{k_y \triangle y}{2})\lambda^2 - 2\lambda + 1 = 0. \qquad (3.47)$$

Let $d = 4\mu_x^2 \sin^2 \frac{k_x \triangle x}{2} + 4\mu_y^2 \sin^2 \frac{k_y \triangle y}{2}$. By the quadratic formula we have

$$\lambda = \frac{2 \pm \sqrt{2^2 - 4(1+d)}}{2(1+d)} = \frac{1 \pm \sqrt{-d}}{1+d} = \frac{1 \pm i\sqrt{d}}{1+d},$$

which leads to

$$|\lambda|^2 = \frac{1}{(1+d)^2} \cdot (1+d) = \frac{1}{1+d} < 1,$$

i.e., the scheme is unconditionally stable.

Assume that the rectangular domain is divided by $0 = x_0 < x_1 < \cdots < x_{J_x} = 1$ and $0 = y_0 < y_1 < \cdots < y_{J_y} = 1$, and denote the solution vector $\mathbf{u}^{n+1} = (u_{1,1}, \cdots, u_{J_x-1,1}, u_{1,2}, \cdots, u_{J_x-1,2}, \cdots, u_{1,J_y-1}, \cdots, u_{J_x-1,J_y-1})'$. To implement the scheme (3.46), at each time step we have to solve a linear system $A\mathbf{u}^{n+1} = \mathbf{b}$ for some known vector $\mathbf{b}$, where the matrix A has dimension $[(J_x - 1) \times (J_y - 1)] \times [(J_x - 1) \times (J_y - 1)]$. This is a big matrix. Think about a case with $J_x = J_y = 101$, i.e., 100 partitions in each dimension, we end up with a $10^4 \times 10^4$ matrix. ☐

Hence, in practice, we usually do not use implicit schemes for high-dimensional hyperbolic problems. Instead, similar ADI schemes as those of parabolic equations are popular and have been developed by many researchers (see [6] and references therein). Below we give three exemplary ADI schemes for 2-D hyperbolic equations.

Example 3.11

In [7], Lees presented two ADI schemes. The first one is given as follows: for any parameter $\eta \in (0,1)$,

$$u_{ij}^* - 2u_{ij}^n + u_{ij}^{n-1} = \frac{(\triangle t)^2}{(\triangle x)^2} [\eta \delta_x^2 u_{ij}^* + (1 - 2\eta)\delta_x^2 u_{ij}^n + \eta \delta_x^2 u_{ij}^{n-1}]$$

$$+ \frac{(\triangle t)^2}{(\triangle y)^2} [(1 - 2\eta)\delta_y^2 u_{ij}^n + 2\eta \delta_y^2 u_{ij}^{n-1}], \qquad (3.48)$$

$$u_{ij}^{n+1} - u_{ij}^* = \frac{(\triangle t)^2}{(\triangle y)^2} \cdot \eta(\delta_y^2 u_{ij}^{n+1} - \delta_y^2 u_{ij}^{n-1}). \qquad (3.49)$$

This scheme is easy to be implemented, since u^* can be obtained by solving a tridiagonal system from (3.48), and u^{n+1} can be obtained by solving another tridiagonal system from (3.49).

Substituting u^* from (3.49) into (3.48), we have

$$
u_{ij}^{n+1} - 2u_{ij}^n + u_{ij}^{n-1} = \frac{(\triangle t)^2}{(\triangle y)^2} \cdot \eta(\delta_y^2 u_{ij}^{n+1} - \delta_y^2 u_{ij}^{n-1})
$$

$$
+ \frac{(\triangle t)^2}{(\triangle x)^2} \eta \delta_x^2 [u_{ij}^{n+1} - \frac{(\triangle t)^2}{(\triangle y)^2} \cdot \eta(\delta_y^2 u_{ij}^{n+1} - \delta_y^2 u_{ij}^{n-1})]
$$

$$
+ \frac{(\triangle t)^2}{(\triangle x)^2} [(1 - 2\eta)\delta_x^2 u_{ij}^n + \eta \delta_x^2 u_{ij}^{n-1}] + \frac{(\triangle t)^2}{(\triangle y)^2} [(1 - 2\eta)\delta_y^2 u_{ij}^n + 2\eta \delta_y^2 u_{ij}^{n-1}].
$$

Dividing both sides by $(\triangle t)^2$ and collecting like terms, we obtain

$$
\frac{1}{(\triangle t)^2} \delta_t^2 u_{ij}^n - (\frac{1}{(\triangle x)^2} \delta_x^2 u_{ij}^n + \frac{1}{(\triangle y)^2} \delta_y^2 u_{ij}^n) - \eta \delta_t^2 (\frac{1}{(\triangle x)^2} \delta_x^2 u_{ij}^n + \frac{1}{(\triangle y)^2} \delta_y^2 u_{ij}^n)
$$

$$
+ (\triangle t)^2 \cdot \eta^2 \frac{1}{(\triangle x)^2} \cdot \frac{1}{(\triangle y)^2} (u_{ij}^{n+1} - u_{ij}^{n-1}) = 0. \tag{3.50}
$$

We see that (3.50) is a perturbation of the forward Euler scheme (3.45). It is easy to prove that the truncation error of this scheme is $O((\triangle t)^2 + (\triangle x)^2 + (\triangle y)^2)$. Furthermore, it is proved [7] that the ADI scheme (3.48)-(3.49) satisfies an energy inequality if $4\eta > 1$, which implies that this ADI scheme is unconditionally stable if $4\eta > 1$. ☐

Example 3.12
The second ADI scheme proposed by Lees [7] is given as follows: for any parameter $\eta \in (0, 1)$,

$$
u_{ij}^* - 2u_{ij}^n + u_{ij}^{n-1} = \frac{(\triangle t)^2}{(\triangle x)^2} [\eta \delta_x^2 u_{ij}^* + (1 - 2\eta)\delta_x^2 u_{ij}^n + \eta \delta_x^2 u_{ij}^{n-1}]
$$

$$
+ \frac{(\triangle t)^2}{(\triangle y)^2} \delta_y^2 u_{ij}^n, \tag{3.51}
$$

$$
u_{ij}^{n+1} - 2u_{ij}^n + u_{ij}^{n-1} = \frac{(\triangle t)^2}{(\triangle x)^2} [\eta \delta_x^2 u_{ij}^* + (1 - 2\eta)\delta_x^2 u_{ij}^n + \eta \delta_x^2 u_{ij}^{n-1}]
$$

$$
+ \frac{(\triangle t)^2}{(\triangle y)^2} [\eta \delta_y^2 u_{ij}^{n+1} + (1 - 2\eta)\delta_y^2 u_{ij}^n + \eta \delta_y^2 u_{ij}^{n-1}]. \tag{3.52}
$$

This scheme can be easily implemented also, since u^* can be obtained by solving a tridiagonal system from (3.51), and u^{n+1} can be obtained by solving another tridiagonal system from (3.52).

Subtracting (3.51) from (3.52), we obtain

$$
u_{ij}^{n+1} = u_{ij}^* + \frac{(\triangle t)^2}{(\triangle y)^2} \eta \delta_y^2 (u_{ij}^{n+1} - 2u_{ij}^n + u_{ij}^{n-1}). \tag{3.53}
$$

Substituting u^* from (3.53) into (3.52), we have

$$\delta_t^2 u_{ij}^n = \frac{(\triangle t)^2}{(\triangle x)^2} \cdot \eta \delta_x^2 [u_{ij}^{n+1} - \frac{(\triangle t)^2}{(\triangle y)^2} \cdot \eta \delta_y^2 (u_{ij}^{n+1} - 2u_{ij}^n + u_{ij}^{n-1})]$$

$$+ \frac{(\triangle t)^2}{(\triangle x)^2} [(1 - 2\eta)\delta_x^2 u_{ij}^n + \eta \delta_x^2 u_{ij}^{n-1}]$$

$$+ \frac{(\triangle t)^2}{(\triangle y)^2} [\eta \delta_y^2 u_{ij}^{n+1} + (1 - 2\eta)\delta_y^2 u_{ij}^n + \eta \delta_y^2 u_{ij}^{n-1}].$$

Dividing both sides by $(\triangle t)^2$ and collecting like terms, we obtain

$$\frac{1}{(\triangle t)^2}\delta_t^2 u_{ij}^n - (\frac{1}{(\triangle x)^2}\delta_x^2 u_{ij}^n + \frac{1}{(\triangle y)^2}\delta_y^2 u_{ij}^n) - \eta\delta_t^2(\frac{1}{(\triangle x)^2}\delta_x^2 u_{ij}^n + \frac{1}{(\triangle y)^2}\delta_y^2 u_{ij}^n)$$

$$+ (\triangle t)^2 \cdot \eta^2 \frac{1}{(\triangle x)^2} \cdot \frac{1}{(\triangle y)^2} \delta_x^2 \delta_y^2 \delta_t^2 u_{ij}^n = 0. \qquad (3.54)$$

We see that (3.54) reduces to the forward Euler scheme (3.45) if we drop the last two high-order perturbed terms. The truncation error of this scheme is $O((\triangle t)^2 + (\triangle x)^2 + (\triangle y)^2)$. □

Example 3.13
Finally, we like to mention a sixth-order accurate (in space) ADI scheme developed by Fairweather and Mitchell [1]:

$$u_{ij}^* = 2u_{ij}^n - u_{ij}^{n-1} - \frac{1}{12}(1 - \mu^2)\delta_x^2 [u_{ij}^* - \frac{2(1 + 5\mu^2)}{1 - \mu^2}u_{ij}^n + u_{ij}^{n-1}]$$

$$- \mu^2 \cdot \frac{1 + \mu^2}{1 - \mu^2}\delta_y^2 u_{ij}^n, \qquad (3.55)$$

$$u_{ij}^{n+1} = u_{ij}^* - \frac{1 - \mu^2}{12}\delta_y^2 [u_{ij}^{n+1} - \frac{2(1 + 10\mu^2 + \mu^4)}{(1 - \mu^2)^2}u_{ij}^n + u_{ij}^{n-1}], \quad (3.56)$$

if $\mu \neq 1$. Furthermore, Fairweather and Mitchell [1] proved that the local truncation error is

$$-\frac{1}{180}h^6\mu^2[(\mu^4 - \frac{3}{4})(u_{x^6} + u_{y^6}) + \frac{7}{4}(\mu^4 - \frac{29}{21})(u_{x^4y^2} + u_{x^2y^4})] = O(h^6)$$

and the scheme is stability under the condition [1, Eq. (4.2)]

$$\mu = \frac{\triangle t}{h} \leq \sqrt{3} - 1.$$

□

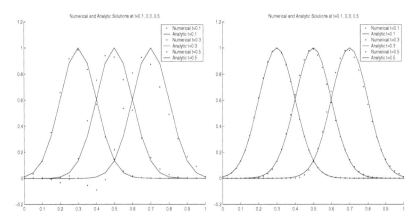

FIGURE 3.1
Solutions for the 1-D hyperbolic problem.

3.6 Numerical examples with MATLAB codes

Here we show an example using the Lax-Wendroff scheme to solve the hyperbolic equation

$$u_t + u_x = 0 \quad 0 \le x \le 1, 0 < t < 0.5,$$

with proper boundary condition at $x = 0$ and initial condition:

$$u(x, 0) = \exp(-c(x - 0.5)^2), \quad 0 \le x \le 1,$$

such that the analytic solution is a smooth Gaussian wave

$$u(x, t) = \exp(-c(x - t - 0.5)^2),$$

where $c > 0$ is a constant determining the narrowness of the wave. The larger c is, the narrower the wave will be.

Examplary numerical and analytic solutions are presented in Fig. 3.1 for $c = 50$ solved by different numbers of grid points: the left graphs are obtained on the 20×50 grid; the right graphs are obtained on the 40×50 grid.

From Fig. 3.1, we can see that when the grid is not fine enough, the numerical solution cannot approximate the analytic solution well. Use of finer mesh has reduced both the amplitude and phase errors significantly. Furthermore, we can see that the numerical wave always lags behind the analytic wave, which is consistent with our theoretical phase error analysis carried out earlier.

The MATLAB source code *hyper1d.m* is listed below.

```
%----------------------------------------------------------------
% hyper1d.m:
%     use Lax-Wendroff scheme to solve the hyperbolic equation
%        u_t(x,t) + u_x(x,t) = 0,          xl < x < xr, 0 < t < tf
%        u(x, 0) = f(x),                   xl < x < xr
%        u(0, t) = g(t),                   0  < t < tf
%
% A special case is choosing f and g properly such that the
% The analytic solution is:
%        u(x,t)= f(x-t)=e^(-10(x-t-0.2)^2)
%----------------------------------------------------------------
clear all;                    % clear all variables in memory

xl=0; xr=1;                   % x domain [xl,xr]
J = 40;                       % J: number of division for x
dx = (xr-xl) / J;             % dx: mesh size
tf = 0.5;                     % final simulation time
Nt = 50;                      % Nt: number of time steps
dt = tf/Nt;
c = 50;                       % parameter for the solution

mu = dt/dx;

if mu > 1.0       % make sure dt satisfy stability condition
    error('mu should < 1.0!')
end

% Evaluate the initial conditions
x = xl : dx : xr;             % generate the grid point
f = exp(-c*(x-0.2).^2);       % dimension f(1:J+1)

% store the solution at all grid points for all time steps
u = zeros(J+1,Nt);

% Find the approximate solution at each time step
for n = 1:Nt
    t = n*dt;                 % current time
    gl = exp(-c*(xl-t-0.2)^2);   % BC at left side
    gr = exp(-c*(xr-t-0.2)^2);   % BC at right side
    if n==1                   % first time step
        for j=2:J             % interior nodes
        u(j,n) = f(j) - 0.5*mu*(f(j+1)-f(j-1)) + ...
                       0.5*mu^2*(f(j+1)-2*f(j)+f(j-1));
```

```
      end
      u(1,n) = gl;          % the left-end point
      u(J+1,n) = gr;        % the right-end point
   else
      for j=2:J             % interior nodes
         u(j,n) = u(j,n-1) - 0.5*mu*(u(j+1,n-1)-u(j-1,n-1)) + ...
                  0.5*mu^2*(u(j+1,n-1)-2*u(j,n-1)+u(j-1,n-1));
      end
      u(1,n) = gl;          % the left-end point
      u(J+1,n) = gr;        % the right-end point
   end

   % calculate the analytic solution
   for j=1:J+1
      xj = xl + (j-1)*dx;
      u_ex(j,n)=exp(-c*(xj-t-0.2)^2);
   end

end

% plot the analytic and numerical solution at different times
figure;
hold on;
n=10;
plot(x,u(:,n),'r.',x,u_ex(:,n),'r-');    % r for red
n=30;
plot(x,u(:,n),'g.',x,u_ex(:,n),'g-');
n=50;
plot(x,u(:,n),'b.',x,u_ex(:,n),'b-');

legend('Numerical t=0.1','Analytic t=0.1',...
       'Numerical t=0.3','Analytic t=0.3',...
       'Numerical t=0.5','Analytic t=0.5');
title('Numerical and Analytic Solutions at t=0.1, 0.3, 0.5');
```

3.7 Bibliographical remarks

In this chapter we introduced some classic difference methods for solving hyperbolic equations including conservation laws. More advanced difference methods and their analysis can be found in other more focused books such as [2, 4, 5, 8, 10].

3.8   Exercises

1. For the hyperbolic equation

$$u_t + au_x = 0, \quad a = \text{constant} > 0,$$

we can construct an implicit scheme

$$\frac{u_j^{n+1} - u_j^n}{\Delta t} + a\frac{u_{j+1}^{n+1} - u_{j-1}^{n+1}}{2\Delta x} = 0. \tag{3.57}$$

Use the von Neumann technique to prove that the amplification factor for the scheme (3.57) is $\lambda = 1/(1+i\mu \sin k\Delta x)$, where $\mu = a\frac{\Delta t}{\Delta x}$. Hence the scheme is unconditionally stable. Study the corresponding dissipation and dispersion errors for the scheme.

2. Modify the code *hyper1d.m* and investigate what happens with increasingly larger C (say $C = 100$)? What do you observe for nonsmooth solution cases such as if the initial condition is a sequence pause:

$$u(x,0) = 1, \quad \text{if } 0.1 \le x \le 0.3; \ 0, \quad \text{elsewhere.}$$

3. Consider a system of conservation laws in two space dimensions

$$\frac{\partial u}{\partial t} + \frac{\partial f(u)}{\partial x} + \frac{\partial g(u)}{\partial y} = 0. \tag{3.58}$$

The unstaggered leap-frog scheme can be obtained immediately

$$\frac{u_{ij}^{n+1} - u_{ij}^{n-1}}{2\Delta t} + \frac{f(u_{i+1,j}^n) - f(u_{i-1,j}^n)}{2\Delta x} + \frac{g(u_{i,j+1}^n) - g(u_{i,j-1}^n)}{2\Delta y} = 0. \tag{3.59}$$

Show that for the scalar case $f(u) = au, g(u) = bu$, the von Neumann technique leads to

$$\lambda^2 + i(2\mu_x \sin k_x\Delta x + 2\mu_y \sin k_y\Delta y)\lambda - 1 = 0, \tag{3.60}$$

for the amplification factor λ, where $\mu_x = a\frac{\Delta t}{\Delta x}, \mu_y = b\frac{\Delta t}{\Delta y}$. Hence the scheme (3.59) is stable under the condition $|a\frac{\Delta t}{\Delta x}| + |b\frac{\Delta t}{\Delta y}| \le 1$.

4. Implement the explicit scheme (3.45) for solving the problem (3.41)-(3.44) with corresponding initial and boundary conditions such that the analytic solution is

$$u(x, y, t) = \sin \pi x \sin \pi y \cos \sqrt{2}\pi t.$$

Let $h = \frac{1}{10}$ and $t_F = 0.5$. Solve the problem with $\Delta t = 0.5h$ and $\Delta t = 0.8h$. Plot the numerical and analytic solutions at t_F. What do you observe? What happens if you try a smaller mesh size, say $h = \frac{1}{20}$?

5. For the hyperbolic problem

$$u_{tt} = a^2 u_{xx}, \quad \forall \, (x,t) \in (0,1) \times (0,t_F),$$
$$u(x,0) = f(x), \; u_t(x,0) = g(x), \quad \forall \, 0 \le x \le 1,$$
$$u(0,t) = u(1,t) = 0, \quad \forall \, t \in (0,t_F),$$

where $a > 0$ is a constant. Prove that the energy

$$E(t) = \int_0^1 (a^2 u_x^2(x,t) + u_t^2(x,t)) dx$$

is conserved for all time if u is a smooth solution. Hint: prove that $\frac{dE(t)}{dt} = 0$.

6. For the damped wave equation

$$u_{tt} + u_t = a^2 u_{xx}, \quad \forall \, (x,t) \in (0,1) \times (0,t_F),$$
$$u(x,0) = f(x), \; u_t(x,0) = g(x), \quad \forall \, 0 \le x \le 1,$$
$$u(0,t) = u(1,t) = 0, \quad \forall \, t \in (0,t_F).$$

Prove that the energy is decreasing, i.e.,

$$E(t) \le E(0) \quad \text{for any } t \ge 0,$$

where

$$E(t) = \int_0^1 (a^2 u_x^2(x,t) + u_t^2(x,t)) dx.$$

Use the von Neumann method to perform a stability analysis for the scheme

$$\frac{u_j^{n+1} - 2u_j^n + u_j^{n-1}}{(\Delta t)^2} + \frac{u_j^{n+1} - u_j^{n-1}}{2\Delta t} = a^2 \frac{u_{j+1}^n - 2u_j^n + u_{j-1}^n}{(\Delta x)^2}.$$

7. Let u be a sufficiently smooth solution to the initial value problem

$$v_t + f'(v)v_x = 0, \quad x \in \mathcal{R}, \; t > 0,$$
$$v(x,0) = v_0(x), \quad x \in \mathcal{R}.$$

Prove that v satisfies

$$v(x,t) = v_0(x - f'(v(x,t))t), \quad x \in \mathcal{R}, \; t > 0.$$

8. For the conservation law

$$u_t + (f(u))_x = 0, \tag{3.61}$$

the so-called three-point conservative difference scheme [10, Ch. 9] is in the form

$$u_k^{n+1} = u_k^n - \frac{\Delta t}{\Delta x}(h_{k+\frac{1}{2}}^n - h_{k-\frac{1}{2}}^n), \tag{3.62}$$

where the numerical flux

$$h^n_{k+\frac{1}{2}} = h(u^n_k, u^n_{k+1}), \quad h^n_{k-\frac{1}{2}} = h(u^n_{k-1}, u^n_k).$$

To make sure the difference scheme (3.62) is consistent with the conservation law (3.61), the flux h must satisfy

$$h(u, u) = f(u). \tag{3.63}$$

Prove that the Lax-Wendroff scheme (3.26) can be written as (3.62) with

$$h^n_{k+\frac{1}{2}} = \frac{1}{2}[f(u^n_{k+1}) + f(u^n_k)] - \frac{\triangle t}{2 \triangle x} a(u^n_{k+\frac{1}{2}})(f(u^n_{k+1}) - f(u^n_k))$$

where $a(u^n_{k+\frac{1}{2}}) = f'(\frac{1}{2}(u^n_k + u^n_{k+1}))$. Hence the Lax-Wendroff scheme is a consistent conservative scheme.

9. Prove that the Lax-Friedrichs scheme

$$u^{n+1}_k = \frac{1}{2}(u^n_{k-1} + u^n_{k+1}) - \frac{\triangle t}{2 \triangle x}(f(u^n_{k+1}) - f(u^n_{k-1}))$$

is also consistent and conservative for (3.61) by setting

$$h^n_{k+\frac{1}{2}} = \frac{1}{2}(f(u^n_k) + f(u^n_{k+1})) - \frac{\triangle x}{2 \triangle t}(u^n_{k+1} - u^n_k).$$

References

[1] G. Fairweather and A.R. Mitchell. A high accuracy alternating direction method for the wave equation. *J. Inst. Math. Appl.*, 1:309–316, 1965.

[2] E. Godlewski and P.-A. Raviart. *Numerical Approximation of Hyperbolic Systems of Conservation Laws*. Springer, New York, NY, 1996.

[3] B. Gustafsson, H.-O. Kreiss and J. Oliger. *Time Dependent Problems and Difference Methods*. Wiley, New York, NY, 1995.

[4] B.S. Jovanović and E. Süli. *Analysis of Finite Difference Schemes*. Springer-Verlag, London, UK, 2014.

[5] D. Kröner. *Numerical Schemes for Conservation Laws*. Wiley, Teubner, 1997.

[6] L. Lapidus and G.F. Pinder. *Numerical Solution of Partial Differential Equations in Science and Engineering*. John Wiley & Sons, New York, NY, 1982.

[7] M. Lees, Alternating direction methods for hyperbolic differential equations. *Journal of the Society for Industrial and Applied Mathematics*, 10(4):610–616, 1962.

[8] R.J. LeVeque. *Numerical Methods for Conservation Laws*. Birkhäuser, Basel, 2nd Edition, 2006.

[9] K.W. Morton and D.F. Mayers. *Numerical Solution of Partial Differential Equations*. Cambridge University Press, Cambridge, UK, 1994.

[10] J.W. Thomas. *Numerical Partial Differential Equations: Conservation Laws and Elliptic Equations*. Springer, New York, NY, 1999.

4

Finite Difference Methods for Elliptic Equations

In this chapter, we switch the discussion of finite difference method (FDM) to time-independent problems. In Sec. 4.1, we construct a difference scheme for solving the two-dimensional Poisson equation. Then in Sec. 4.2, we briefly introduce both direct methods and iterative methods for solving a general system of linear equations, such as that resulting from the difference scheme for the Poisson equation. Sec. 4.3 is devoted to the error analysis of the difference method for solving elliptic equations. In Sec. 4.4, we extend the construction of difference methods to some other elliptic problems. Finally, in Sec. 4.5 we present an examplary MATLAB code for solving the Poisson equation.

4.1 Introduction

Suppose that Ω is a bounded domain of R^2 with boundary $\partial\Omega$. The equation

$$a(x,y)\frac{\partial^2 u}{\partial x^2} + 2b(x,y)\frac{\partial^2 u}{\partial x\partial y} + c(x,y)\frac{\partial^2 u}{\partial y^2} = f(x,y,u,\frac{\partial u}{\partial x},\frac{\partial u}{\partial y}), \qquad (4.1)$$

is said to be elliptic if $b^2 - ac < 0$ for all points $(x,y) \in \Omega$. Proper boundary conditions (such as Dirichlet, Neumann, or Robin type) are needed for (4.1).

To illustrate how difference methods can be used to solve elliptic equations, let us start with the Poisson equation

$$-(u_{xx} + u_{yy}) = f(x,y) \quad \forall\ (x,y) \in \Omega \equiv (0,1)^2, \qquad (4.2)$$

$$u|_{\partial\Omega} = g(x,y) \quad \forall\ (x,y) \in \partial\Omega. \qquad (4.3)$$

As before, we assume that Ω is covered by a uniform grid

$$x_i = ih,\ y_j = jh,\ \ 0 \leq i,j \leq J,\ h = \frac{1}{J}.$$

The approximate solution at point (x_i, y_j) is denoted by u_{ij}. Using the second-order central difference

$$\delta_x^2 u_{ij} = \frac{u_{i+1,j} - 2u_{ij} + u_{i-1,j}}{h^2},\ \delta_y^2 u_{ij} = \frac{u_{i,j+1} - 2u_{ij} + u_{i,j-1}}{h^2}$$

to approximate the derivatives u_{xx} and u_{yy} in (4.2), respectively, we obtain the five-point difference scheme:

$$\frac{u_{i+1,j} - 2u_{ij} + u_{i-1,j}}{h^2} + \frac{u_{i,j+1} - 2u_{ij} + u_{i,j-1}}{h^2} + f_{ij} = 0, \quad 1 \leq i, j \leq J - 1,$$
(4.4)

or

$$u_{i+1,j} + u_{i-1,j} + u_{i,j+1} + u_{i,j-1} - 4u_{ij} = -h^2 f_{ij}. \tag{4.5}$$

The boundary condition (4.3) is approximated directly. For example, on boundary $y = 0$, we have

$$u_{i0} = g(x_i, 0) \equiv g_{i0}, \quad \forall\, 0 \leq i \leq J.$$

Define the local truncation error

$$T_{ij} \equiv \frac{1}{h^2} [u(x_{i+1}, y_j) + u(x_{i-1}, y_j) + u(x_i, y_{j+1}) + u(x_i, y_{j-1}) - 4u(x_i, y_j)]$$
$$+ f(x_i, y_j). \tag{4.6}$$

By Taylor expansion, we find that

$$T_{ij} = (u_{xx} + u_{yy})(x_i, y_j) + \frac{h^2}{12}(u_{x^4} + u_{y^4})(x, y) + f(x_i, y_j),$$

from which we obtain

$$|T_{ij}| \leq \frac{h^2}{12} \max_{(x,y) \in [0,1]^2} (|u_{x^4}| + |u_{y^4}|) \equiv T, \tag{4.7}$$

i.e., the truncation error is $O(h^2)$ when the derivatives of u are continuous up to order four in both x and y.

Denote the $(J-1)^2$ dimensional vector

$$U = [u_{1,1} \cdots u_{J-1,1}; u_{1,2} \cdots u_{J-1,2}; \cdots ; u_{1,J-1} \cdots u_{J-1,J-1}]^T,$$

where T means the transpose. With this notation, the scheme (4.5) ends up with a linear system

$$AU = F, \tag{4.8}$$

where A is a matrix of order $(J-1)^2$ given by

$$A = \begin{bmatrix} B & -I & & & & 0 \\ -I & B & -I & & & \\ & & \cdot & \cdot & \cdot & \\ & & \cdot & \cdot & \cdot & \\ & & & \cdot & \cdot & \cdot \\ & & & & -I & B & -I \\ 0 & & & & & -I & B \end{bmatrix}$$

with I being the identity matrix of order $J - 1$, and B being a matrix of order $J - 1$ given by

$$
B = \begin{bmatrix}
4 & -1 & & & & 0 \\
-1 & 4 & -1 & & & \\
 & \cdot & \cdot & \cdot & & \\
 & & \cdot & \cdot & \cdot & \\
 & & & \cdot & \cdot & \cdot \\
 & & & -1 & 4 & -1 \\
0 & & & & -1 & 4
\end{bmatrix} \tag{4.9}
$$

Furthermore, F is a $(J - 1)^2$ dimensional vector. Detailed elements are listed below:

$$
\begin{aligned}
F_{1,1} &= h^2 f_{1,1} + g_{1,0} + g_{0,1}, \\
F_{i,1} &= h^2 f_{i,1} + g_{i,0}, \quad 2 \leq i \leq J - 2, \\
F_{J-1,1} &= h^2 f_{J-1,1} + g_{J-1,0} + g_{J,1}, \\
F_{1,2} &= h^2 f_{1,2} + g_{0,2}, \\
F_{i,2} &= h^2 f_{i,2}, \quad 2 \leq i \leq J - 2, \\
F_{J-1,2} &= h^2 f_{J-1,2} + g_{J,2}, \\
&\cdots \\
F_{1,J-1} &= h^2 f_{1,J-1} + g_{1,J} + g_{0,J-1}, \\
F_{i,J-1} &= h^2 f_{i,J-1} + g_{i,J}, \quad 2 \leq i \leq J - 2, \\
F_{J-1,J-1} &= h^2 f_{J-1,J-1} + g_{J-1,J} + g_{J,J-1}.
\end{aligned}
$$

4.2 Numerical solution of linear systems

In this section, we present those popular solvers for linear systems classified into three categories: direct methods; iterative methods; and more advanced iterative methods.

4.2.1 Direct methods

We have seen that application of the finite difference method to a linear elliptic problem leads to a linear system of equations

$$
Ax = b, \tag{4.10}
$$

where $A = (a_{ij})$ is a nonsingular $n \times n$ matrix, and $b \in R^n$.

The basic direct method for solving (4.10) is the Gaussian elimination method, which is equivalent to decomposing the matrix A into the so-called LU-factorization:

$$
A = LU, \tag{4.11}
$$

where $L = (l_{ij})$ is a lower triangular $n \times n$ matrix (i.e., $l_{ij} = 0$ if $j > i$), and $U = (u_{ij})$ is an upper triangular matrix (i.e., $u_{ij} = 0$ if $j < i$). With the factorization (4.11), we can easily solve the system (4.10) by using forward and backward substitution to solve the triangular systems:

$$Ly = b, \quad Ux = y. \tag{4.12}$$

Note that the LU decomposition is achieved through the n-step procedure

$$A = A^{(1)} \to A^{(2)} \to \cdots \to A^{(n)} = U = L^{-1}A.$$

The basic step from $A^{(k)}$ to $A^{(k+1)}$ is to transform the matrix $A^{(k)}$ such that the elements of the kth column under the diagonal of the new matrix $A^{(k+1)}$ become zero.

The Gaussian elimination algorithm can be described as follows:

$$A^{(1)} = A;$$

For $k = 1 : n - 1$

    For $i = k + 1 : n$

$$l_{ik} = -a_{ik}^{(k)}/a_{kk}^{(k)}$$

        For $j = k + 1 : n$

$$a_{ij}^{(k+1)} = a_{ij}^{(k)} + l_{ik}a_{kj}^{(k)}$$

        End j

    End i

End k

The elements of matrix $L = (l_{ij})$ are given as follows:

$$l_{ii} = 1, \quad i = 1, \cdots, n,$$
$$l_{ik} = -a_{ik}^{(k)}/a_{kk}^{(k)}, \quad i = k + 1, \cdots, n, \ k = 1, \cdots, n,$$
$$l_{ik} = 0, \quad \text{if } i < k.$$

The Gaussian elimination method can terminate prematurely if one of the pivots $a_{kk}^{(k)}$ becomes zero. Fortunately, it is proven [4] that for a positive definite matrix A, all $a_{kk}^{(k)} > 0$, i.e., A has a unique LU-factorization.

When A is symmetric positive definite, we have the Cholesky decomposition

$$A = LL^T, \tag{4.13}$$

where L is a lower triangular matrix with positive diagonal entries. The elements of L can be obtained as follows:

For $k = 1 : n$

$$l_{kk} = (a_{kk} - \sum_{j=1}^{k-1} l_{kj}^2)^{1/2}$$

For $i = k + 1 : n$

$$l_{ik} = (a_{ik} - \sum_{j=1}^{k-1} l_{ij}l_{kj})/l_{kk}$$

End i

End k

Note that both the Gaussian elimination method and the Cholesky decomposition require $O(n^3)$ operations. Fortunately, in many practical applications, the matrix A is sparse (i.e., with many zero elements), in which case, it is possible to greatly reduce the number of operations. A special case is that if the matrix A is a band matrix, i.e., there is an integer d (the bandwidth), such that

$$a_{ij} = 0 \quad \text{if } |i - j| > d.$$

For such a band matrix, both the Gaussian elimination algorithm and Cholesky decomposition only require $O(nd^2)$ operations.

4.2.2 Simple iterative methods

In this subsection, we will introduce some classical iterative methods. Let the matrix A be represented as

$$A = L + D + U, \tag{4.14}$$

where D is the diagonal of A, and L, U are its lower and upper triangular parts.

The Jacobi method is obtained as follows:

$$Dx^{k+1} = -(L + U)x^k + b$$

or

$$x^{k+1} = -D^{-1}(L + U)x^k + D^{-1}b. \tag{4.15}$$

Componentwise, the algorithm becomes

$$x_i^{k+1} = (b_i - \sum_{j=1, j \neq i}^{n} a_{ij}x_j^k)/a_{ii}, \quad i = 1, \cdots, n. \tag{4.16}$$

Let the error vector $e^k = x - x^k$. Hence we have the error equation

$$e^{k+1} = B_J e^k, \quad B_J = -D^{-1}(L + U). \tag{4.17}$$

Note that the iterates x^k converge to the exact solution x of (4.10) for any initial vector x^0 if and only if the spectral radius of the iteration matrix B_J satisfies

$$\rho(B_J) < 1. \tag{4.18}$$

The condition (4.18) is satisfied for a strictly diagonally dominant matrix A, i.e.,

$$|a_{ii}| > \sum_{j=1, j \neq i}^{n} |a_{ij}| \quad \text{for } i = 1, \cdots, n. \tag{4.19}$$

We like to remark that the Jacobi method can be divergent even if A is symmetric positive definite [14, p. 111].

The Gauss-Seidel method is formulated as follows:

$$(L + D)x^{k+1} = -Ux^k + b,$$

or

$$x^{k+1} = -(L + D)^{-1}Ux^k + (L + D)^{-1}b, \tag{4.20}$$

i.e., the corresponding iteration matrix is $B_{GS} = -(L+D)^{-1}U$. The iteration (4.20) can be written componentwisely as

$$x_i^{k+1} = (b_i - \sum_{j=1}^{i-1} a_{ij}x_j^{k+1} - \sum_{j=i+1}^{n} a_{ij}x_j^{k})/a_{ii}, \quad i = 1, \cdots, n.$$

The Gauss-Seidel method is proved to be convergent either if A is symmetric positive definite or A is strictly diagonally dominant. Furthermore, when A is a tridiagonal matrix, we have [12, 14]

$$\rho(B_{GS}) = (\rho(B_J))^2,$$

which yields that the Gauss-Seidel method converges if and only if the Jacobi method converges.

To speed up the Gauss-Seidel method, we can construct the so-called successive over-relaxation (SOR) method:

$$x_i^{k+1} = (1-\omega)x_i^k + \omega(b_i - \sum_{j=1}^{i-1} a_{ij}x_j^{k+1} - \sum_{j=i+1}^{n} a_{ij}x_j^{k})/a_{ii}, \quad i = 1, \cdots, n, \tag{4.21}$$

where ω is a positive parameter.

Note that the SOR method (4.21) can be written in vector form

$$(D + \omega L)x^{k+1} = [(1 - \omega)D - \omega U]x^k + \omega b, \tag{4.22}$$

which has the iteration matrix

$$B_{SOR} = (D + \omega L)^{-1}[(1 - \omega)D - \omega U], \tag{4.23}$$

and coincides with B_{GS} when $\omega = 1$.

From (4.23), we have

$$|det(B_{SOR})| = \Pi_{i=1}^{n}|\lambda_i| = |det(D|^{-1} \cdot |1 - \omega|^n \cdot |det(D)| \leq (\rho(B_{SOR}))^n,$$

which yields that

$$\rho(B_{SOR}) \geq |\omega - 1|. \tag{4.24}$$

Therefore, a necessary condition for convergence of the SOR method is

$$0 < \omega < 2. \tag{4.25}$$

Actually, it can be proved that (4.25) becomes a sufficient condition if A is symmetric positive definite.

Now we want to give a simple example for applying these classical iteration methods to our difference equation (4.5), in which case, the Jacobi iterative method becomes:

$$u_{ij}^{n+1} = \frac{1}{4}(u_{i+1,j}^n + u_{i-1,j}^n + u_{i,j+1}^n + u_{i,j-1}^n) + \frac{h^2}{4}f_{ij}, \tag{4.26}$$

where u_{ij}^{n+1} denotes the $(n+1)$th estimate of the solution at point (x_i, y_j). To start the process, we need an initial guess u_{ij}^0, which usually takes the average of the boundary values, i.e.,

$$u_{ij}^0 = \frac{1}{(J+1)^2} \sum_{i=0}^{J} \sum_{j=0}^{J} g_{ij}.$$

To terminate the iteration, we need a stop criterion such as when the change between successive estimates at all interior points falls below an error tolerance $\epsilon > 0$, e.g.,

$$\max_{1 \leq i,j \leq J-1} |u_{ij}^{n+1} - u_{ij}^n| \leq \epsilon.$$

We can improve the convergence speed by using the new values of u_{ij} on the right-hand side of (4.26) as soon as they become available, which results in the Gauss-Seidel iterative method

$$u_{ij}^{n+1} = \frac{1}{4}(u_{i+1,j}^n + u_{i-1,j}^{n+1} + u_{i,j+1}^{n+1} + u_{i,j-1}^n) + \frac{h^2}{4}f_{ij}, \tag{4.27}$$

Faster convergence can be realized by using the SOR method:

$$u_{ij}^{n+1} = (1 - \omega)u_{i,j}^n + \omega u_{ij}^*, \tag{4.28}$$

where

$$u_{ij}^* = \frac{1}{4}(u_{i+1,j}^n + u_{i-1,j}^{n+1} + u_{i,j+1}^{n+1} + u_{i,j-1}^n) + \frac{h^2}{4}f_{ij}. \tag{4.29}$$

Here the parameter ω lies in the range $0 < \omega < 2$. The optimal choice for ω is based on the study of eigenvalues of the iteration matrix and is given by the formula

$$\omega_{opt} = \frac{2}{1 + \sqrt{1 - \cos^2 \frac{\pi}{J}}} \tag{4.30}$$

for the case of a rectangular region and Dirichlet boundary condition.

4.2.3   Modern iterative methods

Here we first consider two iterative methods for solving (4.10) when A is symmetric positive definite.

Example 4.1

The first one is the simple gradient method: Given an initial approximation $x^0 \in R^n$, find successive approximations $x^k \in R^n$ of the form

$$x^{k+1} = x^k + \alpha d^k, \quad k = 0, 1, \cdots, \tag{4.31}$$

where the search direction $d^k = -(Ax^k - b)$ and the step length α is a suitably chosen positive constant.

Using (4.10) and (4.31), we obtain the error equation

$$e^{k+1} = (I - \alpha A)e^k, \quad k = 0, 1, \cdots, \tag{4.32}$$

which leads to

$$|e^{k+1}| \le |I - \alpha A| \cdot |e^k|. \tag{4.33}$$

Here we denote $|\eta|$ for the usual Euclidean norm $|\eta| = (\sum_{i=1}^{n} \eta_i^2)^{1/2}$ for any vector $\eta \in R^n$, and the matrix norm

$$|B| = \max_{0 \neq \eta \in R^n} \frac{|B\eta|}{|\eta|}. \tag{4.34}$$

From linear algebra, if B is symmetric with eigenvalues $\lambda_1, \cdots, \lambda_n$, then we have

$$|B| = \max_i |\lambda_i|. \tag{4.35}$$

To guarantee the convergence of the gradient method, we like to have that

$$|I - \alpha A| \equiv \gamma < 1, \tag{4.36}$$

i.e., the error will be reduced by a factor γ at each iterative step. Note that the smaller γ is, the more rapid is the convergence. Since our matrix A is positive definite, we have

$$|I - \alpha A| = \max_i |1 - \alpha \lambda_i|.$$

Hence $|I - \alpha A| < 1$ is equivalent to

$$-1 < 1 - \alpha \lambda_i < 1, \quad i = 1, \cdots, n,$$

i.e., α has to be chosen such that

$$\alpha \lambda_{max} < 2. \tag{4.37}$$

A common choice is $\alpha = 1/\lambda_{max}$, in which case we have

$$|I - \alpha A| = 1 - \frac{\lambda_{min}}{\lambda_{max}} = 1 - \frac{1}{K(A)}, \qquad (4.38)$$

where $K(A) = \frac{\lambda_{max}}{\lambda_{min}}$ is the condition number for a symmetric matrix A.
From (4.33) and (4.38), we obtain

$$|e^k| \le (1 - \frac{1}{K(A)})|e^{k-1}| \le \cdots \le (1 - \frac{1}{K(A)})^k |e^0|, \quad k = 1, 2, \cdots .$$

Hence to reduce the error by a factor $\epsilon > 0$, the number of iteration step k needs to satisfy the inequality

$$(1 - \frac{1}{K(A)})^k \le \epsilon.$$

Using the inequality $1 - x \le e^{-x}$ for any $x \ge 0$, we obtain

$$(1 - \frac{1}{K(A)})^k \le e^{-k/K(A)},$$

which shows that to reduce the error by a factor ϵ, we need the number of iteration step k satisfying the following:

$$k \ge K(A) \log \frac{1}{\epsilon}. \qquad (4.39)$$

□

Example 4.2
A more efficient iterative method for (4.10) is the conjugate gradient method, whose step size α is chosen to be optimal and the search directions d^k are conjugate, i.e.,

$$d^i \cdot Ad^j = 0, \quad i \ne j. \qquad (4.40)$$

The conjugate gradient method can be stated as follows: Given $x^0 \in R^n$ and $d^0 = -r^0$, find x^k and d^k, $k = 1, 2, \cdots$, such that

$$x^{k+1} = x^k + \alpha_k d^k, \qquad (4.41)$$

$$\alpha_k = -\frac{r^k \cdot d^k}{d^k \cdot Ad^k}, \qquad (4.42)$$

$$d^{k+1} = -r^{k+1} + \beta_k d^k, \qquad (4.43)$$

$$\beta_k = \frac{r^{k+1} \cdot Ad^k}{d^k \cdot Ad^k}, \qquad (4.44)$$

where we denote $r^k = Ax^k - b$.

It can be shown that the conjugate gradient algorithm gives, in the absence of round-off error, the exact solution after at most n steps [7, p. 133]. Furthermore, for a given error tolerance $\epsilon > 0$, in order to satisfy

$$(x - x^k, A(x - x^k)) \leq \epsilon(x - x^0, A(x - x^0)),$$

it is sufficient to choose the iteration step k such that [7]

$$k \geq \frac{1}{2}\sqrt{K(A)} \cdot \log\frac{2}{\epsilon}. \tag{4.45}$$

Compared to (4.39), for large $K(A)$, the conjugate gradient method is much more efficient than the gradient method.   ▯

For non-symmetric matrices, there are many interesting iterative algorithms developed since the 1980s. Examples include the Generalized Minimal Residual (GRMES) method, the Bi-Conjugate Gradient (Bi-CG) method, the Conjugate Gradient-Squared (CGS) method, and the Bi-Conjugate Gradient Stabilized method (Bi-CGSTAB) method, among others. Details about these methods and their implementation in C/C++, Fortran 77 and MATLAB can be consulted in the book [1]. The most efficient method for solving a linear system of equations (4.10) may be the so-called multigrid method, whose required number of operations is of the order $O(n)$. Due to its complexity, interested readers can consult special books on the multigrid method [5, 3, 13] and references therein.

4.3   Error analysis with a maximum principle

Assume that a general elliptic equation

$$Lu = f$$

is approximated by a difference scheme

$$L_h U_p \equiv \alpha_p U_p - \sum_{i=1}^{k} \alpha_i U_i = f_p, \quad \forall\, p \in J_\Omega,$$

where $i = 1, \cdots, k$ denotes the neighbors of point p, f_p is the value of function f evaluated at point p, and J_Ω denotes the set of interior points of Ω.

We assume that the coefficients satisfy the following conditions:
(i) For all points $p \in J_\Omega, \alpha_p > 0, \alpha_i \geq 0, \ i = 1, \cdots, k$.
(ii) For all points $p \in J_\Omega, \sum_{i=1}^{k} \alpha_i \leq \alpha_p$. Strict inequality must be true for at least one p.

LEMMA 4.1

(Maximum Principle) *Suppose that the difference operator L_h satisfy the conditions (i) and (ii), and*

$$L_h V_p \leq 0, \quad \text{for all interior points } p \in \Omega,$$

then V cannot achieve its non-negative maximum at an interior point, i.e.,

$$\max_{p \in J_\Omega} V_p \leq \max\{0, \max_{Q \in J_{\partial\Omega}} V_Q\},$$

where $J_{\partial\Omega}$ denotes the boundary points of Ω.

Proof. Suppose that the non-negative maximum of V occurs at some interior point $\tilde{p}$, i.e.,

$$V_{\tilde{p}} = \max_{p \in J_\Omega} V_p \quad \text{and} \quad V_{\tilde{p}} > \max_{Q \in J_{\partial\Omega}} V_Q. \tag{4.46}$$

Note that $L_h V_{\tilde{p}} \leq 0$ and conditions (i)-(ii) lead to

$$\alpha_{\tilde{p}} V_{\tilde{p}} \leq \sum_{i=1}^{k} \alpha_i V_i \leq (\sum_{i=1}^{k} \alpha_i) V_{\tilde{p}} \leq \alpha_{\tilde{p}} V_{\tilde{p}},$$

which implies that all $V_i = V_{\tilde{p}}$ (including boundary points), which contradicts the assumption (4.46). □

In the following, we shall use the Maximum Principle to derive the error estimate for our model problem (4.2)-(4.3). For any interior point $p(x_i, y_j)$, we denote the difference operator

$$L_h u_{ij} \equiv \frac{4}{h^2} u_{ij} - \frac{1}{h^2} (u_{i-1,j} + u_{i+1,j} + u_{i,j-1} + u_{i,j+1}).$$

It is easy to see that

$$L_h u_{ij} = f_{ij} \quad \forall\, 1 \leq i, j \leq J - 1,$$

and conditions (i) and (ii) are satisfied. The strict inequality of (ii) holds true for those interior points with some neighboring boundary points.

Denote the pointwise error

$$e_{ij} = u_{ij} - u(x_i, y_j),$$

which along with (4.6) leads to the error equation

$$L_h e_{ij} = f_{ij} - L_h u(x_i, y_j) = T_{ij}. \tag{4.47}$$

Let us introduce the grid function

$$\psi_{ij} = e_{ij} + \frac{1}{2} T \phi_{ij}, \tag{4.48}$$

where T is the maximum truncation error defined by (4.7), and ϕ_{ij} is defined as $\phi_{ij} = x_i^2$.

By the definitions of L_h and ϕ_{ij}, we have

$$L_h\phi_{ij} = \frac{4}{h^2}x_i^2 - \frac{1}{h^2}[(x_i - h)^2 + (x_i + h)^2 + x_i^2 + x_i^2] = -2,$$

from which, (4.47), and the definition of T, we obtain

$$L_h\psi_{ij} = L_h e_{ij} + \frac{1}{2}TL_h\phi_{ij} = T_{ij} - T \leq 0, \forall\, 1 \leq i, j \leq J - 1.$$

Furthermore, the maximum of ψ_{ij} on the boundary points are

$$\max_{(i,j)\in J_{\partial\Omega}} \psi_{ij} = 0 + \frac{1}{2}T \max_{(i,j)\in J_{\partial\Omega}} \phi_{ij} = \frac{1}{2}T.$$

Hence, by Lemma 4.1, we see that

$$\max_{(i,j)\in J_\Omega} \psi_{ij} \leq \frac{1}{2}T = \frac{h^2}{24} \max_{(x,y)\in[0,1]^2}(|u_{x^4}| + |u_{y^4}|),$$

which, along with the fact $\phi_{ij} \geq 0$, leads to

$$e_{ij} \leq \frac{h^2}{24} \max_{(x,y)\in[0,1]^2}(|u_{x^4}| + |u_{y^4}|). \tag{4.49}$$

Repeating the above analysis with the function

$$\psi_{ij} = -e_{ij} + \frac{1}{2}T\phi_{ij}$$

shows that

$$-e_{ij} \leq \frac{h^2}{24} \max_{(x,y)\in[0,1]^2}(|u_{x^4}| + |u_{y^4}|), \tag{4.50}$$

which combining with (4.49) proves that

$$|u_{ij} - u(x_i, y_j)| \leq \frac{h^2}{24} \max_{(x,y)\in[0,1]^2}(|u_{x^4}| + |u_{y^4}|).$$

In conclusion, the solution of the five-point scheme (4.4) converges to the exact solution (4.2)-(4.3) in order of $O(h^2)$.

4.4 Some extensions

In this section, we first show how to develop finite difference schemes for elliptic problems with mixed boundary conditions and conserved forms. Then we show how to construct higher-order difference schemes.

4.4.1 Mixed boundary conditions

Now let us consider the Poisson equation with mixed boundary conditions:

$$-(u_{xx} + u_{yy}) = f(x, y) \quad \forall \ (x, y) \in \Omega \equiv (0, 1)^2, \tag{4.51}$$
$$u(0, y) = u^L(y), \quad u(1, y) = u^R(y) \quad \forall \ y \in (0, 1), \tag{4.52}$$
$$u_y(x, 0) = g^B(x), \quad u_y(x, 1) = g^T(x) \quad \forall \ x \in (0, 1), \tag{4.53}$$

where some corner compatibility conditions are implied in order to guarantee that the solution has enough regularity. For example, at corner $(0, 0)$, we have $\frac{\partial u^L}{\partial y}(0) = g^B(0)$.

We discretize (4.51) by the five-point scheme (4.4), i.e.,

$$u_{i+1,j} + u_{i-1,j} + u_{i,j+1} + u_{i,j-1} - 4u_{i,j} = -h^2 f_{i,j}, \quad \forall \ 1 \le i, j \le J-1. \tag{4.54}$$

Approximating (4.52) directly gives

$$u_{0,j} = u_j^L, \quad u_{J,j} = u_j^R, \quad 0 \le j \le J. \tag{4.55}$$

Approximating (4.53) by central difference, we have

$$u_{i,1} - u_{i,-1} = 2hg_i^B, \quad u_{i,J+1} - u_{i,J-1} = 2hg_i^T, \quad 0 \le i \le J. \tag{4.56}$$

Eliminating $u_{i,-1}$ from (4.56) and (4.54) with $j = 0$, we obtain

$$u_{i+1,0} + u_{i-1,0} + 2u_{i,1} - 4u_{i,0} = 2hg_i^B - h^2 f_{i,0}. \quad \forall \ 1 \le i \le J-1. \tag{4.57}$$

Similarly, eliminating $u_{i,J+1}$ from (4.56) and (4.54) with $j = J$, we obtain

$$u_{i+1,J} + u_{i-1,J} + 2u_{i,J-1} - 4u_{i,J} = -2hg_i^T - h^2 f_{i,J}. \quad \forall \ 1 \le i \le J-1. \tag{4.58}$$

Define the vector

$$U = [u_{1,0} \cdots u_{J-1,0}; u_{1,1} \cdots u_{J-1,1}; \cdots ; u_{1,J} \cdots u_{J-1,J}]^T,$$

which forms the unknown approximate solution for the model problem (4.51)-(4.53).

Combining (4.54), (4.55), (4.57) and (4.58), we obtain the linear system

$$AU = F, \tag{4.59}$$

where the matrix A is of order $((J-1)(J+1))^2$ or $(J^2-1)^2$, the vector F has $J^2 - 1$ elements which is composed of function values of f, u^L, u^R, g^B and g^T.

By careful calculation, the coefficient matrix A is formed by $(J+1) \times (J+1)$ blocks of $(J-1) \times (J-1)$ submatrices:

$$A = \begin{bmatrix} -B & 2I & & & 0 \\ I & -B & I & & \\ & & \ddots & \ddots & \ddots \\ & & \ddots & \ddots & \ddots \\ & & & I & -B & I \\ 0 & & & & 2I & -B \end{bmatrix}$$

where the matrix B is presented by (4.9).

4.4.2   Self-adjoint problems

Now we consider a self-adjoint elliptic equation

$$-[(au_x)_x + (bu_y)_y] + cu = f(x,y), \quad \forall\,(x,y) \in (0,1)^2, \tag{4.60}$$

where $a(x,y), b(x,y) > 0, c(x,y) \geq 0$.
Approximating

$$(au_x)_x|_{i,j} \approx (a_{i+\frac{1}{2},j}\frac{u_{i+1,j} - u_{i,j}}{h} - a_{i-\frac{1}{2},j}\frac{u_{i,j} - u_{i-1,j}}{h})/h$$

and

$$(bu_y)_y|_{i,j} \approx (b_{i,j+\frac{1}{2}}\frac{u_{i,j+1} - u_{i,j}}{h} - b_{i,j-\frac{1}{2}}\frac{u_{i,j} - u_{i,j-1}}{h})/h,$$

we obtain the difference scheme for (4.60):

$$\alpha_1 u_{i+1,j} + \alpha_2 u_{i,j+1} + \alpha_3 u_{i-1,j} + \alpha_4 u_{i,j-1} - \alpha_0 u_{i,j} = -h^2 f_{i,j}, \tag{4.61}$$

where we denoted

$$\alpha_1 = a_{i+\frac{1}{2},j}, \alpha_2 = b_{i,j+\frac{1}{2}}, \alpha_3 = a_{i-\frac{1}{2},j}, \alpha_4 = b_{i,j-\frac{1}{2}}, \alpha_0 = \sum_{k=1}^{4} \alpha_k + h^2 c_{i,j}.$$

It is easy to see that the scheme (4.61) satisfies Lemma 4.1.

4.4.3   A fourth-order scheme

Finally, we like to mention how to construct a fourth-order scheme for (4.2)-(4.3) using nine points (cf., Fig. 4.1). For simplicity, we assume a uniform mesh $\Delta x = \Delta y = h$. To construct the scheme, we approximate

$$(u_{xx} + u_{yy})(x_0, y_0) \approx \alpha_0 u_0 + \alpha_1 (u_1 + u_3) + \alpha_2 (u_2 + u_4) + \alpha_3 (u_5 + u_6 + u_7 + u_8), \tag{4.62}$$

and

$$f(x_0, y_0) \approx \beta_0 f_0 + \beta_1 (f_1 + f_2 + f_3 + f_4), \tag{4.63}$$

where u_i represents the approximate solution of u at point i, while α_i and β_i are unknown coefficients.
By Taylor expansion, we have

$$u_1 = u_0 + h\frac{\partial u_0}{\partial x} + \frac{h^2}{2!}\frac{\partial^2 u_0}{\partial x^2} + \frac{h^3}{3!}\frac{\partial^3 u_0}{\partial x^3} + \frac{h^4}{4!}\frac{\partial^4 u_0}{\partial x^4} + \frac{h^5}{5!}\frac{\partial^5 u_0}{\partial x^5} + O(h^6),$$

$$u_3 = u_0 - h\frac{\partial u_0}{\partial x} + \frac{h^2}{2!}\frac{\partial^2 u_0}{\partial x^2} - \frac{h^3}{3!}\frac{\partial^3 u_0}{\partial x^3} + \frac{h^4}{4!}\frac{\partial^4 u_0}{\partial x^4} - \frac{h^5}{5!}\frac{\partial^5 u_0}{\partial x^5} + O(h^6),$$

$$u_2 = u_0 + h\frac{\partial u_0}{\partial y} + \frac{h^2}{2!}\frac{\partial^2 u_0}{\partial y^2} + \frac{h^3}{3!}\frac{\partial^3 u_0}{\partial y^3} + \frac{h^4}{4!}\frac{\partial^4 u_0}{\partial y^4} + \frac{h^5}{5!}\frac{\partial^5 u_0}{\partial y^5} + O(h^6),$$

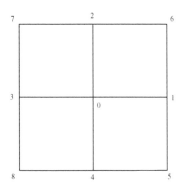

FIGURE 4.1
Grid point location.

$$u_4 = u_0 - h\frac{\partial u_0}{\partial y} + \frac{h^2}{2!}\frac{\partial^2 u_0}{\partial y^2} - \frac{h^3}{3!}\frac{\partial^3 u_0}{\partial y^3} + \frac{h^4}{4!}\frac{\partial^4 u_0}{\partial y^4} - \frac{h^5}{5!}\frac{\partial^5 u_0}{\partial y^5} + O(h^6),$$

which lead to

$$\alpha_1(u_1 + u_3) = \alpha_1[2u_0 + h^2\frac{\partial^2 u_0}{\partial x^2} + \frac{h^4}{12}\frac{\partial^4 u_0}{\partial x^4}] + O(h^6), \tag{4.64}$$

$$\alpha_2(u_2 + u_4) = \alpha_2[2u_0 + h^2\frac{\partial^2 u_0}{\partial y^2} + \frac{h^4}{12}\frac{\partial^4 u_0}{\partial y^4}] + O(h^6). \tag{4.65}$$

On the other hand, we have the Taylor expansion

$$u_6 = u_0 + h(\frac{\partial u_0}{\partial x} + \frac{\partial u_0}{\partial y}) + \frac{h^2}{2!}(\frac{\partial^2 u_0}{\partial x^2} + 2\frac{\partial^2 u_0}{\partial x \partial y} + \frac{\partial^2 u_0}{\partial y^2})$$
$$+ \frac{h^3}{3!}(\frac{\partial^3 u_0}{\partial x^3} + 3\frac{\partial^3 u_0}{\partial x^2 \partial y} + 3\frac{\partial^3 u_0}{\partial x \partial y^2} + \frac{\partial^3 u_0}{\partial y^3})$$
$$+ \frac{h^4}{4!}(\frac{\partial^4 u_0}{\partial x^4} + 4\frac{\partial^4 u_0}{\partial x^3 \partial y} + 6\frac{\partial^4 u_0}{\partial x^2 \partial y^2} + 4\frac{\partial^4 u_0}{\partial x \partial y^3} + \frac{\partial^4 u_0}{\partial y^4})$$
$$+ \frac{h^5}{5!}(\frac{\partial^5 u_0}{\partial x^5} + 5\frac{\partial^5 u_0}{\partial x^4 \partial y} + 10\frac{\partial^5 u_0}{\partial x^3 \partial y^2} + 10\frac{\partial^5 u_0}{\partial x^2 \partial y^3} + 5\frac{\partial^5 u_0}{\partial x \partial y^4} + \frac{\partial^5 u_0}{\partial y^5}).$$

Similar Taylor expansions for u_5, u_7 and u_8 can be obtained. It is not difficult to see that

$$\alpha_3(u_5 + u_6 + u_7 + u_8)$$
$$= \alpha_3[4u_0 + \frac{h^2}{2!} \cdot 4(\frac{\partial^2 u_0}{\partial x^2} + \frac{\partial^2 u_0}{\partial y^2})$$
$$+ \frac{h^4}{4!} \cdot 4(\frac{\partial^4 u_0}{\partial x^4} + 6\frac{\partial^4 u_0}{\partial x^2 \partial y^2} + \frac{\partial^4 u_0}{\partial y^4})] + O(h^6). \tag{4.66}$$

From (4.2), we have

$$f_0 = -(\frac{\partial^2 u_0}{\partial x^2} + \frac{\partial^2 u_0}{\partial y^2}),$$

which leads further to

$$\frac{\partial^2 f_0}{\partial x^2} = -(\frac{\partial^4 u_0}{\partial x^4} + \frac{\partial^4 u_0}{\partial x^2 \partial y^2}), \quad \frac{\partial^2 f_0}{\partial y^2} = -(\frac{\partial^4 u_0}{\partial x^2 \partial y^2} + \frac{\partial^4 u_0}{\partial y^4}).$$

Hence, we can obtain

$$\beta_0 f_0 = \beta_0[-(\frac{\partial^2 u_0}{\partial x^2} + \frac{\partial^2 u_0}{\partial y^2})], \tag{4.67}$$

and

$$
\begin{aligned}
&\beta_1(f_1 + f_2 + f_3 + f_4) \\
&= \beta_1\{4f_0 + \frac{h^2}{2!} \cdot 2(\frac{\partial^2 f_0}{\partial x^2} + \frac{\partial^2 f_0}{\partial y^2}) + O(h^4)\} \\
&= -\beta_1[4(\frac{\partial^2 u_0}{\partial x^2} + \frac{\partial^2 u_0}{\partial y^2}) + h^2(\frac{\partial^4 u_0}{\partial x^4} + 2\frac{\partial^4 u_0}{\partial x^2 \partial y^2} + \frac{\partial^4 u_0}{\partial y^4})] + O(h^4)
\end{aligned} \tag{4.68}
$$

Substituting (4.64)-(4.66) into (4.62), (4.67)-(4.68) into (4.63), and comparing the same order for (4.62) and (4.63), we obtain

$$u_0 : \quad \alpha_0 + 2\alpha_1 + 2\alpha_2 + 4\alpha_3 = 0,$$

$$\frac{\partial^2 u_0}{\partial x^2} : \quad h^2(\alpha_1 + 2\alpha_3) = -\beta_0 - 4\beta_1,$$

$$\frac{\partial^2 u_0}{\partial y^2} : \quad h^2(\alpha_2 + 2\alpha_3) = -\beta_0 - 4\beta_1,$$

$$\frac{\partial^4 u_0}{\partial x^4} : \quad h^2(\alpha_1 + 2\alpha_3) = -12\beta_1,$$

$$\frac{\partial^4 u_0}{\partial y^4} : \quad h^2(\alpha_2 + 2\alpha_3) = -12\beta_1,$$

$$\frac{\partial^4 u_0}{\partial x^2 \partial y^2} : \quad h^2 \alpha_3 = -2\beta_1.$$

Solving the above linear system, we have

$$\alpha_1 = \alpha_2 = -8\beta_1/h^2, \quad \alpha_3 = -2\beta_1/h^2, \quad \alpha_0 = -40\beta_1/h^2, \quad \beta_0 = 8\beta_1,$$

substituting which into (4.62) and (4.63), we obtain an $O(h^4)$ difference scheme:

$$4(u_1 + u_2 + u_3 + u_4) + (u_5 + u_6 + u_7 + u_8) - 20u_0 = -\frac{h^2}{2}(8f_0 + f_1 + f_2 + f_3 + f_4)$$

i.e., for any interior points $1 \le i, j \le J - 1$, the difference scheme becomes:

$$4(u_{i+1,j} + u_{i,j+1} + u_{i-1,j} + u_{i,j-1})$$
$$+(u_{i+1,j-1} + u_{i+1,j+1} + u_{i-1,j+1} + u_{i-1,j-1}) - 20u_{i,j}$$
$$= -\frac{h^2}{2}(8f_{i,j} + f_{i+1,j} + f_{i,j+1} + f_{i-1,j} + f_{i,j-1}). \tag{4.69}$$

It is shown [2] that on a square mesh, no other 9-point scheme for the Laplace equation can have greater order of accuracy than (4.69). More high-order schemes can be found in [10].

4.5 Numerical examples with MATLAB codes

Here we demonstrate a MATLAB code for solving (4.2) with a homogeneous Dirichlet boundary condition over the domain $\Omega = (0,1)^2$. The right-hand side function f is chosen as $f(x,y) = 2\pi^2 \sin(\pi x) \sin(\pi y)$ such that the exact solution is given by

$$u(x,y) = \sin(\pi x) \sin(\pi y).$$

The listed code *ellip2d.m* solves the problem (4.2) using the five-point difference scheme (4.5). We solve the problem using various mesh sizes, and the results justify the second-order convergence rate. With the number of division $J = 5, 10, 20, 40$, the maximum errors are as follows

$$0.0304, 0.0083, 0.0021, 5.1420e - 004.$$

An exemplary numerical solution obtained with $J = 20$ is shown is Fig. 4.2. The main function *ellip2d.m* is listed below:

```
%------------------------------------------------------------
% ellip2d.m: solve the Poisson equation on unit rectangle.
%
% Need 2 functions: exU.m, srcF.m.
%
% O(h^2) convergence rate is observed by using J=5,10,20,40,
% which gives max errors as follows:
% 0.0304, 0.0083, 0.0021, 5.1420e-004.
%------------------------------------------------------------
clear all;

xl=0; xr=1;    % x domain
yl=0; yr=1;    % y domain
```

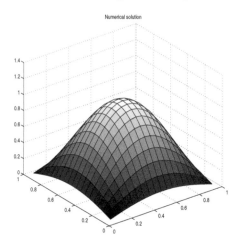

FIGURE 4.2
The numerical solution for the 2-D elliptic problem.

```
J=20;          % number of points in both x- and y-directions
h = (xr-xl)/J;   % mesh size

% build up the coefficient matrix
nr = (J-1)^2;    % order of the matrix
matA = zeros(nr,nr);

% can set J=3,4,5 etc to check the coefficient matrix
for i=1:nr
    matA(i,i)=4;
    if i+1 <= nr & mod(i,J-1) ~= 0
        matA(i,i+1)=-1;
    end
    if i+J-1 <= nr
        matA(i,i+J-1)=-1;
    end
    if i-1 >= 1 & mod(i-1,J-1) ~= 0
        matA(i,i-1)=-1;
    end
    if i-(J-1) >= 1
        matA(i,i-(J-1))=-1;
    end
end
```

```
% build up the right-hand side vector
for j=1:J-1
    y(j) = j*h;
    for i=1:J-1
        x(i) = i*h;
        [fij]=feval(@srcF,x(i),y(j));  % evaluate f(xi,yj)
        vecF((J-1)*(j-1)+i)=h^2*fij;
        [uij]=feval(@exU,x(i),y(j));  % evaluate exact solution
        ExU(i,j)=uij;
    end
end

% solve the system
vecU = matA\vecF';       % vecU is of order (J-1)^2

for j=1:J-1
    for i=1:J-1
        U2d(i,j)=vecU((J-1)*(j-1)+i);  % change into 2-D array
    end
end

% display max error so that we can check convergence rate
disp('Max error ='), max(max(U2d-ExU)),

figure(1);
surf(x,y,U2d);
title('Numerical solution');
figure(2);
surf(x,y,ExU);
title('Exact solution');
```

The two auxiliary functions *exU.m* and *srcF.m* are as follows:

```
% The exact solution of the governing PDE
function uu=exU(x,y)

uu=sin(pi*x)*sin(pi*y);
```

```
% The RHS function f for the governing PDE
function uu=srcF(x,y)

uu=2*pi^2*sin(pi*x)*sin(pi*y);
```

4.6   Bibliographical remarks

In this chapter, we introduced some classic difference schemes for solving elliptic problems. More schemes and detailed discussions on elliptic problems can be found in other books such as [6, 8, 9, 10, 11].

4.7   Exercises

1. For the three-dimensional Poisson problem

$$-(u_{xx} + u_{yy} + u_{zz}) = f(x, y, z) \quad \text{in } \Omega = (0, 1)^3, \tag{4.70}$$
$$u|_{\partial\Omega} = 0. \tag{4.71}$$

We can derive the seven-point scheme

$$\frac{6}{h^2}u_{ijk} - \frac{1}{h^2}(u_{i+1,j,k} + u_{i-1,j,k} + u_{i,j+1,k} + u_{i,j-1,k} + u_{i,j,k+1} + u_{i,j,k-1}) = f_{ijk}. \tag{4.72}$$

Show that the Maximum Principle still applies and we have the convergence result

$$|u_{ijk} - u(x_i, y_j, z_k)| \le \frac{h^2}{24} \max_{(x,y,z)\in[0,1]^3} (|u_{x^4}| + |u_{y^4}| + |u_{z^4}|).$$

2. Implement the SOR method (4.28)-(4.30) and solve the same problem as *ellip2d.m*. Compare your SOR implementation with *ellip2d.m* using $J = 20$ and 40. Which method is more efficient? In this case, you can use the following stop criterion

$$\max_{1 \le i,j \le J-1} |u_{ij}^{n+1} - u(x_i, y_j)| \le \epsilon,$$

where $u(x_i, y_j)$ is the exact solution at point (x_i, y_j).

3. Modify the code *ellip2d.m* to solve the self-adjoint elliptic problem

$$-[(1 + x^2 + y^2)u_x)_x + (e^{xy}u_y)_y] + u = f(x, y) \quad \forall \ (x, y) \in \Omega = (0, 1)^2,$$
$$u|_{\partial\Omega} = 0,$$

where $f(x, y)$ is chosen accordingly such that the exact solution

$$u(x, y) = x(1 - x)y(1 - y).$$

Estimate the convergence rate by running different mesh sizes.

4. Assume that u is harmonic in Ω (i.e., $u_{xx} + u_{yy} = 0$). Then prove that

$$\min_{(x,y)\in\partial\Omega} u(x,y) \leq \max_{(x,y)\in\overline{\Omega}} u(x,y) \leq \max_{(x,y)\in\partial\Omega} u(x,y).$$

Hint: Consider $v_\epsilon = u \pm \epsilon(x^2 + y^2)$ and use the contradiction technique.

5. Consider the eigenvalue problem

$$-\Delta u = \lambda u, \quad \text{in } \Omega \equiv (0,1)^2,$$
$$u = 0 \quad \text{on } \partial\Omega,$$

where the eigenfunction $u \neq 0$. Prove that the eigenvalues and the corresponding eigenfunctions are

$$\lambda_{jk} = (j\pi)^2 + (k\pi)^2, \quad u_{jk}(x,y) = \sin(j\pi x)\sin(k\pi y), \quad j,k = 1,2,\cdots.$$

Develop the five-point scheme to solve this eigenvalue problem.

6. Consider the Poisson equation in the first quadrant of the unit circle $\Omega = \{(r,\theta) : 0 < r < 1, 0 < \theta < \frac{\pi}{2}\}$:

$$\frac{1}{r}\frac{\partial}{\partial r}\left(r\frac{\partial u}{\partial r}\right) + \frac{1}{r^2}\frac{\partial^2 u}{\partial\theta^2} = f(r,\theta), \quad (r,\theta) \in \Omega,$$
$$\frac{\partial u}{\partial\theta} = 0 \quad \text{for } \theta = 0 \text{ and } \theta = \frac{\pi}{2},$$
$$u = g(\theta) \quad \text{for } r = 1.$$

Use uniform grid $r = i\Delta r, \theta = j\Delta\theta, 0 \leq i,j \leq J$, where $\Delta r = \frac{1}{J}, \Delta\theta = \frac{\pi}{2J}$. Derive a five-point scheme to solve the problem. Implement the scheme and solve the problem with the exact solution $u = \frac{1}{4}r^4\sin^2 2\theta$, in which case,

$$f(r,\theta) = 2r^2, \quad g(\theta) = \frac{1}{4}\sin^2 2\theta.$$

Run your code with $J = 5, 10, 20$ to estimate the convergence rate.

7. Develop a second-order difference scheme to solve the biharmonic problem

$$\frac{\partial^4 u}{\partial x^4} + 2\frac{\partial^4 u}{\partial x^2\partial y^2} + \frac{\partial^4 u}{\partial y^4} = 0, \quad (x,y) \in \Omega = (0,1)^2,$$
$$u = f(x,y), \quad \frac{\partial u}{\partial n} = g(x,y), \quad \text{on } \partial\Omega.$$

Implement your scheme and test it with proper f and g such that the exact solution $u = x^3 y^3$.

8. Consider the problem

$$a u_{xx} - 2b u_{xy} + c u_{yy} = f(x,y), \quad (x,y) \in (0,1)^2,$$
$$u(x,0) = g_1(x), \quad u(x,1) = g_2(x), \quad x \in (0,1),$$

$$u(0,y) = g_3(y), \quad u(1,y) = g_4(y), \quad y \in (0,1),$$

solved by the following difference scheme

$$\frac{a}{(\triangle x)^2}\delta_x^2 u_{ij} - \frac{2b}{4\triangle x \triangle y}\delta_x\delta_y u_{ij} + \frac{c}{(\triangle y)^2}\delta_y^2 u_{ij} = f_{ij},$$

$$1 \le i \le J_x - 1, 1 \le j \le J_y - 1,$$

$$u_{i0} = g_1(i\triangle x), \quad u_{i,J_y} = g_2(i\triangle x), \quad 0 \le i \le J_x,$$

$$u_{0j} = g_3(j\triangle y), \quad u_{J_x,j} = g_4(j\triangle y), \quad 0 \le j \le J_y.$$

Here a, b, c are constants satisfying $ac > b^2$, and δ_x denotes the central difference operator. Use the SOR iteration with $J_x = 10, J_y = 20$ to solve the problem with $a = c = 1, b = 0.5$ and proper f and g such that the exact solution $u = x^2 y^2$.

9. Consider a matrix $A = \begin{bmatrix} 1 & a & a \\ a & 1 & a \\ a & a & 1 \end{bmatrix}$. Prove that the eigenvalues of A are

$1 - a, 1 - a$ and $1 + 2a$. Furthermore, prove that the corresponding Jacobi matrix B_J has eigenvalues a, a and $-2a$. Hence, when $a = 0.8$, even though matrix A is positive definite, the Jacobi method fails to converge.

References

[1] R. Barrett, M.W. Berry, T.F. Chan, J. Demmel, J. Donato, J. Dongarra, V. Eijkhout, R. Pozo, C. Romine and H. van der Vorst. *Templates for the Solution of Linear Systems: Building Blocks for Iterative Methods.* SIAM, Philadelphia, PA, 1993.

[2] G. Birkhoff and S. Gulati. Optimal few-point discretizations of linear source problems. *SIAM J. Numer. Anal.*, 11:700–728, 1974.

[3] J.H. Bramble. *Multigrid Methods.* Longman Scientific & Technical, Harlow, 1993.

[4] G.H. Golub and C.F. Van Loan. *Matrix Computations.* Johns Hopkins University Press, 1996.

[5] W. Hackbusch. *Multi-Grid Methods and Applications.* Springer-Verlag, New York, NY, 1985.

[6] W. Hackbusch. *Elliptic Differential Equations: Theory and Numerical Treatment.* 2nd Ed. Springer-Verlag, Berlin, 2017.

[7] C. Johnson. *Numerical Solution of Partial Differential Equations by the Finite Element Method.* Cambridge University Press, Cambridge, UK, 1987.

[8] B.S. Jovanović and E. Süli. *Analysis of Finite Difference Schemes.* Springer-Verlag, London, UK, 2014.

[9] P. Knabner and L. Angerman. *Numerical Methods for Elliptic and Parabolic Partial Differential Equations.* Springer, New York, NY, 2003.

[10] L. Lapidus and G.F. Pinder. *Numerical Solution of Partial Differential Equations in Science and Engineering.* Wiley-Interscience, New York, NY, 1999.

[11] S. Larsson and V. Thomee. *Partial Differential Equations with Numerical Methods.* Springer, New York, NY, 2003.

[12] R.S. Varga. *Matrix Iterative Analysis.* Prentice-Hall, Englewood Cliffs, NJ, 1962.

[13] J. Xu. Iterative methods by space decomposition and subspace correction. *SIAM Review,* 34:581–613, 1992.

[14] D.M. Young. *Iterative Solution of Large Linear Systems.* Academic Press, New York, NY, 1971.

5

High-Order Compact Difference Methods

In many application areas, such as aeroacoustics [12, 13, 27, 31] and electromagnetics [25], the propagation of acoustic and electromagnetic waves
needs to be accurately simulated over very long periods of time and far
distances. To reduce the accumulation of errors, the numerical algorithm
must be highly accurate. To accomplish this goal, high-order compact finite
difference schemes have been developed for wave propagation applications
[2, 12, 15, 25, 26, 30, 31] and surface water modeling [22, 23].

High-order finite difference schemes can be classified into two main categories: explicit schemes and Pade-type or compact schemes. Explicit schemes
compute the numerical derivatives directly at each grid by using large stencils,
while compact schemes obtain all the numerical derivatives along a grid line
using smaller stencils and solving a linear system of equations. Experience has
shown that compact schemes are much more accurate than the corresponding
explicit scheme of the same order.

In this chapter, we first introduce the basic idea for developing compact
difference schemes in Sec. 5.1, where detailed construction of the scheme and
MATLAB exemplary code are presented. Then we extend the discussion to
high-dimensional problems in Sec. 5.2. Finally, in Sec. 5.3, we introduce
some other ways of constructing compact difference schemes.

5.1   One-dimensional problems

Here we consider 1-D time-dependent problems. First, we show how to construct high-order approximations for spatial derivatives from given function
values. This is followed by the dispersive error analysis for various approximations. Then we present the standard time-discretization Runge-Kutta
schemes. To suppress the numerical instability, we introduce the low-pass
spatial filter. Finally, we demonstrate the effectiveness of the high-order compact difference schemes by numerical examples.

5.1.1   Spatial discretization

In the high-order compact difference methods, the spatial derivatives in the
governing PDEs are not approximated directly by some finite differences.
They are evaluated by some compact difference schemes. More specifically,

given scalar pointwise values f, the derivatives of f are obtained by solving a tridiagonal or pentadiagonal system.

To illustrate the idea, we consider solving a linear dispersive equation

$$u_t + c^{-2} u_{xxx} = 0 \quad x \in (XL, XR), \quad t \in (0, T),$$

with proper initial condition and periodic boundary conditions. To solve this problem, we subdivide the physical domain $[XL, XR]$ by a uniform 1-D mesh with mesh size $h = x_{i+1} - x_i$, which consists of N points:

$$XL = x_1 < x_2 < \cdots < x_{i-1} < x_i < x_{i+1} < \cdots < x_N = XR.$$

The time domain $[0, T]$ is subdivided by a uniform grid point $t_n = n\tau, n = 0, \cdots, M$, where $\tau = T/M$.

A simple forward Euler scheme for solving the above equation is: For all $n = 0, \cdots, M - 1$,

$$\frac{u_i^{n+1} - u_i^n}{\tau} + c^{-2} u_{xxx}|_i^n = 0 \quad i = 2, \cdots, N - 1,$$

where the spatial derivative $u_{xxx}|_i^n$ is evaluated by some compact difference scheme shown below. To develop general compact difference schemes for various differential equations, below we construct general formulas for approximating various order derivatives.

Example 5.1
Formula for first derivatives at interior nodes. To evaluate the first spatial derivatives at interior nodes, we assume that they can be obtained by solving the following tridiagonal system

$$\alpha f'_{i-1} + f'_i + \alpha f'_{i+1} = b\frac{f_{i+2} - f_{i-2}}{4h} + a\frac{f_{i+1} - f_{i-1}}{2h}, \quad i = 2, \cdots, N-1, \quad (5.1)$$

where the unknown coefficients α, a and b are obtained by matching the Taylor expansion up to $O(h^4)$.

By Taylor expansion, we can have

$$f(x_{i+1}) - f(x_{i-1})$$
$$= 2[hf'(x_i) + \frac{h^3}{3!}f_i^{(3)} + \frac{h^5}{5!}f_i^{(5)}] + O(h^7), \quad (5.2)$$

$$f(x_{i+2}) - f(x_{i-2})$$
$$= 2[2hf'(x_i) + \frac{(2h)^3}{3!}f_i^{(3)} + \frac{(2h)^5}{5!}f_i^{(5)}] + O(h^7), \quad (5.3)$$

and

$$f'(x_{i+1}) + f'(x_{i-1})$$
$$= 2[f'(x_i) + \frac{h^2}{2!}f_i^{(3)} + \frac{h^4}{4!}f_i^{(5)}] + O(h^6). \quad (5.4)$$

Let us denote the truncation error for the scheme (5.1)

$$R \equiv a(f'(x_{i+1}) + f'(x_{i-1})) + f'(x_i)$$
$$- \frac{b}{4h}(f(x_{i+2}) - f(x_{i-2})) - \frac{a}{2h}(f(x_{i+1}) - f(x_{i-1})). \qquad (5.5)$$

Substituting (5.2)-(5.4) into (5.5), we obtain

$$R = 2\alpha f'(x_i) + \alpha h^2 f_i^{(3)} + \frac{\alpha h^4}{12} f_i^{(5)}$$
$$+ f'(x_i) - b[f'(x_i) + \frac{4h^2}{6} f_i^{(3)} + \frac{16 h^4}{5!} f_i^{(5)}]$$
$$- a[f'(x_i) + \frac{h^2}{6} f_i^{(3)} + \frac{h^4}{5!} f_i^{(5)}] + O(h^6)$$
$$= (2\alpha + 1 - a - b)f'(x_i) + (\alpha - \frac{4b}{6} - \frac{a}{6})h^2 f_i^{(3)}$$
$$+ (10\alpha - 16b - a)\frac{h^4}{5!} f_i^{(5)} + O(h^6). \qquad (5.6)$$

Letting the coefficients of $f'(x_i)$ and $f_i^{(3)}$ be zero, we obtain an α-family of fourth-order tridiagonal scheme (5.1) with

$$a = \frac{2}{3}(\alpha + 2), \quad b = \frac{1}{3}(4\alpha - 1).$$

Also, the truncation error of (5.1) is $-\frac{4}{5!}(3\alpha - 1)h^4 f^{(5)}$. Note that $\alpha = 0$ gives the explicit fourth-order scheme for the first derivative without solving a linear system. When $\alpha = \frac{1}{3}$, the scheme becomes sixth-order accurate, in which case

$$a = \frac{14}{9}, \quad b = \frac{1}{9}.$$

□

Example 5.2
Formula for second derivatives at interior nodes. Similarly, to evaluate the second derivatives at interior nodes, we can derive the formula [15]

$$\alpha f''_{i-1} + f''_i + \alpha f''_{i+1} = b\frac{f_{i+2} - 2f_i + f_{i-2}}{4h^2} + a\frac{f_{i+1} - 2f_i + f_{i-1}}{h^2}, \qquad (5.7)$$

which provides an α-family of fourth-order tridiagonal schemes with

$$a = \frac{4}{3}(1 - \alpha), \quad b = \frac{1}{3}(-1 + 10\alpha).$$

The special case $\alpha = 0$ gives the explicit fourth-order scheme for the second derivative. The truncation error of (5.7) is $\frac{-4}{6!}(11\alpha - 2)h^4 f^{(6)}$ [15, p. 19].

When $\alpha = \frac{2}{11}$, the scheme becomes sixth-order accurate, in which case

$$a = \frac{12}{11}, \quad b = \frac{3}{11}.$$

☐

Example 5.3
Formula for third derivatives at interior nodes. For the third derivatives at interior nodes, we have the formula [15]

$$\begin{aligned}
&\alpha f'''_{i-1} + f'''_i + \alpha f'''_{i+1}\\
&= b\frac{f_{i+3} - 3f_{i+1} + 3f_{i-1} - f_{i-3}}{8h^3} + a\frac{f_{i+2} - 2f_{i+1} + 2f_{i-1} - f_{i-2}}{2h^3}, \quad (5.8)
\end{aligned}$$

which provides an α-family of fourth-order tridiagonal schemes with

$$a = 2, \quad b = 2\alpha - 1.$$

$\alpha = 0$ gives the explicit fourth-order scheme for the third derivative. A simple sixth-order tridiagonal scheme is given by the coefficients

$$\alpha = \frac{7}{16}, \quad a = 2, \quad b = -\frac{1}{8}.$$

☐

For near-boundary nodes, approximation formulas for the derivatives of non-periodic problems can be derived by one-sided schemes. There are detailed listings for the first and second derivatives in [9, 15].

Example 5.4
Formula for first derivatives at near-boundary nodes. For example, at boundary point 1, a sixth-order formula is [9]

$$f'_1 + \alpha f'_2 = (c_1 f_1 + c_2 f_2 + c_3 f_3 + c_4 f_4 + c_5 f_5 + c_6 f_6 + c_7 f_7)/h,$$

where

$$\alpha = 5, c_1 = \frac{-197}{60}, c_2 = \frac{-5}{12}, c_3 = 5, c_4 = \frac{-5}{3}, c_5 = \frac{5}{12}, c_6 = \frac{-1}{20}, c_7 = 0.$$

At boundary point 2, the sixth-order formula is [9]

$$\alpha f'_1 + f'_2 + \alpha f'_3 = (c_1 f_1 + c_2 f_2 + c_3 f_3 + c_4 f_4 + c_5 f_5 + c_6 f_6 + c_7 f_7)/h,$$

where

$$\alpha = \frac{2}{11}, c_1 = \frac{-20}{33}, c_2 = \frac{-35}{132}, c_3 = \frac{34}{33}, c_4 = \frac{-7}{33}, c_5 = \frac{2}{33}, c_6 = \frac{-1}{132}, c_7 = 0.$$

At boundary point $N - 1$, the sixth-order formula is [9]

$$\alpha f'_{N-2} + f'_{N-1} + \alpha f'_N$$
$$= (c_1 f_N + c_2 f_{N-1} + c_3 f_{N-2} + c_4 f_{N-3} + c_5 f_{N-4} + c_6 f_{N-5} + c_7 f_{N-6})/h,$$

where $\alpha = \frac{2}{11}$. The remaining coefficients are the opposite of those given for point 2 (i.e., the signs are reversed).

At boundary point N, the sixth-order formula is [9]

$$\alpha f'_{N-1} + f'_N$$
$$= (c_1 f_N + c_2 f_{N-1} + c_3 f_{N-2} + c_4 f_{N-3} + c_5 f_{N-4} + c_6 f_{N-5} + c_7 f_{N-6})/h,$$

where $\alpha = 5$. The remaining coefficients are the opposite of those given for point 1 (i.e., the signs are reversed). ☐

For those near-boundary points, Lele [15] has presented the third-order compact scheme for the second derivatives. Below we show the reader how to derive some sixth-order compact formulas at those near-boundary points.

Example 5.5

Formula for third derivatives at near-boundary nodes. At boundary point 1, we construct the sixth-order formula

$$f''_1 + \alpha f''_2 = (c_1 f_1 + c_2 f_2 + c_3 f_3 + c_4 f_4 + c_5 f_5 + c_6 f_6 + c_7 f_7)/h^2,$$

where the coefficients can be found by matching the Taylor series expansions up to the order of h^7, which gives us the following linear system

$$c_1 + c_2 + c_3 + c_4 + c_5 + c_6 + c_7 = 0$$
$$c_2 + 2c_3 + 3c_4 + 4c_5 + 5c_6 + 6c_7 = 0$$
$$c_2 + 2^2 c_3 + 3^2 c_4 + 4^2 c_5 + 5^2 c_6 + 6^2 c_7 = (2!)(1 + \alpha)$$
$$c_2 + 2^3 c_3 + 3^3 c_4 + 4^3 c_5 + 5^3 c_6 + 6^3 c_7 = (3!)\alpha$$
$$c_2 + 2^4 c_3 + 3^4 c_4 + 4^4 c_5 + 5^4 c_6 + 6^4 c_7 = \frac{4!}{2!}\alpha$$
$$c_2 + 2^5 c_3 + 3^5 c_4 + 4^5 c_5 + 5^5 c_6 + 6^5 c_7 = \frac{5!}{3!}\alpha$$
$$c_2 + 2^6 c_3 + 3^6 c_4 + 4^6 c_5 + 5^6 c_6 + 6^6 c_7 = \frac{6!}{4!}\alpha$$
$$c_2 + 2^7 c_3 + 3^7 c_4 + 4^7 c_5 + 5^7 c_6 + 6^7 c_7 = \frac{7!}{5!}\alpha$$

The solution to the above system is

$$c_1 = 2077/157, c_2 = -2943/110, c_3 = 573/44, c_4 = 167/99, \quad (5.9)$$
$$c_5 = -18/11, c_6 = 57/110, c_7 = -131/1980, \alpha = 126/11. \quad (5.10)$$

At boundary point 2, we can construct the sixth-order formula

$$\alpha f_1'' + f_2'' + \alpha f_3'' = (c_1 f_1 + c_2 f_2 + c_3 f_3 + c_4 f_4 + c_5 f_5 + c_6 f_6 + c_7 f_7)/h^2.$$

Matching the Taylor series expansions up to the order of h^7 gives us the following linear system

$$c_1 + c_2 + c_3 + c_4 + c_5 + c_6 + c_7 = 0$$
$$c_2 + 2c_3 + 3c_4 + 4c_5 + 5c_6 + 6c_7 = 0$$
$$c_2 + 2^2 c_3 + 3^2 c_4 + 4^2 c_5 + 5^2 c_6 + 6^2 c_7 = (2!)(1 + 2\alpha)$$
$$c_2 + 2^3 c_3 + 3^3 c_4 + 4^3 c_5 + 5^3 c_6 + 6^3 c_7 = (3!)(1 + 2\alpha)$$
$$c_2 + 2^4 c_3 + 3^4 c_4 + 4^4 c_5 + 5^4 c_6 + 6^4 c_7 = \frac{4!}{2!}(1 + 2^2 \alpha)$$
$$c_2 + 2^5 c_3 + 3^5 c_4 + 4^5 c_5 + 5^5 c_6 + 6^5 c_7 = \frac{5!}{3!}(1 + 2^3 \alpha)$$
$$c_2 + 2^6 c_3 + 3^6 c_4 + 4^6 c_5 + 5^6 c_6 + 6^6 c_7 = \frac{6!}{4!}(1 + 2^4 \alpha)$$
$$c_2 + 2^7 c_3 + 3^7 c_4 + 4^7 c_5 + 5^7 c_6 + 6^7 c_7 = \frac{7!}{5!}(1 + 2^5 \alpha)$$

which has the solution

$$c_1 = 585/512, c_2 = -141/64, c_3 = 459/512, c_4 = 9/32, \qquad (5.11)$$
$$c_5 = -81/512, c_6 = 3/64, c_7 = -3/512, \alpha = 11/128. \qquad (5.12)$$

Similarly, at boundary point $N - 1$, the sixth-order formula is

$$\alpha f_{N-2}'' + f_{N-1}'' + \alpha f_N''$$
$$= (c_1 f_N + c_2 f_{N-1} + c_3 f_{N-2} + c_4 f_{N-3} + c_5 f_{N-4} + c_6 f_{N-5} + c_7 f_{N-6})/h^2,$$

where the coefficients are given by (5.11)-(5.12). And at boundary point N, the sixth-order formula is

$$\alpha f_{N-1}'' + f_N''$$
$$= (c_1 f_N + c_2 f_{N-1} + c_3 f_{N-2} + c_4 f_{N-3} + c_5 f_{N-4} + c_6 f_{N-5} + c_7 f_{N-6})/h^2,$$

where the coefficients are given by (5.9)-(5.10). ▯

5.1.2   Dispersive error analysis

In many real applications, the governing PDEs often contain some second or higher derivatives. How to approximate those derivatives accurately and efficiently becomes quite interesting. For example, we can approximate the second derivatives directly by formula (5.7) or by applying the first derivative formula (5.1) twice (see Visbal et al. [30, 31, 32]). To approximate a third derivative, there are more choices: using formula (5.8) directly; using the first derivative operator (5.1) three times; using the first derivative operator (5.1) twice, followed by the second derivative operator (5.7) once.

To compare the dispersive errors introduced by different approximation methods, let us consider a periodic function $f(x)$ over the domain $[0, 2\pi]$, i.e.,

$f(0) = f(2\pi)$. Furthermore, we let N be an even positive integer and denote $h = 2\pi/N$.

By Fourier analysis, f can be approximated by its truncated Fourier series

$$f(x) \approx \sum_{k=-\frac{N}{2}}^{\frac{N}{2}} \hat{f}_k e^{ikx}, \qquad (5.13)$$

where the Fourier coefficients

$$\hat{f}_k = \frac{1}{2\pi} \int_0^{2\pi} f(x) e^{-ikx} dx = \frac{1}{2\pi} \int_0^{2\pi} f(x)(\cos kx - i \sin kx) dx.$$

Since we assume that f is a real-valued function,

$$\hat{f}_{-k} = \hat{f}_k{}^* \quad 0 \le k \le \frac{N}{2},$$

where $\hat{f}_k{}^*$ is the complex conjugate of $\hat{f}_k$.

For convenience, we introduce a scaled wavenumber $\omega_k = kh = \frac{2\pi k}{N}$ and a scaled coordinate $s = \frac{x}{h}$. Hence the domain of the scaled wavenumber $\omega_k \in [0, \pi]$, and (5.13) can be written as

$$f(x(s)) \approx \sum_{k=-\frac{N}{2}}^{\frac{N}{2}} \hat{f}_k e^{i\omega_k s}. \qquad (5.14)$$

Taking the derivative of (5.14) with respect to s twice generates the exact Fourier coefficients

$$\hat{f}_k'' = (i\omega_k)^2 \hat{f}_k = -\omega_k^2 \hat{f}_k \qquad (5.15)$$

for the second derivative function.

On the other hand, we can assume that a truncated Fourier series for $f''(x)$ be represented as

$$f''(x(s)) \approx \sum_{k=-\frac{N}{2}}^{\frac{N}{2}} \hat{f}_k{}'' e^{i\omega_k s},$$

where $\hat{f}_k{}''$ can be obtained from a difference scheme such as (5.7).

If considering just one specific mode ω_k, and substituting

$$f|_{x_i} = \hat{f}_k e^{i\omega_k s_i}$$

and

$$\frac{d^2 f}{dx^2}\Big|_{x_i} = \frac{1}{h^2} \frac{d^2 f}{ds^2}\Big|_{s_i} = \frac{1}{h^2} \hat{f}_k{}'' e^{i\omega_k s_i}$$

into (5.7), then we obtain the left-hand side of (5.7) as

$$LHS = \frac{1}{h^2} \hat{f}_k{}'' e^{i\omega_k s_i} [\alpha e^{-i\omega_k} + 1 + \alpha e^{i\omega_k}]$$

$$= \frac{1}{h^2} \hat{f}_k{}'' e^{i\omega_k s_i} [2\alpha \cos \omega_k + 1],$$

and

$$RHS = \hat{f}_k e^{i\omega_k s_i}[\frac{b}{4h^2}(e^{i2\omega_k} - 2 + e^{-i2\omega_k}) + \frac{a}{h^2}(e^{i\omega_k} - 2 + e^{-i\omega_k})]$$
$$= \hat{f}_k e^{i\omega_k s_i}[\frac{b}{4h^2}(2\cos 2\omega_k - 2) + \frac{a}{h^2}(2\cos\omega_k - 2)],$$

which leads to

$$\hat{f}_k{}'' = \hat{f}_k \cdot [\frac{b}{2}(\cos 2\omega_k - 1) + 2a(\cos\omega_k - 1)]/(2\alpha\cos\omega_k + 1). \qquad (5.16)$$

Comparing (5.16) with (5.15), we obtain

$$w''_{ex}(\omega) = \omega^2, \qquad (5.17)$$
$$w''_{appr}(\omega) = [\frac{b}{2}(1 - \cos 2\omega) + 2a(1 - \cos\omega)]/(1 + 2\alpha\cos\omega), \qquad (5.18)$$

where we dropped the subscript k.

Hence for the second derivative, the resolution formula of (5.7) is:

$$w''_{dir2}(\omega) = \frac{2a(1 - \cos(\omega)) + (b/2)(1 - \cos(2\omega))}{1 + 2\alpha\cos(\omega)}.$$

The resolution characteristics of exact differentiation for the second derivative is given by

$$w''_{ex2}(\omega) = \omega^2.$$

Similar analysis can be carried out for the first derivative. Taking the derivative of (5.14) with respect to S generates the exact Fourier coefficients

$$\hat{f}'_k = i\omega_k \hat{f}_k \qquad (5.19)$$

for the first derivative function.

On the other hand, we can assume that a truncated Fourier series for $f'(x)$ is represented as

$$f'(x(s)) \approx \sum_{k=-\frac{N}{2}}^{\frac{N}{2}} \hat{f}_k{}' e^{i\omega_k s},$$

where $\hat{f}_k{}'$ can be obtained from a difference scheme such as (5.1).

Substituting one specific mode ω_k solution

$$f|_{x_i} = \hat{f}_k e^{i\omega_k s_i}$$

and

$$\frac{df}{dx}|_{x_i} = \frac{1}{h}\frac{df}{ds}|_{s_i} = \frac{1}{h}\hat{f}_k{}' e^{i\omega_k s_i}$$

into (5.1), then we obtain the left-hand side of (5.1) as

$$LHS = \frac{1}{h}\hat{f_k}' e^{i\omega_k s_i}[\alpha e^{-i\omega_k} + 1 + \alpha e^{i\omega_k}]$$
$$= \frac{1}{h}\hat{f_k}' e^{i\omega_k s_i}[2\alpha\cos\omega_k + 1],$$

and

$$RHS = \hat{f_k} e^{i\omega_k s_i}[\frac{b}{4h}(e^{i2\omega_k} - e^{-i2\omega_k}) + \frac{a}{2h}(e^{i\omega_k} - e^{-i\omega_k})]$$
$$= \hat{f_k} e^{i\omega_k s_i}[\frac{b}{4h}2i\sin 2\omega_k + \frac{a}{2h}(2i\sin\omega_k)],$$

which leads to

$$\hat{f_k}' = i\hat{f_k}\cdot[\frac{b}{2}\sin 2\omega_k + a\sin\omega_k]/(2\alpha\cos\omega_k + 1). \qquad (5.20)$$

Comparing (5.20) with (5.19), we obtain

$$w_{ex}'(\omega) = \omega, \qquad (5.21)$$
$$w_{appr}'(\omega) = [\frac{b}{2}\sin 2\omega + a\sin\omega]/(1 + 2\alpha\cos\omega), \qquad (5.22)$$

where we dropped the subscript k.

Hence two consecutive applications of the first derivative operator (5.1) gives

$$w_{ind2}''(\omega) = (\frac{a\sin(\omega) + (b/2)\sin(2\omega)}{1 + 2\alpha\cos(\omega)})^2.$$

By the same technique, the resolution characteristics of the third derivative scheme (5.8) can be derived:

$$w_{dir3}'''(\omega) = \frac{a(2\sin(\omega) - \sin(2\omega)) + (b/4)(3\sin(\omega) - \sin(3\omega))}{1 + 2\alpha\cos(\omega)},$$

where the corresponding coefficients are given in Section 5.1.1. While three consecutive applications of the first derivative operator (5.1) give

$$w_{ind3}'''(\omega) = (\frac{a\sin(\omega) + (b/2)\sin(2\omega)}{1 + 2\alpha\cos(\omega)})^3.$$

Note that the resolution characteristics of exact differentiation for the third derivative is given by

$$w_{ex3}'''(\omega) = \omega^3.$$

For completeness, we present the resolution characteristics for the sixth-order explicit scheme for approximating the third derivatives. We construct the sixth-order explicit scheme as

$$h^3 f_i''' = a(f_{i+1} - f_{i-1}) + b(f_{i+2} - f_{i-2}) + c(f_{i+3} - f_{i-3}) + d(f_{i+4} - f_{i-4}). \qquad (5.23)$$

By matching the Taylor series coefficients of various orders, we obtain the following linear system

$$a + 2b + 3c + 4d = 0,$$
$$a + 2^3 b + 3^3 c + 4^3 d = 3,$$
$$a + 2^5 b + 3^5 c + 4^5 d = 0,$$
$$a + 2^7 b + 3^7 c + 4^7 d = 0,$$

which gives the solution

$$a = -488/240, \quad b = 338/240, \quad c = -72/240, \quad d = 7/240. \qquad (5.24)$$

The truncation error is $\frac{2}{9!}(a + 2^9 b + 3^9 c + 4^9 d)h^6 f^{(9)} \approx 0.0136 h^6 f^{(9)}$. Note that the truncation error for the sixth-order tridiagonal scheme [15, p. 35] is $\frac{36}{9!} h^6 f^{(9)} \approx 0.0000992 h^6 f^{(9)}$, which is much smaller than the corresponding sixth-order explicit scheme. By performing the similar Fourier error analysis as above, we can obtain the resolution characteristics for (5.23)

$$w'''_{ex6}(\omega) = -2[a \sin(\omega) + b \sin(2\omega) + c \sin(3\omega) + d \sin(4\omega)],$$

where $a, b, c,$ and d are given by (5.24).

Plots of the above-modified wavenumbers w'' and w''' against wavenumber ω for a variety of schemes are presented in Figs. 5.1 and 5.2, respectively. From Figs. 5.1 and 5.2, it is not difficult to see that using the direct way to obtain the approximations of the second and third derivatives should be better than using the indirect way.

An example for reconstruction of the third derivatives of a given function $u(x) = \cos(2\pi x) + \cos(4\pi x)$ is carried out by using both direct and indirect ways. The approximated third derivatives and the corresponding errors are plotted in Figs. 5.3 through 5.4 for $N = 50$ and $N = 100$ uniform intervals over $x \in [0, 1]$, respectively. These figures show clearly that reconstruction by the direct way always performs better than the indirect way: the direct way is at least one order of magnitude better than the indirect way.

5.1.3   Temporal discretization

In our implementation, the governing PDEs are often integrated in time with the classical second-order two-stage Runge-Kutta (RK2) method or the fourth-order, four-stage Runge-Kutta (RK4) method. Assuming that the governing equation is

$$\frac{\partial U}{\partial t} = R(U),$$

where $R(U)$ denotes the residual.

The classical RK4 method integrates from time t_0 (step n) to $t_0 + \triangle t$ (step $n + 1$) through the operations

$$U_0 = u(x, t_0), \quad k_0 = \triangle t R(U_0),$$

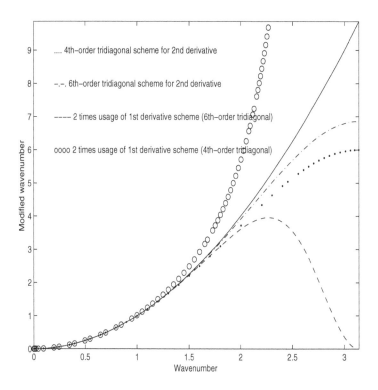

FIGURE 5.1

Differencing error for second derivative vs. wavenumber by different algorithms: the solid line is for the exact differentiation. (From Fig. 1 of [18] with kind permission of Springer Science and Business Media.)

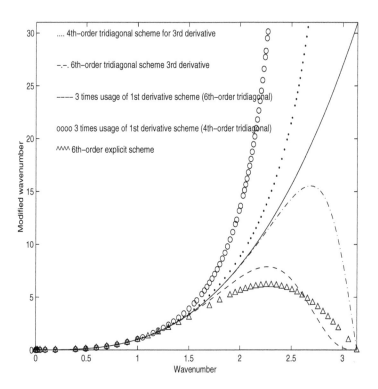

FIGURE 5.2

Differencing error for third derivative vs. wavenumber by different algorithms: the solid line is for the exact differentiation. (From Fig. 2 of [18] with kind permission of Springer Science and Business Media.)

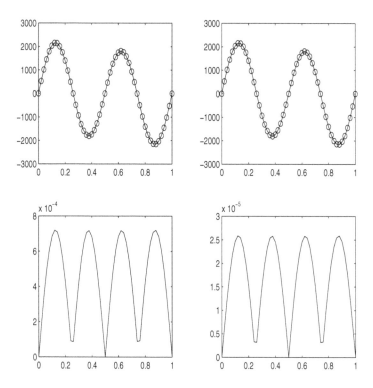

FIGURE 5.3

The reconstructed third derivative (row one) and the pointwise error (row two) with 50 intervals. Columns 1 are obtained by three consecutive applications of the sixth-order implicit first derivative scheme; columns 2 are obtained by the sixth-order implicit third derivative scheme. (From Fig. 4 of [18] with kind permission of Springer Science and Business Media.)

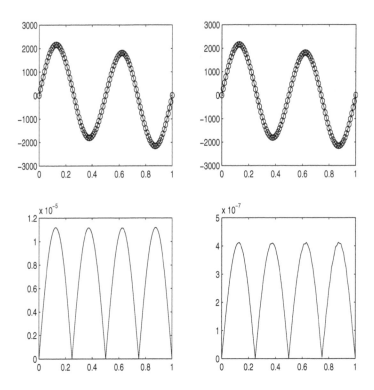

FIGURE 5.4

The reconstructed third derivative (row one) and the pointwise error (row two) with 100 intervals. Columns 1 are obtained by three consecutive applications of the sixth-order implicit first derivative scheme; columns 2 are obtained by the sixth-order implicit third derivative scheme. (From Fig. 4 of [18] with kind permission of Springer Science and Business Media.)

$$U_1 = U_0 + k_0/2, \quad k_1 = \Delta t R(U_1),$$
$$U_2 = U_0 + k_1/2, \quad k_2 = \Delta t R(U_2),$$
$$U_3 = U_0 + k_2, \quad k_3 = \Delta t R(U_3),$$
$$U^{n+1} = U_0 + \frac{1}{6}(k_0 + 2k_1 + 2k_2 + k_3).$$

In order to save computational cost, sometimes the low-order accurate RK2 scheme is preferred. The classical RK2 method integrates from time t_0 (step n) to $t_0 + \Delta t$ (step $n + 1$) through the operations

$$U_0 = u(x, t_0), \quad k_0 = \Delta t R(U_0), \tag{5.25}$$
$$U_1 = U_0 + k_0, \quad k_1 = \Delta t R(U_1), \tag{5.26}$$
$$U_2 = U_1 + k_1, \tag{5.27}$$
$$U^{n+1} = \frac{1}{2}(U_0 + U_2) = U_0 + \frac{\Delta t}{2}(R(U_0) + R(U_1)). \tag{5.28}$$

5.1.4 Low-pass spatial filter

Like other centered schemes, compact difference discretizations are nondissipative and are therefore susceptible to numerical instabilities originating from a variety of sources, such as mesh nonuniformity, truncated boundaries, and nonlinearity. Previous experiences [15, 30, 31, 32] show that a low-pass filter [29] is ideal for suppressing the amplitude of undesirable high-frequency components and does not affect the remaining components of the solution.

A high-order tridiagonal filter can be formulated as [9, 15]

$$\alpha_f \hat{\phi}_{i-1} + \hat{\phi}_i + \alpha_f \hat{\phi}_{i+1} = \sum_{n=0}^{N} \frac{a_n}{2}(\phi_{i+n} + \phi_{i-n}), \tag{5.29}$$

where ϕ_i denotes the given value at point i, and $\hat{\phi}_i$ denotes the value after filtering. The spectral function (or frequency response) of the operator is [9]

$$SF(\omega) = (\sum_{n=0}^{N} a_n \cos(n\omega))/(1 + 2\alpha_f \cos(\omega)).$$

To obtain the unknown coefficients, we insist that the highest frequency mode be eliminated by enforcing the condition $SF(\pi) = 0$. For flexibility, we retain α_f as a free parameter. Then the remaining N equations can be derived by matching Taylor series coefficients of the left and right sides. By doing this, (5.29) provides a $2N$-th order formula on a $2N + 1$ point stencil. Note that $SF(\omega)$ is real, hence the filter only modifies the amplitude of each wave component without affecting the phase.

Coefficients of a family of tenth-order filters are derived in [9], which are

$$a_0 = \frac{193 + 126\alpha_f}{256}, a_1 = \frac{105 + 302\alpha_f}{256}, a_2 = \frac{15(-1 + 2\alpha_f)}{64}, \tag{5.30}$$

$$a_3 = \frac{45(1 - 2\alpha_f)}{512}, a_4 = \frac{5(-1 + 2\alpha_f)}{256}, a_5 = \frac{1 - 2\alpha_f}{512}. \tag{5.31}$$

Here α_f is a free parameter which satisfies the inequality $-0.5 < \alpha_f \le 0.5$. Extensive numerical experience [30, 31, 32] suggests that values between 0.3 and 0.5 for α_f are appropriate.

Special formulas are needed at near boundary points due to the large stencil of the filter. Very detailed high-order one-sided formulas are provided in [9]. Readers can find many specific formulas there.

5.1.5 Numerical examples with MATLAB codes

In this section, two numerical examples with MATLAB source codes are provided to show the superior performance of high-order compact schemes. We use the sixth-order tridiagonal scheme and the classic explicit fourth-order Runge-Kutta method in all our examples. Below we denote $\triangle x$ as the grid size, and $\triangle t$ as the time step size.

Example 1. First we compute the solution of the linear dispersive equation

$$u_t + c^{-2}u_{xxx} = 0 \tag{5.32}$$

with an initial condition $u(x,0) = \sin(cx)$ and a periodic boundary condition such that the exact solution is given by $u(x,t) = \sin(c(x + t))$ for $x \in [0, 2\pi]$ and $t \in [0, 1]$. This example is a generalized problem presented in [33], where only $c = 1$ is discussed. Note that the exact solution is a left-moving wave.

To show the superior performance of compact schemes compared to the same order explicit scheme, we first solved the problem (5.32) with $c = 8$ using fourth-order explicit and compact schemes (with $\alpha = 15/32$), and sixth-order explicit and compact schemes (with $\alpha = 7/32$). This problem is solved with 41 uniformly distributed points (i.e., $\triangle x = 2\pi/40$) and time step size $\triangle t = a(dx)^3, a = 0.125$. We want to remark that $c = 8$ is a challenging problem, where we use only five points per wavelength and the total time steps are 2064.

The obtained solution and pointwise error at time $t = 0.3, 0.7$ and 1 are graphed in Fig. 5.5, from which we see that the L_∞ errors for the sixth-order explicit scheme and the sixth-order compact scheme are 0.35 and 6×10^{-4}, respectively, i.e., the compact scheme is much more accurate than the corresponding same order explicit scheme.

Next we solved (5.32) with $c = 1$ using both the sixth-order explicit and compact schemes. The obtained solution and L_∞ error are shown in Fig. 5.6 at $t = 0, 0.25, 0.5, 0.75$, and 1. Due to the nice property of the solution, this time all schemes solve the problem very well. But the compact scheme still achieves better accuracy than the corresponding same order explicit scheme as shown by the fact that the errors obtained for the sixth-order compact scheme and the sixth-order explicit scheme are 8×10^{-10} and 2.5×10^{-7}, respectively. Note that these results are much better than the best L_∞ error 7.2362×10^{-7}

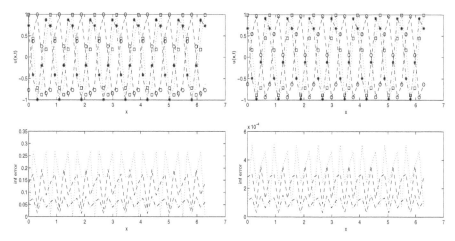

FIGURE 5.5

Linear dispersive wave with $c = 8$ at $t = 0.3, 0.7, 1$. The left column is obtained by the sixth-order explicit scheme, while the right column is obtained by the sixth-order compact scheme. Top row: $*, \circ, \square$ represent the numerical solutions at $t = 0.3, 0.7, 1$, respectively. Bottom row: $-\cdot, --, \cdots$ represent the pointwise errors at $t = 0.3, 0.7, 1$, respectively. (From Fig. 6 of [18] with kind permission of Springer Science and Business Media.)

achieved by the third-order local discontinuous Galerkin (LDG) method with the same number of points $N = 41$ [33, Table 1.1]. We want to emphasize that no filters are used in any of the computations for this example.

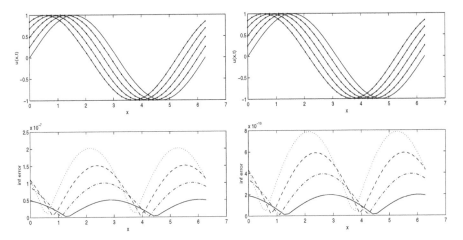

FIGURE 5.6

Linear dispersive wave with $c = 1$. The left column is obtained by the sixth-order explicit scheme, while the right column is obtained by the sixth-order compact scheme. Top row shows the calculated (dotted) and exact (solid line) left moving wave at $t = 0, 0.25, 0.5, 0.75, 1$. Bottom row shows the corresponding errors increasing with time. (From Fig. 8 of [18] with kind permission of Springer Science and Business Media.)

The MATLAB source code *disp1d.m* and its auxiliary codes *JSC1.m, resJSC1.m, reconuxxxp.m* are listed below.

```
%-----------------------------------------------------------------
% disp1d.m: solve the linear dispersive equation
%               u_t + c^{-2}u_{xxx}=0
%        by high order compact difference method.
%
% Used functions:
%     JSC1.m: the analytic solution
%     reconuxxxp.m: reconstrcution u_{xxx} from u
%     resJSC1.m: the residual function
%-----------------------------------------------------------------
clear all
%%%%%%%%%%%%%%%%%%%%%%%%%%%%%%%%%%%%%%
% generate 1D mesh

Nx=41; XL=0; XR=2*pi;
dx=(XR-XL)/(Nx-1);

for i=1:Nx
    XX(i)=XL+(i-1)*dx;
end

%%%%%%%%%march-in-time%%%%%%%%%%%%%%
% cfl has to be small!
cfl=0.125; dt=cfl*dx^3; tlast=1;
Nstep=tlast/dt;
cc=1;    % the bigger, the larger frequency

IRK=4;

% get initial condition
[u0ex]=feval(@JSC1,XX,0,cc);
uold=u0ex;

subplot(2,1,1), plot(XX,u0ex, 'y-');
            hold on
% we use 4th-order RK scheme:IRK=2,4
for k=1:Nstep
  if IRK==4
    u0x=reconuxxxp(uold,Nx,dx);    % reconstruct ux from u_n
    k0=dt*resJSC1(uold,u0x,cc);    % k0=dt*R(dummy,uxxx,dummy)
    u1=uold+k0/2;
```

```
    u1x=reconuxxxp(u1,Nx,dx);
    k1=dt*resJSC1(uold,u1x,cc);
    u2=uold+k1/2;

    u2x=reconuxxxp(u2,Nx,dx);
    k2=dt*resJSC1(uold,u2x,cc);
    u3=uold+k2;

    u3x=reconuxxxp(u3,Nx,dx);
    k3=dt*resJSC1(uold,u3x,cc);

    unew=uold+(k0+2*k1+2*k2+k3)/6.;   % finish one-time step
  elseif IRK==2
    u0x=reconuxp(uold,Nx,dx);          % reconstruct ux from u_n
    u0xx=reconuxp(u0x,Nx,dx);
    u0xxx=reconuxp(u0xx,Nx,dx);   % obtain u_xxx
    u2x=reconuxp(uold.^2,Nx,dx);  % reconstruct u^2
    k0=dt*resJSC1(u2x,u0xxx,cc);  % k0=dt*R((u_n)_x)
    u1=uold+k0;

    u1x=reconuxp(u1,Nx,dx);
    uxx=reconuxp(u1x,Nx,dx);
    uxxx=reconuxp(uxx,Nx,dx);    % obtain u_xxx
    u2x=reconuxp(u1.^2,Nx,dx);   % reconstruct u^2
    k1=dt*resJSC1(u2x,uxxx,cc);
    u2=u1+k1;

    unew=(uold+u2)/2;
  end

    uold=unew;     % update for next time step
    eps=0.3*dt;    % error tolerance
    % plot the solution at some specific times
    if   abs(k*dt-0.25) < eps | abs(k*dt-0.5) < eps ...
         | abs(k*dt-0.75) < eps | abs(k*dt-1) < eps
       disp(k),

       [u0ex]=feval(@JSC1,XX,k*dt,cc);
       subplot(2,1,1),plot(XX,unew, 'y-',XX,u0ex,'g.');
       xlabel('x'); ylabel('u(x,t)');
       hold on
       subplot(2,1,2),
       if abs(k*dt-0.25) < eps, plot(XX,abs(u0ex-unew), 'r-'); end
       if abs(k*dt-0.5) < eps, plot(XX,abs(u0ex-unew), 'r-.'); end
```

```
        if abs(k*dt-0.75) < eps, plot(XX,abs(u0ex-unew), 'r--'); end
        if abs(k*dt-1) < eps, plot(XX,abs(u0ex-unew), 'r:'); end

        xlabel('x'); ylabel('inf error');
        hold on
    end

end


%----------------------------------------------------
% The analytic solution
function uu=JSC1(x,ap,cc)
uu=sin(cc*(x+ap));


%-----------------------------------------------
% the residual R=-0.5*(u^2)_x-cc*(u_xxx)
% i.e., u1=(u^2)_x, u2=u_xxx
function uu=resJSC1(u1,u2,cc)
uu=-cc*u2;



%-----------------------------------------------------------
% reconstruction the 3rd derivative for periodic function
%-----------------------------------------------------------
function ux=reconuxxxp(u,N,h)

IEX=0;    % 4 means explicte 4th-order reconstruction
if IEX==4
  a=2; b=-1;
  for i=1:N
    if i==1
      tmp = b*(u(i+3)-3*u(i+1)+3*u(N-1)-u(N-3))/4 ...
          + a*(u(i+2)-2*u(i+1)+2*u(N-1)-u(N-2));
    elseif i==2
      tmp = b*(u(i+3)-3*u(i+1)+3*u(i-1)-u(N-2))/4 ...
          + a*(u(i+2)-2*u(i+1)+2*u(i-1)-u(N-1));
    elseif i==3
      tmp = b*(u(i+3)-3*u(i+1)+3*u(i-1)-u(N-1))/4 ...
          + a*(u(i+2)-2*u(i+1)+2*u(i-1)-u(i-2));
    elseif i== (N-2)
      tmp = b*(u(2)-3*u(i+1)+3*u(i-1)-u(i-3))/4 ...
          + a*(u(i+2)-2*u(i+1)+2*u(i-1)-u(i-2));
```

```
   elseif i== (N-1)
     tmp = b*(u(3)-3*u(i+1)+3*u(i-1)-u(i-3))/4 ...
         + a*(u(2)-2*u(i+1)+2*u(i-1)-u(i-2));
   elseif i==N
     tmp = b*(u(4)-3*u(2)+3*u(i-1)-u(i-3))/4 ...
         + a*(u(3)-2*u(2)+2*u(i-1)-u(i-2));
   else
     tmp = b*(u(i+3)-3*u(i+1)+3*u(i-1)-u(i-3))/4 ...
         + a*(u(i+2)-2*u(i+1)+2*u(i-1)-u(i-2));
   end
     ux(i)=tmp/(2*h^3);
 end
 return
end


  % 6 means explict 6th-order reconstruction
if IEX==6
 a=-488/240; b=338/240; c=-72/240; d=7/240;
 for i=1:N
   if i==1
     tmp = a*(u(i+1)-u(N-1))+b*(u(i+2)-u(N-2)) ...
         + c*(u(i+3)-u(N-3))+d*(u(i+4)-u(N-4));
   elseif i==2
     tmp = a*(u(i+1)-u(i-1))+b*(u(i+2)-u(N-1)) ...
         + c*(u(i+3)-u(N-2))+d*(u(i+4)-u(N-3));
   elseif i==3
     tmp = a*(u(i+1)-u(i-1))+b*(u(i+2)-u(i-2)) ...
         + c*(u(i+3)-u(N-1))+d*(u(i+4)-u(N-2));
   elseif i==4
     tmp = a*(u(i+1)-u(i-1))+b*(u(i+2)-u(i-2)) ...
         + c*(u(i+3)-u(i-3))+d*(u(i+4)-u(N-1));
   elseif i==(N-3)
     tmp = a*(u(i+1)-u(i-1))+b*(u(i+2)-u(i-2)) ...
         + c*(u(i+3)-u(i-3))+d*(u(2)-u(i-4));
   elseif i== (N-2)
     tmp = a*(u(i+1)-u(i-1))+b*(u(i+2)-u(i-2)) ...
         + c*(u(2)-u(i-3))+d*(u(3)-u(i-4));
   elseif i== (N-1)
     tmp = a*(u(i+1)-u(i-1))+b*(u(2)-u(i-2)) ...
         + c*(u(3)-u(i-3))+d*(u(4)-u(i-4));
   elseif i==N
     tmp = a*(u(2)-u(i-1))+b*(u(3)-u(i-2)) ...
         + c*(u(4)-u(i-3))+d*(u(5)-u(i-4));
   else
     tmp = a*(u(i+1)-u(i-1))+b*(u(i+2)-u(i-2)) ...
```

```
                + c*(u(i+3)-u(i-3))+d*(u(i+4)-u(i-4));
      end
        ux(i)=tmp/(h^3);
   end
   return
end

% below are implicit reconstruction

ISC=6;      % order of the scheme
% for 4th-order compact scheme: alfa cannot be 1/2
if ISC==4
   alfa=15/32; aa=2; bb=2*alfa-1;
% for 6th-order compact scheme
elseif ISC==6
   alfa=7/16; aa=2; bb=-1.0/8;
end

amat=zeros(N,N);
B=zeros(N,1);

for i=1:N
    if i==1
        amat(1,1)=1; amat(1,2)=alfa; amat(1,N-1)=alfa;
        B(i)=bb*(u(i+3)-3*u(i+1)+3*u(N-1)-u(N-3))/4 ...
            +aa*(u(i+2)-2*u(i+1)+2*u(N-1)-u(N-2));
    elseif i==2
        amat(2,1)=alfa; amat(2,2)=1; amat(2,3)=alfa;
        B(i)=bb*(u(i+3)-3*u(i+1)+3*u(i-1)-u(N-2))/4 ...
            +aa*(u(i+2)-2*u(i+1)+2*u(i-1)-u(N-1));
    elseif i==3
        amat(i,i-1)=alfa; amat(i,i)=1; amat(i,i+1)=alfa;
        B(i)=bb*(u(i+3)-3*u(i+1)+3*u(i-1)-u(N-1))/4 ...
            +aa*(u(i+2)-2*u(i+1)+2*u(i-1)-u(i-2));
    elseif i==N-2
        amat(i,i-1)=alfa; amat(i,i)=1; amat(i,i+1)=alfa;
        B(i)=bb*(u(2)-3*u(i+1)+3*u(i-1)-u(i-3))/4 ...
            +aa*(u(i+2)-2*u(i+1)+2*u(i-1)-u(i-2));
    elseif i==N-1
        amat(i,i-1)=alfa; amat(i,i)=1; amat(i,i+1)=alfa;
        B(i)=bb*(u(3)-3*u(i+1)+3*u(i-1)-u(i-3))/4 ...
            +aa*(u(2)-2*u(i+1)+2*u(i-1)-u(i-2));
    elseif i==N
        amat(N,1)=-1; amat(N,N)=1;
        B(i)=0;
```

```
      else      % i>=4 & i <=N-3
            amat(i,i-1)=alfa; amat(i,i)=1; amat(i,i+1)=alfa;
            B(i)=bb*(u(i+3)-3*u(i+1)+3*u(i-1)-u(i-3))/4 ...
                  +aa*(u(i+2)-2*u(i+1)+2*u(i-1)-u(i-2));
      end
      B(i)=B(i)/(2*h^3);
end

% call trisys.m
ux=(amat\B)';
```

Example 2. In order to see how our method works for a nonlinear problem with a small coefficient for the third derivative term, we compute the classical soliton solutions of the KdV equation [33]

$$u_t + 0.5(u^2)_x + \epsilon u_{xxx} = 0.$$

Here we present the MATLAB code for the interesting double soliton collision case, which has the initial condition

$$u_0(x) = 3c_1 \text{sech}^2(k_1(x - x_1)) + 3c_2 \text{sech}^2(k_2(x - x_2))$$

with $c_1 = 0.3, c_2 = 0.1, x_1 = 0.4, x_2 = 0.8, k_i = 0.5\sqrt{c_i/\epsilon}$ for $i = 1, 2$, and $\epsilon = 4.84 \times 10^{-4}$. The solution is modeled for $x \in [0, 2]$ and $t \in [0, 4]$ with periodic boundary conditions and 101 uniform points. Fig. 5.7 and Fig. 5.8 show the result obtained with $\Delta t = 0.5(\Delta x)^2$ (so the total time steps are 20,000) with and without the tenth-order filter (using $\alpha = 0.4$ every 1000 steps), respectively. Fig. 5.7 is obtained without filter, and we see some numerical oscillations. But the tenth-order filter eliminates those high-frequency components very well (see Fig. 5.8).

By simple modification of the code *soliton.m*, we can solve for the triple soliton splitting problem, which has the initial condition

$$u_0(x) = \frac{2}{3}\text{sech}^2(\frac{x-1}{\sqrt{108\epsilon}})$$

with $\epsilon = 10^{-4}$. The solution is computed for $x \in [0, 3]$ and $t \in [0, 4]$ with periodic boundary conditions, 151 uniform points and $\Delta t = 0.5(\Delta x)^2$. The computed solutions at $t = 0, 1, 2$ and the solution contour up to $t = 4$ is shown in Fig. 5.9. Here to eliminate the high-frequency oscillations, we use the tenth-order periodic filter with $\alpha = 0.4$ every 100 steps. Note that our tests showed that filtering every 1000 steps would not eliminate the small oscillations very well. More examples can be found in our paper [18].

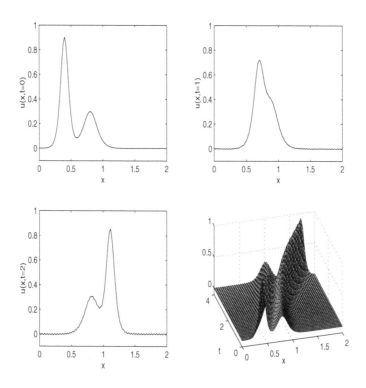

FIGURE 5.7

Double soliton collision obtained without filter. (From Fig. 12 of [18] with kind permission of Springer Science and Business Media.)

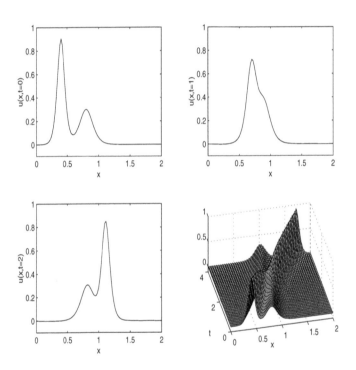

FIGURE 5.8
Double soliton collision obtained with the tenth-order filter. (From Fig. 12 of
[18] with kind permission of Springer Science and Business Media.)

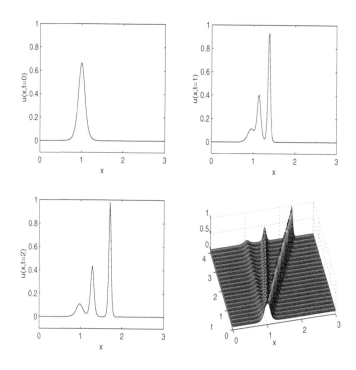

FIGURE 5.9

Triple soliton splitting obtained with tenth-order filter. (From Fig. 13 of [18]
with kind permission of Springer Science and Business Media.)

The MATLAB source code *soliton.m* and its auxiliary codes *ushu51.m*, *resshu5.m*, *reconuxp.m*, and *filterLI.m* are listed below.

```
%-----------------------------------------------------
% soliton.m: solve the double solition problem
%
%  Functions used:
%     ushu51.m: the exact solution
%     resshu5.m: define the residual function
%     reconuxxxp.m: reconstruct u_{xxx} from u
%     reconuxp.m: reconstruct u_x from u
%     filterLI.m: our 10th order filter
%-----------------------------------------------------
clear all
%%%%%%%%%%%%%%%%%%%%%%%%%%%%%%%%%%%%%%%%
% generate 1D mesh
% for ex 4.5 (4.9) & (4.10)
Nx=101; XL=0; XR=2;
% for ex 4.5 (4.11)
%Nx=151; XL=0; XR=3;
dx=(XR-XL)/(Nx-1);

for i=1:Nx
    XX(i)=XL+(i-1)*dx;
end

%%%%%%%%march-in-time%%%%%%%%%%%%
% cfl has to be small!
cfl=0.5; dt=cfl*dx^2; tlast=4;
Nstep=tlast/dt;
IRK=2;
%ap=5*10^(-4);   % coefficient eps for the PDE, (4.9)
ap=4.84*10^(-4);   % for (4.10) of Shu
%ap=10^(-4);      % for (4.11)

% calculate init cond.: stored at u0ex
[u0ex]=feval(@ushu51,XX,ap);
% for ex.4.1., ap likes t
%[u0ex]=feval(@ushu51,XX,0);
uold=u0ex;

IT=1; % indicator for storing solution uall(1:Nx,1:NT)
uall(:,1)=u0ex(:);   % for contour plot

% we use 4th-order RK scheme:IRK=2,4
```

```
for k=1:Nstep
  if IRK==4
  % u0x=reconux(uold,Nx,dx);      % reconstruct ux from u_n
  % k0=dt*resshu5(u0x);           % k0=dt*R((u_n)_x)
  % u1=uold+k0/2;

  % u1x=reconux(u1,Nx,dx);
  % k1=dt*resshu5(u1x);
  % u2=u1+k1/2;

  % u2x=reconux(u2,Nx,dx);
  % k2=dt*resshu5(u2x);
  % u3=u2+k2;

  % u3x=reconux(u3,Nx,dx);
  % k3=dt*resshu5(u3x);

  % unew=uold+(k0+2*k1+2*k2+k3)/6.;   % finish one-time step
  elseif IRK==2
      % method 1: use 1st derivative construction three times
    %u0x=reconuxp(uold,Nx,dx);    % reconstruct ux from u_n
    %u0xx=reconuxp(u0x,Nx,dx);
    %u0xxx=reconuxp(u0xx,Nx,dx);  % obtain u_xxx
    % method 2: construct directly
    u0xxx=reconuxxxp(uold,Nx,dx);
    u2x=reconuxp(uold.^2,Nx,dx);  % reconstruct u^2
    k0=dt*resshu5(u2x,u0xxx,ap);  % k0=dt*R((u_n)_x)
    u1=uold+k0;

    %u1x=reconuxp(u1,Nx,dx);
    %uxx=reconuxp(u1x,Nx,dx);
    %uxxx=reconuxp(uxx,Nx,dx);    % obtain u_xxx
    uxxx=reconuxxxp(u1,Nx,dx);
    u2x=reconuxp(u1.^2,Nx,dx);    % reconstruct u^2
    k1=dt*resshu5(u2x,uxxx,ap);
    u2=u1+k1;

    unew=(uold+u2)/2;
  end
    uold=unew;     % update for next time step

    % for (4.11)
    subplot(2,2,1), plot(XX,u0ex,'g-');
    axis([XL XR -0.1 1]); xlabel('x'); ylabel('u(x,t=0)');
    if abs(k*dt-1) < eps
```

```
            subplot(2,2,2),
            unew=filterLI(unew,Nx);
            plot(XX,unew,'g-');      % plot t=1
            axis([XL XR -0.1 1]); xlabel('x'); ylabel('u(x,t=1)');
        elseif abs(k*dt-2) < eps
            subplot(2,2,3),
            unew=filterLI(unew,Nx);
            plot(XX,unew,'g-');      % plot t=2
            axis([XL XR -0.1 1]); xlabel('x'); ylabel('u(x,t=2)');
        end

        % save some vaules for contour plot
        if mod(k,500)==0
            disp(k), IT=IT+1;
            uall(:,IT)=unew(:);
        end
    end

% do contour plot
ht=tlast/(IT-1);
[X,Y]=meshgrid(XL:dx:XR, 0:ht:tlast);
subplot(2,2,4), mesh(X,Y,uall');
xlabel('x'); ylabel('t');
axis([XL XR 0 tlast -0.1 1]);
```

```
%-----------------------------------------------------------------
```

```
function uu=ushu51(x,ap)
% ex 4.5.: single solition
%c=0.3; x0=0.5; k=sqrt(c/ap)/2;
%uu=3*c*(sech(k*(x-x0))).^2;

% ex 4.5: double solition
c1=0.3; c2=0.1;x1=0.4;x2=0.8;
k1=0.5*sqrt(c1/ap); k2=0.5*sqrt(c2/ap);
uu=3*c1*(sech(k1*(x-x1))).^2+3*c2*(sech(k2*(x-x2))).^2;
% ex 4.5: triple soliton
%tmp=sqrt(108*ap);
%uu=2/3*(sech((x-1)/tmp)).^2;      %.^ array power
% below for ex4.1. of Shu, so ap serves as t
%uu=sin(x+ap);
```

```
% work on kdv equ of Shu's ex4.5
% the residual R=-0.5*(u^2)_x-ap*(u_xxx)
```

```
% i.e., u1=(u^2)_x, u2=u_xxx
function uu=resshu5(u1,u2,ap)
uu=-0.5*u1-ap*u2;
```

```
% Reconstruction for periodic function

function ux=reconuxp(u,N,h)

IEX=0;    % 1 for explicit 4th-order scheme
if IEX==1
  alfa=0; aa=2/3*(alfa+2); bb=(4*alfa-1)/3;
  for i=1:N
    if i==1
        tmp =bb*(u(i+2)-u(N-2))/2+aa*(u(i+1)-u(N-1));
    elseif i==2
        tmp =bb*(u(i+2)-u(N-1))/2+aa*(u(i+1)-u(i-1));
    elseif i==(N-1)
        tmp=bb*(u(2)-u(i-2))/2+aa*(u(i+1)-u(i-1));
    elseif i==N
        tmp=bb*(u(3)-u(i-2))/2+aa*(u(2)-u(i-1));
    else
        tmp=bb*(u(i+2)-u(i-2))/2+aa*(u(i+1)-u(i-1));
    end
    ux(i)=tmp/(2*h);
  end
  return
end

% below are implicit scheme
ISC=6;    % order of the scheme
% for C4 scheme
if ISC==4
   alfa=1.0/4.0; aa=3.0/2; bb=0;
% for C6 scheme
elseif ISC==6
   alfa=1.0/3; aa=14.0/9; bb=1.0/9;
end

amat=zeros(N,N);
B=zeros(N,1);

for i=1:N
    if i==1
```

```
          amat(1,1)=1; amat(1,2)=alfa; amat(1,N-1)=alfa;
          B(i)=bb*(u(3)-u(N-2))/4+aa*(u(2)-u(N-1))/2;
      elseif i==2
          amat(2,1)=alfa; amat(2,2)=1; amat(2,3)=alfa;
          B(i)=bb*(u(4)-u(N-1))/4+aa*(u(3)-u(1))/2;
      elseif i==N-1
          amat(i,i-1)=alfa; amat(i,i)=1; amat(i,i+1)=alfa;
          B(i)=bb*(u(2)-u(N-3))/4+aa*(u(N)-u(N-2))/2;
      elseif i==N
          amat(N,1)=-1; amat(N,N)=1;
          B(i)=0;
      else     % i>=3 & i <=N-2
          amat(i,i-1)=alfa; amat(i,i)=1; amat(i,i+1)=alfa;
          B(i)=bb*(u(i+2)-u(i-2))/4+aa*(u(i+1)-u(i-1))/2;
      end
      B(i)=B(i)/h;
end

% call trisys.m
ux=(amat\B)';
```

```
%-------------------------------------------------------------
% implement 10th order filter for periodic function
%-------------------------------------------------------------
function uf=filterLI(u,N)

ID=5;     %10th-order filter
af=0.1;

a(1)=(193+126*af)/256;
a(2)=(105+302*af)/256;
a(3)=15*(-1+2*af)/64;
a(4)=45*(1-2*af)/512;
a(5)=5*(-1+2*af)/256;
a(6)=(1-2*af)/512;

amat=zeros(N,N);
B=zeros(N,1);

for i=1:N
    if i==1
        amat(1,1)=1; amat(1,2)=af; amat(1,N-1)=af;
    elseif i==N
        amat(N,1)=-1; amat(N,N)=1;
```

```
        else     % special for the 1st and last rows
            amat(i,i-1)=af; amat(i,i)=1; amat(i,i+1)=af;
        end
          B(i)=0;
          for k=0:ID
                % try to keep i+k and i-k into the range of [1,N]
                k1=i+k;
                if k1 > N
                    k1=mod(k1,N-1);
                end
                k2=i-k;
                if k2 <=0
                    k2=k2+(N-1);
                end
                %disp(k1), disp(k2),
                B(i)=B(i)+a(k+1)*(u(k1)+u(k2));
            end
            %disp('----------');
          B(i)=B(i)/2;
end

B(N)=0;      % the last row is special
% call linear system solver
uf=(amat\B)';
```

5.2 High-dimensional problems

Here we consider sixth-order compact schemes coupled with alternating direction implicit (ADI) methods and apply them to 2-D and 3-D parabolic problems *.

5.2.1 Temporal discretization for 2-D problems

Considering the efficiency of ADI methods for solving 2-D or 3-D problems, we will use the ADI method in our implementation. Though ADI coupled with fourth-order compact schemes have been investigated in [1, 6, 14, 19], our derivations are quite different and much simpler as shown below. For

*This section is reprinted from *Computers & Mathematics with Application*, Vol.52, Jichun Li, Yitung Chen and Guoqing Liu, High-order compact ADI methods for parabolic equations, pp. 1343-1356, Copyright(2006), with permission from Elsevier.

clarity and generality, we present the algorithm for the following 2-D parabolic equation:

$$u_t = \nu(u_{xx} + u_{yy}) + F(x, y, t), \quad (x, y, t) \in \Omega \times (0, T], \tag{5.33}$$

$$u(x, y, 0) = G(x, y), \quad (x, y) \in \Omega, \tag{5.34}$$

$$u(x, y, t)|_{\partial\Omega} = H(x, y), \quad t \in [0, T], \tag{5.35}$$

where the diffusion coefficient ν is a positive constant, $\Omega \equiv [0, 1]^2$, and $\partial\Omega$ is the boundary of the domain.

To develop our high-order compact scheme, Ω is divided into a uniform mesh in each direction, i.e.,

$$x_i = (i - 1)\triangle x, i = 1, \cdots, N_x; \quad y_j = (j - 1)\triangle y, j = 1, \cdots, N_y,$$

where $\triangle x, \triangle y$ are the mesh sizes in the x- and y-directions, respectively. We denote u_{ij}^n the approximate solution of $u(i\triangle x, j\triangle y, n\triangle t)$, where $\triangle t$ is the time step.

By applying the Peaceman-Rachford ADI method [24] to (5.33), we have

$$\frac{u_{ij}^{n+1/2} - u_{ij}^n}{0.5\triangle t} = \nu[(u_{xx})_{ij}^{n+1/2} + (u_{yy})_{ij}^n] + (F)_{ij}^{n+1/2}, \tag{5.36}$$

$$\frac{u_{ij}^{n+1} - u_{ij}^{n+1/2}}{0.5\triangle t} = \nu[(u_{xx})_{ij}^{n+1/2} + (u_{yy})_{ij}^{n+1}] + (F)_{ij}^{n+1/2}, \tag{5.37}$$

where $(F)_{ij}^{n+1/2} = F(x_i, y_j, (n + \frac{1}{2})\triangle t)$. Then all the derivatives in (5.36)-(5.37) are approximated by the sixth-order compact formulas developed in §2.1. For example, we can write

$$(u_{xx})_{,j} = \frac{1}{(\triangle x)^2} A^{-1} B u_{,j}, \tag{5.38}$$

where A and B are the corresponding $N_x \times N_x$ triangular and sparse matrices, $u_{,j} = (u_{1,j}, u_{2,j}, \cdots, u_{N_x,j})'$ is the solution vector at the j-th row. Substituting (5.38) into (5.36) gives us

$$(I_x - \frac{1}{2}\nu\frac{\triangle t}{(\triangle x)^2} A^{-1} B)u_{,j}^{n+1/2} = u_{,j}^n + \frac{1}{2}\triangle t[\nu(u_{yy})_{,j}^n + F_{,j}^{n+1/2}], \tag{5.39}$$

where I_x denotes the $N_x \times N_x$ identity matrix.

Similarly for equation (5.37) in the y-direction, we can obtain

$$(I_y - \frac{1}{2}\nu\frac{\triangle t}{(\triangle y)^2} C^{-1} D)u_{i,}^{n+1} = u_{i,}^{n+1/2} + \frac{1}{2}\triangle t[\nu(u_{xx})_{i,}^{n+1/2} + F_{i,}^{n+1/2}], \tag{5.40}$$

by using

$$(u_{yy})_{i,} = \frac{1}{(\triangle y)^2} C^{-1} D u_{i,},$$

where C and D are the corresponding $N_y \times N_y$ triangular and sparse matrices, $u_i = (u_{i,1}, u_{i,2}, \cdots, u_{i,N_y})'$ is the solution vector at the i-th column, and I_y denotes the $N_y \times N_y$ identity matrix.

We like to mention that the above scheme has a truncation error $O((\Delta t)^2, (\Delta x)^6, (\Delta y)^6)$. Note that the coefficient matrices of (5.39) and (5.40) are time-independent, hence we can store the inverse of those coefficient matrices before the time-marching in the implementation for computational efficiency.

5.2.2 Stability analysis

To study the stability of our scheme, we use the von Neumann stability analysis. For simplicity, we assume $F \equiv 0$ in (5.36)-(5.37), and u is periodic in both x and y.

For the sixth-order compact scheme (5.7) for a periodic problem, the matrices A and B have the property

$$Au_i = \alpha u_{i-1} + u_i + \alpha u_{i+1}, \quad Bu_i = \frac{b}{4}(u_{i+2} - 2u_i + u_{i-2}) + a(u_{i+1} - 2u_i + u_{i-1}).$$

Let

$$u_{ij}^n = \xi^n e^{I(w_x i + w_y j)}, \quad I = \sqrt{-1}$$

be the solution of (5.36)-(5.37), where $w_x = \frac{2\pi \Delta x}{l_x}$ and $w_y = \frac{2\pi \Delta y}{l_y}$ are phase angles with wavelengths l_x and l_y, respectively. It is easy to verify that

$$Ae^{Iw_x i} = e^{Iw_x i}(\alpha e^{Iw_x} + 1 + \alpha e^{-Iw_x}) = e^{Iw_x i}(2\alpha \cos w_x + 1),$$

$$Be^{Iw_x i} = e^{Iw_x i}[\frac{b}{4}(e^{I2w_x} - 2 + e^{-I2w_x}) + a(e^{Iw_x} - 2 + e^{-Iw_x})]$$

$$= e^{Iw_x i}[-b\sin^2 w_x - 4a\sin^2 \frac{w_x}{2})].$$

Therefore, we have

$$(u_{xx})_{ij}^{n+1/2} = \frac{1}{(\Delta x)^2} A^{-1} B u_{ij}^{n+1/2}$$

$$= \frac{u_{ij}^{n+1/2}}{(\Delta x)^2} \frac{(-b\sin^2 w_x - 4a\sin^2 \frac{w_x}{2})}{(2\alpha \cos w_x + 1)}. \quad (5.41)$$

We denote

$$m_x = \frac{1}{2} \frac{\nu \Delta t}{(\Delta x)^2}, \quad m_y = \frac{1}{2} \frac{\nu \Delta t}{(\Delta y)^2},$$

and

$$\gamma_x = \frac{(-b\sin^2 w_x - 4a\sin^2 \frac{w_x}{2})}{(2\alpha \cos w_x + 1)}, \quad \gamma_y = \frac{(-b\sin^2 w_y - 4a\sin^2 \frac{w_y}{2})}{(2\alpha \cos w_y + 1)}.$$

Hence (5.41) can be written as

$$(u_{xx})_{ij}^{n+1/2} = \frac{1}{(\triangle x)^2} A^{-1} B u_{ij}^{n+1/2} = \frac{1}{(\triangle x)^2} \gamma_x u_{ij}^{n+1/2}. \tag{5.42}$$

Similarly, it is easy to find that

$$(u_{yy})_{ij}^{n+1} = \frac{1}{(\triangle y)^2} C^{-1} D u_{ij}^{n+1} = \frac{1}{(\triangle y)^2} \gamma_y u_{ij}^{n+1}. \tag{5.43}$$

Substituting (5.41) and (5.42) into (5.36) with $F = 0$, we obtain

$$(1 - m_x \gamma_x) u_{ij}^{n+1/2} = (1 + m_y \gamma_y) u_{ij}^n. \tag{5.44}$$

In the same way, from (5.37) we can easily obtain

$$(1 - m_y \gamma_y) u_{ij}^{n+1} = (1 + m_x \gamma_x) u_{ij}^{n+1/2}. \tag{5.45}$$

From (5.44)-(5.45), we see that the amplification factor is

$$|\xi| = |\frac{u_{ij}^{n+1}}{u_{ij}^{n+1/2}}| \cdot |\frac{u_{ij}^{n+1/2}}{u_{ij}^n}| = |\frac{1 + m_x \gamma_x}{1 - m_y \gamma_y} \cdot \frac{1 + m_y \gamma_y}{1 - m_x \gamma_x}| \le 1$$

whenever $\gamma_x \le 0$ and $\gamma_y \le 0$ hold true. It is easy to see that for our special sixth-order scheme (5.7) with

$$\alpha = \frac{2}{11}, \quad a = \frac{12}{11}, \quad b = \frac{3}{11}, \tag{5.46}$$

$\gamma_x \le 0$ and $\gamma_y \le 0$ hold true, which means that our scheme (5.36)-(5.37) is unconditionally stable in this case.

5.2.3   Extensions to 3-D compact ADI schemes

The above compact ADI scheme can be extended directly to the 3-D case, such as

$$u_t = \nu(u_{xx} + u_{yy} + u_{zz}) + F(x, y, z, t), \quad (x, y, z, t) \in \Omega \times (0, T]. \tag{5.47}$$

By applying the Douglas ADI method [7] to (5.47), we have

$$\frac{u_{ijk}^{n+1/3} - u_{ijk}^n}{\triangle t} = \nu[\frac{1}{2}((u_{xx})_{ijk}^{n+1/3} + (u_{xx})_{ijk}^n) + (u_{yy})_{ijk}^n + (u_{zz})_{ijk}^n]$$

$$+ (F)_{ijk}^{n+1/2}, \tag{5.48}$$

$$\frac{u_{ijk}^{n+2/3} - u_{ijk}^{n+1/3}}{0.5\triangle t} = \nu[(u_{yy})_{ijk}^{n+2/3} - (u_{yy})_{ijk}^n], \tag{5.49}$$

$$\frac{u_{ijk}^{n+1} - u_{ijk}^{n+2/3}}{0.5\triangle t} = \nu[(u_{zz})_{ijk}^{n+1} - (u_{zz})_{ijk}^n], \tag{5.50}$$

where $(F)_{ijk}^{n+1/2} = F(x_i, y_j, z_k, (n + \frac{1}{2})\triangle t)$, and u_{ijk}^n denotes the approximate solution of $u(x_i, y_j, z_k, t_n)$. We approximate all the second derivatives in (5.48)-(5.50) by those sixth-order compact formulas developed previously.

Applying the similar technique used for 2-D, we can prove unconditional stability of our scheme (5.48)-(5.50) with periodic boundary conditions. Details can be found in our paper [17]. Similar analysis can be carried out for other ADI schemes (e.g., [8]), and references can be found in books such as [11]. The stability for non-periodic cases can be pursued using other methods such as [3, 21], which is much more complicated and shall be further studied.

5.2.4 Numerical examples with MATLAB codes

Here we solve the 2-D diffusion problem

$$u_t = u_{xx} + u_{yy} + f(x, y, t) \quad (x, y) \in \Omega \equiv (0, 1)^2, \quad 0 < t < 1 \tag{5.51}$$

with periodic boundary conditions in both x and y directions, $f = (16\pi^2 - 1)e^{-t}(\sin(4\pi x) + \sin(4\pi y))$ and a properly selected initial condition so that the exact solution is given by

$$u(x, y, t) = e^{-t}(\sin(4\pi x) + \sin(4\pi y)). \tag{5.52}$$

This problem was solved by the ADI method (5.36)-(5.37) using the sixth-order scheme (5.7) to approximate those derivatives. The theoretical accuracy $O((\triangle t)^2, (\triangle x)^6, (\triangle y)^6)$ was confirmed by solving the problem with various mesh sizes and time step sizes. The results presented in Table 5.1 (with a fixed mesh size) show the convergence rate of $O((\triangle t)^2)$ very well. The results presented in Table 5.2 (with a fixed small time step) demonstrate the convergence rate of $O((\triangle x)^6)$ very well. An examplary numerical solution and maximum error are plotted in Fig. 5.10 for $T = 1$ solved by the listed MATLAB code *hoc2d.m* with $Nx = Ny = 51, dt = \frac{1}{80}$. The computation was performed under MATLAB 6.0 installed on a laptop with Intel Pentium 3 processor at 533MHz and 224MB RAM.

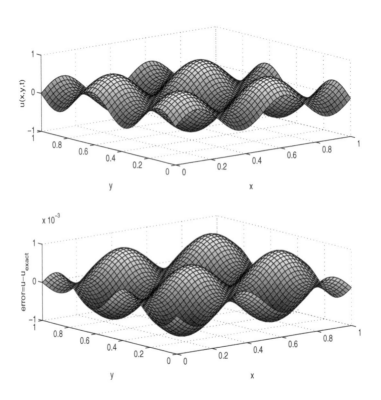

FIGURE 5.10

The numerical solutions (top) and pointwise errors (bottom) at $t = 1.0$.

TABLE 5.1
Maximum error at $t = 1$ with fixed mesh size $\triangle x = \triangle y = 1/50$

Time step size	Maximum error	Convergence rate $O((\triangle t)^r)$	CPU time (seconds)
$\triangle t = 1/10$	7.2469e-4		4.90s
$\triangle t = 1/20$	3.1364e-4	1.2083	9.45s
$\triangle t = 1/40$	5.7551e-5	2.4462	17.40s
$\triangle t = 1/80$	1.4346e-5	2.0042	33.95s
$\triangle t = 1/160$	3.5438e-6	2.0173	67.05s
$\triangle t = 1/320$	8.4330e-7	2.0712	133.80s

TABLE 5.2
Maximum error at $t = 1$ with fixed time step size $\triangle t = 10^{-4}$

Mesh size	Maximum error	Convergence rate $O((\triangle x)^k)$	CPU time (seconds)
$\triangle x = \triangle y = 1/10$	9.2351e-4		218.90s
$\triangle x = \triangle y = 1/20$	1.3497e-5	5.9470	746.02s
$\triangle x = \triangle y = 1/40$	2.1695e-7	5.9591	2695.10s

The MATLAB source code *hoc2d.m* and its auxiliary functions *F2dconeF.m*, *F2dcone.m*, *reconux2p.m* are listed below.

```
%-----------------------------------------------------------------
% hoc2d.m: solve 2D parabolic equ:
%          u_t=u_{xx}+u_{yy}+F(x,y,t), (x,y)\in (0,1)^2
% by high-order compact ADI scheme.
%
% Other functions needed:
%       F2dconeF.m: source term F
%       F2dcone.m: the exact solution u
%       reconux2p.m: generate the matrix used by reconstruction
%-----------------------------------------------------------------
clear all;

    T=1;   %hr
    XL=0; XR=1;  % left and right end points
    Nx=51;     % total number of points
    dx=(XR-XL)/(Nx-1);
    YL=0; YR=1;  % left and right end points
    Ny=51;     % total number of points
    dy=(YR-YL)/(Ny-1);

    dt=1/80;
    uold=zeros(Nx,Ny);

Nstep=round(T/dt);

dtx=dt/dx; dty=dt/dy;

for i=1:Nx
    XX(i)=XL+(i-1)*dx;
end
for i=1:Ny
    YY(i)=YL+(i-1)*dy;
end

% set up initial cond
for i=1:Nx
    for j=1:Ny
        uold(i,j)=F2dcone(XX(i),YY(j),0);
    end
end

%%%%%%%%march-in-time%%%%%%%%%%%%
```

```
matx=zeros(Nx,Nx);  % define matrix for reconstructing u_xx
maty=zeros(Ny,Ny);  % define matrix for reconstructing u_yy

matx=reconux2p(Nx,dx);   % reconstruction in x-direction
maty=reconux2p(Ny,dy);   % reconstruction in y-direction

t1=cputime;

Ainv=inv(eye(Nx,Nx)-0.5*dt*matx);
Binv=inv(eye(Ny,Ny)-0.5*dt*maty);

for k=1:Nstep
  % reconstruct u_{yy}
  for i=1:Nx
      tmp=(uold(i,1:Ny))';     % has to transpose
      tmp2=maty*tmp;           % construct u_yy
      uyy(i,1:Ny)=tmp2';
  end

    % start ADI scheme
  for j=1:Ny
      for i=1:Nx
        TMPF(i,j)=0.5*dt*F2dconeF(XX(i),YY(j),(k-0.5)*dt);
        rhs(i)=uold(i,j)+0.5*dt*uyy(i,j)+TMPF(i,j);
      end
      uHaf(1:Nx,j)=Ainv*rhs';
  end

  % reconstruct u_{xx}
  for j=1:Ny
      uxx(1:Nx,j)=matx*uHaf(1:Nx,j);   % reconstruct u_{xx}
  end

  for i=1:Nx
      for j=1:Ny
         rhs(j)=uHaf(i,j)+0.5*dt*uxx(i,j)+TMPF(i,j);
      end
      unew(i,1:Ny)=(Binv*rhs')';
  end

  uold=unew;    % update for next time step
end

    % draw numerical solution and max errors
    if k == Nstep
```

```
            disp('k='), k
            errL2=0;
            for i=1:Nx
                for j=1:Ny
                    % get the exact solution
                    u2d(i,j)=feval(@F2dcone,XX(i),YY(j),dt*k);
                    errL2=errL2+(u2d(i,j)-unew(i,j))^2;
                end
            end
            errL2=sqrt(errL2/(Nx*Ny));

            subplot(2,1,1), surf(XX,YY,unew');
            xlabel('x'); ylabel('y'); zlabel('u(x,y,t)');
            subplot(2,1,2), surf(XX,YY,(unew-u2d)');
            xlabel('x'); ylabel('y'); zlabel('error=u-u_{exact}');
            err=max(max(abs(unew-u2d)));
            disp('L2 error='),errL2,
            disp('absolute max error='), err,
            disp('relative max error='), err/max(max(abs(u2d))),
    end

t_used=cputime-t1;     % get time CPU time used in seconds
disp('CPU time used='), disp(t_used),
```

```
% Calculate the RHS function

function rhsF=F2dconeF(x,y,t)
ID=4;
if ID==1     % rotating cone model
   rt=(2+sin(pi*t))/4;
   st=(2+cos(pi*t))/4;
   tmp=exp(-80*((x-rt)^2+(y-st)^2));
   ut=0.8*tmp*40*pi*((x-rt)*cos(pi*t)-(y-st)*sin(pi*t));
   uxx=0.8*tmp*(160^2*(x-rt)^2-160);
   uyy=0.8*tmp*(160^2*(y-st)^2-160);
   rhsF=ut-uxx-uyy;
elseif ID==2     % periodic function model
   rhsF=0;
elseif ID==4
    rhsF=(-1+16*pi*pi)*exp(-t)*(sin(4*pi*x)+sin(4*pi*y));
end
```

```
%-----------------------------------------------------
% Calculate the exact solution of the given PDE
```

```
function u2d=F2dcone(x,y,t)
ID=4;
if ID==1
  rt=(2+sin(pi*t))/4;
  st=(2+cos(pi*t))/4;
  u2d=0.8*exp(-80*((x-rt)^2+(y-st)^2));
elseif ID==2
  u2d=exp(-8*pi*pi*t)*sin(2*pi*x)*sin(2*pi*y);
 elseif ID==3
  u2d=exp(-t)*sin(4*pi*x)*sin(4*pi*y);
 elseif ID==4
  u2d=exp(-t)*(sin(4*pi*x)+sin(4*pi*y));
end
```

```
%------------------------------------------------------------
% Ref to S.K. Lele's 1992 paper
% Reconstruct the 2nd derivative of a periodic function
%------------------------------------------------------------
function matAB=reconux2p(N,h)

matA=zeros(N,N); matB=zeros(N,N);
% for interior point
% for 6th-order scheme (2.2.7)
alfa=2.0/11.0; aa=4./3.*(1-alfa); bb=1./3.*(-1+10*alfa);

for i=3:N-2
    matA(i,i)=1;
    matA(i,i+1)=alfa;
    matA(i,i-1)=alfa;
    matB(i,i-2)=bb/4;
    matB(i,i-1)=aa;
    matB(i,i)=-bb/2-2*aa;
    matB(i,i+1)=aa;
    matB(i,i+2)=bb/4;
end

%%%%%%%%%%%%%%%%%%%%%%%%%%%%%%%%%%%%
% for boundary point 1: 6th-order at boundary nodes!!
for i=1:1
    matA(i,i)=1;
    matA(i,i+1)=alfa;
    matA(i,N-1)=alfa;

    matB(i,N-2)=bb/4;
```

```
    matB(i,N-1)=aa;
    matB(i,i)=-bb/2-2*aa;
    matB(i,i+1)=aa;
    matB(i,i+2)=bb/4;
end
%%%%%%%%%%%%%%%%%%%%%%%%%%%%
% for boundry point 2
for i=2:2
    matA(i,i)=1;
    matA(i,i+1)=alfa;
    matA(i,i-1)=alfa;

    matB(i,N-1)=bb/4;
    matB(i,i-1)=aa;
    matB(i,i)=-bb/2-2*aa;
    matB(i,i+1)=aa;
    matB(i,i+2)=bb/4;
end
%%%%%%%%%%%%%%%%%%%%%%%%%%%%
% for boundary point N-1
for i=N-1:N-1
    matA(i,i)=1;
    matA(i,i+1)=alfa;
    matA(i,i-1)=alfa;

    matB(i,i-2)=bb/4;
    matB(i,i-1)=aa;
    matB(i,i)=-bb/2-2*aa;
    matB(i,i+1)=aa;
    matB(i,2)=bb/4;
end
%%%%%%%%%%%%%%%%%%%%%%%%%%%%
% for boundary point N
for i=N:N
    matA(i,i)=1;
    matA(i,2)=alfa;
    matA(i,i-1)=alfa;

    matB(i,i-2)=bb/4;
    matB(i,i-1)=aa;
    matB(i,i)=-bb/2-2*aa;
    matB(i,2)=aa;
    matB(i,3)=bb/4;
end
%%%%%%%%%%%%%%%%%%%%%%%%%%%%
```

```
matB=matB./(h*h);

%matA, matB,

matAB=inv(matA)*matB;    % the matrix for u_xx=A^(-1)B*u
```

5.3   Other high-order compact schemes

Here we discuss another popular way to construct high-order compact difference schemes (e.g., [26] and references cited therein). The basic idea is to
apply central differences to the governing PDE and then repeatedly replace
those higher-order derivatives in the truncation error by low-order derivatives
using the PDE.

5.3.1   One-dimensional problems

To demonstrate the technique, let us start with a one-dimensional steady-state
problem

$$-a\frac{d^2u}{dx^2} + b\frac{du}{dx} = s(x), \quad x \in \Omega, \tag{5.53}$$

with corresponding Dirichlet boundary conditions. For simplicity, we assume
that a and b are constants, and a uniform mesh of Ω is used, which has a
constant mesh size h.

From Taylor expansions

$$u_{i+1} = u_i + hu_i' + \frac{h^2}{2}u_i'' + \frac{h^3}{6}u_i^{(3)} + \frac{h^4}{24}u_i^{(4)} + \frac{h^5}{120}u_i^{(5)} + O(h^6),$$

and

$$u_{i-1} = u_i - hu_i' + \frac{h^2}{2}u_i'' - \frac{h^3}{6}u_i^{(3)} + \frac{h^4}{24}u_i^{(4)} - \frac{h^5}{120}u_i^{(5)} + O(h^6),$$

we have

$$u_i' = \delta_x u_i - \frac{h^2}{6}u_i^{(3)} - \frac{h^4}{120}u_i^{(5)} + O(h^6), \tag{5.54}$$

$$u_i'' = \delta_x^2 u_i - \frac{h^2}{12}u_i^{(4)} - \frac{h^4}{360}u_i^{(6)} + O(h^6), \tag{5.55}$$

where we denote the standard first and second central difference operators

$$\delta_x u_i = \frac{u_{i+1} - u_{i-1}}{2h}, \quad \delta_x^2 u_i = \frac{u_{i+1} - 2u_i + u_{i-1}}{h^2}. \tag{5.56}$$

Substituting (5.54) and (5.55) into (5.53), we obtain

$$-a(\delta_x^2 u_i - \frac{h^2}{12}u_i^{(4)}) + b(\delta_x u_i - \frac{h^2}{6}u_i^{(3)}) = s_i + O(h^4). \tag{5.57}$$

Differentiating (5.53) with respect to x once and twice, respectively, we have

$$u^{(3)} = \frac{1}{a}(bu'' - s') = \frac{b}{a}u'' - \frac{1}{a}s', \tag{5.58}$$

$$u^{(4)} = \frac{1}{a}(bu^{(3)} - s'') = \frac{b^2}{a^2}u'' - \frac{b}{a^2}s' - \frac{1}{a}s''. \tag{5.59}$$

Then replacing the second derivative u'' in (5.58) and (5.59) by (5.55) and substituting into (5.57), we obtain

$$-a[\delta_x^2 u_i - \frac{h^2}{12}(\frac{b^2}{a^2}u_i'' - \frac{b}{a^2}s_i' - \frac{1}{a}s_i'')]$$
$$+b[\delta_x u_i - \frac{h^2}{6}(\frac{b}{a}u_i'' - \frac{1}{a}s_i')] = s_i + O(h^4),$$

which leads to a fourth-order compact difference scheme for (5.53):

$$-a(1 + \frac{b^2 h^2}{12a^2})\delta_x^2 u_i + b\delta_x u_i = s_i - \frac{bh^2}{12a}s_i' + \frac{h^2}{12}s_i''. \tag{5.60}$$

The above technique can be easily extended to the transient convection-diffusion equation

$$\frac{\partial u}{\partial t} + b\frac{\partial u}{\partial x} - a\frac{\partial^2 u}{\partial x^2} = s(x), \quad x \in \Omega, \tag{5.61}$$

with proper initial condition and Dirichlet boundary conditions. Replacing s by $s - \frac{\partial u}{\partial t}$ in (5.60), we have

$$\frac{\partial u}{\partial t}|_i + \frac{h^2}{12a}\frac{\partial}{\partial t}(au_i'' - bu_i') - a(1 + \frac{b^2 h^2}{12a^2})\delta_x^2 u_i + b\delta_x u_i = s_i - \frac{bh^2}{12a}s_i' + \frac{h^2}{12}s_i'',$$

which, along with (5.54) and (5.55), leads to the fourth-order compact semi-discrete scheme for (5.61):

$$\frac{\partial u}{\partial t}|_i + \frac{h^2}{12a}\frac{\partial}{\partial t}(a\delta_x^2 u_i - b\delta_x u_i) - a(1 + \frac{b^2 h^2}{12a^2})\delta_x^2 u_i + b\delta_x u_i = s_i - \frac{bh^2}{12a}s_i' + \frac{h^2}{12}s_i''. \tag{5.62}$$

Note that (5.62) is a tridiagonal semi-discrete system at each grid point i. To get a fully discrete scheme, we can consider either the Runge-Kutta scheme or θ-scheme used for time-dependent problems. For example, the Crank-Nicolson scheme at point i at time level n for (5.62) can be written as

$$\frac{u_i^{n+1} - u_i^n}{\Delta t} + \frac{h^2}{12a\Delta t}[a\delta_x^2(u_i^{n+1} - u_i^n) - b\delta_x(u_i^{n+1} - u_i^n)]$$

$$-a(1 + \frac{b^2 h^2}{12a^2}) \cdot \frac{1}{2}\delta_x^2(u_i^{n+1} + u_i^n) + b \cdot \frac{1}{2}\delta_x(u_i^{n+1} + u_i^n)$$

$$= \frac{1}{2}(s_i^{n+1} + s_i^n) - \frac{bh^2}{24a} \cdot \frac{\partial}{\partial x}(s_i^{n+1} + s_i^n) + \frac{h^2}{24} \cdot \frac{\partial^2}{\partial x^2}(s_i^{n+1} + s_i^n). \quad (5.63)$$

By the von Neumann stability technique, it is easy to see that this scheme is unconditionally stable.

5.3.2 Two-dimensional problems

The above approach can be used for problems in higher spatial dimensions. For example, let us consider the 2-D transient diffusion problem:

$$\frac{\partial u}{\partial t} - a(\frac{\partial^2 u}{\partial x^2} + \frac{\partial^2 u}{\partial y^2}) = s(x, y) \quad \text{in } \Omega, \quad (5.64)$$

$$u|_{t=0} = f(x, y), \quad \text{in } \Omega, \quad (5.65)$$

$$u|_{\partial\Omega} = g(x, y, t). \quad (5.66)$$

Repeating the above procedure for $b = 0$, it is easy to obtain the fourth-order compact Crank-Nicolson scheme for (5.64):

$$\frac{u_{ij}^{n+1} - u_{ij}^n}{\triangle t} + \frac{h^2}{12\triangle t}[\delta_x^2(u_{ij}^{n+1} - u_{ij}^n) + \delta_y^2(u_{ij}^{n+1} - u_{ij}^n)]$$

$$- \frac{a}{2}[\delta_x^2(u_{ij}^{n+1} + u_{ij}^n) + \delta_y^2(u_{ij}^{n+1} + u_{ij}^n) + \frac{h^2}{6}\delta_x^2\delta_y^2(u_i^{n+1} + u_i^n)]$$

$$= \frac{1}{2}(s_{ij}^{n+1} + s_{ij}^n) + \frac{h^2}{24}\frac{\partial^2}{\partial x^2}(s_{ij}^{n+1} + s_{ij}^n) + \frac{h^2}{24}\frac{\partial^2}{\partial y^2}(s_{ij}^{n+1} + s_{ij}^n). \quad (5.67)$$

Substituting $s_{ij}^n = \lambda^n e^{\sqrt{-1}k_x ih} e^{\sqrt{-1}k_y jh}$ into (5.67) with $s = 0$, we obtain the amplification factor

$$\lambda = \frac{1 + (\frac{1}{12} + \frac{a\triangle t}{2h^2})(-4\sin^2\frac{1}{2}k_x h - 4\sin^2\frac{1}{2}k_y h) + \frac{a\triangle t}{12h^2}(4\sin^2\frac{1}{2}k_x h)(4\sin^2\frac{1}{2}k_y h)}{1 + (\frac{1}{12} - \frac{a\triangle t}{2h^2})(-4\sin^2\frac{1}{2}k_x h - 4\sin^2\frac{1}{2}k_y h) - \frac{a\triangle t}{12h^2}(4\sin^2\frac{1}{2}k_x h)(4\sin^2\frac{1}{2}k_y h)}$$

which is equivalent to

$$\lambda = \frac{1 - \frac{1}{3}(\mu_x + \mu_y) - \frac{a\triangle t}{h^2}(2\mu_x + 2\mu_y - \frac{4}{3}\mu_x\mu_y)}{1 - \frac{1}{3}(\mu_x + \mu_y) + \frac{a\triangle t}{h^2}(2\mu_x + 2\mu_y + \frac{4}{3}\mu_x\mu_y)}$$

where we denote $\mu_x = \sin^2\frac{1}{2}k_x h$ and $\mu_y = \sin^2\frac{1}{2}k_y h$.
It is easy to check that

$$-1 \leq \lambda \leq 1$$

holds true for any $\mu_x, \mu_y, \triangle t$ and h. Hence the scheme (5.67) is also unconditionally stable.

Finally, let us consider a variable coefficient problem such as the 2-D convection-diffusion equation

$$-(\frac{\partial^2 u}{\partial x^2} + \frac{\partial^2 u}{\partial y^2}) + a(x,y)\frac{\partial u}{\partial x} + b(x,y)\frac{\partial u}{\partial y} = s(x,y) \tag{5.68}$$

with Dirichlet boundary condition. For simplicity, we assume that mesh sizes in both x- and y-directions are h.

Using (5.54) and (5.55) for (5.68), we have

$$-(\delta_x^2 u_{ij} + \delta_y^2 u_{ij}) + \frac{h^2}{12}(\frac{\partial^4 u_{ij}}{\partial x^4} + \frac{\partial^4 u_{ij}}{\partial y^4}) + a_{ij}(\delta_x u_{ij} - \frac{h^2}{6}\frac{\partial^3 u_{ij}}{\partial x^3})$$

$$+b_{ij}(\delta_y u_{ij} - \frac{h^2}{6}\frac{\partial^3 u_{ij}}{\partial y^3}) = s_{ij} + O(h^4). \tag{5.69}$$

To obtain a fourth-order compact scheme, we need to replace those third and fourth derivatives in (5.69) by first and second central differences. Differentiating (5.68) with respect to x once and twice, respectively, we obtain

$$-\frac{\partial^3 u}{\partial x^3} = \frac{\partial^3 u}{\partial x \partial y^2} - \frac{\partial a}{\partial x}\frac{\partial u}{\partial x} - a\frac{\partial^2 u}{\partial x^2} - \frac{\partial b}{\partial x}\frac{\partial u}{\partial y} - b\frac{\partial^2 u}{\partial x \partial y} + \frac{\partial s}{\partial x}, \tag{5.70}$$

and

$$\frac{\partial^4 u}{\partial x^4} = -\frac{\partial^4 u}{\partial x^2 \partial y^2} + \frac{\partial^2 a}{\partial x^2}\frac{\partial u}{\partial x} + 2\frac{\partial a}{\partial x}\frac{\partial^2 u}{\partial x^2} + a\frac{\partial^3 u}{\partial x^3}$$

$$+\frac{\partial^2 b}{\partial x^2}\frac{\partial u}{\partial y} + 2\frac{\partial b}{\partial x}\frac{\partial^2 u}{\partial x \partial y} + b\frac{\partial^3 u}{\partial x^2 \partial y} + \frac{\partial^2 s}{\partial x^2}. \tag{5.71}$$

Combining (5.70) and (5.71), and using (5.54) and (5.55) again, we have

$$(\frac{\partial^4 u}{\partial x^4} - 2a\frac{\partial^3 u}{\partial x^3})_{ij} = [-\frac{\partial^4 u}{\partial x^2 \partial y^2} + \frac{\partial^2 a}{\partial x^2}\frac{\partial u}{\partial x} + 2\frac{\partial a}{\partial x}\frac{\partial^2 u}{\partial x^2} + \frac{\partial^2 b}{\partial x^2}\frac{\partial u}{\partial y}$$

$$+2\frac{\partial b}{\partial x}\frac{\partial^2 u}{\partial x \partial y} + b\frac{\partial^3 u}{\partial x^2 \partial y} + \frac{\partial^2 s}{\partial x^2} + a\frac{\partial^3 u}{\partial x \partial y^2} - a\frac{\partial a}{\partial x}\frac{\partial u}{\partial x}$$

$$-a^2\frac{\partial^2 u}{\partial x^2} - a\frac{\partial b}{\partial x}\frac{\partial u}{\partial y} - ab\frac{\partial^2 u}{\partial x \partial y} + a\frac{\partial s}{\partial x}]_{ij}$$

$$= -\delta_x^2 \delta_y^2 u_{ij} + \frac{\partial^2 a_{ij}}{\partial x^2}\delta_x u_{ij} + 2\frac{\partial a_{ij}}{\partial x}\delta_x^2 u_{ij} + \frac{\partial^2 b_{ij}}{\partial x^2}\delta_y u_{ij}$$

$$+2\frac{\partial b_{ij}}{\partial x^2}\delta_x \delta_y u_{ij} + b_{ij}\delta_x^2 \delta_y u_{ij} + \frac{\partial^2 s_{ij}}{\partial x^2} + a_{ij}\delta_x \delta_y^2 u_{ij}$$

$$-a_{ij}\frac{\partial a_{ij}}{\partial x}\delta_x u_{ij} - a_{ij}^2 \delta_x^2 u_{ij} - a_{ij}\frac{\partial b_{ij}}{\partial x}\delta_y u_{ij}$$

$$-a_{ij}b_{ij}\delta_x \delta_y u_{ij} + a_{ij}\frac{\partial s_{ij}}{\partial x} + O(h^2). \tag{5.72}$$

Similarly, we have

$$-\frac{\partial^3 u}{\partial y^3} = \frac{\partial^3 u}{\partial y \partial x^2} - \frac{\partial a}{\partial y}\frac{\partial u}{\partial x} - a\frac{\partial^2 u}{\partial y \partial x} - \frac{\partial b}{\partial y}\frac{\partial u}{\partial y} - b\frac{\partial^2 u}{\partial y^2} + \frac{\partial s}{\partial y},$$

$$\frac{\partial^4 u}{\partial y^4} = -\frac{\partial^4 u}{\partial x^2 \partial y^2} + \frac{\partial^2 a}{\partial y^2}\frac{\partial u}{\partial x} + 2\frac{\partial a}{\partial y}\frac{\partial^2 u}{\partial y \partial x} + a\frac{\partial^3 u}{\partial y^2 \partial x}$$
$$+ \frac{\partial^2 b}{\partial y^2}\frac{\partial u}{\partial y} + 2\frac{\partial b}{\partial y}\frac{\partial^2 u}{\partial y \partial y^2} + b\frac{\partial^3 u}{\partial y^3} + \frac{\partial^2 s}{\partial y^2},$$

and

$$\left(\frac{\partial^4 u}{\partial y^4} - 2b\frac{\partial^3 u}{\partial y^3}\right)_{ij} = \Big[-\frac{\partial^4 u}{\partial x^2 \partial y^2} + \frac{\partial^2 a}{\partial y^2}\frac{\partial u}{\partial x} + 2\frac{\partial a}{\partial y}\frac{\partial^2 u}{\partial y \partial x} + a\frac{\partial^3 u}{\partial y^2 \partial x}$$
$$+ \frac{\partial^2 b}{\partial y^2}\frac{\partial u}{\partial y} + 2\frac{\partial b}{\partial y}\frac{\partial^2 u}{\partial y \partial y^2} + \frac{\partial^2 s}{\partial y^2} + b\frac{\partial^3 u}{\partial y \partial x^2}$$
$$- b\frac{\partial a}{\partial y}\frac{\partial u}{\partial x} - ba\frac{\partial^2 u}{\partial y \partial x} - b\frac{\partial b}{\partial y}\frac{\partial u}{\partial y} - b^2\frac{\partial^2 u}{\partial y^2} + b\frac{\partial s}{\partial y}\Big]_{ij}$$
$$= -\delta_x^2 \delta_y^2 u_{ij} + \frac{\partial^2 a_{ij}}{\partial y^2}\delta_x u_{ij} + 2\frac{\partial a_{ij}}{\partial y}\delta_y \delta_x u_{ij} + a_{ij}\delta_y^2 \delta_x u_{ij}$$
$$+ \frac{\partial^2 b_{ij}}{\partial y^2}\delta_y u_{ij} + 2\frac{\partial b_{ij}}{\partial y}\delta_y^2 u_{ij} + \frac{\partial^2 s_{ij}}{\partial y^2} + b_{ij}\delta_y \delta_x^2 u_{ij}$$
$$- b_{ij}\frac{\partial a_{ij}}{\partial y}\delta_x u_{ij} - b_{ij}a_{ij}\delta_y \delta_x u_{ij}$$
$$- b_{ij}\frac{\partial b_{ij}}{\partial y}\delta_y u_{ij} - b_{ij}^2 \delta_y^2 u_{ij} + b_{ij}\frac{\partial s_{ij}}{\partial y} + O(h^2). \quad (5.73)$$

Substituting (5.72) and (5.73) into (5.69) gives us a fourth-order compact scheme for (5.68). This scheme only involves 9 points.

In the special case of $a = b =$ constant, the fourth-order compact scheme (5.73) can be simplified to

$$-(1 + \frac{h^2 a^2}{12})\delta_x^2 u_{ij} - (1 + \frac{h^2 b^2}{12})\delta_y^2 u_{ij}$$
$$+ \frac{h^2}{6}[-\delta_x^2 \delta_y^2 u_{ij} + a\delta_x \delta_y^2 u_{ij} + b\delta_y \delta_x^2 u_{ij} - ab\delta_x \delta_y u_{ij}] + a\delta_x u_{ij} + b\delta_y u_{ij}$$
$$= s_{ij} - \frac{h^2}{12}\Big[\frac{\partial^2 s_{ij}}{\partial x^2} + \frac{\partial^2 s_{ij}}{\partial y^2} + a\frac{\partial s_{ij}}{\partial x} + b\frac{\partial s_{ij}}{\partial y}\Big]. \quad (5.74)$$

For the special case of $a = b = 0$ (i.e., the diffusion problem), the scheme (5.74) can be reduced further to

$$-(\delta_x^2 u_{ij} + \delta_y^2 u_{ij} + \frac{h^2}{6}\delta_x^2 \delta_y^2 u_{ij}) = s_{ij} - \frac{h^2}{12}\left(\frac{\partial^2 s_{ij}}{\partial x^2} + \frac{\partial^2 s_{ij}}{\partial y^2}\right). \quad (5.75)$$

Multiplying both sides of (5.75) by $6h^2$ and expanding it, we see that (5.75) becomes

$$
\begin{aligned}
-[6(&u_{i+1,j} - 2u_{ij} + u_{i-1,j}) + 6(u_{i,j+1} - 2u_{ij} + u_{i,j-1}) \\
+(&u_{i+1,j+1} - 2u_{i,j+1} + u_{i-1,j+1}) - 2(u_{i+1,j} - 2u_{ij} + u_{i-1,j}) \\
+(&u_{i+1,j-1} - 2u_{i,j-1} + u_{i-1,j-1})] \\
= 6h^2 s_{ij} &- \frac{h^2}{2}[(s_{i+1,j} - 2s_{ij} + s_{i-1,j}) + (s_{i,j+1} - 2s_{ij} + s_{i,j-1})],
\end{aligned}
$$

which can be simplified to the 9-point fourth-order scheme (4.69) obtained in Chapter 4.

5.4 Bibliographical remarks

Extensions of compact schemes to nonuniform mesh is trivial but very technical, and interested readers can find more details in papers such as [10, 34] and references cited therein. To increase accuracy, the so-called combined compact difference scheme was proposed [4, 20], which solves for first and second derivatives synchronously. To our best knowledge, books specifically covering high-order difference methods are not many [5, 11, 28]. Gustafsson et al. [11] provide solid mathematical theory on both the Fourier method and the energy method for building stable difference schemes. Some fourth-order schemes are analyzed for hyperbolic and parabolic equations. Cohen [5] introduces some classical fourth-order schemes in time and space for the wave equations. Detailed numerical analysis on dispersion and stability of the schemes is provided in his book. Tolstykh introduces high-order schemes using compact upwind differencing in his book [28], where arbitrary odd- (e.g., third-, fifth-, seventh-) order schemes are constructed and used for hyperbolic systems and convection-diffusion equations. Recently, we [16] developed and analyzed fourth-order compact schemes for time-dependent Maxwell's equations in metamaterials. More references on high-order compact difference schemes for Maxwell's equations can be found there.

5.5 Exercises

1. Using the Taylor expansion, prove that with $a = \frac{2}{3}(\alpha+2)$ and $b = \frac{1}{3}(4\alpha-1)$, the scheme (5.1) is fourth-order accurate. Find out the leading two terms for the truncation error.

2. Use the Taylor expansion to prove that with coefficients (5.30)-(5.31), the filter (5.29) is indeed tenth-order.

3. Modify the code *hoc2d.m* and its auxiliary functions to solve the problem

$$u_t = u_{xx} + u_{yy} + f(x,y,t) \quad (x,y) \in \Omega \equiv (0,1)^2, \quad 0 < t < 2 \qquad (5.76)$$

with properly selected initial condition and Dirichlet boundary condition so that the exact solution is given by

$$u(x,y,t) = 0.8\exp(-80[(x - r(t))^2 + (y - s(t))^2]) \qquad (5.77)$$

where

$$r(t) = \frac{1}{4}(2 + \sin \pi t), \quad s(t) = \frac{1}{4}(2 + \cos \pi t).$$

Note that the solution is a cone that is initially centered at $(\frac{1}{2}, \frac{3}{4})$ and rotates around the center of the domain Ω in a clockwise direction.

4. Consider the wave equation

$$\frac{\partial^2 u}{\partial t^2} - c^2 \frac{\partial^2 u}{\partial x^2} = 0, \quad (x,t) \in (0,1) \times (0,t_F), \qquad (5.78)$$

where $c > 0$ is a constant. Prove that the semi-discrete scheme

$$\frac{1}{c^2}\frac{d^2 u_j}{dt^2} - \frac{4}{3}\frac{u_{j+1} - 2u_j + u_{j-1}}{h^2} + \frac{1}{3}\frac{u_{j+2} - 2u_j + u_{j-2}}{4h^2} = 0, \qquad (5.79)$$

has truncation error $O(h^4)$, where u_j is the approximate solution of $u(x_j, t)$ at points $x_j = jh, 0 \le j \le J, h = \frac{1}{J}$. Then use the Taylor expansion and the governing equation (5.78) to derive the following fourth-order scheme (in both time and space):

$$\frac{u_j^{n+1} - 2u_j^n + u_j^{n-1}}{\tau^2} - \frac{4c^2}{3}\frac{u_{j+1}^n - 2u_j^n + u_{j-1}^n}{h^2} + \frac{c^2}{3}\frac{u_{j+2}^n - 2u_j^n + u_{j-2}^n}{4h^2}$$
$$- \frac{c^4\tau^2}{12}\frac{u_{j+2}^n - 4u_{j+1}^n + 6u_j^n - 4u_{j-1}^n + u_{j-2}^n}{h^4} = 0,$$

where τ is the time step. Hint: Start with the Taylor expansion

$$\frac{u_j^{n+1} - 2u_j^n + u_j^{n-1}}{\tau^2} = \frac{\partial^2 u}{\partial t^2}(x_j, n\tau) + \frac{\tau^2}{12}\frac{\partial^4 u}{\partial t^4}(x_j, n\tau) + O(\tau^4).$$

5. Extend the idea of Exercise 4 to the wave equation in an inhomogeneous medium

$$\frac{\partial^2 u}{\partial t^2} - \frac{\partial}{\partial x}(\mu(x)\frac{\partial u}{\partial x}) = 0$$

by developing a five-point fourth-order scheme in both time and space.

6. Show that the two-stage Runge-Kutta (RK2) method (5.25)-(5.28) can be written as

$$U^{n+1} = U^n + \frac{\Delta t}{2}[R(U^n) + R(U^n + \Delta t R(U^n))].$$

Then use the Taylor expansion to prove that its truncation error is $O(\triangle t^2)$.

7. To approximate the fourth derivatives, we can construct the difference scheme

$$
\begin{aligned}
&\alpha f_{i-1}^{(4)} + f_i^{(4)} + \alpha f_{i+1}^{(4)} \\
&= \frac{a}{h^4}(f_{i+2} - 4f_{i+1} + 6f_i - 4f_{i-1} + f_{i-2}) \\
&\quad + \frac{b}{6h^4}(f_{i+3} - 9f_{i+1} + 16f_i - 9f_{i-1} + f_{i-3}).
\end{aligned}
\tag{5.80}
$$

Prove that a family of fourth-order schemes can be defined by

$$
a = 2(1 - \alpha), \quad b = 4\alpha - 1,
$$

and the truncation error of this scheme is $\frac{7-26\alpha}{240}h^4 f^{(8)}$.

8. Prove that the resolution characteristics of the scheme (5.80) is

$$
\begin{aligned}
\omega_{appr}^{(4)} &= [2a(\cos 2\omega - 4\cos\omega + 3) \\
&\quad + \frac{b}{3}(\cos 3\omega - 9\cos\omega + 8)]/(1 + 2\alpha\cos\omega).
\end{aligned}
$$

9. Derivatives can be evaluated using staggered grids on which the function values are prescribed. For example, we can construct the following scheme

$$
\begin{aligned}
&\alpha f_{i-1}' + f_i' + \alpha f_{i+1}' \\
&= \frac{a}{h}(f_{i+\frac{1}{2}} - f_{i-\frac{1}{2}}) + \frac{b}{3h}(f_{i+\frac{3}{2}} - f_{i-\frac{3}{2}}).
\end{aligned}
$$

Prove that a family of fourth-order schemes can be obtained by

$$
a = \frac{3}{8}(3 - 2\alpha), \quad b = \frac{1}{8}(22\alpha - 1),
$$

and the truncation error of this scheme is $\frac{9-62\alpha}{1920}h^4 f^{(5)}$.

10. Prove that the following combined compact difference schemes

$$
7f_{i-1}' + 16f_i' + 7f_{i+1}' + h(f_{i-1}'' - f_{i+1}'') = \frac{15}{h}(f_{i+1} - f_{i-1})
$$

and

$$
-f_{i-1}'' + 8f_i'' - f_{i+1}'' + \frac{9}{h}(f_{i+1}' - f_{i-1}') = \frac{24}{h^2}(f_{i+1} - 2f_i + f_{i-1})
$$

are sixth-order accurate. Find the leading term for the truncation error of each scheme.

References

[1] Y. Adam. Highly accurate compact implicit methods and boundary conditions. *J. Comp. Phys.*, 24:10–22, 1977.

[2] M.H. Carpenter, D. Gottlieb and S. Abarbanel. Time-stable boundary conditions for finite-difference schemes solving hyperbolic systems: methodology and application to high-order compact system. *J. Comp. Phys.*, 111:220–236, 1994.

[3] M.H. Carpenter, D. Gottlieb and S. Abarbanel. Stable and accurate boundary treatments for compact high order finite difference schemes. *Appl. Numer. Math.*, 12:55–87, 1993.

[4] P.C. Chu and C. Fan. A three-point combined compact difference scheme. *J. Comp. Phys.*, 140:370–399, 1998.

[5] G.C. Cohen. *Higher-Order Numerical Methods for Transient Wave Equations*. Springer-Verlag, Berlin, 2002.

[6] W. Dai and R. Nassar. Compact ADI method for solving parabolic differential equations. *Numer. Methods for Partial Differential Equations*, 18:129–142, 2002.

[7] J. Douglas. Alternating direction methods for three space variables. *Numer. Math.*, 4:41–63, 1962.

[8] J. Douglas and J.E. Gunn. A general formulation of alternating direction methods – Part I. parabolic and hyperbolic problems. *Numer. Math.*, 6:428–453, 1964.

[9] D.V. Gaitonde and M.R. Visbal. High-order schemes for Navier-Stokes equations: algorithms and implementation into FDL3DI. *Technical Report AFRL-VA-WP-TR-1998-3060*, Air Force Research Laboratory, Wright-Patterson AFB, Ohio, 1998.

[10] W.J. Goedheer and J.H.M. Potters. A compact finite difference scheme on a nonequidistance mesh. *J. Comp. Phys.*, 61:269–279, 1985.

[11] B. Gustafsson, H.-O. Kreiss and J. Oliger. *Time Dependent Problems and Difference Methods*. Wiley, New York, 1995.

[12] R. Hixon and E. Turkel. Compact implicit MacCormack-type schemes with high accuracy. *J. Comp. Phys.*, 158:51–70, 2000.

[13] F.Q. Hu, M.Y. Hussaini and J.L. Manthey. Low-dissipation and low-dispersion Runge-Kutta schemes for computational acoustics. *J. Comp. Phys.*, 124:177–191, 1996.

[14] S. Karaa and J. Zhang. High order ADI method for solving unsteady convection-diffusion problems. *J. Comp. Phys.*, 198:1–9, 2004.

[15] S.K. Lele. Compact finite difference schemes with spectral-like solution. *J. Comp. Phys.*, 103:16–42, 1992.

[16] J. Li, M. Chen and M. Chen. Developing and analyzing fourth-order difference methods for the metamaterial Maxwell's equations. *Adv. Comput. Math.*, (2019) 45:213–241, 2019.

[17] J. Li, Y. Chen and G. Liu. High-order compact ADI methods for parabolic equations. *Comp. Math. Appl.*, 52:1343–1356, 2006.

[18] J. Li and M.R. Visbal. High-order compact schemes for nonlinear dispersive waves. *J. Sci. Comput.*, 26(1):1–23, 2006.

[19] W. Liao, J. Zhu and A.Q.M. Khaliq. An efficient high-order algorithm for solving systems of reaction-diffusion equations. *Numer. Methods for Partial Differential Equations*, 18:340–354, 2002.

[20] K. Mahesh. A family of high order finite difference schemes with good spectral resolution. *J. Comp. Phys.*, 145:332–358, 1998.

[21] K. Mattsson and J. Nordstrom. Summation by parts operators for finite difference approximations of second derivatives. *J. Comp. Phys.*, 199:503–540, 2004.

[22] I.M. Navon and H.A. Riphagen. An implicit compact fourth order algorithm for solving the shallow-water equations in conservation-law form. *Monthly Weather Review*, 107:1107–1127, 1979.

[23] I.M. Navon and H.A. Riphagen. SHALL4 – An implicit compact fourth-order Fortran program for solving the shallow-water equations in conservation-law form. *Computers & Geosciences*, 12:129–150, 1986.

[24] D.W. Peaceman and H.H. Rachford. The numerical solution of parabolic and elliptic differential equations. *J. Soc. Indust. Appl. Math.*, 3:28–41, 1955.

[25] J.S. Shang. High-order compact-difference schemes for time-dependent Maxwell equations. *J. Comp. Phys.*, 153:312–333, 1999.

[26] W.F. Spotz and G.F. Carey. Extension of high order compact schemes to time dependent problems. *Numer. Methods for Partial Differential Equations*, 17:657–672, 2001.

[27] C.K.W. Tam and J.C. Webb. Dispersion-relation-preserving finite difference schemes for computational acoustics. *J. Comp. Phys.*, 107:262–281, 1993.

[28] A.I. Tolstykh. *High Accuracy Non-Centered Compact Difference Schemes for Fluid Dynamics Applications*. World Scientific, Hackensack, NJ, 1994.

[29] R. Vichnevetsky and J.B. Bowles. *Fourier Analysis of Numerical Approximations of Hyperbolic Equations*. SIAM, Philadelphia, 1982.

[30] M.R. Visbal and D.V. Gaitonde. High-order-accurate methods for complex unsteady subsonic flows. *AIAA J.*, 37:1231–1239, 1999.

[31] M.R. Visbal and D.V. Gaitonde. Very high-order spatially implicit schemes for computational acoustics on curvilinear meshes. *J. Comput. Acoustics*, 9:1259–1286, 2001.

[32] M.R. Visbal and D.V. Gaitonde. On the use of higher-order finite-difference schemes on curvilinear and deforming meshes. *J. Comp. Phys.*, 181:155–185, 2002.

[33] J. Yan and C.-W. Shu. A local discontinuous Galerkin method for KdV type equations. *SIAM J. Numer. Anal.*, 40:769–791, 2002.

[34] X. Zhong and M. Tatineni. High-order non-uniform grid schemes for numerical simulation of hypersonic boundary-layer stability and transition. *J. Comp. Phys.*, 190:419–458, 2003.

6

Finite Element Methods: Basic Theory

It is known that finite difference methods are not very good at handling irregularly shaped domains. The finite element method can overcome this disadvantage. It is arguably that the most robust and popular method for solving differential equations is the finite element method (FEM), which was first conceived by Courant [19], who used a group of triangular elements to study the St. Venant torsion problem. Then engineers independently re-invented the method in the early 1950s. The early contributions are attributed to Argyris [5], and Turner, Clough, Martin and Topp [44], etc. The term "finite element" was proposed by Clough [18]. Today, FEMs have become the mainstream numerical methods for solving all kinds of PDEs as evidenced by the wide use of many advanced commercial packages.

In this chapter, we will introduce the finite element method and related fundamental theory. In Sec. 6.1 and Sec. 6.2, we illustrate the finite element method through some 1-D and 2-D examples, respectively. Then we discuss some fundamental theory needed for analyzing the finite element method in Sec. 6.3. In Sec. 6.4 and Sec. 6.5, we present some commonly used conforming and nonconforming finite element spaces, respectively. Basic finite element interpolation theory is introduced in Sec. 6.6. Sec. 6.7 is devoted to the error analysis for elliptic problems. Both conforming and nonconforming elements are discussed. Finally, in Sec. 6.8 we discuss the finite element method for parabolic equations. Practical programming of the finite element method will be introduced in the next chapter.

6.1 Introduction to one-dimensional problems

Here we illustrate the basics of using finite element method to solve 1-D problems through two examples.

6.1.1 The second-order equation

As an introduction, we start with a two-point boundary value problem

$$-\frac{d^2u}{dx^2} = f(x), \quad 0 < x < 1, \tag{6.1}$$

$$u(0) = u(1) = 0, \tag{6.2}$$

where f is a given real-valued piecewise continuous bounded function.

Before we introduce FEM, we need to define the notation

$$(v, w) = \int_0^1 v(x)w(x)dx,$$

for real-valued piecewise continuous bounded functions. We also need the linear space

$$V = \{v : v \text{ is a continuous function on } [0,1], \frac{dv}{dx}$$
$$\text{is piecewise continuous and bounded, and } v(0) = v(1) = 0\}. \quad (6.3)$$

Multiplying (6.1) by an arbitrary function $v \in V$, a so-called test function, and integrating over the interval $(0, 1)$, we have

$$-(\frac{d^2u}{dx^2}, v) = (f, v).$$

Then integrating by parts and using the fact that $v(0) = v(1) = 0$, we obtain a variational problem

$$(\frac{du}{dx}, \frac{dv}{dx}) = (f, v) \quad \forall\, v \in V. \quad (6.4)$$

Equation (6.4) is called a weak form of (6.1)-(6.2). On the other hand, if we assume that $\frac{d^2u}{dx^2}$ exists and is piecewise continuous, then we can integrate the left-hand side by parts and use the fact that $v(0) = v(1) = 0$, we have

$$-(\frac{d^2u}{dx^2} + f, v) = 0 \quad \forall\, v \in V,$$

which yields

$$(\frac{d^2u}{dx^2} + f)(x) = 0, \quad 0 < x < 1,$$

under the assumption that $\frac{d^2u}{dx^2}$ and f are piecewise continuous and bounded on $[0, 1]$. Hence the original problem (6.1)-(6.2) is equivalent to the variational problem (6.4) under proper regularity assumptions.

Since the space V is of infinite dimension, we want to approximate it by a finite dimensional subspace V_h. To that end, we divide the interval $[0, 1]$ into subintervals

$$I_j = [x_{j-1}, x_j], \quad 1 \leq j \leq N + 1,$$

with length $h_j = x_j - x_{j-1}$, where N is a positive integer, and

$$0 = x_0 < x_1 < \cdots < x_N < x_{N+1} = 1. \quad (6.5)$$

We denote $h = \max_{1 \leq j \leq N+1} h_j$. Note that the mesh size h is used to measure how fine the partition is.

FIGURE 6.1
The hat function: p_1 basis function in 1-D.

Now we can define the finite element space

$$V_h = \{v : v \text{ is a continuous function on } [0,1],$$
$$v \text{ is linear on each subinterval } I_j, \text{ and } v(0) = v(1) = 0\}. \quad (6.6)$$

Comparing (6.3) and (6.6), we see that $V_h \subset V$, i.e., V_h is indeed a subspace of V. Let us now introduce the linear basis function $\phi_j(x) \in V_h, 1 \le j \le N$, which satisfies the property

$$\phi_j(x_i) = \begin{cases} 1 \text{ if } i = j, \\ 0 \text{ if } i \ne j, \end{cases}$$

i.e., $\phi_j(x)$ is piecewise continuous on $[0,1]$ and its value is one at node j and zero at other nodes. More specifically, $\phi_j(x)$ is given by

$$\phi_j(x) = \begin{cases} \frac{x - x_{j-1}}{h_j}, & \text{if } x \in [x_{j-1}, x_j], \\ \frac{x_{j+1} - x}{h_{j+1}}, & \text{if } x \in [x_j, x_{j+1}], \\ 0 & \text{elsewhere.} \end{cases} \quad (6.7)$$

Since the shape of ϕ_j looks like a hat, $\phi_j(x)$ is often called a hat function (see Figure 6.1).

Note that any function $v \in V_h$ has a unique representation

$$v(x) = \sum_{j=1}^{N} v_j \phi_j(x), \quad x \in [0,1], \quad (6.8)$$

where $v_i = v(x_i)$, i.e., V_h is a linear space of dimension N with basis function $\{\phi_j\}_{j=1}^{N}$.

With all the above preparations, we can now formulate the finite element method for the problem (6.1)-(6.2) as follows: Find $u_h \in V_h$ such that

$$(\frac{du_h}{dx}, \frac{dv}{dx}) = (f, v) \quad \forall\, v \in V_h. \quad (6.9)$$

From the unique representation (6.8), we can set the finite element solution u_h of (6.9) as

$$u_h(x) = \sum_{j=1}^{N} u_j \phi_j(x), \quad u_j = u_h(x_j).$$

Substituting $u_h(x)$ into (6.9) and choosing $v = \phi_i(x)$ in (6.9) for each i, we obtain

$$\sum_{j=1}^{N} (\frac{d\phi_j}{dx}, \frac{d\phi_i}{dx}) u_j = (f, \phi_i) \quad 1 \le i \le N, \tag{6.10}$$

which is a linear system of N equations in N unknowns u_j. The system can be written in matrix form

$$Au = F, \tag{6.11}$$

where $A = (a_{ij})$ is an $N \times N$ matrix with elements $a_{ij} = (\frac{d\phi_j}{dx}, \frac{d\phi_i}{dx})$, $u = (u_1, \cdots, u_N)^T$ is an N-dimensional vector, and $F = (F_1, \cdots, F_N)^T$ is an N-dimensional vector with elements $F_i = (f, \phi_i)$.

The matrix A is called the stiffness matrix and F the load vector. From (6.7), we can explicitly calculate the matrix A. From (6.7), we easily have

$$(\frac{d\phi_j}{dx}, \frac{d\phi_j}{dx}) = \int_{x_{j-1}}^{x_j} \frac{1}{h_j^2} dx + \int_{x_j}^{x_{j+1}} \frac{1}{h_{j+1}^2} dx = \frac{1}{h_j} + \frac{1}{h_{j+1}}, \quad 1 \le j \le N,$$

$$(\frac{d\phi_j}{dx}, \frac{d\phi_{j-1}}{dx}) = (\frac{d\phi_{j-1}}{dx}, \frac{d\phi_j}{dx}) = \int_{x_{j-1}}^{x_j} \frac{-1}{h_j^2} dx = -\frac{1}{h_j}, \quad 2 \le j \le N,$$

$$(\frac{d\phi_j}{dx}, \frac{d\phi_i}{dx}) = 0 \quad \text{if } |j - i| > 1.$$

Thus the matrix A is tri-diagonal, i.e., only elements in the main diagonal and the two adjacent diagonals to the main diagonal are non-zero.

Note that

$$\sum_{i,j=1}^{N} v_j (\frac{d\phi_j}{dx}, \frac{d\phi_i}{dx}) v_i = (\sum_{j=1}^{N} v_j \frac{d\phi_j}{dx}, \sum_{i=1}^{N} v_i \frac{d\phi_i}{dx}) \ge 0,$$

and the equality holds only if $\frac{dv}{dx} \equiv 0$, where we denote $v(x) = \sum_{j=1}^{N} v_j \phi_j(x)$. Noting that $v(0) = 0$, hence $\frac{dv}{dx} \equiv 0$ is equivalent to $v(x) \equiv 0$, or $v_j = 0$ for all $j = 1, \cdots, N$. Therefore, the matrix A is symmetric and positive definite, which guarantees that A is nonsingular, i.e., the linear system (6.11) has a unique solution.

6.1.2 The fourth-order equation

In this subsection, we consider a fourth-order problem

$$\frac{d^4 u}{dx^4} = f(x), \quad 0 < x < 1, \tag{6.12}$$

$$u(0) = u(1) = \frac{du}{dx}(0) = \frac{du}{dx}(1) = 0, \tag{6.13}$$

where f is a given real-valued piecewise continuous bounded function. For this problem, we need the linear space

$$V = \{v : v \text{ and } \frac{dv}{dx} \text{ are continuous on } [0,1], \frac{d^2v}{dx^2} \text{ is piecewise continuous,}$$

$$\text{and } v(0) = v(1) = \frac{dv}{dx}(0) = \frac{dv}{dx}(1) = 0\}. \tag{6.14}$$

Multiplying (6.12) by a test function $v \in V$, integrating over the interval $(0,1)$ and using the boundary condition (6.13), we can obtain

$$(\frac{d^2u}{dx^2}, \frac{d^2v}{dx^2}) = (f, v) \quad \forall v \in V. \tag{6.15}$$

Similar to the previous subsection, it can be proved that (6.12)-(6.13) is equivalent to the weak form (6.15).

Using the same partition (6.5) as in the previous subsection, we can define the finite element space

$$V_h = \{v : v \text{ and } \frac{dv}{dx} \text{ are continuous on } [0,1],$$

$$v \text{ is a polynomial of degree 3 on each subinterval } I_j,$$

$$\text{and } v(0) = v(1) = \frac{dv}{dx}(0) = \frac{dv}{dx}(1) = 0\}, \tag{6.16}$$

which is a subspace of V, i.e., we have $V_h \subset V$.

On each subinterval $I_j = [x_{j-1}, x_j]$, any polynomial of degree 3 (i.e., $v(x) \in P_3(I_j)$) can be uniquely determined by the values

$$v(x_{j-1}), v(x_j), \frac{dv}{dx}(x_{j-1}), \frac{dv}{dx}(x_j).$$

The basis function corresponding to the value $v(x_{j-1})$ is a cubic polynomial v such that

$$v(x_{j-1}) = 1, v(x_j) = \frac{dv}{dx}(x_{j-1}) = \frac{dv}{dx}(x_j) = 0. \tag{6.17}$$

It is easy to see that this basis function can be represented as

$$\phi_{j-1}(x) = (x - x_j)^2(ax + b). \tag{6.18}$$

Letting $\phi_{j-1}(x_{j-1}) = 1$ and $\frac{d\phi_{j-1}}{dx}(x_{j-1}) = 0$, we obtain

$$h_j^2(ax_{j-1} + b) = 1,$$

$$-2h_j(ax_{j-1} + b) + ah_j^2 = 0,$$

which lead to $a = \frac{2}{h_j^3}, b = \frac{1}{h_j^3}(h_j - 2x_{j-1})$. Substituting a and b into (6.18), we obtain the basis function

$$\phi_{j-1}(x) = \frac{1}{h_j^3}(x - x_j)^2[h_j + 2(x - x_{j-1})], \quad x_{j-1} \le x \le x_j. \tag{6.19}$$

Similarly, the basis function corresponding to the value $v(x_j)$ must satisfy

$$v(x_j) = 1, v(x_{j-1}) = \frac{dv}{dx}(x_{j-1}) = \frac{dv}{dx}(x_j) = 0. \tag{6.20}$$

Skipping the details, we can derive the basis function as

$$\phi_j(x) = \frac{1}{h_j^3}(x - x_{j-1})^2[h_j - 2(x - x_j)], \quad x_{j-1} \le x \le x_j. \tag{6.21}$$

In the same way, the basis function corresponding to the value $\frac{dv}{dx}(x_{j-1})$ must satisfy

$$\frac{dv}{dx}(x_{j-1}) = 1, v(x_{j-1}) = v(x_j) = \frac{dv}{dx}(x_j) = 0.$$

It can be checked that this basis function is given by

$$\psi_{j-1}(x) = \frac{1}{h_j^2}(x - x_j)^2(x - x_{j-1}). \tag{6.22}$$

Finally, the basis function corresponding to the value $\frac{dv}{dx}(x_j)$ must satisfy

$$\frac{dv}{dx}(x_j) = 1, v(x_{j-1}) = v(x_j) = \frac{dv}{dx}(x_{j-1}) = 0.$$

This basis function can be derived similarly as above and is given by

$$\psi_j(x) = \frac{1}{h_j^2}(x - x_{j-1})^2(x - x_j), \quad x_{j-1} \le x \le x_j. \tag{6.23}$$

From the definition (6.16), we know that any $v \in V_h$ can be uniquely represented as

$$v(x) = \sum_{j=1}^{N}[v_j\phi_j(x) + v_j'\psi_j(x)], \quad v_j = v(x_j), v_j'(x) = \frac{dv}{dx}(x_j). \tag{6.24}$$

Now we can formulate the finite element method for the problem (6.12)-(6.13): Find $u_h \in V_h$ such that

$$(\frac{d^2u_h}{dx^2}, \frac{d^2v}{dx^2}) = (f, v) \quad \forall\, v \in V_h. \tag{6.25}$$

Substituting the finite element solution u_h of (6.25)

$$u_h(x) = \sum_{j=1}^{N} [u_j \phi_j(x) + u'_j \psi_j(x)], \tag{6.26}$$

into (6.25) and choosing $v = \phi_i(x)$ and $\psi_i(x)$, respectively, in (6.25), we obtain

$$\sum_{j=1}^{N} (\frac{d^2\phi_j}{dx^2}, \frac{d^2\phi_i}{dx^2}) u_j + \sum_{j=1}^{N} (\frac{d^2\psi_j}{dx^2}, \frac{d^2\phi_i}{dx^2}) u'_j = (f, \phi_i), \quad 1 \le i \le N, \tag{6.27}$$

$$\sum_{j=1}^{N} (\frac{d^2\phi_j}{dx^2}, \frac{d^2\psi_i}{dx^2}) u_j + \sum_{j=1}^{N} (\frac{d^2\psi_j}{dx^2}, \frac{d^2\psi_i}{dx^2}) u'_j = (f, \psi_i), \quad 1 \le i \le N, \tag{6.28}$$

which is a linear system of $2N$ equations in unknowns u_j and u'_j.

The above system can be written in matrix form

$$\begin{pmatrix} A & B \\ B^T & C \end{pmatrix} \begin{pmatrix} u \\ u' \end{pmatrix} = \begin{pmatrix} F_\phi \\ F_\psi \end{pmatrix}, \tag{6.29}$$

where $A = (a_{ij})$ is an $N \times N$ matrix with element $a_{ij} = (\frac{d^2\phi_j}{dx^2}, \frac{d^2\phi_i}{dx^2})$, $B = (b_{ij})$ is an $N \times N$ matrix with element $b_{ij} = (\frac{d^2\psi_j}{dx^2}, \frac{d^2\phi_i}{dx^2})$, $C = (c_{ij})$ is an $N \times N$ matrix with element $c_{ij} = (\frac{d^2\psi_j}{dx^2}, \frac{d^2\psi_i}{dx^2})$, B^T is the transpose of B. Furthermore, we denote the N-dimensional vectors

$$u = (u_1, \cdots, u_N)^T, \quad u' = (u'_1, \cdots, u'_N)^T,$$

and

$$F_\phi = ((f, \phi_1), \cdots, (f, \phi_N))^T, \quad F_\psi = ((f, \psi_1), \cdots, (f, \psi_N))^T.$$

The coefficient matrix of (6.29) can be proved to be symmetric and positive definite, which guarantees that the solution of our finite element method (6.25) exists and is unique. To prove the positive definiteness, multiplying the left-hand side of (6.27) by u_i and sum up i, multiplying the left-hand side of (6.28) by u'_i and sum up i, and adding them together, we have

$$(u, u') \cdot \begin{pmatrix} A & B \\ B^T & C \end{pmatrix} \begin{pmatrix} u \\ u' \end{pmatrix} = (\frac{d^2 u_h(x)}{dx^2}, \frac{d^2 u_h(x)}{dx^2}) \ge 0,$$

with equality only if $\frac{d^2 u_h(x)}{dx^2} \equiv 0$. Considering the boundary conditions $\frac{du_h}{dx}(0) = \frac{du_h}{dx}(1) = 0$, we see that

$$\frac{du_h(x)}{dx} \equiv 0,$$

which, along with the boundary conditions $u_h(0) = u_h(1) = 0$, yields $u_h(x) \equiv 0$. Hence, the positive definiteness of matrix $\begin{pmatrix} A & B \\ B^T & C \end{pmatrix}$ is proved.

6.2   Introduction to two-dimensional problems

Here we use two examples to illustrate the basic techniques for developing the finite element method for 2-D PDEs.

6.2.1   The Poisson equation

In this subsection, we consider the Poisson equation:

$$-\Delta u = f(x_1, x_2) \quad \text{in} \ \Omega, \tag{6.30}$$

$$u = 0 \quad \text{on} \ \partial\Omega, \tag{6.31}$$

where Ω is a bounded domain in the plane with boundary $\partial\Omega$, f is a given real-values piecewise continuous bounded function in Ω, and the Laplacian operator Δ is defined as

$$\Delta u = \frac{\partial^2 u}{\partial x_1^2} + \frac{\partial^2 u}{\partial x_2^2}.$$

Recall that in two dimensions, we have the divergence theorem

$$\int_\Omega \nabla \cdot b\, dx = \int_{\partial\Omega} b \cdot n\, ds, \tag{6.32}$$

where $b = (b_1, b_2)$ is a vector-valued function defined in Ω, the divergence operator

$$\nabla \cdot b = \frac{\partial b_1}{\partial x_1} + \frac{\partial b_2}{\partial x_2},$$

and $n = (n_1, n_2)$ is the outward unit normal to $\partial\Omega$. Here dx denotes the element of area, and ds the element of arc length along $\partial\Omega$.

If we apply the divergence theorem to $b = (w\frac{\partial v}{\partial x_1}, 0)$ and $b = (0, w\frac{\partial v}{\partial x_2})$, respectively, we find that

$$\int_\Omega (w\frac{\partial^2 v}{\partial x_1^2} + \frac{\partial v}{\partial x_1}\frac{\partial w}{\partial x_1})dx = \int_{\partial\Omega} w\frac{\partial v}{\partial x_1} n_1 ds, \tag{6.33}$$

and

$$\int_\Omega (w\frac{\partial^2 v}{\partial x_2^2} + \frac{\partial v}{\partial x_2}\frac{\partial w}{\partial x_2})dx = \int_{\partial\Omega} w\frac{\partial v}{\partial x_2} n_2 ds. \tag{6.34}$$

Denote by ∇v the gradient of v, i.e., $\nabla v = (\frac{\partial v}{\partial x_1}, \frac{\partial v}{\partial x_2})$. Adding (6.33) and (6.34) together, we obtain the following Green's formula:

$$\int_\Omega (w\Delta v + \nabla v \cdot \nabla w)dx = \int_{\partial\Omega} w \cdot \frac{\partial v}{\partial n} ds, \tag{6.35}$$

where we denote by $\frac{\partial v}{\partial n}$ the normal derivative

$$\frac{\partial v}{\partial n} = \frac{\partial v}{\partial x_1} n_1 + \frac{\partial v}{\partial x_2} n_2.$$

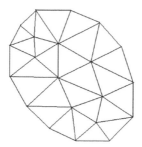

FIGURE 6.2

An exemplary finite element triangulation.

Before continuing, we introduce the linear space

$$V = \{v : v \text{ is continuous in } \Omega, \; \frac{\partial v}{\partial x_1} \text{ and } \frac{\partial v}{\partial x_2}$$
$$\text{are piecewise continuous in } \Omega, \text{ and } v = 0 \text{ on } \partial \Omega\}. \tag{6.36}$$

Multiplying (6.30) by a test function $v \in V$, integrating over Ω, and using the Green's formula (6.35) and the homogeneous boundary condition, we obtain the variational formulation of problem (6.30)-(6.31): Find $u \in V$ such that

$$a(u, v) = (f, v) \quad \forall \, v \in V, \tag{6.37}$$

where

$$a(u, v) = \int_{\Omega} \nabla u \cdot \nabla v dx, \quad (f, v) = \int_{\Omega} f v dx.$$

As in the 1-D case, we can prove that variational problem (6.37) is equivalent to the original problem (6.30)-(6.31).

We now construct a finite dimensional subspace V_h of V. For simplicity, we assume that Ω is a polygonal domain (i.e., $\partial \Omega$ is a polygonal curve). Let us now make a partition of Ω, called a triangulation, by subdividing Ω into a set T_h of non-overlapping triangles K_i, i.e.,

$$\Omega = \cup_{K \in T_h} K = K_1 \cup K_2 \cup \cdots \cup K_m,$$

such that no vertex of one triangle lies in the interior of the edge of another triangle (cf. Fig. 6.2).

We introduce the mesh notation

$$h = \max_{K \in T_h} \text{diam}(K), \quad \text{diam}(K) = \text{longest side of } K,$$

and the finite element space

$$V_h = \{v : v \text{ is continuous in } \Omega, v$$
$$\text{is linear on each triangle } K \in T_h, \text{ and } v = 0 \text{ on } \partial\Omega\}, \quad (6.38)$$

from which and (6.36), we see that $V_h \subset V$.

Assume that T_h contains N interior vertices $\boldsymbol{x}_i, 1 \le i \le N$; then we can define a basis function $\phi_j \in V_h$ for each interior vertex satisfying

$$\phi_j(\boldsymbol{x}_i) = \delta_{ij} \equiv \begin{cases} 1 \text{ if } i = j, \\ 0 \text{ if } i \ne j. \end{cases}$$

Notice that the support of ϕ_j (the set of point $\boldsymbol{x}$ such that $\phi_j(\boldsymbol{x}) \ne 0$) consists of the triangles with the common vertex $\boldsymbol{x}_j$. Furthermore, any function $v \in V_h$ has the unique representation

$$v(\boldsymbol{x}) = \sum_{j=1}^{N} v_j \phi_j(\boldsymbol{x}), \quad v_j = v(\boldsymbol{x}_j). \quad (6.39)$$

With all the above preparations, now we can formulate the finite element method for (6.30)-(6.31): Find $u_h \in V_h$ such that

$$a(u_h, v) = (f, v) \quad \forall\, v \in V_h. \quad (6.40)$$

Exactly as in the 1-D case, we can show that (6.40) is equivalent to a linear system
$$Au = F,$$
where $A = (a_{ij})$ is an $N \times N$ matrix with element $a_{ij} = (\nabla\phi_j, \nabla\phi_i)$, and $F = (F_i)$ is an N-dimensional vector with element $F_i = (F, \phi_i)$. Furthermore, the matrix A is symmetric positive definite so that the solution u_h of (6.40) exists and is unique.

6.2.2   The biharmonic problem

Let us now consider a fourth-order problem in 2-D, namely the biharmonic problem

$$\triangle^2 u = f(x_1, x_2) \quad \text{in } \Omega, \quad (6.41)$$
$$u = \frac{\partial u}{\partial n} = 0 \quad \text{on } \partial\Omega, \quad (6.42)$$

where $\frac{\partial}{\partial n}$ denotes the normal derivative. For this problem, we introduce the linear space

$$V = \{v : v, \frac{\partial v}{\partial x_1}, \frac{\partial v}{\partial x_2} \text{ are continuous on } \Omega, \frac{\partial^2 v}{\partial x_1^2}, \frac{\partial^2 v}{\partial x_2^2}, \frac{\partial^2 v}{\partial x_1 \partial x_2}$$

$$\text{are piecewise continuous, and } v = \frac{\partial v}{\partial n} = 0 \text{ on } \partial\Omega\}. \qquad (6.43)$$

Multiplying (6.41) by a test function $v \in V$, integrating over Ω, and using Green's formula (6.35) twice with $v = \frac{\partial v}{\partial n} =)$ on $\partial\Omega$, we obtain

$$\int_\Omega f v dx = \int_\Omega \triangle\triangle u \cdot v dx$$

$$= \int_{\partial\Omega} v \cdot \frac{\partial}{\partial n}(\triangle u) ds - \int_\Omega \nabla\triangle u \cdot \nabla v dx$$

$$= -\int_\Omega \nabla\triangle u \cdot \nabla v dx = -\int_{\partial\Omega} \triangle u \cdot \frac{\partial v}{\partial n} ds + \int_\Omega \triangle u \cdot \triangle v dx,$$

which leads to the variational problem of the biharmonic problem (6.41)-(6.42): Find $u \in V$ such that

$$a(u, v) = (f, v) \quad \forall v \in V, \qquad (6.44)$$

where

$$a(u, v) = \int_\Omega \triangle u \cdot \triangle v dx, \quad (f, v) = \int_\Omega f v dx.$$

Let T_h be a triangulation of Ω into triangles as in the last subsection. Now we can define the finite element space

$$V_h = \{v : v, \nabla v \text{ are continuous on } \Omega, v \text{ is a polynomial}$$

$$\text{of degree 5 on each triangle, and } v = \frac{\partial v}{\partial n} = 0 \text{ on } \partial\Omega\}. \quad (6.45)$$

The elements in this space are known as the Argyris triangles, and it is true that $V_h \subset V$. We will elaborate on the Argyris triangle in late section.

The finite element method for the biharmonic problem (6.41)-(6.42) can be formulated as: Find $u_h \in V_h$ such that

$$a(u_h, v) = (f, v) \quad \forall v \in V_h. \qquad (6.46)$$

The existence and uniqueness of the solution for (6.46) can be proved similar to the 1-D case.

6.3 Abstract finite element theory

In this section, we first present the general results for proving the existence and uniqueness of a solution to an abstract variational problem. Then we present two fundamental results used for proving the stability and error estimate of the finite element method.

6.3.1   Existence and uniqueness

Consider the abstract variational problem: Find $u \in V$ such that

$$A(u, v) = F(v) \quad \forall \, v \in V, \tag{6.47}$$

where V denotes a real Hilbert space and $F \in V'$. The following lemma provides the existence and uniqueness theory for this problem. Detailed proof can be found in many books (e.g., [17, p. 8]).

THEOREM 6.1

(Lax-Milgram lemma) Let V be a real Hilbert space with norm $||\cdot||_V$, $A(\cdot, \cdot) : V \times V \to R$ a bilinear form, and $F(\cdot) : V \to R$ a linear continuous functional. Furthermore, suppose that $A(\cdot, \cdot)$ is bounded, i.e.,

$$\exists \beta > 0: \quad |A(w, v)| \leq \beta ||w||_V ||v||_V \quad \text{for all } w, v \in V,$$

and coercive, i.e.,

$$\exists \alpha > 0: \quad |A(v, v)| \geq \alpha ||v||_V^2 \quad \text{for all } v \in V.$$

Then, there exists a unique solution $u \in V$ to (6.47) and

$$||u||_V \leq \frac{1}{\alpha} ||F||_{V'},$$

where V' denotes the dual space of V.

Similar results hold true for a more general problem: Find $u \in W$ such that

$$A(u, v) = F(v) \quad \forall \, v \in V, \tag{6.48}$$

where W can be different from V.

THEOREM 6.2

(Generalized Lax-Milgram lemma) Let W and V be two real Hilbert spaces with norms $||\cdot||_W$ and $||\cdot||_V$ respectively, $A(\cdot, \cdot) : W \times V \to R$ a bilinear form, and $F(\cdot) : V \to R$ a linear continuous functional. Furthermore, assume that

(i) $\exists \beta > 0: \quad |A(w, v)| \leq \beta ||w||_W ||v||_V \quad \text{for all } w \in W, v \in V,$

(ii) $\exists \alpha > 0: \quad \inf\limits_{w \in W, ||w||_W = 1} \sup\limits_{v \in V, ||v||_V \leq 1} |A(w, v)| \geq \alpha,$

(iii) $\sup\limits_{w \in W} |A(w, v)| > 0 \quad \text{for every } 0 \neq v \in V.$

Then, there exists a unique solution $u \in W$ to (6.48) and

$$||u||_W \leq \frac{1}{\alpha} ||F||_{V'}.$$

For a detailed proof of this lemma, readers can consult Babuska and Aziz's classic paper [6].

6.3.2 Stability and convergence

Now we assume that a family of finite dimensional subspaces V_h is constructed to approximate the infinite dimensional space V, i.e.,

$$\inf_{v_h \in V_h} ||v - v_h||_V \to 0 \quad \text{as} \quad h \to 0, \quad \text{for all } v \in V.$$

The Galerkin FEM for solving (6.47) is: Find $u_h \in V_h$ such that

$$A(u_h, v_h) = F(v_h) \quad \forall\, v_h \in V_h. \tag{6.49}$$

We have the following proven stability and convergence theory.

THEOREM 6.3
(Céa lemma) Under the assumption of Theorem 6.1, there exists a unique solution u_h to (6.49) and

$$||u_h||_V \le \frac{1}{\alpha} ||F||_{V'}.$$

Furthermore, if u denotes the solution to (6.47), then

$$||u - u_h||_V \le \frac{\beta}{\alpha} \inf_{v_h \in V_h} ||u - v_h||_V,$$

i.e., u_h converges to u as $h \to 0$.

Proof. From (6.47) and (6.49), we obtain

$$A(u - u_h, v_h) = 0 \quad \forall\, v_h \in V_h, \tag{6.50}$$

which, along with the coercivity and continuity of $A(\cdot, \cdot)$ yields

$$\alpha ||u - u_h||_V^2 \le A(u - u_h, u - u_h) = A(u - u_h, u - v_h) \le \beta ||u - u_h||_V ||u - v_h||_V,$$

which concludes the proof. $\square$

A more general technique than the Galerkin FEM is the so-called Petrov-Galerkin method for (6.47): Find $u_h \in W_h$ such that

$$A_h(u_h, v_h) = F_h(v_h) \quad \forall\, v_h \in V_h, \tag{6.51}$$

where W_h and V_h are two families of finite dimensional subspaces of W and V, respectively. In general, $W_h \ne V_h$ but $\dim W_h = \dim V_h$. Here $A_h : W_h \times V_h \to R$ and $F_h : V_h \to R$ are some approximations to A and F, respectively. For problem (6.51), we have the following result (e.g., [6]).

THEOREM 6.4
Under the assumptions of Theorem 6.2 and the same properties as Theorem 6.2 for the discrete functional F_h and bilinear form A_h with W replaced by

W_h, V replaced by V_h, and the constants β and α replaced by β_h and α_h, respectively. Then, there exists a unique solution u_h to (6.51) and

$$||u_h||_W \leq \frac{1}{\alpha_h} \sup_{v_h \in V_h, v_h \neq 0} \frac{|F_h(v_h)|}{||v_h||_V}.$$

Furthermore, if u denotes the solution to (6.49), then

$$||u - u_h||_W$$
$$\leq \frac{1}{\alpha_h} \sup_{v_h \in V_h, v_h \neq 0} \frac{|F(v_h) - F_h(v_h)|}{||v_h||_V}$$
$$+ \inf_{w_h \in W_h} [(1 + \frac{\beta_h}{\alpha_h})||u - w_h||_W + \frac{1}{\alpha_h} \sup_{v_h \in V_h, v_h \neq 0} \frac{|A(u, v_h) - A_h(u_h, v_h)|}{||v_h||_V}].$$

Proof. Let a function $w_h \in W_h$. By the coercivity and continuity of A_h, we have

$$||u - u_h||_W$$
$$\leq ||u - w_h||_W + ||w_h - u_h||_W$$
$$\leq ||u - w_h||_W + \frac{1}{\alpha_h} \sup_{v_h \in V_h, v_h \neq 0} \frac{|A_h(w_h - u_h, v_h)|}{||v_h||_V}$$
$$\leq ||u - w_h||_W + \frac{1}{\alpha_h} \sup_{v_h \in V_h, v_h \neq 0} \frac{|A_h(w_h - u, v_h) + A_h(u - u_h, v_h)|}{||v_h||_V}$$
$$\leq ||u - w_h||_W + \frac{\beta_h}{\alpha_h}||w_h - u||_W + \frac{1}{\alpha_h} \sup_{v_h \in V_h, v_h \neq 0} \frac{|A_h(u - u_h, v_h)|}{||v_h||_V} \quad (6.52)$$

Using the definitions of A and A_h, we obtain

$$A_h(u - u_h, v_h) = A_h(u, v_h) - A_h(u_h, v_h)$$
$$= A_h(u, v_h) - A(u, v_h) + A(u, v_h) - A_h(u_h, v_h)$$
$$= F_h(v_h) - F(v_h) + A(u, v_h) - A_h(u_h, v_h),$$

substituting which into (6.52), we complete the proof. □

6.4   Examples of conforming finite element spaces

For simplicity, we assume that the physical domain $\Omega \subset R^d (d = 2, 3)$ is a polygonal domain, which is triangulated by d-simplex (i.e., triangle in 2-D, or tetrahedron in 3-D) or d-rectangles, i.e., $\overline{\Omega} = \cup_{K \in T_h} K$.

Denote P_k $(k \geq 0)$ as the space of polynomials of degree less than or equal to k in variables $x_1, \cdots, x_d$, and Q_k as the space of polynomials of degree less than or equal to k in each variable $x_1, \cdots, x_d$, i.e.,

$$P_k(\boldsymbol{x}) = \sum_{0 \leq \alpha_i, \alpha_1 + \cdots + \alpha_d \leq k} c_{\alpha_1 \alpha_2 \cdots \alpha_d} x_1^{\alpha_1} x_2^{\alpha_2} \cdots x_d^{\alpha_d}$$

and

$$Q_k(\boldsymbol{x}) = \sum_{0 \leq \alpha_1, \cdots, \alpha_d \leq k} c_{\alpha_1 \alpha_2 \cdots \alpha_d} x_1^{\alpha_1} x_2^{\alpha_2} \cdots x_d^{\alpha_d}$$

for proper coefficients $c_{\alpha_1 \alpha_2 \cdots \alpha_d}$. Furthermore, the dimensions of the spaces P_k and Q_k are given by

$$\dim P_k = \frac{(k+d) \cdots (k+1)}{d!} = C_{k+d}^k, \quad \dim Q_k = (k+1)^d.$$

According to Ciarlet [17], a finite element space V_h is composed of three parts $(K, P_K, \sum_K)$, where K defines the element domain/type (e.g., a triangle or tetrahedron), P_K defines the shape functions on each finite element $K \in \Omega, \sum_K$ denotes the set of degrees of freedom (i.e., the parameters which uniquely define the function P_K on each element K). Below we will discuss some widely used finite element spaces.

6.4.1 Triangular finite elements

Let us first look at the triangle element. Consider a triangle K with three vertices $a_i(x_i, y_i), i = 1, 2, 3$, ordered in a counterclockwise direction. Let the local basis function of $P_1(K)$ at each vertex a_i be λ_i, which are defined by

$$\lambda_i(a_j) = \delta_{ij}, \quad 1 \leq i, j \leq 3.$$

It is not difficult to check that

$$\lambda_i = \frac{1}{2A}(\alpha_i + \beta_i x + \gamma_i y), \quad i = 1, 2, 3,$$

where A is the area of the triangle, i.e.,

$$A = \frac{1}{2} \begin{vmatrix} 1 & x_1 & y_1 \\ 1 & x_2 & y_2 \\ 1 & x_3 & y_3 \end{vmatrix}.$$

Furthermore, the constants α_i, β_i and γ_i are defined as

$$\alpha_i = x_j y_k - x_k y_j, \quad \beta_i = y_j - y_k, \quad \gamma_i = -(x_j - x_k),$$

where $i \neq j \neq k$, and i, j and k permute naturally.

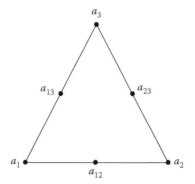

FIGURE 6.3
The P_2 element.

Note that λ_i are often called the barycentric coordinates of the triangle. Now any function $v \in P_1(K)$ can be uniquely represented by

$$v(x, y) = \sum_{i=1}^{3} v(a_i)\lambda_i \quad \forall \ (x, y) \in K.$$

Hence the degrees of freedom $\sum_K$ consist of the function values at the vertices. It is easy to see that for any v such that $v|_K \in P_1(K), K \in T_h$, if it is continuous at all internal vertices, then $v \in C^0(\overline{\Omega})$. Therefore, we can define the continuous piecewise linear finite element space

$$V_h = \{v \in C^0(\overline{\Omega}) : v|_K \in P_1(K), \ \forall \ K \in T_h\}.$$

Similarly, we can define the continuous piecewise quadratic finite element space

$$V_h = \{v \in C^0(\overline{\Omega}) : v|_K \in P_2(K), \ \forall \ K \in T_h\},$$

where the degrees of freedom on each triangle K consist of the values of v at the vertices a_i and the midpoints a_{ij} of edges (cf. Fig. 6.3).

LEMMA 6.1
A function $v \in P_2(K)$ is uniquely determined by the values of $v(a_i)$, $1 \leq i \leq 3$, and $v(a_{ij})$, $1 \leq i < j \leq 3$.

Proof. Since the number of degrees of freedom is equal to the dimension of $P_2(K)$, we have only to show that if $v(a_i) = v(a_{ij}) = 0$, then $v \equiv 0$. Note that the restriction of $v \in P_2(K)$ to edge a_2a_3 is a quadratic function of one variable vanishing at three distinct points, hence $v(\boldsymbol{x})$ must be zero over that edge, i.e., $v(\boldsymbol{x})$ should include a factor $\lambda_1(\boldsymbol{x})$.

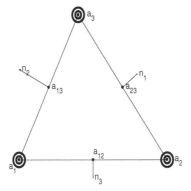

FIGURE 6.4
The Argyris triangle.

By the same argument, v must be zero over edges a_1a_2 and a_1a_3, i.e., $v(\boldsymbol{x})$ also contains factors $\lambda_2(\boldsymbol{x})$ and $\lambda_3(\boldsymbol{x})$. Hence we can write

$$v(\boldsymbol{x}) = c\lambda_1(\boldsymbol{x})\lambda_2(\boldsymbol{x})\lambda_3(\boldsymbol{x}).$$

But $v(\boldsymbol{x}) \in P_2$, it follows that $c = 0$, which concludes the proof. $\square$

Let us remark that this choice of degrees of freedom guarantees that $v \in C^0(\overline{\Omega})$ since each side of the element uniquely identifies the restriction of v on that side. More specifically, any $v \in P_2(K)$ has the representation [17, p. 47]

$$v(\boldsymbol{x}) = \sum_{i=1}^{3} v(a_i)\lambda_i(2\lambda_i - 1) + \sum_{i<j,i,j=1}^{3} 4v(a_{ij})\lambda_i\lambda_j.$$

Similarly, continuous higher-order finite element spaces on triangular element can be defined. More examples can be found in [17].

For a fourth-order problem, the conforming finite elements are C^1 elements (i.e., we need continuity of derivatives). Constructions of such elements are quite complicated, one example is the so-called Argyris triangle, which is a fifth-degree polynomial involving 21 degrees of freedom on a triangular element. The Argyris triangle finite element space is given as

$$V_h = \{v \in C^1(\overline{\Omega}) : v|_K \in P_5(K), \ \forall \ K \in T_h\},$$

where the degrees of freedom on each triangle K consist of the values of v at the vertices a_i, the first and second derivatives at the vertices a_i, and the normal derivatives at the midpoints a_{ij} of edges (cf. Fig. 6.4).

LEMMA 6.2

A function $v \in P_5(K)$ is uniquely determined by the following values:

$$v(a_i),\ \frac{\partial v}{\partial x}(a_i),\ \frac{\partial v}{\partial y}(a_i),\ \frac{\partial^2 v}{\partial x^2}(a_i),\ \frac{\partial^2 v}{\partial x \partial y}(a_i),\ \frac{\partial^2 v}{\partial y^2}(a_i),\ i = 1, 2, 3,$$

$$\frac{\partial v}{\partial n_1}(a_{23}),\ \frac{\partial v}{\partial n_2}(a_{13}),\ \frac{\partial v}{\partial n_3}(a_{12}),$$

where n_i denote the outward unit normal vectors shown in Fig. 6.4.

Proof. Since this is a finite dimensional problem, the uniqueness is equivalent to the existence of zero solution when all the degrees of freedom are zero.

First, note that the restriction of $v \in P_5(K)$ to line $a_2 a_3$ is a fifth-order polynomial in one variable with triple zeros at a_2 and a_3. Hence $v = 0$ on this line. Similarly, we can prove that $v = 0$ on lines $a_1 a_3$ and $a_1 a_2$. Therefore, we can write $v = Q \cdot \lambda_1 \lambda_2 \lambda_3$, where λ_i are the standard barycentric coordinates, and Q is a polynomial of degree two.

Recall that the directional derivative of a function f at (x_0, y_0) in the direction of a unit vector $\boldsymbol{u} = (u_1, u_2)$ is defined as follows:

$$\partial_{\boldsymbol{u}} f(x_0, y_0) = \lim_{h \to 0} \frac{f(x_0 + u_1 h, y_0 + u_2 h) - f(x_0, y_0)}{h},$$

which leads to

$$\partial_{\boldsymbol{u}} f(x_0, y_0) = u_1 \partial_x f(x_0, y_0) + u_2 \partial_y f(x_0, y_0) = \boldsymbol{u} \cdot \nabla f(x_0, y_0).$$

It is easy to see that $\partial_{\lambda_i} \lambda_i = 0$, where for simplicity ∂_{λ_i} denotes the directional derivative along line $\lambda_i = 0$. By the assumptions that all second derivatives are zero at a_3, we have $(\partial_{\lambda_1} \partial_{\lambda_2} v)(a_3) = 0$.

Using the fact that $\partial_{\lambda_2} \lambda_2 = 0$, we have

$$\partial_{\lambda_2} v = \partial_{\lambda_2} Q \cdot \lambda_1 \lambda_2 \lambda_3 + Q \lambda_2 (\lambda_1 \partial_{\lambda_2} \lambda_3 + \lambda_3 \partial_{\lambda_2} \lambda_1).$$

Furthermore, using the fact that $\partial_{\lambda_i} \lambda_j$ ($i \neq j$) is a constant, we obtain

$$\partial_{\lambda_1} \partial_{\lambda_2} v = \partial_{\lambda_1} \partial_{\lambda_2} Q \cdot \lambda_1 \lambda_2 \lambda_3 + \partial_{\lambda_2} Q \cdot \lambda_1 (\lambda_2 \partial_{\lambda_1} \lambda_3 + \lambda_3 \partial_{\lambda_1} \lambda_2)$$
$$+ \partial_{\lambda_1} Q \cdot (\lambda_2 \lambda_1 \partial_{\lambda_2} \lambda_3 + \lambda_2 \lambda_3 \partial_{\lambda_2} \lambda_1)$$
$$+ Q \cdot (\lambda_1 \partial_{\lambda_1} \lambda_2 \partial_{\lambda_2} \lambda_3 + \lambda_3 \partial_{\lambda_1} \lambda_2 \partial_{\lambda_2} \lambda_1 + \lambda_2 \partial_{\lambda_1} \lambda_3 \partial_{\lambda_2} \lambda_1),$$

from which and the facts that $\lambda_1(a_3) = \lambda_2(a_3) = 0$ and $\lambda_3(a_3) = 1$, we have

$$0 = \partial_{\lambda_1} \partial_{\lambda_2} v(a_3) = Q(a_3) \partial_{\lambda_1} \lambda_2 \partial_{\lambda_2} \lambda_1,$$

this leads to $Q(a_3) = 0$. Similarly, we can obtain $Q(a_1) = Q(a_2) = 0$.

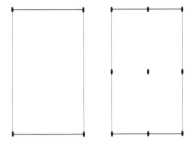

FIGURE 6.5
The Q_1 and Q_2 elements.

On the other hand, note that

$$\frac{\partial v}{\partial n_1}(a_{23}) = (\frac{\partial Q}{\partial n_1}\lambda_1\lambda_2\lambda_3 + Q\lambda_2\lambda_3\frac{\partial\lambda_1}{\partial n_1} + Q\lambda_1\lambda_3\frac{\partial\lambda_2}{\partial n_1} + Q\lambda_1\lambda_2\frac{\partial\lambda_3}{\partial n_1})(a_{23})$$

$$= Q(a_{23})\lambda_2(a_{23})\lambda_3(a_{23})\frac{\partial\lambda_1}{\partial n_1}, \qquad (6.53)$$

where we used the properties that $\lambda_1(a_{23}) = 0$ and $\frac{\partial\lambda_1}{\partial n_1}$ is a constant.

Using the assumption that $\frac{\partial v}{\partial n_1}(a_{23}) = 0$, from (6.53) we have $Q(a_{23}) = 0$. By the same argument, we have $Q(a_{13}) = Q(a_{12}) = 0$.

Hence the polynomial Q of degree two in two variables satisfies

$$Q(a_1) = Q(a_2) = Q(a_3) = Q(a_{23}) = Q(a_{13}) = Q(a_{12}) = 0,$$

which leads to $Q \equiv 0$. This completes the proof. $\square$

6.4.2 Rectangular finite elements

Without loss of generality, we assume that the rectangular element $K = [0, 1]^2$. We can define the continuous rectangular finite element spaces

$$V_h = \{v \in C^0(\overline{\Omega}) : \ v|_K \in Q_r(K), \quad K \in T_h\}, \quad \forall\, r \geq 1,$$

where $r = 1$ and $r = 2$ correspond to the so-called bilinear and biquadratic finite element spaces, respectively.

It will be shown in Lemma 6.3 that for $r = 1$ the degrees of freedom are the values at the vertices of the rectangle; for $r = 2$ the degrees of freedom are the values at the vertices, midpoints of each side, and at the center of the rectangle. We plot the degrees of freedom in Fig. 6.5.

LEMMA 6.3
If $v \in Q_r(K), r = 1, 2$, vanishes at the nodes shown in Fig. 6.5, then $v = 0$.

Proof. For $r = 1$, the restriction of v to each side of $K = [0,1]^2$ is a linear function of one variable. Hence v vanishes on each side, i.e., v can be represented as

$$v = cx(1-x)y(1-y), \tag{6.54}$$

following which $c = 0$, since $v \in Q_1(K)$.

Similarly, when $r = 2$, v also vanishes on each side, i.e., v still has the form (6.54), from which we obtain the value at the center of K as

$$v\left(\frac{1}{2}, \frac{1}{2}\right) = \frac{c}{16}.$$

But by assumption $v(\frac{1}{2}, \frac{1}{2}) = 0$, we see that $c = 0$, therefore $v \equiv 0$, which concludes the proof for $r = 2$. $\square$

6.5 Examples of nonconforming finite elements

In a previous section, we develop finite element space V_h, which is a subspace of $V \equiv H^1(\Omega)$ for a second-order elliptic problem. In practice, we can release this constraint so that V_h should not be a subspace of V. Here we will introduce some classic nonconforming finite elements.

6.5.1 Nonconforming triangular elements

Let Ω be a polygonal domain in the plane, and T_h be the triangulation of Ω. Consider an arbitrary triangle $K \in T_h$ with vertices a_i and midpoints of edges a_{ij} (see Fig. 6.3).

We claim that a linear function v on each K is uniquely determined by the function values at the three midpoints of the edges. In fact, let the basis function at a_{23} be

$$\phi_1(\boldsymbol{x}) = c_1\lambda_1(\boldsymbol{x}) + c_2\lambda_2(\boldsymbol{x}) + c_3\lambda_3(\boldsymbol{x}), \tag{6.55}$$

where $\lambda_i(\boldsymbol{x})$ are the barycentric coordinates defined in a previous section. The unknown coefficients c_1, c_2, and c_3 are determined by the following conditions

$$1 = \phi_1(a_{23}), \quad i.e., \quad \frac{1}{2}c_2 + \frac{1}{2}c_3 = 1,$$

$$0 = \phi_1(a_{12}), \quad i.e., \quad \frac{1}{2}c_1 + \frac{1}{2}c_2 = 0,$$

$$0 = \phi_1(a_{13}), \quad i.e., \quad \frac{1}{2}c_1 + \frac{1}{2}c_3 = 0,$$

which leads to the solution $c_1 = -1$, $c_2 = c_3 = 1$. Hence the basis function at node a_{23} is

$$\phi_1 = -\lambda_1 + \lambda_2 + \lambda_3 = 1 - 2\lambda_1,$$

where we used the fact that $\lambda_1 + \lambda_2 + \lambda_3 = 1$. Similarly, we can derive the basis functions at nodes a_{13} and a_{12}.

Therefore, any linear function v on K can be uniquely represented by

$$v(x,y) = v(a_{23})(1 - 2\lambda_1) + v(a_{13})(1 - 2\lambda_2) + v(a_{12})(1 - 2\lambda_3), \quad \forall\, (x,y) \in K.$$

Now we can define the nonconforming P_1 space on triangles

$$\boldsymbol{V}_h = \{v \in L^2(\Omega) : v|_K \text{ is linear for all } K \in T_h, v$$
$$\text{is continuous at the midpoints of the interior edges}\},$$

which is the linear Crouzeix-Raviart element. Note that the function $v \in \boldsymbol{V}_h$ is continuous only at the midpoints of interior edges, so $\boldsymbol{V}_h \not\subset \boldsymbol{V} \equiv H^1(\Omega)$. General high-order nonconforming elements on triangles are discussed in [4].

As we showed above, the conforming finite elements for solving a fourth-order PDE are C^1 elements and their constructions such as the Argyris triangle are quite complicated. To reduce this burden, nonconforming elements are especially popular for fourth-order problems. One classic example is the Morley element on triangles (cf. Fig. 6.6), which is defined by

$$\boldsymbol{V}_h = \{v \in L^2(\Omega) : v|_K \in P_2(K) \text{ for all } K \in T_h, v$$
$$\text{is continuous at the interior vertices, the normal derivative}$$
$$\frac{\partial v}{\partial n} \text{ is continuous at the midpoints of interior edges}\},$$

i.e., on each triangle, $v \in \boldsymbol{V}_h$ is uniquely defined by 6 degrees of freedom (function values at the vertices, and the values of the first normal derivatives at the midpoints of the edges). Hence the Morley element uses much fewer degrees of freedom compared to the Argyris element. Note that the Morley element is not even of C^0, thus $\boldsymbol{V}_h \not\subset \boldsymbol{V} \equiv H^2(\Omega)$. Detailed error estimates for the Morley element can consult the paper by Shi [39].

6.5.2 Nonconforming rectangular elements

We now consider the case where the partition T_h of Ω is formed by rectangles such that edges of the rectangles are parallel to the x- and y- axes, respectively.

The simplest nonconforming rectangular element space is the so-called rotated Q_1 element [36], which can be defined as

$$\boldsymbol{V}_h = \{v \in L^2(\Omega) : v|_K = a_1 + a_2 x + a_3 y + a_4(x^2 - y^2) \text{ for all rectangles } K \in T_h,$$
$$v \text{ is continuous at the midpoints of interior vertices}, a_i \in R\},$$

i.e., the degrees of freedom are the function values at the midpoints of edges (cf. Fig. 6.7).

Another popular element is Wilson's rectangle, which is defined by

$$\boldsymbol{V}_h = \{v \in L^2(\Omega) : v|_K \in P_2(K) \text{ for all rectanlges } K \in T_h,$$

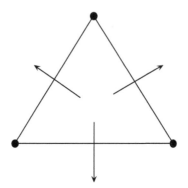

FIGURE 6.6
The degrees of freedom for the Morley element.

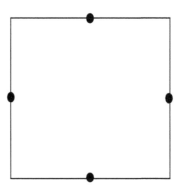

FIGURE 6.7
The degrees of freedom for the rotated Q_1 element.

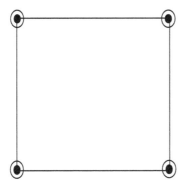

FIGURE 6.8
The degrees of freedom for the Adini element.

v is determined by its values at the vertices of K and the mean values
of its second derivatives over K : $\dfrac{1}{|K|} \displaystyle\int_K \dfrac{\partial^2 v}{\partial x^2}\,d\boldsymbol{x}$ and $\dfrac{1}{|K|} \displaystyle\int_K \dfrac{\partial^2 v}{\partial y^2}\,d\boldsymbol{x}\}$.

The nonconforming quadrilateral Wilson element is discussed in [40].

For a fourth-order problem, the Adini element is often used, which is defined
by

$$\boldsymbol{V}_h = \{v \in L^2(\Omega) : v|_K \in P_3(K) \oplus \operatorname{span}\{x^3 y, xy^3\} \text{ for all rectanlges } K \in T_h,$$
$$v, \frac{\partial v}{\partial x} \text{ and } \frac{\partial v}{\partial y} \text{ are continuous at the interior vertices}\}.$$

Hence the functions in the Adini element are $C^0(\Omega)$, but not $C^1(\Omega)$. The
degrees of freedom of the Adini element are the function values and the first
derivatives with respect to x and y at the vertices of the rectangle (cf. Fig. 6.8).

6.6 Finite element interpolation theory

Before we discuss the finite element interpolation theory, we need to introduce
some important spaces for theoretical analysis of partial differential equations.

6.6.1 Sobolev spaces

Since its introduction around 1950 by S.L. Sobolev, Sobolev spaces have become a cornerstone for analyzing partial differential equations. For a compre-
hensive discussion, see the book by Adams [1].

Let Ω be an open bounded domain in R^d with a boundary $\partial\Omega$. We define the Lebesgue spaces

$$L^p(\Omega) = \{v : ||v||_{L^p(\Omega)} < \infty\},$$

where the associated norm is

$$||v||_{L^p(\Omega)} = (\int_\Omega |v(x)|^p dx)^{1/p} \quad \text{for } 1 \le p < \infty,$$

or

$$||v||_{L^\infty(\Omega)} = \sup\{|v(x)| : x \in \Omega\} \quad \text{for } p = \infty.$$

Note that $L^2(\Omega)$ is a Hilbert space (i.e., a complete inner product space), endowed with scalar product $(u, v) = \int_\Omega u(x)v(x)dx$. Recall that a vector space endowed with a norm $|| \cdot ||$ is called a normed linear space. While a normed linear space V with norm $|| \cdot ||$ is called a Banach space if it is complete with respect to $||\cdot||$, i.e., any Cauchy sequence $\{v_j\}$ of V has a limit $v \in V$.

It is known that [11, p. 23]: For any $1 \le p \le \infty$, $L^p(\Omega)$ is a Banach space.

Let $C_0^\infty(\Omega)$ (or $D(\Omega)$) be the space of infinitely differentiable functions with compact support in Ω. We usually use the multi-index notation to represent the derivatives of v, i.e., let $\alpha = (\alpha_1, \cdots, \alpha_d)$ be an d-tuple of nonnegative integers and denote the length of α by $|\alpha| = \sum_{i=1}^d \alpha_i$. Then the α-th partial derivative of v is defined by

$$D^\alpha v = \frac{\partial^{\alpha_1+\alpha_2+\cdots+\alpha_d} v}{\partial x_1^{\alpha_1} \partial x_2^{\alpha_2} \cdots \partial x_d^{\alpha_d}} = \frac{\partial^{|\alpha|} v}{\partial x_1^{\alpha_1} \partial x_2^{\alpha_2} \cdots \partial x_d^{\alpha_d}}.$$

Now we will define derivatives for a class of not-so-smooth functions. To do that, let us first consider the set of locally integrable functions

$$L_{loc}^1(\Omega) = \{v : v \in L^1(K), \text{ any compact } K \subset \text{ interior } \Omega\}.$$

If for a given $v \in L_{loc}^1(\Omega)$, there is a function $w \in L_{loc}^1(\Omega)$ such that

$$(w, \phi) = (-1)^{|\alpha|}(v, D^\alpha \phi) \quad \forall \phi \in D(\Omega),$$

then we say w is the weak derivative of v, denoted as $D^\alpha v = w$. It is easy to check that for any $v \in C^{|\alpha|}(\Omega)$, the weak derivative $D^\alpha v$ exists and equals the ordinary pointwise derivative.

With all the above preparations, we are ready to define the Sobolev spaces

$$W^{k,p}(\Omega) = \{v \in L_{loc}^1(\Omega) : ||v||_{W^{k,p}(\Omega)} < \infty\},$$

equipped with the corresponding Sobolev norm

$$||v||_{W^{k,p}(\Omega)} = (\sum_{|\alpha|\le k} ||D^\alpha v||_{L^p(\Omega)}^p)^{1/p} \quad \forall 1 \le p < \infty,$$

and

$$||v||_{W^{k,\infty}(\Omega)} = \max_{|\alpha|\le k} ||D^\alpha v||_{L^\infty(\Omega)}.$$

It is proved that $W^{k,p}(\Omega)$ is a Banach space (see e.g., [11, p. 28]). Furthermore, we denote $W_0^{k,p}(\Omega)$ as the closure of $C_0^\infty(\Omega)$ with respect to the norm $||\cdot||_{W^{k,p}(\Omega)}$. When $p = 2$, we write $H^k(\Omega)$ (or $H_0^k(\Omega)$) instead of $W^{k,2}(\Omega)$ or $(W_0^{k,2}(\Omega))$. Similarly, we can define the seminorms

$$|v|_{W^{k,p}(\Omega)} = (\sum_{|\alpha|=k} ||D^\alpha v||_{L^p(\Omega)}^p)^{1/p}, \quad |v|_{W^{k,\infty}(\Omega)} = \max_{|\alpha|=k} ||D^\alpha v||_{L^\infty(\Omega)}.$$

Sobolev spaces provide a way for quantifying the degree of smoothness of functions. Before we do that, we have to explain the word "embedding." A normed space U is said to be embedded in a normed space V, denoted as $U \hookrightarrow V$, if
(i) U is a linear subspace of V;
(ii) the injection of U into V is continuous, i.e., there exists a constant $C > 0$ such that $||u||_V \le C||u||_U \ \forall\, u \in U$.

THEOREM 6.5
(Sobolev embedding theorem) Suppose that Ω is an open set of R^d with a Lipschitz continuous boundary, and $1 \le p < \infty$. Then the following embeddings exist:
(i) If $0 \le kp < d$, then $W^{k,p}(\Omega) \hookrightarrow L^{p_*}$ for $p_* = \frac{dp}{d-kp}$.
(ii) If $kp = d$, then $W^{k,p}(\Omega) \hookrightarrow L^q$ for any q such that $p \le q < \infty$.
(iii) If $kp > d$, then $W^{k,p}(\Omega) \hookrightarrow C^0(\overline{\Omega})$.

6.6.2 Interpolation theory

First we define a local k-th order interpolation operator in a finite element K:

$$\Pi_K^k(v) = \sum_i v(a_i)\phi_i \quad \forall\, v \in C^0(K),$$

where a_i are the nodes in K, ϕ_i are the shape functions. Then we can define a corresponding global interpolation operator

$$\Pi_h^k(v)|_K = \Pi_K^k(v|_K) \quad \forall\, v \in C^0(\overline{\Omega}), K \in T_h.$$

We assume further that each element K of T_h can be obtained as an affine mapping of a reference element $\hat{K}$, i.e.,

$$K = F_K(\hat{K}), \quad F_K(\hat{x}) = B_K\hat{x} + b_K,$$

where B_K is a $d \times d$ nonsingular matrix. For example, we can map a reference element with vertices

$$(\xi_1, \eta_1) = (0,0), \quad (\xi_2, \eta_2) = (1,0), \quad (\xi_3, \eta_3) = (0,1)$$

to an arbitrary triangle with vertices $(x_i, y_i), 1 \leq i \leq 3$, by the following affine mapping:

$$\begin{pmatrix} x \\ y \end{pmatrix} = \begin{pmatrix} x_2 - x_1 & x_3 - x_1 \\ y_2 - y_1 & y_3 - y_1 \end{pmatrix} \begin{pmatrix} \xi \\ \eta \end{pmatrix} + \begin{pmatrix} x_1 \\ y_1 \end{pmatrix}.$$

The estimate of interpolation error will be obtained by the following four lemmas.

LEMMA 6.4
Denote $\hat{v} = v(F_K(\hat{x}))$ for any $v \in H^m(K), m \geq 0$. Then $\hat{v} \in H^m(\hat{K})$, and there exists a constant $C = C(m, d)$ such that

$$|v|_{m,K} \leq C||B_K^{-1}||^m |det B_K|^{1/2} |\hat{v}|_{m,\hat{K}}, \quad \forall \, \hat{v} \in H^m(\hat{K}), \quad (6.56)$$

and

$$|\hat{v}|_{m,\hat{K}} \leq C||B_K||^m |det B_K|^{-1/2} |v|_{m,K}, \quad \forall \, v \in H^m(K), \quad (6.57)$$

where $||\cdot||$ denotes the matrix norm associated to the Euclidean norm in R^d.

Proof. Since $C^\infty(K)$ is dense in $H^m(K)$, it is sufficient to prove (6.56) for a smooth v. By definition, we have

$$|v|_{m,K}^2 = \sum_{|\alpha|=m} \int_K |D^\alpha v|^2. \quad (6.58)$$

Using the chain rule and the mapping F_K, we have

$$||D^\alpha v||_{0,K}^2 \leq C||B_K^{-1}||^{2m} \sum_{|\beta|=m} ||\hat{D}^\beta \hat{v}||_{0,K}^2$$

$$\leq C||B_K^{-1}||^{2m} \sum_{|\beta|=m} ||\hat{D}^\beta \hat{v}||_{0,\hat{K}}^2 \cdot (det B_K), \quad (6.59)$$

which along with (6.58) completes the proof of (6.56). The other proof can be pursued similarly. $\square$

For any element K, we denote

$$h_K = diam(K), \quad \rho_K = \sup\{diam(S) : S \text{ is a sphere inscribed in } K\}.$$

Similarly, $h_{\hat{K}}$ and $\rho_{\hat{K}}$ can be defined on the reference element $\hat{K}$.

LEMMA 6.5
The following estimates hold

$$||B_K^{-1}|| \leq \frac{h_{\hat{K}}}{\rho_K}, \quad ||B_K|| \leq \frac{h_K}{\rho_{\hat{K}}}.$$

Proof. By definition, we have

$$\|B_K^{-1}\| = \frac{1}{\rho_K} \sup_{|\xi|=\rho_K} |B_K^{-1}\xi|. \tag{6.60}$$

For any ξ satisfying $|\xi| = \rho_K$, we can always find two points $x, y \in K$ such that $x - y = \xi$. Note that $|B_K^{-1}\xi| = |B_K^{-1}(x-y)| = |\hat{x} - \hat{y}| \leq h_{\hat{K}}$, substituting which into (6.60) completes the proof of the first inequality. The other one can be proved in a similar way. □

LEMMA 6.6
(Bramble-Hilbert lemma) Let the linear continuous mapping

$$\hat{L} \in \mathcal{L}(H^s(\hat{K}); H^m(\hat{K})), m \geq 0, s \geq 0,$$

such that

$$\hat{L}(\hat{p}) = 0 \quad \text{for all } \hat{p} \in P_l, l \geq 0.$$

Then

$$|\hat{L}(\hat{v})|_{m,\hat{K}} \leq \|\hat{L}\|_{\mathcal{L}(H^s;H^m)} \inf_{\hat{p}\in P_l} \|\hat{v} + \hat{p}\|_{s,\hat{K}}, \forall\, \hat{v} \in H^s(\hat{K}).$$

Proof. The proof follows from

$$|\hat{L}(\hat{v})|_{m,\hat{K}} = |\hat{L}(\hat{v} + \hat{p})|_{m,\hat{K}} \leq \|\hat{L}\|_{\mathcal{L}(H^s;H^m)}\|\hat{v} + \hat{p}\|_{s,\hat{K}}.$$

□

LEMMA 6.7
[35, p. 88] There exists a constant $C(\hat{K})$ such that

$$\inf_{\hat{p}\in P_k(\hat{K})} \|\hat{v} + \hat{p}\|_{k+1,\hat{K}} \leq C(\hat{K})|\hat{v}|_{k+1,\hat{K}} \quad \forall\, \hat{v} \in H^{k+1}(\hat{K}).$$

THEOREM 6.6
Let $m = 0, 1, l = \min(k, s-1) \geq 1$. Then there exists a constant C such that

$$|v - \Pi_K^k v|_{m,K} \leq C\frac{h_K^{l+1}}{\rho_K^m}|v|_{l+1,K} \quad \forall\, v \in H^s(K).$$

Proof. Denote $\widehat{\Pi_K^k v} = \Pi_K^k v \circ F_K$. Noting that the shape functions $\hat{\phi}_i$ in $\hat{K}$ are given by $\hat{\phi}_i = \phi_i \circ F_K$, we obtain

$$\widehat{\Pi_K^k v} = \Pi_h^k v \circ F_K = \sum_i v(a_i)(\phi_i \circ F_K) = \sum_i v(F_K(\hat{a}_i))\hat{\phi}_i = \Pi_{\hat{K}}^k \hat{v},$$

from which and Lemmas 6.4-6.7, we have

$$|v - \Pi_K^k v|_{m,K} \leq C \cdot \frac{1}{\rho_K^m} |det B_K|^{1/2} |\hat{v} - \widehat{\Pi_K^k v}|_{m,\hat{K}}$$

$$\leq \frac{C}{\rho_K^m} |det B_K|^{1/2} |\hat{v} - \Pi_{\hat{K}}^k \hat{v}|_{m,\hat{K}}$$

$$\leq \frac{C}{\rho_K^m} |det B_K|^{1/2} \|I - \Pi_{\hat{K}}^k\|_{\mathcal{L}(H^s;H^m)} \inf_{\hat{p} \in P_l} \|\hat{v} + \hat{p}\|_{s,\hat{K}} \quad (6.61)$$

If $s - 1 < k$, then $l = s - 1$, i.e., $s = l + 1$. Therefore, by Lemmas 6.4-6.5, we have

$$\inf_{\hat{p} \in P_l} \|\hat{v} + \hat{p}\|_{s,\hat{K}} \leq C|\hat{v}|_{l+1,\hat{K}} \leq C\|B_K\|^{l+1} |det B_K|^{-1/2} |v|_{l+1,K}$$

$$\leq C h_K^{l+1} |det B_K|^{-1/2} |v|_{l+1,K},$$

substituting which into (6.61) yields

$$|v - \Pi_K^k v|_{m,K} \leq C \frac{h_K^{l+1}}{\rho_K^m} |v|_{l+1,K}.$$

If $s - 1 \geq k$, then $l = k$, in which case the proof completes by noting that

$$|\hat{v} - \Pi_{\hat{K}}^k \hat{v}|_{m,\hat{K}} \leq C\|I - \Pi_{\hat{K}}^k\|_{\mathcal{L}(H^{k+1};H^k)} \inf_{\hat{p} \in P_l} \|\hat{v} + \hat{p}\|_{k+1,\hat{K}} \leq C|\hat{v}|_{k+1,\hat{K}}.$$

☐

In order to consider the global interpolation error over Ω, we assume that the triangulation T_h of Ω is regular, i.e., there exists a constant $\sigma \geq 1$ such that

$$\frac{h_K}{\rho_K} \leq \sigma \quad \text{for any } K \in T_h.$$

Noting that $|v - \Pi_h^k v|_{m,\Omega}^2 = \sum_{K \in T_h} |v - \Pi_h^k v|_{m,K}^2$, which along with Theorem 6.6, and the regularity assumption, we have the following global interpolation error estimate.

THEOREM 6.7
Let $m = 0, 1, l = \min(k, s-1) \geq 1$, and T_h be a regular family of triangulations of Ω. Then there exists a constant C, independent of $h = \max_{K \in T_h} h_K$, such that

$$|v - \Pi_h^k v|_{m,\Omega} \leq C h^{l+1-m} |v|_{l+1,\Omega} \quad \forall v \in H^s(\Omega).$$

REMARK 6.1 By performing a similar proof as above (cf. [17]), we have both local and global L^∞ interpolation error estimates: for $m = 0, 1, l = \min(k, s-1) \geq 1$,

$$|v - \Pi_K^k v|_{m,\infty,K} \leq C \frac{h_K^{l+1}}{\rho_K^m} |v|_{l+1,\infty,K} \quad \forall v \in W^{s,\infty}(K),$$

and

$$|v - \Pi_h^k v|_{m,\infty,\Omega} \leq Ch^{l+1-m}|v|_{l+1,\infty,\Omega} \quad \forall\, v \in W^{s,\infty}(\Omega).$$

6.7 Finite element analysis of elliptic problems

In this section, we will present the basic finite element error analysis for elliptic problems. First we will consider the conforming finite elements, then we extend the analysis to nonconforming finite elements.

Without loss of generality, we consider the Dirichlet problem for the Laplace equation

$$-\triangle u = f \quad \text{in } \Omega, \tag{6.62}$$

$$u = 0 \quad \text{on } \partial\Omega. \tag{6.63}$$

6.7.1 Analysis of conforming finite elements

Multiplying (6.62) by a test function $v \in H_0^1(\Omega)$ and using the Green's formula

$$-\int_\Omega \triangle u v dx = -\int_{\partial\Omega} \frac{\partial u}{\partial n} v ds + \int_\Omega \sum_{i=1}^d \frac{\partial u}{\partial x_i}\frac{\partial v}{\partial x_i}, \quad \forall\, u \in H^2(\Omega), v \in H^1(\Omega),$$
$$\tag{6.64}$$

where $\frac{\partial}{\partial n} = \sum_{i=1}^d n_i \frac{\partial}{\partial x_i}$ is the normal derivative operator, we obtain the variational problem: Find $u \in H_0^1(\Omega)$ such that

$$A(u,v) \equiv (\nabla u, \nabla v) = (f,v), \quad \forall\, v \in H_0^1(\Omega). \tag{6.65}$$

Application of the Cauchy-Schwarz inequality shows that

$$|(\nabla u, \nabla v)| \leq ||\nabla u||_0 ||\nabla v||_0 \leq ||u||_1 ||v||_1,$$

which reveals that $A(\cdot,\cdot)$ is continuous on $H_0^1(\Omega) \times H_0^1(\Omega)$.

Using the Poincaré inequality [17, p. 12]

$$\int_\Omega v^2(x)dx \leq C_\Omega \int_\Omega |\nabla v(x)|^2 dx, \quad \forall\, v \in H_0^1(\Omega), \tag{6.66}$$

we obtain

$$A(v,v) = ||\nabla v||_0^2 \geq \frac{1}{1+C_\Omega}||v||_1^2,$$

which proves the coercivity of A. Therefore, by the Lax-Milgram lemma, the variational problem (6.65) has a unique solution $u \in H_0^1(\Omega)$.

To solve (6.65) by the finite element method, we construct a finite dimensional subspace V_h of $H_0^1(\Omega)$ using continuous piecewise basis functions, i.e.,

$$V_h = \{v_h \in C^0(\overline{\Omega}) : v_h|_K \in P_k, \forall\, K \in T_h\}, \tag{6.67}$$

where T_h is a regular family of triangulations of Ω.

The finite element approximation $u_h \in V_h$ of (6.64) is: Find $u_h \in V_h$ such
that

$$(\nabla u_h, \nabla v_h) = (f, v_h), \quad \forall\, v_h \in V_h.$$

By the Céa lemma, for any $u \in H^s(\Omega) \cup H_0^1(\Omega), s \geq 2$, we have

$$||u - u_h||_1 \leq (1 + C_\Omega) \inf_{v_h \in V_h} ||u - v_h||_1 \leq (1 + C_\Omega)||u - \Pi_h u||_1 \leq Ch^l||u||_{l+1},$$

where $l = \min(k, s - 1)$.

Hence if u is very smooth, then the rate of convergence in the H^1-norm will
be $O(h^k), k = 1, 2, \cdots$, which is optimal in the sense that it is the highest
possible rate of convergence allowed by the polynomial degree k.

To derive error estimates in the L^2-norm, we need to use a duality argument
(also called the Aubin-Nitsche technique). Let the error $e = u - u_h$, and w
be the solution of

$$-\triangle w = e \text{ in } \Omega, \quad w = 0 \text{ on } \partial\Omega, \tag{6.68}$$

whose variational formulation is: Find $w \in H_0^1(\Omega)$ such that

$$A(v, w) = (e, v) \quad \forall\, v \in H_0^1(\Omega). \tag{6.69}$$

Hence, we have

$$||u - u_h||_0^2 = (u - u_h, u - u_h) = A(u - u_h, w) = A(u - u_h, w - \Pi_h w)$$
$$\leq ||u - u_h||_1 ||w - \Pi_h w||_1 \leq Ch^l||u||_{l+1} \cdot h||w||_2. \tag{6.70}$$

Note that the adjoint problem (6.68) satisfies the property

$$||w||_2 \leq C||e||_0, \tag{6.71}$$

substituting which into (6.70), we have

$$||u - u_h||_0 \leq Ch^{l+1}||u||_{l+1},$$

which is optimal in the L^2-norm.

To summarize, we have the following convergence results for the Laplace
problem (6.62)-(6.63).

THEOREM 6.8
*Let Ω be a polygonal domain of $R^d, d = 2, 3$, with Lipschitz boundary, and T_h
be a regular family of triangulations of Ω. Let V_h be defined in (6.67). If the
exact solution $u \in H^s(\Omega) \cap H_0^1(\Omega), s \geq 2$, the error estimate holds*

$$||u - u_h||_1 \leq Ch^l||u||_{l+1}, \quad l = \min(k, s - 1), k \geq 1.$$

Suppose, furthermore, for each $e \in L^2(\Omega)$, the solution w of (6.69) belongs to $H^2(\Omega)$ and satisfies

$$||w||_2 \leq C||e||_0 \quad \forall\, e \in L^2(\Omega).$$

Then we have

$$||u - u_h||_0 \leq Ch^{l+1}||u||_{l+1}, \quad l = \min(k, s-1), k \geq 1.$$

Using more sophisticated weighted-norm techniques, the following L^∞-error estimates hold true [17, p. 165].

THEOREM 6.9
Under the assumption of $u \in H_0^1(\Omega) \cap W^{k+1,\infty}(\Omega), k \geq 1$, we have

$$||u - u_h||_{\infty,\Omega} + h||\nabla(u - u_h)||_{\infty,\Omega} \leq Ch^2|\ln h|\,|u|_{2,\infty,\Omega}, \quad \text{for } k = 1,$$

and

$$||u - u_h||_{\infty,\Omega} + h||\nabla(u - u_h)||_{\infty,\Omega} \leq Ch^{k+1}|u|_{k+1,\infty,\Omega}, \quad \text{for } k \geq 2.$$

6.7.2 Analysis of nonconforming finite elements

Here we consider the error analysis for the nonconforming finite element method for solving the Laplace problem (6.62)-(6.63). For simplicity, we consider the linear Crouzeix-Raviart element

$$V_h = \{v \in L^2(\Omega) : v|_K \text{ is linear for all } K \in T_h, v$$
$$\text{is continuous at the midpoints of the interior edges,}$$
$$\text{and } v = 0 \text{ at the midpoints of } \partial\Omega\},$$

in which case the discrete problem becomes: Find $u_h \in V_h$ such that

$$A_h(u_h, v) = (f, v) \quad \forall\, v \in V_h, \tag{6.72}$$

where the bilinear form $A_h(\cdot, \cdot)$ is defined by

$$A_h(v, w) = \sum_{K \in T_h} \int_K \nabla v \cdot \nabla w d\mathbf{x} = \sum_{K \in T_h} (\nabla v, \nabla w)_K, \quad v, w \in V_h,$$

with associated norm $||v||_h \equiv \sqrt{A_h(v, v)}$.

Note that $||v||_h$ is indeed a norm, since $A_h(v, v) = 0$ implies that v is piecewise constant, which leads to $v \equiv 0$ due to the zero boundary condition and the continuity at midpoints.

THEOREM 6.10
Suppose that Ω is a convex polygonal domain. Let $f \in L^2(\Omega)$, u and u_h are the solutions to (6.65) and (6.72), respectively. Then the following error estimate holds

$$\|u - u_h\|_h \leq Ch|u|_{H^2(\Omega)}.$$

Proof. It is easy to see that

$$|A_h(v_h, w_h)| \leq \|v_h\|_h \|w_h\|_h$$

and $A_h(v_h, v_h) = \|v_h\|_h^2$. Hence we can use Theorem 6.4.

Note that

$$
\begin{aligned}
A_h(u - u_h, v) &= \sum_{K \in T_h} \int_K \nabla u \cdot \nabla v d\boldsymbol{x} - \int_\Omega f v d\boldsymbol{x} \\
&= \sum_{K \in T_h} \left(\int_{\partial K} \frac{\partial u}{\partial n} v ds - \int_K \triangle u \cdot v d\boldsymbol{x} \right) - \int_\Omega f v d\boldsymbol{x} \\
&= \sum_{K \in T_h} \int_{\partial K} \frac{\partial u}{\partial n} v ds.
\end{aligned}
\tag{6.73}
$$

For each edge $e \in \partial K$, we define the mean value of v on e

$$\bar{v} = \frac{1}{|e|} \int_e v|_K ds.$$

Note that each interior edge appears twice, hence we can rewrite (6.73) as

$$A_h(u - u_h, v) = \sum_{K \in T_h} \sum_{e \in \partial K} \int_e \left(\frac{\partial u}{\partial n_K} - \overline{\frac{\partial u}{\partial n_K}} \right)(v - \bar{v})ds. \tag{6.74}$$

Consider a reference triangle $\hat{K}$, for an edge $\hat{e} \in \partial \hat{K}$, we have

$$\|w - \bar{w}\|_{L^2(\hat{e})} \leq C|w|_{H^1(\hat{K})},$$

from which and the standard scaling argument, we obtain

$$\|w - \bar{w}\|_{L^2(e)} \leq Ch^{1/2}|w|_{H^1(K)}. \tag{6.75}$$

Applying (6.75) to (6.74), we have

$$
\begin{aligned}
|A_h(u - u_h, v)| &\leq \sum_{K \in T_h} \sum_{e \in \partial K} Ch \left| \frac{\partial u}{\partial n} \right|_{H^1(K)} |v|_{H^1(K)} \\
&\leq Ch|u|_{H^2(\Omega)} \|v\|_h,
\end{aligned}
$$

which along with (6.52) and the standard interpolation estimate

$$\|u - \Pi u\|_h \leq Ch|u|_{H^2(\Omega)},$$

completes the proof. □

Applying the duality argument, we can obtain the following optimal error estimate in the L^2-norm [15, p.112]:

$$||u - \Pi u||_0 \leq Ch^2 |u|_{H^2(\Omega)}.$$

6.8 Finite element analysis of time-dependent problems

Here we first give a brief introduction to time-dependent problems discretized by the finite element method. Then we illustrate the development and analysis of both semi-discrete and fully-discrete finite element methods applied to a parabolic problem.

6.8.1 Introduction

For time-dependent problems, we will consider the space-time functions $v(x, t)$, $(x, t) \in \Omega \times (0, T)$, for which we need the space

$$L^q(0, T; W^{k,p}(\Omega))$$

$$= \{v : (0, T) \to W^{k,p}(\Omega) : \int_0^T ||v(t)||_{W^{k,p}(\Omega)}^q dt < \infty\}, \ 1 \leq q < \infty,$$

endowed with norm

$$||v||_{L^q(0,T;W^{k,p}(\Omega))} = (\int_0^T ||v(t)||_{W^{k,p}(\Omega)}^q dt)^{1/q}.$$

Similarly, when $q = \infty$, we can define the space $L^\infty(0, T; W^{k,p}(\Omega))$ with norm

$$||v||_{L^\infty(0,T;W^{k,p}(\Omega))} = \max_{0 \leq t \leq T} ||v(\cdot, t)||_{W^{k,p}(\Omega)}.$$

Another popular Sobolev space for time-dependent problems is

$$H^1(0, T; V) = \{v \in L^2(0, T; V) : \frac{\partial v}{\partial t} \in L^2(0, T; V)\}$$

where V can be any Banach space such as $W^{k,p}(\Omega)$.

Let us consider an abstract time-dependent problem

$$\frac{\partial u}{\partial t} + Lu = f \quad \text{in } (0, T) \times \Omega,$$
$$u(x, t) = 0 \quad \text{on } (0, T) \times \partial\Omega, \qquad (6.76)$$
$$u(x, 0) = u_0(x) \quad \text{on } \Omega,$$

where L is a linear differential operator.

The weak formulation of (6.76) is: Find $u \in L^2(0, T; V) \cap C^0(0, T; H)$ such that

$$\frac{d}{dt}(u(t), v) + A(u(t), v) = (f(t), v) \quad \forall \, v \in V, \qquad (6.77)$$

and $u = u_0$ at $t = 0$. Here the continuous bilinear form $A(\cdot, \cdot)$ is defined on $V \times V$, and V and H are two Hilbert spaces.

The problem (6.77) needs to be discretized with respect to both time and space variables. If the discretization is carried out only for the space variable, we obtain the so-called semi-discrete approximation. For example, the semi-discretization of (6.77) is: For each $t \in [0, T]$, find $u_h(\cdot, t) \in V_h \subset V$ such that

$$\frac{d}{dt}(u_h(t), v_h) + A(u_h(t), v_h) = (f(t), v_h) \quad \forall \, v_h \in V_h, \qquad (6.78)$$

with $u_h(0) = u_{0,h}$ where $u_{0,h}$ usually is the interpolant or L^2-projection of u_0.

In semi-discrete approximation, the solution u_h is sought as

$$u_h(x, t) = \sum_{j=1}^{N} u_j(t)\phi_j(x), \qquad (6.79)$$

where ϕ_j are the basis functions of V_h. Substituting (6.79) into (6.78), we end up with a system of ordinary differential equations for unknown vectors $\vec{u} = (u_1, \cdots, u_N)^T$:

$$M\frac{du(t)}{dt} + Su(t) = F(t), \quad u(0) = u_0, \qquad (6.80)$$

where the mass matrix M, the stiffness matrix S, and the right-hand side vector F are given as follows:

$$M = [M_{ij}] = [(\phi_j, \phi_i)], \; S = [S_{ij}] = [S(\phi_j, \phi_i)], \; F = (F_i) = (f(t), \phi_i).$$

Note that (6.80) is not fully solved, hence we need to further discrete it with respect to the time variable. To this purpose, we partition the time interval $[0, T]$ into N equal subintervals $[t_k, t_{k+1}]$ of length $\triangle t = \frac{T}{N}$. A simple fully discrete scheme is the θ-scheme:

$$\frac{1}{\triangle t}(u_h^{k+1} - u_h^k, v_h) + A(\theta u_h^{k+1} + (1-\theta)u_h^k, v_h)$$
$$= \theta(f(t_{t_{k+1}}), v_h) + (1-\theta)(f(t_{t_k}), v_h), \quad \forall \, v_h \in V_h, \qquad (6.81)$$

for $k = 0, 1, \cdots, N-1$. Here u_h^k is the approximation to $u(t)$ at time $t_k = k \cdot \triangle t$. Note that $\theta = 0$ refers to the forward Euler scheme, $\theta = 1$ refers to the backward Euler scheme, $\theta = \frac{1}{2}$ corresponds to the Crank-Nicolson scheme.

6.8.2   FEM for parabolic equations

Here we consider the parabolic equation

$$u_t - \nabla \cdot (a(\boldsymbol{x})\nabla u) = f(\boldsymbol{x}, t) \quad (\boldsymbol{x}, t) \in \Omega \times (0, T], \qquad (6.82)$$
$$u(\boldsymbol{x}, 0) = u_0(\boldsymbol{x}) \quad \boldsymbol{x} \in \Omega, \qquad (6.83)$$

$$u(\boldsymbol{x}, t) = 0 \quad (\boldsymbol{x}, t) \in \partial\Omega \times (0, T] \tag{6.84}$$

where we assume that

$$0 < a_0 \le a(\boldsymbol{x}) \le a_1. \tag{6.85}$$

The weak formulation of (6.82)-(6.84) is: Find $u \in H_0^1(\Omega)$ such that

$$(u_t, v) + (a\nabla u, \nabla v) = (f, v) \quad \forall\, v \in H_0^1(\Omega),\ t \in (0, T], \tag{6.86}$$
$$u(\boldsymbol{x}, 0) = u_0(\boldsymbol{x}) \quad \boldsymbol{x} \in \Omega. \tag{6.87}$$

6.8.2.1 The semi-discrete scheme and its analysis

Let the finite dimensional space $V_h \subset H_0^1(\Omega)$ be defined by (6.67). Then we can define the semi-discrete finite element method for (6.86)-(6.87): Find $u_h : [0, T] \to V_h$ such that

$$(\partial_t u_h, v) + (a\nabla u_h, \nabla v) = (f, v) \quad \forall\, v \in V_h, \tag{6.88}$$
$$u_h(\cdot, 0) = \Pi_h u_0, \tag{6.89}$$

where Π_h is the elliptic projection operator [47] into V_h, i.e., for any $w \in H^1(\Omega)$, $\Pi_h w \in V_h$ satisfies the equation

$$(a\nabla(w - \Pi_h w), \nabla v) = 0 \quad \forall\, v \in V_h. \tag{6.90}$$

Furthermore, if T_h is a regular family of triangulations and for each $r \in L^2(\Omega)$ the adjoint problem

$$(a\nabla v, \nabla\phi(r)) = (r, v) \quad \forall\, v \in H^1(\Omega) \tag{6.91}$$

has a solution $\phi(r) \in H^2(\Omega)$, then we have ([47], see also [35, p. 376])

$$||v - \Pi_h v||_0 + h||v - \Pi_h v||_1 \le Ch^{k+1}|v|_{k+1}, \quad \forall\, v \in H^{k+1}(\Omega). \tag{6.92}$$

To obtain the energy stability and error estimate for time-dependent problems, we usually need the following Gronwall inequality. Its proof is given as an exercise of this chapter.

LEMMA 6.8
Let f and g be piecewise continuous non-negative functions defined on $0 \le t \le T$, g being non-decreasing. If for each $t \in [0, T]$ and any constant $c > 0$,

$$f(t) \le g(t) + c \int_0^t f(\tau)d\tau, \tag{6.93}$$

then $f(t) \le e^{ct}g(t)$.

First, we can establish the following discrete stability for the semi-discrete scheme (6.88)–(6.89).

THEOREM 6.11

For the solution $u_h(t)$ of (6.88)–(6.89), we have

$$||u_h(t)||_0^2 + 2a_0 \int_0^t ||\nabla u_h(s)||_0^2 ds \le e^t \left(||u_h(0)||_0^2 + \int_0^t ||f(s)||_0^2 ds \right),$$

and

$$||u_h(t)||_0^2 + a_0 \int_0^t ||\nabla u_h(s)||_0^2 ds \le ||u_h(0)||_0^2 + \frac{C_\Omega}{a_0} \int_0^t ||f(s)||_0^2 ds,$$

where the constant C_Ω comes from the Poincaré inequality (6.66).

Proof. (I) Choosing $v = u_h$ in (6.88) and using Young's inequality

$$\xi\eta \le \frac{\epsilon\xi^2}{2} + \frac{\eta^2}{2\epsilon}, \quad \forall \xi, \eta \in \mathcal{R}, \ \forall \ \epsilon > 0, \tag{6.94}$$

we have

$$\frac{1}{2} \cdot \frac{d}{dt} ||u_h||_0^2 + ||a^{\frac{1}{2}}\nabla u_h||_0^2 = (f, u_h) \le \frac{1}{2}||u_h||_0^2 + \frac{1}{2}||f||_0^2. \tag{6.95}$$

Integrating (6.95) in time from 0 to t, and using the assumption (6.85), we obtain

$$\frac{1}{2}||u_h(t)||_0^2 + a_0 \int_0^t ||\nabla u_h(s)||_0^2 ds$$

$$\le \frac{1}{2}||u_h(0)||_0^2 + \frac{1}{2}\int_0^t ||u_h(s)||_0^2 ds + \frac{1}{2}\int_0^t ||f(s)||_0^2 ds. \tag{6.96}$$

Using the Gronwall inequality given in Lemma 6.8, we obtain the following discrete stability for the solution of (6.88):

$$||u_h(t)||_0^2 + 2a_0 \int_0^t ||\nabla u_h(s)||_0^2 ds \le e^t \left(||u_h(0)||_0^2 + \int_0^t ||f(s)||_0^2 ds \right). \tag{6.97}$$

(II) The stability can be proved without using the Gronwall inequality. Choosing $v = u_h$ in (6.88) and using Young's inequality (6.94), we obtain

$$\frac{1}{2} \cdot \frac{d}{dt} ||u_h||_0^2 + ||a^{\frac{1}{2}}\nabla u_h||_0^2 \le ||f||_0 ||u_h||_0 \le \frac{a_0}{2C_\Omega}||u_h||_0^2 + \frac{C_\Omega}{2a_0}||f||_0^2$$

$$\le \frac{a_0}{2}||\nabla u_h||_0^2 + \frac{C_\Omega}{2a_0}||f||_0^2, \tag{6.98}$$

where in the last step we used the Poincaré inequality (6.66): $||u_h||_0 \le C_\Omega^{\frac{1}{2}}||\nabla u_h||_0$.

Using the assumption (6.85) and simplifying (6.98), we obtain

$$\frac{d}{dt}||u_h(t)||_0^2 + a_0||\nabla u_h||_0^2 \le \frac{C_\Omega}{a_0}||f||_0^2.$$

Integrating in time, we have the following improved stability:

$$||u_h(t)||_0^2 + a_0 \int_0^t ||\nabla u_h(s)||_0^2 ds \le ||u_h(0)||_0^2 + \frac{C_\Omega}{a_0} \int_0^t ||f(s)||_0^2 ds. \quad (6.99)$$

☐

By following exactly the same technique as developed above, we can obtain the following a priori energy estimates for the solution u of (6.86)-(6.87): For any $t > 0$,

$$||u(t)||_0^2 + 2a_0 \int_0^t ||\nabla u(s)||_0^2 ds \le e^t \left(||u(0)||_0^2 + \int_0^t ||f(s)||_0^2 ds \right), \quad (6.100)$$

and

$$||u(t)||_0^2 + a_0 \int_0^t ||\nabla u(s)||_0^2 ds \le ||u(0)||_0^2 + \frac{C_\Omega}{a_0} \int_0^t ||f(s)||_0^2 ds. \quad (6.101)$$

Finally, we present the following optimal error estimate in the L^2 norm.

THEOREM 6.12
Let u and u_h be the solutions to (6.86)-(6.87) and (6.88)-(6.89), respectively, where the finite element space $V_h \subset H_0^1(\Omega)$ contains piecewise polynomials of degree less than or equal to k. Then the following error estimate holds true

$$||u - u_h||_{L^\infty(0,T;L^2(\Omega))}$$
$$\le Ch^{k+1}(||u||_{L^\infty(0,T;H^{k+1}(\Omega))} + ||u_t||_{L^2(0,T;H^{k+1}(\Omega))}), \quad (6.102)$$

where the constant $C > 0$ is independent of h.

Proof. Subtracting (6.88) from (6.86) with $v = v_h \in V_h$, we have

$$((\Pi_h u - u_h)_t, v) + (a\nabla(\Pi_h u - u_h), \nabla v)$$
$$= ((\Pi_h u - u)_t, v) + (a\nabla(\Pi_h u - u), \nabla v). \quad (6.103)$$

Choosing $v = \Pi_h u - u_h$ and using (6.85) and (6.90), we can reduce (6.103) to

$$\frac{1}{2}\frac{d}{dt}||\Pi_h u - u_h||_0^2 + a_0||\nabla(\Pi_h u - u_h)||_0^2$$
$$\le ||(\Pi_h u - u)_t||_0||\Pi_h u - u_h||_0$$
$$\le Ch^{k+1}|u_t|_{k+1}||\Pi_h u - u_h||_0. \quad (6.104)$$

Dropping the a_0 term, integrating both sides of (6.104) with respect to t and using (6.89), we obtain

$$\frac{1}{2}||(\Pi_h u - u_h)(t)||_0^2$$

$$\leq Ch^{2(k+1)} \int_0^t |u_t(\tau)|_{k+1}^2 d\tau + \frac{1}{2} \int_0^t ||(\Pi_h u - u_h)(\tau)||_0^2 d\tau. \quad (6.105)$$

By Gronwall's inequality, we have

$$||(\Pi_h u - u_h)(t)||_0^2 \leq Ch^{2(k+1)} \left(\int_0^t |u_t(\tau)|_{k+1}^2 d\tau \right) e^t,$$

which reduces to

$$||(\Pi_h u - u_h)(t)||_0 \leq Ch^{k+1} \left(\int_0^t ||u_t(\tau)||_{k+1}^2 d\tau \right)^{1/2}, \quad \forall\, t \in [0, T],$$

where we observed the dependence on T into the generic constant C.

Using the triangle inequality and the estimate (6.92), we obtain

$$||(u - u_h)(t)||_0 \leq Ch^{k+1} \left(||u(t)||_{k+1} + \left(\int_0^t ||u_t(\tau)||_{k+1}^2 d\tau \right)^{1/2} \right),$$

which concludes the proof of (6.102). ☐

6.8.2.2   The full-discrete schemes and their analysis

The fully discrete finite element method, such as the Crank-Nicolson scheme, for (6.86)-(6.87) can be defined as: For any $v \in V_h$, find $u_h^{n+1} \in V_h$, $n = 0, 1, \cdots, N - 1$, such that

$$\left(\frac{u_h^{n+1} - u_h^n}{\Delta t}, v \right) + \left(a\nabla \frac{u_h^{n+1} + u_h^n}{2}, \nabla v \right) = \left(\frac{f^{n+1} + f^n}{2}, v \right), \quad (6.106)$$

$$u_h^0 = \Pi_h u_0. \quad (6.107)$$

First, we can prove the following discrete stability, which has the similar form as the continuous one (6.101) and shows that the Crank-Nicolson scheme (6.106)-(6.107) is unconditionally stable.

THEOREM 6.13
For any $m \geq 1$, the solution of (6.106)-(6.107) satisfies

$$||u_h^m||_0^2 + a_0 \Delta t \sum_{n=0}^{m-1} ||\nabla \left(\frac{u_h^{n+1} + u_h^n}{2} \right)||_0^2$$

$$\leq ||u_h^0||_0^2 + \frac{C_\Omega}{a_0} \sum_{n=0}^{m-1} \Delta t ||\frac{f^{n+1} + f^n}{2}||_0^2. \quad (6.108)$$

Proof. Taking $v = \triangle t \cdot \frac{u_h^{n+1} + u_h^n}{2}$ in (6.106), we have

$$\frac{1}{2}(\|u_h^{n+1}\|_0^2 - \|u_h^n\|_0^2) + \triangle t \|a^{\frac{1}{2}} \nabla \left(\frac{u_h^{n+1} + u_h^n}{2} \right)\|_0^2$$

$$\leq \triangle t \|\frac{f^{n+1} + f^n}{2}\|_0 \|\frac{u_h^{n+1} + u_h^n}{2}\|_0$$

$$\leq \frac{a_0 \cdot \triangle t}{2 C_\Omega} \|\frac{u_h^{n+1} + u_h^n}{2}\|_0^2 + \frac{C_\Omega \cdot \triangle t}{2 a_0} \|\frac{f^{n+1} + f^n}{2}\|_0^2$$

$$\leq \frac{a_0 \cdot \triangle t}{2} \|\nabla \left(\frac{u_h^{n+1} + u_h^n}{2} \right)\|_0^2 + \frac{C_\Omega \cdot \triangle t}{2 a_0} \|\frac{f^{n+1} + f^n}{2}\|_0^2,$$

which leads to

$$\|u_h^{n+1}\|_0^2 - \|u_h^n\|_0^2 + a_0 \cdot \triangle t \|\nabla \left(\frac{u_h^{n+1} + u_h^n}{2} \right)\|_0^2 \leq \frac{C_\Omega \cdot \triangle t}{a_0} \|\frac{f^{n+1} + f^n}{2}\|_0^2.$$

Summing over n from 0 to $m - 1$, we complete the proof. □

Finally, we like to present the error estimate for the Crank-Nicolson scheme (6.106)-(6.107).

THEOREM 6.14
For any $m \geq 1$, the solution of (6.106)-(6.107) satisfies the error estimate

$$\|u(t_m) - u_h^m\|_0^2 + a_0 \triangle t \sum_{n=0}^{m-1} \|\nabla \left(\frac{u(t_{n+1}) - u_h^{n+1} + u(t_n) - u_h^n}{2} \right)\|_0^2$$

$$\leq C \left(h^{2k} + (\triangle t)^4 \right), \tag{6.109}$$

where the positive constant C is independent of h and $\triangle t$, and k is the degree of the finite element basis function in V_h. Furthermore, we have the optimal error estimate in the L^2 norm:

$$\max_{0 \leq n \leq N} \|u_h^n - u(t_n)\|_0 \leq C(h^{k+1} + (\triangle t)^2). \tag{6.110}$$

Proof. To make the proof easy to follow, we introduce the error between the finite element solution u_h^n and the elliptic projection $\Pi_h u(t_n)$:

$$e_h^n = u_h^n - \Pi_h u(t_n).$$

By using the definition of e_h^n, the weak formulation (6.86)-(6.87), and the definition of the elliptic projection, we obtain: For any $v \in V_h$,

$$\left(\frac{e_h^{n+1} - e_h^n}{\triangle t}, v \right) + \left(a \nabla \left(\frac{e_h^{n+1} + e_h^n}{2} \right), \nabla v \right)$$

$$= \left(\frac{f^{n+1} + f^n}{2}, v \right) - \left(\frac{\Pi_h(u(t_{n+1}) - u(t_n))}{\triangle t}, v \right) - \left(a \nabla \Pi_h \left(\frac{u(t_{n+1}) + u(t_n)}{2} \right), \nabla v \right)$$

$$= (\partial_t \left(\frac{u(t_{n+1}) + u(t_n)}{2} \right) - \frac{u(t_{n+1}) - u(t_n)}{\Delta t}, v)$$

$$+ ((I - \Pi_h) \frac{u(t_{n+1}) - u(t_n)}{\Delta t}, v) := (\delta_{n+\frac{1}{2}}, v). \tag{6.111}$$

Note that the error equation has the same form as (6.106). Hence, by adjusting the discrete stability (6.108), we have: For any $m \geq 1$,

$$||e_h^m||_0^2 + a_0 \Delta t \sum_{n=0}^{m-1} ||\nabla \left(\frac{e_h^{n+1} + e_h^n}{2} \right)||_0^2$$

$$\leq ||e_h^0||_0^2 + \frac{C_\Omega}{a_0} \sum_{n=0}^{m-1} \Delta t ||\delta_{n+\frac{1}{2}}||_0^2. \tag{6.112}$$

By the choice of initial condition (6.107), we see that $e_h^0 = 0$. Hence we just need to estimate $||\delta_{n+\frac{1}{2}}||_0$. By the projection error estimate (6.92), we have

$$||(I - \Pi_h) \frac{u(t_{n+1}) - u(t_n)}{\Delta t}||_0^2 = (\frac{1}{\Delta t})^2 || \int_{t_n}^{t_{n+1}} (I - \Pi_h)\partial_t u(s)ds||_0^2$$

$$\leq \frac{1}{\Delta t} \int_{t_n}^{t_{n+1}} ||(I - \Pi_h)\partial_t u(s)||_0^2 ds$$

$$\leq \frac{1}{\Delta t} \int_{t_n}^{t_{n+1}} ch^{2(k+1)}|\partial_t u|_{k+1}^2 ds. \tag{6.113}$$

Applying the following integral identity

$$\frac{w(t_{n+1}) + w(t_n)}{2} - \frac{1}{\Delta t} \int_{t_n}^{t_{n+1}} w(s)ds = \frac{1}{2\Delta t} \int_{t_n}^{t_{n+1}} (s - t_n)(t_{n+1} - s)w_{t^2}(s)ds$$

to $w(s) = \partial_t u(s)$, we obtain

$$||\partial_t \left(\frac{u(t_{n+1}) + u(t_n)}{2} \right) - \frac{u(t_{n+1}) - u(t_n)}{\Delta t}||_0^2$$

$$\leq \frac{1}{4(\Delta t)^2} || \int_{t_n}^{t_{n+1}} (s - t_n)(t_{n+1} - s)u_{t^3}(s)ds||_0^2$$

$$\leq \frac{1}{4(\Delta t)^2} (\int_{t_n}^{t_{n+1}} (s - t_n)^2(t_{n+1} - s)^2 ds)(\int_{t_n}^{t_{n+1}} ||u_{t^3}(s)||_0^2 ds)$$

$$\leq \frac{(\Delta t)^3}{4} \int_{t_n}^{t_{n+1}} ||u_{t^3}(s)||_0^2 ds. \tag{6.114}$$

Substituting the estimates (6.113) and (6.114) into (6.112), we obtain

$$||e_h^m||_0^2 + a_0 \Delta t \sum_{n=0}^{m-1} ||\nabla \left(\frac{e_h^{n+1} + e_h^n}{2} \right)||_0^2$$

$$\leq \frac{C_\Omega}{a_0} \sum_{n=0}^{m-1} [ch^{2(k+1)} \int_{t_n}^{t_{n+1}} |\partial_t u|_{k+1}^2 ds + \frac{(\triangle t)^4}{4} \int_{t_n}^{t_{n+1}} ||u_{t^3}(s)||_0^2 ds]$$

$$\leq C[h^{2(k+1)} \int_0^{t_m} |\partial_t u|_{k+1}^2 ds + (\triangle t)^4 \int_0^{t_m} ||u_{t^3}(s)||_0^2 ds]. \tag{6.115}$$

Using the triangle inequality and the projection error estimate (6.92), we obtain

$$||u(t_m) - u_h^m||_0^2 + a_0 \triangle t \sum_{n=0}^{m-1} ||\nabla \left(\frac{u(t_{n+1}) - u_h^{n+1} + u(t_n) - u_h^n}{2} \right) ||_0^2$$

$$\leq C \left(h^{2k} + (\triangle t)^4 \right),$$

where the constant $C > 0$ depends on u but not on h and $\triangle t$. This completes the proof of (6.109).

The optimal L^2 error estimate is easy to obtain by using the triangle inequality, the estimate (6.115) and the projection error estimate (6.92):

$$||u(t_m) - u_h^m||_0 \leq ||e_h^m||_0 + ||(I - \Pi_h)u(t_m)||_0 \leq C \left(h^{k+1} + (\triangle t)^2 \right).$$

Taking the maximum of m completes the proof of (6.110). □

Finally, we like to mention the general θ scheme for (6.86)-(6.87): For any $v \in V_h$, find $u_h^{n+1} \in V_h$, $n = 0, 1, \cdots, N - 1$, such that

$$(\frac{u_h^{n+1} - u_h^n}{\triangle t}, v) + (a\nabla(\theta u_h^{n+1} + (1 - \theta)u_h^n), \nabla v) = (\theta f^{n+1} + (1 - \theta)f^n, v), \tag{6.116}$$

$$u_h^0 = \Pi_h u_0, \tag{6.117}$$

where as usual $\theta \in [0, 1]$. When $\theta = 0$, we obtain the Forward Euler scheme; when $\theta = 1$, we obtain the Backward Euler scheme; and when $\theta = \frac{1}{2}$, we obtain the Crank-Nicolson scheme (6.106)-(6.107). Similar stability and error estimate can be derived for the θ scheme. We leave the proof to the interested reader as exercises. Details can be found in books such as [22, 35, 43].

6.9 Bibliographical remarks

Here we only present the very basic theory for analyzing a finite element method. More complete and advanced theory can be consulted in many classic books in this area (e.g., [10, 11, 17, 21, 22, 24, 26, 34, 42]). There are other more specialized books on different areas of finite element methods. For example, readers can find detailed discussions on mixed finite element methods in [9, 12, 23]; finite elements on anisotropic meshes [3]; hp finite element methods in [20, 32, 38, 41]; analysis for time-dependent problems in [22, 43]; finite element superconvergence theory in [14, 28, 30, 31, 46]; adaptive finite element methods in [2, 7, 45]; finite element methods for singularly perturbed problems [32, 37]; least-squares finite element methods and applications [8, 25]; finite element applications in fluids and electromagnetics [15, 29, 33]; and finite elements for reservoir simulation [16, 13].

6.10   Exercises

1. Let $\Omega \subset R^2$ be a bounded polygonal domain. Use the definition of the Sobolev space to prove that

$$|v|^2_{H^2(\Omega)} = \int_\Omega \{|\frac{\partial^2 v}{\partial x_1^2}|^2 + |\frac{\partial^2 v}{\partial x_2^2}|^2 + 2|\frac{\partial^2 v}{\partial x_1 \partial x_2}|^2\}dx_1 dx_2, \quad \forall\, v \in H^2(\Omega).$$

Furthermore, for any $v \in H_0^2(\Omega) \equiv \{w \in H^2(\Omega) : w = \frac{\partial w}{\partial n} = 0 \text{ on } \partial\Omega\}$, where $\frac{\partial}{\partial n}$ is the outward normal derivative, prove the relation [17, p. 15]:

$$|\triangle v|_{L^2(\Omega)} = |v|_{H^2(\Omega)},$$

where we denote $\triangle v = \frac{\partial^2 v}{\partial x_1^2} + \frac{\partial^2 v}{\partial x_2^2}$.

2. Assume that $f(x) \in C^1[a, b]$, and $f(a) = 0$. Prove the one-dimensional Poincaré inequality:

$$||f||_{L^2} \leq \frac{b - a}{\sqrt{2}}||f'||_{L^2}.$$

3. Let Ω be a bounded domain which has a Lipschitz boundary. It is known that the so-called trace theorem holds true [17, p. 13]:

$$||v||_{L^2(\partial\Omega)} \leq C_\Omega||v||_{H^1(\Omega)}, \quad \forall\, v \in H^1(\Omega), \tag{6.118}$$

where $C_\Omega > 0$ is a constant depending on the domain Ω. Prove that (6.118) is true for $u \in C^1(\Omega)$, where Ω is the unit disk in R^2:

$$\Omega = \{(x, y) : x^2 + y^2 = 1\} = \{(r, \theta) : 0 \leq r < 1,\ 0 \leq \theta < 2\pi\}.$$

Hint: Use $||v||_{L^2(\partial\Omega)} = (\int_0^{2\pi} |v(1, \theta)|^2 d\theta)^{1/2}$, and start with

$$v(1, \theta)^2 = \int_0^1 \frac{\partial}{\partial r}(r^2 v(r, \theta)^2)dr.$$

4. For the boundary value problem

$$-u'' + u' + u = f \quad \text{in } (0, 1),$$

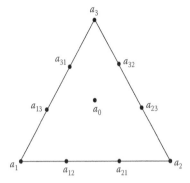

FIGURE 6.9
The P_3 element.

$$u'(0) = u'(1) = 0,$$

we can form the following variational problem: find $u \in H^1(0,1)$ such that

$$A(u,v) = (f,v) \quad \forall\, v \in H^1(0,1),$$

where $A(u,v) = \int_0^1 (u'v' + u'v + uv)dx$. Show that $A(\cdot,\cdot)$ is continuous and coercive.

5. Derive the variational formulation for the problem

$$-\triangle u + au = f \quad \text{in } \Omega,$$
$$\frac{\partial u}{\partial n} = g \quad \text{on } \partial\Omega,$$

where $a(x) \geq a_* > 0$ on Ω, and $f \in L^2(\Omega), g \in L^2(\Omega)$. Check if the conditions of the Lax-Milgram lemma are satisfied.

6. For a triangle K (see Fig. 6.9). Prove that a function $v \in P_3(K)$ is uniquely determined by the values $v(a_i), v(a_{i,j})$ and $v(a_0), 1 \leq i, j \leq 3, i \neq j$.

7. Prove that the problem (6.65) is equivalent to the minimization in $H_0^1(\Omega)$ of the energy functional

$$J(v) = \frac{1}{2}A(v,v) - (f,v),$$

i.e., $u = \min_{v \in H_0^1(\Omega)} J(v)$.

8. Let V_h be the finite element space (6.67), P_h^k is the projection operator from $L^2(\Omega) \rightarrow V_h$, i.e., for any $v \in L^2(\Omega)$,

$$(P_h^k v - v, w_h) = 0, \quad \forall\, w_h \in V_h.$$

Prove that

$$\|P_h^k v - v\|_{L^2(\Omega)} \leq Ch^{l+1}|v|_{H^{l+1}(\Omega)}, \quad 1 \leq l \leq k, \ \forall\, v \in H^{l+1}(\Omega).$$

9. Prove Gronwall's inequality stated in Lemma 6.7. Hint: Let $F(t) = \int_0^t f(\tau)d\tau$ and rewrite (6.93) as $\frac{d}{dt}[e^{-ct}F(t)] \le g(t)e^{-ct}$.

10. Prove that the implicit Backward Euler scheme (6.116)-(6.117) with $\theta = 1$ is unconditionally stable. Furthermore, prove that the following optimal L^2 error estimate holds true: For any $m \ge 1$,

$$\|u(t_m) - u_h^m\|_0 \le C\left(h^{k+1} + (\triangle t)^2\right),$$

where $k \ge 1$ is the degree of the polynomial basis function in V_h.

11. Let $v = a_1 + a_2 x + a_3 y + a_4 xy$, where $a_i \in R$ $(i = 1, 2, 3, 4)$, and K is a rectangle. Is v uniquely determined by its values at the four edge midpoints of K?

12. Let $v = a_1 + a_2 x + a_3 y + a_4(x^2 - y^2)$, where $a_i \in R$ $(i = 1, 2, 3, 4)$, and K is a rectangle. Prove that v is uniquely defined by its values at the four edge midpoints of K.

13. Let $v = a_1 + a_2 x + a_3 y + a_4(x^2 - y^2)$, where $a_i \in R$ $(i = 1, 2, 3, 4)$, and K is a rectangle. Prove that v is uniquely defined by its four integral values $\int_{l_i} v\, dl$ on the edges l_i of K.

14. Let $v = a_1 + a_2 x + a_3 y + a_4 z + a_5(x^2 - y^2) + a_6(x^2 - z^2)$, where $a_i \in R$ $(i = 1, \cdots, 6)$, and K is a cubic element. Show that v is uniquely defined by its values at the centroids of the six faces of K.

15. Let $v = a_1 + a_2 x + a_3 y + a_4 z + a_5(x^2 - y^2) + a_6(x^2 - z^2)$, where $a_i \in R$ $(i = 1, \cdots, 6)$, and K is a cubic element. Show that v is uniquely defined by its mean values $\int_{e_i} v\, ds$ over six faces e_i of K.

References

[1] R.A. Adams. *Sobolev Spaces*. Academic Press, New York, NY, 1975.

[2] M. Ainsworth and J.T. Oden. *A Posteriori Error Estimation in Finite Element Analysis*. Wiley-Interscience, New York, NY, 2000.

[3] T. Apel. *Anisotropic Finite Elements: Local Estimates and Applications*. Teubner, Stuttgart, 1999.

[4] T. Arbogast and Z. Chen. On the implementation of mixed methods as nonconforming methods for second-order elliptic problems. *Math. Comp.*, 64(211):943–972, 1995.

[5] J.H. Argyris. Energy theorems and structural analysis. *Aircraft Engineering*, 26:347–356, 1954.

[6] I. Babuska and A.K. Aziz. Survey lectures on the mathematical foundations of the finite element method, in *The Mathematical Foundations*

of the Finite Element Method with Applications to Partial Differential Equations, A.K. Aziz (ed.), Academic Press, New York, NY, 3–359, 1972.

[7] W. Bangerth and R. Rannacher. *Adaptive Finite Element Methods for Solving Differential Equations*. Birkhäuser, Basel, 2003.

[8] P.B. Bochev and M.D. Gunzburger. *Least-Squares Finite Element Methods*. Springer-Verlag, New York, 2009.

[9] D. Boffi, F. Brezzi and M. Fortin. *Mixed Finite Element Methods and Applications*. Springer-Verlag, Berlin, 2013.

[10] D. Braess. *Finite Elements*. Cambridge University Press, 3rd Edition, Cambridge, UK, 2007.

[11] S.C. Brenner and L.R. Scott. *The Mathematical Theory of Finite Element Methods*. Springer-Verlag, Berlin/Heidelberg, 1994.

[12] F. Brezzi and M. Fortin. *Mixed and Hybrid Finite Element Methods*. Springer-Verlag, Berlin/Heidelberg, 1991.

[13] G. Chavent and J. Jaffré. *Mathematical Models and Finite Elements for Reservoir Simulation*. North-Holland, Amsterdam, 1978.

[14] C.M. Chen and Y.Q. Huang, *High Accuracy Theory of Finite Element Methods* (in Chinese). Hunan Science Press, China, 1995.

[15] Z. Chen. *Finite Element Methods and Their Applications*. Springer-Verlag, Berlin, 2005.

[16] Z. Chen, G. Huan and Yuanle Ma. *Computational Methods for Multiphase Flows in Porous Media*. SIAM, Philadelphia, PA, 2006.

[17] P.G. Ciarlet. *The Finite Element Method for Elliptic Problems*. North-Holland, Amsterdam, 1978.

[18] R.W. Clough. The finite element method in plane stress analysis, in *Proceedings of the Second ASCE Conference on Electronic Computation*. Pittsburgh, Pennsylvania, 1960.

[19] R. Courant. Variational methods for the solution of problems of equilibrium and vibrations. *Bull. Amer. Math. Soc.*, 49:1–23, 1943.

[20] L. Demkowicz. *Computing with hp-Adaptive Finite Elements*. Chapman & Hall/CRC, Boca Raton, FL, 2006.

[21] K. Eriksson, D. Estep, P. Hansbo, C. Johnson. *Computational Differential Equations*. Cambridge University Press, Cambridge, UK, 1996.

[22] G. Fairweather. *Finite Element Galerkin Methods for Differential Equations*. Marcel Dekker, New York-Basel, 1978.

[23] V. Girault and P.A. Raviart. *Finite Element Methods for Navier-Stokes Equations – Theory and Algorithms.* Springer-Verlag, Berlin, 1986.

[24] J.-L. Guermond and A. Ern. *Theory and Practice of Finite Elements.* Springer-Verlag, New York, 2004.

[25] B.N. Jiang. *The Least-Squares Finite Element Method: Theory and Applications in Fluid Dynamics and Electromagnetics.* Springer-Verlag, Berlin, 1998.

[26] C. Johnson. *Numerical Solution of Partial Differential Equations by the Finite Element Method.* Cambridge University Press, New York, 1988.

[27] M. Krizek, P. Neittaanmaki and R. Stenberg (eds.). *Finite Element Methods: Fifty Years of the Courant Element.* Marcel Dekker, New York, 1994.

[28] M. Krizek, P. Neittaanmaki and R. Stenberg (eds.). *Finite Element Methods: Superconvergence, Post-Processing A Posteriori Estimates.* Marcel Dekker, New York, 1998.

[29] W. Layton. *Introduction to the Numerical Analysis of Incompressible Viscous Flows.* SIAM, Philadelphia, PA, 2008.

[30] Q. Lin and J. Lin. *Finite Element Methods: Accuracy and Improvement.* Science Press, Beijing, 2007.

[31] Q. Lin and N. Yan. *The Construction and Analysis of High Accurate Finite Element Methods* (in Chinese). Hebei University Press, Hebei, China, 1996.

[32] J.M. Melenk. *hp-Finite Element Methods for Singular Perturbations.* Springer, New York, NY, 2002.

[33] P. Monk. *Finite Element Methods for Maxwell's Equations.* Oxford University Press, Oxford, UK, 2003.

[34] J.T. Oden and G.F. Carey. *Finite Elements – Mathematical Aspects,* Vol. IV. Prentice-Hall, Englewood Cliffs, NJ, 1983.

[35] A. Quarteroni and A. Valli. *Numerical Approximation of Partial Differential Equations.* Springer-Verlag, Berlin, 1994.

[36] R. Rannacher and S. Turek. Simple nonconforming quadrilateral Stokes element. *Numer. Methods Partial Differential Equations,* 8(2):97–111, 1992.

[37] H.-G. Roos, M. Stynes and L. Tobiska. *Numerical Methods for Singularly Perturbed Differential Equations: Convection-Diffusion and Flow Problems.* Springer-Verlag, Berlin, 1996.

[38] C. Schwab. *p- and hp- Finite Element Methods, Theory and Applications to Solid and Fluid Mechanics*. Oxford University Press, New York, NY, 1998.

[39] Z.C. Shi. Error estimates of Morley element. *Chinese J. Numer. Math. & Appl.*, 12:102–108, 1990.

[40] Z.C. Shi. A convergence condition for the quadrilateral Wilson element. *Numer. Math.*, 44(3):349–361, 1984.

[41] P. Solin, K. Segeth and I. Dolezel. *Higher-Order Finite Element Methods*. Chapman & Hall/CRC Press, Boca Raton, FL, 2003.

[42] G. Strang and G.J. Fix. *An Analysis of the Finite Element Method*. Prentice-Hall, Englewood Cliffs, NJ, 1973.

[43] V. Thomee. *Galerkin Finite Element Methods for Parabolic Problems*. Springer-Verlag, Berlin/New York, 1997.

[44] M.J. Turner, R.W. Clough, H.C. Martin and L.J. Topp. Stiffness and deflection analysis of complex structures. *J. Aero. Sci.*, 23:805–823, 1956.

[45] R. Verfürth. *A Review of a Posteriori Error Estimation and Adaptive Mesh-Refinement Techniques*. Wiley, Teubner, 1996.

[46] L. Wahlbin. *Superconvergence in Galerkin Finite Element Methods*. Springer-Verlag, Berlin/Heidelberg, 1995.

[47] M.F. Wheeler. A priori L_2 error estimates for Galerkin approximations to parabolic partial differential equations. *SIAM J. Numer. Anal.*, 10:723–759, 1973.

7

Finite Element Methods: Programming

In this chapter, we will demonstrate the practical implementation of finite element methods for a 2-D convection-diffusion problem:

$$-\triangle u + \vec{v} \cdot \nabla u = s \quad \text{in } \Omega, \tag{7.1}$$

$$u = u_{exact} \quad \text{on } \partial\Omega. \tag{7.2}$$

The basic procedure for solving partial differential equations (PDEs) is as follows:

1. Discretize the physical domain into finite elements;
2. Rewrite the PDE in its weak formulation;
3. Calculate element matrices for each finite element;
4. Assemble element matrices to form a global linear system;
5. Implement the boundary conditions by modifying the global linear system;
6. Solve the linear system and postprocess the solution.

In the rest of this chapter, we will discuss details of implementation of the above steps. More specifically, in Sec. 7.1, we present a simple grid-generation algorithm and its implementation. Sec. 7.2 formulates the finite element equation for the convection-diffusion model. In Sec. 7.3, we discuss how to calculate those elementary matrices. Then we discuss the finite element assembly procedure and how to implement the Dirichlet boundary condition in Sec. 7.4. In Sec. 7.5, we present the complete MATLAB code for the P_1 element and an example problem solved by our code. Finally, we present a MATLAB code for the Q_1 element and demonstrate how to use it to solve an elliptic problem.

7.1 FEM mesh generation

For simplicity, we assume that the physical domain Ω is a polygon, which can be discretized into N_E triangular elements T_h. Here we describe a simple algorithm: implementation starting with a very coarse mesh, then generating different levels of finer meshes by uniform subdivision, i.e., each triangle is subdivided into four smaller triangles [13].

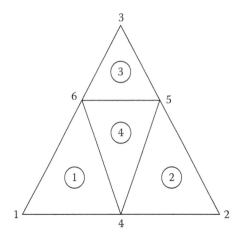

FIGURE 7.1
Uniform refinement of a given triangle 123.

For example (see Fig. 7.1), a triangle 123 is subdivided into four smaller triangles $146, 425, 653$ and 456, which are labeled as subtriangles $1, 2, 3$, and 4, respectively. Note that vertices of each triangle are labeled counterclockwise.

An FEM mesh generally contains much more information than the finite difference method. First, we generate the nodal coordinates $(x_i, y_i), 1 \leq i \leq N_G$, where N_G denotes the total number of nodes in the mesh. Then we need a connectivity matrix $conn(i, j)$ to describe the relation between local nodes and global nodes. For a linear triangular grid, $conn(i, j), j = 1, 2, 3, i = 1, \cdots, N_E$, denotes the global label of the jth node of the ith element, where N_E is the total number of elements.

To implement boundary conditions, we need to know which nodes are on the boundary, and what types of boundary conditions are there. In our implementation, we introduce a global boundary condition indicator $gbc(i, j), 1 \leq i \leq N_G, 1 \leq j \leq 2$. For the ith global node, we define

$$gbc(i, 1) = \begin{cases} 0 & \text{if it is an interior node} \\ 1 & \text{if it is a Dirichlet node} \\ 2 & \text{if it is a Neumann node} \end{cases}$$

Furthermore, when $gbc(i, 1) \neq 0$, we use $gbc(i, 2)$ to specify the corresponding boundary value.

In order to generate $gbc(i, j)$ easily, we let $gbc(i, j)$ inherit directly from the element boundary condition indicator $efl(i, j), 1 \leq i \leq N_E, 1 \leq j \leq 3$. Note that $efl(i, j)$ can be generated from the initial coarse mesh as follows: when the jth local node is a midpoint generated from two vertices of the previous level, it will be an interior node if either of the two vertices is an interior

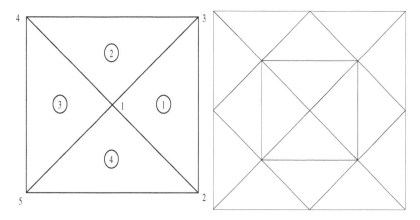

FIGURE 7.2
The initial mesh and the one after one refinement.

node; it will be a Dirichlet (or Neumann) node if both vertices are Dirichlet (or Neumann) nodes; it will be a Neumann node if one is Dirichlet and the other one is Neumann; when the jth local node is a vertex from the previous level, it will have the same type as that vertex.

An example initial coarse mesh and the mesh after one refinement are given in Fig. 7.2, where we label all the nodes and elements for the initial mesh. The connectivity matrix for the initial mesh is:

$$conn(1, 1:3) = 1, 2, 3, \quad conn(2, 1:3) = 1, 3, 4,$$
$$conn(3, 1:3) = 1, 4, 5, \quad conn(4, 1:3) = 1, 5, 2.$$

The MATLAB source code for generating the triangular grid is listed as follows.

```
%------------------------------------------------------------
% generate a triangular grid by uniform refinement.
%
% ne: total number of elements
% np: total number of nodes
% p:  the x- and y-coordinates of all nodes
% conn: connectivity matrix
% gbc: boundary condition indicator for all nodes
% nref>=1: number of refinement level
%------------------------------------------------------------
function [ne,np,p,conn,gbc] = gen_p1grid(nref)

% define an initial coarse mesh
ne = 4;
```

```
x(1,1)= 0.5; y(1,1)= 0.5; efl(1,1)=0;  % 0 for inner node
x(1,2)= 1.0; y(1,2)= 0.0; efl(1,2)=1;  % 1 for Dirichlet node
x(1,3)= 1.0; y(1,3)= 1.0; efl(1,3)=1;

x(2,1)= 0.5; y(2,1)= 0.5; efl(2,1)=0;  % second element
x(2,2)= 1.0; y(2,2)= 1.0; efl(2,2)=1;
x(2,3)= 0.0; y(2,3)= 1.0; efl(2,3)=1;

x(3,1)= 0.5; y(3,1)= 0.5; efl(3,1)=0;  % third element
x(3,2)= 0.0; y(3,2)= 1.0; efl(3,2)=1;
x(3,3)= 0.0; y(3,3)= 0.0; efl(3,3)=1;

x(4,1)= 0.5; y(4,1)= 0.5; efl(4,1)=0;  % fourth element
x(4,2)= 0.0; y(4,2)= 0.0; efl(4,2)=1;
x(4,3)= 1.0; y(4,3)= 0.0; efl(4,3)=1;

% generate finer mesh based on the given coarse mesh
for i=1:nref
  nm = 0; % count the elements generated by each refinement

  for j=1:ne
    % mid-edge nodes of each triangle become new nodes
    x(j,4)=0.5*(x(j,1)+x(j,2)); y(j,4)=0.5*(y(j,1)+y(j,2));
    x(j,5)=0.5*(x(j,2)+x(j,3)); y(j,5)=0.5*(y(j,2)+y(j,3));
    x(j,6)=0.5*(x(j,3)+x(j,1)); y(j,6)=0.5*(y(j,3)+y(j,1));

    % generate mid-node BC indicatior from its parent
    if (efl(j,1)==1 & efl(j,2)==1)
        efl(j,4) = 1;
    else
        efl(j,4)=0;
    end
    if (efl(j,2)==1 & efl(j,3)==1)
        efl(j,5) = 1;
    else
        efl(j,5)=0;
    end
    if (efl(j,3)==1 & efl(j,1)==1)
        efl(j,6) = 1;
    else
        efl(j,6)=0;
    end

  % generate four sub-elements
```

```
nm = nm+1;    %  1st sub-element
xn(nm,1)=x(j,1); yn(nm,1)=y(j,1); efln(nm,1)=efl(j,1);
xn(nm,2)=x(j,4); yn(nm,2)=y(j,4); efln(nm,2)=efl(j,4);
xn(nm,3)=x(j,6); yn(nm,3)=y(j,6); efln(nm,3)=efl(j,6);

nm = nm+1;    %  2nd sub-element
xn(nm,1)=x(j,4); yn(nm,1)=y(j,4); efln(nm,1)=efl(j,4);
xn(nm,2)=x(j,2); yn(nm,2)=y(j,2); efln(nm,2)=efl(j,2);
xn(nm,3)=x(j,5); yn(nm,3)=y(j,5); efln(nm,3)=efl(j,5);

nm = nm+1;    %  3rd sub-element
xn(nm,1)=x(j,6); yn(nm,1)=y(j,6); efln(nm,1)=efl(j,6);
xn(nm,2)=x(j,5); yn(nm,2)=y(j,5); efln(nm,2)=efl(j,5);
xn(nm,3)=x(j,3); yn(nm,3)=y(j,3); efln(nm,3)=efl(j,3);

nm = nm+1;    %  4th sub-element
xn(nm,1)=x(j,4); yn(nm,1)=y(j,4); efln(nm,1)=efl(j,4);
xn(nm,2)=x(j,5); yn(nm,2)=y(j,5); efln(nm,2)=efl(j,5);
xn(nm,3)=x(j,6); yn(nm,3)=y(j,6); efln(nm,3)=efl(j,6);
end % end of loop over current elements

ne = 4*ne;  % increase the number of elements by a factor of four
for k=1:ne     % relabel the new points
   for l=1:3
      x(k,l)=  xn(k,l);         y(k,l)=  yn(k,l);
      efl(k,l)=efln(k,l);
   end
end
end % end of refinement loop

% get rid of redundant mid-edge nodes:
% fix the first element, then loop the rest elements
p(1,1)=x(1,1); p(1,2)=y(1,1); gbc(1,1)=efl(1,1);
p(2,1)=x(1,2); p(2,2)=y(1,2); gbc(2,1)=efl(1,2);
p(3,1)=x(1,3); p(3,2)=y(1,3); gbc(3,1)=efl(1,3);
conn(1,1)=1;   conn(1,2)=2;   conn(1,3)=3;

np=3;    % we already has 3 nodes from 1st element!
% loop over rest elements: Id=0 means a new node
eps = 1.0e-8;
for i=2:ne         % loop over elements
 for j=1:3            % loop over element nodes

 Id=0;
 for k=1:np
```

```
if(abs(x(i,j)-p(k,1)) < eps & abs(y(i,j)-p(k,2)) < eps)
    Id = 1;            % indicate this node has already been used
    conn(i,j) = k;   % jth node of element i = kth global node
  end
end

if(Id==0)  % record the new node
  np = np+1;
  p(np,1)=x(i,j);           p(np,2)=y(i,j);
  gbc(np,1) = efl(i,j);   conn(i,j)=np;
end

end
end  % end of loop over elements

return;
```

7.2   Forming FEM equations

The weak formulation of (7.1)-(7.2) is: Find $u \in H^1(\Omega)$ such that

$$(\nabla u, \nabla \phi) + (\vec{v} \cdot \nabla u, \phi) = (s, \phi), \quad \forall \phi \in H_0^1(\Omega) \tag{7.3}$$

The corresponding finite element method is: Find $u_h \in V_h \subset H^1(\Omega)$ such that

$$(\nabla u_h, \nabla \phi_h) + (\vec{v} \cdot \nabla u_h, \phi_h) = (s, \phi_h), \quad \forall \phi_h \in V_h, \tag{7.4}$$

where the finite element space

$$V_h = \{v : v \text{ is continuous on } \Omega \text{ and } v|_E \in P_1(E), E \in T_h\}.$$

Suppose that u is approximated over a finite element E (here it is a triangle) by

$$u(x,y) \approx u_h^E(x,y) = \sum_{j=1}^{3} u_j^E \psi_j^E(x,y), \tag{7.5}$$

where u_j^E is the value of u_h at the jth node of the element, and ψ_j^E is the Lagrange interpolation function, which satisfies

$$\psi_j^E(x_i, y_i) = \delta_{ij}.$$

Over each element E, we need to calculate some element matrices as follows. Substituting (7.5) into (7.4) with test function $\phi = \psi_i^E, i = 1, 2, 3$, respectively, and approximating the source function s by

$$s(x,y) \approx \sum_{j=1}^{3} s_j \psi_j^E(x,y), \quad s_j = s(x_j, y_j),$$

we obtain the element diffusion matrix (stiffness matrix)

$$A_{ij} \equiv \int_E \nabla \psi_j^E \cdot \nabla \psi_i^E \, dx dy, \quad i,j = 1,2,3, \tag{7.6}$$

the element convection matrix

$$B_{ij} \equiv \int_E (\vec{v} \cdot \nabla \psi_j^E) \psi_i^E \, dx dy, \quad i,j = 1,2,3, \tag{7.7}$$

and the element mass matrix

$$M_{ij} \equiv \int_E \psi_j^E \psi_i^E \, dx dy, \quad i,j = 1,2,3. \tag{7.8}$$

Summing up all elements $E_n, 1 \leq n \leq N_E$, of the mesh T_h, we obtain a system of linear equations for the unknown numerical solution u_j at all nodes:

$$\sum_{n=1}^{N_E} (A_{ij} + B_{ij}) u_j = \sum_{n=1}^{N_E} M_{ij} s_j. \tag{7.9}$$

7.3 Calculation of element matrices

The element matrices are often computed on a reference element. The mapping to an arbitrary triangular element with vertices $(x_i, y_i), 1 \leq i \leq 3$, from a reference element with vertices

$$(\xi_1, \eta_1) = (0,0), \quad (\xi_2, \eta_2) = (1,0), \quad (\xi_3, \eta_3) = (0,1)$$

is given by

$$x = x_1 + (x_2 - x_1)\xi + (x_3 - x_1)\eta = \sum_{j=1}^{3} x_j \hat{\psi}_j(\xi, \eta), \tag{7.10}$$

$$y = y_1 + (y_2 - y_1)\xi + (y_3 - y_1)\eta = \sum_{j=1}^{3} y_j \hat{\psi}_j(\xi, \eta), \tag{7.11}$$

where

$$\hat{\psi}_1(\xi, \eta) = \zeta = 1 - \xi - \eta, \quad \hat{\psi}_2(\xi, \eta) = \xi, \quad \hat{\psi}_3(\xi, \eta) = \eta.$$

Note that (ξ, η, ζ) are the triangle barycentric coordinates.

The Jacobi matrix of the mapping is denoted as

$$J \equiv \begin{bmatrix} \frac{\partial x}{\partial \xi} & \frac{\partial x}{\partial \eta} \\ \frac{\partial y}{\partial \xi} & \frac{\partial y}{\partial \eta} \end{bmatrix},$$

which determinant is $det(J) = \begin{vmatrix} x_2 - x_1 & x_3 - x_1 \\ y_2 - y_1 & y_3 - y_1 \end{vmatrix} = 2A$, where A denotes the area of the triangle.

The interpolation function $\psi_j(x, y)$ over the physical triangle is defined as

$$\psi_j(x, y) = \hat{\psi}_j(\xi(x, y), \eta(x, y)),$$

from which we have

$$\frac{\partial \hat{\psi}_j}{\partial \xi} = \frac{\partial \psi_j}{\partial x}\frac{\partial x}{\partial \xi} + \frac{\partial \psi_j}{\partial y}\frac{\partial y}{\partial \xi}, \qquad (7.12)$$

$$\frac{\partial \hat{\psi}_j}{\partial \eta} = \frac{\partial \psi_j}{\partial x}\frac{\partial x}{\partial \eta} + \frac{\partial \psi_j}{\partial y}\frac{\partial y}{\partial \eta}, \qquad (7.13)$$

i.e.,

$$J^T \cdot \nabla\psi_j = \begin{bmatrix} \frac{\partial \hat{\psi}_j}{\partial \xi} \\ \frac{\partial \hat{\psi}_j}{\partial \eta} \end{bmatrix}, \quad j = 1, 2, 3,$$

where we denote the gradient

$$\nabla\psi_j = \begin{bmatrix} \frac{\partial \psi_j}{\partial x} \\ \frac{\partial \psi_j}{\partial y} \end{bmatrix}.$$

In particular,

$$J^T \cdot \nabla\psi_1 = -\begin{bmatrix} 1 \\ 1 \end{bmatrix}, \quad J^T \cdot \nabla\psi_2 = \begin{bmatrix} 1 \\ 0 \end{bmatrix}, \quad J^T \cdot \nabla\psi_3 = \begin{bmatrix} 0 \\ 1 \end{bmatrix}. \qquad (7.14)$$

By linear algebra, it is easy to prove the following lemma:

LEMMA 7.1

For any 2×2 nonsingular matrix $M = \begin{bmatrix} a & b \\ c & d \end{bmatrix}$,

its inverse $M^{-1} = \frac{1}{det(M)}\begin{bmatrix} d & -b \\ -c & a \end{bmatrix}$.

Using Lemma 7.1, we obtain

$$(J^T)^{-1} = \frac{1}{2A}\begin{bmatrix} y_3 - y_1 & -(y_2 - y_1) \\ -(x_3 - x_1) & x_2 - x_1 \end{bmatrix},$$

substituting which into (7.14) gives us

$$\nabla\psi_1 = \frac{1}{2A}\begin{bmatrix} -(y_3 - y_2) \\ x_3 - x_2 \end{bmatrix}, \nabla\psi_2 = \frac{1}{2A}\begin{bmatrix} -(y_1 - y_3) \\ x_1 - x_3 \end{bmatrix}, \nabla\psi_3 = \frac{1}{2A}\begin{bmatrix} -(y_2 - y_1) \\ x_2 - x_1 \end{bmatrix}.$$

Hence the element diffusion matrix

$$A_{ij} = \int_E \nabla\psi_j \cdot \nabla\psi_i dxdy = 2A \int_{\hat{E}} \nabla\psi_j \cdot \nabla\psi_i d\xi d\eta$$

can be obtained without any numerical integration, since $\nabla\psi_j$ are independent of ξ and η.

LEMMA 7.2

$$A_{11} = \frac{1}{4A}[(y_3 - y_2)^2 + (x_3 - x_2)^2], \tag{7.15}$$

$$A_{12} = \frac{1}{4A}[(y_3 - y_2)(y_1 - y_3) + (x_3 - x_2)(x_1 - x_3)], \tag{7.16}$$

$$A_{13} = \frac{1}{4A}[(y_3 - y_2)(y_2 - y_1) + (x_3 - x_2)(x_2 - x_1)], \tag{7.17}$$

$$A_{23} = \frac{1}{4A}[(y_1 - y_3)(y_2 - y_1) + (x_1 - x_3)(x_2 - x_1)], \tag{7.18}$$

$$A_{22} = \frac{1}{4A}[(y_1 - y_3)^2 + (x_1 - x_3)^2], \tag{7.19}$$

$$A_{33} = \frac{1}{4A}[(y_2 - y_1)^2 + (x_2 - x_1)^2]. \tag{7.20}$$

LEMMA 7.3
[4, p. 201]

$$\int_{\hat{E}} \zeta^l \xi^m \eta^n d\xi d\eta = \frac{l!m!n!}{(l + m + n + 2)!},$$

where $l, m,$ and n are non-negative integers.

Assume that the velocity $\vec{v} = (v_x, v_y)$ is constant, then the element convection matrix can be simplified as

$$B_{ij} = \int_E (\vec{v} \cdot \nabla\psi_j)\psi_i dxdy \tag{7.21}$$

$$= (\vec{v} \cdot \nabla\psi_j)\int_{\hat{E}} \hat{\psi}_i(\xi, \eta) \cdot 2Ad\xi d\eta = \frac{A}{3}(\vec{v} \cdot \nabla\psi_j), \tag{7.22}$$

where we used the fact

$$\int_{\hat{E}} \xi d\xi d\eta = \int_{\hat{E}} \eta d\xi d\eta = \int_{\hat{E}} \zeta d\xi d\eta = \frac{1}{6},$$

which is obtained by Lemma 7.3.

LEMMA 7.4

$$B_{11} = \frac{1}{6}[v_y \cdot (x_3 - x_2) - v_x \cdot (y_3 - y_2)], \quad B_{21} = B_{31} = B_{11}, \quad (7.23)$$

$$B_{12} = \frac{1}{6}[v_y \cdot (x_1 - x_3) - v_x \cdot (y_1 - y_3)], \quad B_{22} = B_{32} = B_{12}, \quad (7.24)$$

$$B_{13} = \frac{1}{6}[v_y \cdot (x_2 - x_1) - v_x \cdot (y_2 - y_1)], \quad B_{23} = B_{33} = B_{13}. \quad (7.25)$$

Similarly, the element mass matrix can be calculated

$$M_{ij} = \int_E \psi_j^E \psi_i^E \, dx \, dy = 2A \int_{\hat{E}} \hat{\psi}_j \hat{\psi}_i \, d\xi \, d\eta,$$

from which and Lemma 7.3, we have

LEMMA 7.5

$$M = 2A \int_{\hat{E}} \begin{bmatrix} \zeta^2 & \zeta\xi & \zeta\eta \\ \xi\zeta & \xi^2 & \xi\eta \\ \eta\zeta & \eta\xi & \eta^2 \end{bmatrix} d\xi\,d\eta = \frac{A}{12} \begin{bmatrix} 2 & 1 & 1 \\ 1 & 2 & 1 \\ 1 & 1 & 2 \end{bmatrix}.$$

Below are the MATLAB functions for calculating the element diffusion, advection and mass matrices.

```
%-----------------------------------------------------------
% evaluate the element diffusion matrix for P1 element
%-----------------------------------------------------------
function [elmA] = locA(x1,y1,x2,y2,x3,y3)

dx23 = x2-x3; dy23 = y2-y3;
dx31 = x3-x1; dy31 = y3-y1;
dx12 = x1-x2; dy12 = y1-y2;

A = 0.5*(dx31*dy12 - dy31*dx12);  % triangle area

elmA(1,1) = 0.25*(dx23*dx23 + dy23*dy23)/A;
elmA(1,2) = 0.25*(dx23*dx31 + dy23*dy31)/A;
elmA(1,3) = 0.25*(dx23*dx12 + dy23*dy12)/A;

elmA(2,1) = 0.25*(dx31*dx23 + dy31*dy23)/A;
elmA(2,2) = 0.25*(dx31*dx31 + dy31*dy31)/A;
elmA(2,3) = 0.25*(dx31*dx12 + dy31*dy12)/A;
```

```
elmA(3,1) = 0.25*(dx12*dx23 + dy12*dy23)/A;
elmA(3,2) = 0.25*(dx12*dx31 + dy12*dy31)/A;
elmA(3,3) = 0.25*(dx12*dx12 + dy12*dy12)/A;

return;
```

```
%---------------------------------------------------------
% calculate the element matrix from convection term
%---------------------------------------------------------
function [elmB] = locB(vx,vy,x1,y1,x2,y2,x3,y3)

 dx32 = x3-x2; dy32 = y3-y2;
 dx13 = x1-x3; dy13 = y1-y3;
 dx21 = x2-x1; dy21 = y2-y1;

 elmB(1,1) = (-vx*dy32 + vy*dx32)/6.0;
 elmB(1,2) = (-vx*dy13 + vy*dx13)/6.0;
 elmB(1,3) = (-vx*dy21 + vy*dx21)/6.0;

 elmB(2,1) = elmB(1,1);    elmB(3,1) = elmB(1,1);
 elmB(2,2) = elmB(1,2);    elmB(3,2) = elmB(1,2);
 elmB(2,3) = elmB(1,3);    elmB(3,3) = elmB(1,3);

return;
```

```
%---------------------------------------------------------------
% calculate the element mass matrix for P1 element
%---------------------------------------------------------------
function [elmM] = locM(x1,y1,x2,y2,x3,y3)

 dx23 = x2-x3; dy23 = y2-y3;
 dx31 = x3-x1; dy31 = y3-y1;
 dx12 = x1-x2; dy12 = y1-y2;

 A = 0.5*(dx31*dy12 - dy31*dx12);
 c_diag = A/6;    % diagonal constant
 c_off = A/12;    % off-diagnonal constant

 for j=1:3
     for i=1:3
         if(i==j)
             elmM(i,j) = c_diag;
         else
```

```
                elmM(i,j) = c_off;
        end
    end
end

return;
```

7.4 Assembly and implementation of boundary conditions

To obtain the global coefficient matrix by assembling the contributions from each element coefficient matrix, we just need to loop through all elements in the mesh, find the corresponding global nodal label for each local node, and put them in the right locations of the global coefficient matrix. A pseudo-code is listed below:

```
for n=1:NE     % loop through all elements
    computer element coefficient matrix
    for i=1:3     % loop through all nodes in the element
        i1 = conn(n,i)
        for j=1:3
            j1=conn(n,j)
            Ag(i1,j1)=Ag(i1,j1)+Aloc(i,j)
        End
    End
End
```

The implementation of the Dirichlet boundary condition can be done after assembly, i.e., we have a linear system

$$Ag \cdot \mathbf{u} = \mathbf{b} \tag{7.26}$$

Suppose we have to specify $u = u(x_k, y_k)$ at the kth global node. We can impose this Dirichlet boundary condition as follows:

1. Replace each entry b_i of $\mathbf{b}$ with $b_i - Ag_{ik}u_k$.
2. Reset all entries in the kth row and kth column of Ag to zero, then the diagonal entry Ag_{kk} to one.
3. Replace the kth entry b_k of $\mathbf{b}$ with u_k.

A pseudo-code is listed below:

```
for k=1:NG     % loop through all global nodes
    If (gbc(k,1) = 1) then   % find a Dirichlet node
        for i=1:NG
```

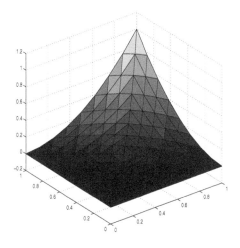

FIGURE 7.3
Numerical solutions obtained with $nref = 3$.

```
        b(i) = b(i) - Ag(i,k)*gbc(k,2)
        Ag(i,k) = 0
        Ag(k,i) = 0
      End
      Ag(k,k) = 1
      b(k) = gbc(k,2)
   End
 End
```

7.5 The MATLAB code for P_1 element

Here we provide our implementation for solving (7.1)-(7.2) with $\vec{v} = (1, 1)$, and $s = -2y^2 - 2x^2 + 2xy^2 + 2yx^2$ such that the exact solution is given by $u = x^2 y^2$ over the domain $\Omega = [0, 1]^2$.

We solved this problem using the P_1 element using several levels of meshes; the maximum errors are

$$0.0044, 0.0014, 4.1030e - 004, 1.2032e - 004,$$

for $nref = 2, 3, 4, 5$ in the MATLAB function *ellip_P1.m*. This assures us that the accuracy of the method is $O(h^2)$, which is consistent with our theoretical result.

An examplary numerical solution with $nref = 3$ is shown in Fig. 7.3.

The MATLAB function *ellip_P1.m* and its auxiliary functions *EXACT.m* and *SRC.m* are shown below.

```
%-----------------------------------------------------------
% ellip.m: finite element code for solving a 2nd-order
%          convecton-diffusion problem using p1 element.
%
% Functions used:
%    get_p1grid.m: generate a triangular grid
%    locA.m, locB.m, locM.m: evaluate element matrices
%    EXACT.m, SRC.m: exact solution and RHS function of the PDE
%
% nref=2,3,4,5 give max error as follows:
%    0.0044, 0.0014, 4.1030e-004, 1.2032e-004.
%-----------------------------------------------------------
clear all;

% velocity components in the governing equation
Vx=1.0;  Vy=1.0;

% generate a triangular grid by uniformly refining a coarse grid
nref = 3;        % level of refinement
[ne,np,p,conn,gbc] = gen_p1grid(nref);

% plot the mesh to see if it is right
figure(1);
trimesh(conn,p(:,1),p(:,2));

pause(2);

% specify the exact solution and use it for Dirichlet BC
for i=1:np
    u_ex(i)=feval(@EXACT,p(i,1),p(i,2));
    if(gbc(i,1)==1)      % indicator for Dirichlet BC
        gbc(i,2) = u_ex(i);
    end
end

% initialize those arrays
Ag = zeros(np,np);
b  = zeros(np,1);

% loop over the elements
for l=1:ne

  j=conn(l,1); x1=p(j,1); y1=p(j,2);
  j=conn(l,2); x2=p(j,1); y2=p(j,2);
  j=conn(l,3); x3=p(j,1); y3=p(j,2);
```

```
  % compute local element matrices
  [elmA] = locA(x1,y1,x2,y2,x3,y3);
  [elmB] = locB(Vx,Vy,x1,y1,x2,y2,x3,y3);
  [elmM] = locM(x1,y1,x2,y2,x3,y3);

   for i=1:3
     i1 = conn(l,i);
     for j=1:3
       j1 = conn(l,j);
       % assemble into the global coefficient matrix
       Ag(i1,j1) = Ag(i1,j1) + elmA(i,j) + elmB(i,j);
       % form the RHS of the FEM equation
       b(i1) = b(i1) + elmM(i,j)*feval(@SRC,p(j1,1),p(j1,2));
     end
   end
end

% impose the Dirichlet BC
for m=1:np
 if(gbc(m,1)==1)
   for i=1:np
     b(i) = b(i) - Ag(i,m) * gbc(m,2);
     Ag(i,m) = 0; Ag(m,i) = 0;
   end
   Ag(m,m) = 1.0; b(m) = gbc(m,2);
  end
end

% solve the linear system
u_fem=Ag\b;

% plot the numerical solution
figure(2);
trisurf(conn,p(:,1),p(:,2),u_fem);

% compare with exact solution
disp('max err='), max(abs(u_fem-u_ex')),
```

```
%-------------------------------------------------------
% exact solution for our model

function val=EXACT (x,y)

val = x^2*y^2;
```

```
%-----------------------------------------------
% source function at the rhs of the equation
% -(u_xx+u_yy)+v_vec*grad u = src
% u=x^2*y^2, v_vec=(1,1),

function val=SRC (x,y)

val = -2*y^2-2*x^2+2*x*y^2+2*y*x^2;
```

7.6   The MATLAB code for the Q_1 element

In this section we present a MATLAB code for solving the elliptic problem

$$-\triangle u + u = f \quad \text{in } \Omega = (0,1)^2, \tag{7.27}$$

$$\frac{\partial u}{\partial n} = 0 \quad \text{on } \partial\Omega, \tag{7.28}$$

using the Q_1 element.

Here we describe some important implementation procedures, since we have already discussed most steps for the triangular P_1 element in previous sections.

First we need to generate a rectangular mesh, which is much easier compared to unstructured triangular mesh. The MATLAB code *getQ1mesh.m* is shown below.

```
% Generate Q1 mesh on rectangle [0, length]x[0, height]
% nx,ny: number of elements in each direction
% x,y: 1-D array for nodal coordinates
% conn(1:ne,1:4): connectivity matrix
% ne, np: total numbers of elements, nodes generated

function[x,y,conn,ne,np] = getQ1mesh(length,height,nx,ny)

ne = nx*ny;
np = (nx+1)*(ny+1);

% create nodal coordinates
dx=length/nx;   dy=height/ny;
for i = 1:(nx+1)
    for j=1:(ny+1)
        x((ny+1)*(i-1)+j) = dx*(i-1);
        y((ny+1)*(i-1)+j) = dy*(j-1);
    end
end
```

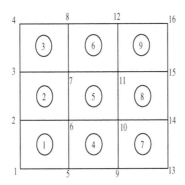

FIGURE 7.4

Example Q_1 element: node and element labelings.

```
% connectivity matrix: countclockwise start at low-left corner
for j=1:nx
   for i=1:ny
      ele = (j-1)*ny + i;
      conn(ele,1) = ele + (j-1);
      conn(ele,2) = conn(ele,1) + ny + 1;
      conn(ele,3) = conn(ele,2) + 1;
      conn(ele,4) = conn(ele,1) + 1;
   end
end
```

A simple 3×3 rectangular mesh generated with this code is shown in Fig. 7.4, where the nodal numbers and element numbers are provided. Note that the nodes of each element are oriented in a counterclockwise manner. For example, the connectivity matrices for element 1 and element 2 are given by:

$$conn(1, 1:4) = 1, 5, 6, 2, \quad conn(2, 1:4) = 2, 6, 7, 3.$$

The finite element method for solving (7.27)-(7.28) is: Find $u_h \in \boldsymbol{V}_h \subset H^1(\Omega)$ such that

$$(\nabla u_h, \nabla \phi_h) + (u_h, \phi_h) = (f, \phi_h) \quad \forall \, \phi_h \in \boldsymbol{V}_h, \tag{7.29}$$

where the Q_1 conforming finite element space

$$\boldsymbol{V}_h = \{v : v \text{ is continuous on} \Omega$$
$$\text{and} v|_E \in Q_1(E), E \in T_h\}.$$

On each rectangular element E, u is approximated by

$$u_h^E(x, y) = \sum_{j=1}^{4} u_j^E \psi_j^E(x, y), \tag{7.30}$$

which leads to the element coefficient matrix of (7.29) as

$$A_{ij} \equiv \int_E \nabla \psi_j^E \cdot \nabla \psi_i^E \, dx\, dy + \int_E \psi_j^E \psi_i^E \, dx\, dy, \quad i, j = 1, \cdots, 4. \tag{7.31}$$

The calculation of A_{ij} is often carried out on a reference rectangle with vertices

$$(\xi_1, \eta_1) = (-1, -1), \ (\xi_2, \eta_2) = (1, -1), \ (\xi_3, \eta_3) = (1, 1), \ (\xi_4, \eta_4) = (-1, 1),$$

whose corresponding shape functions

$$\hat{\psi}_i(\xi, \eta) = \frac{1}{4}(1 + \xi_i \xi)(1 + \eta_i \eta), \quad i = 1, \cdots, 4. \tag{7.32}$$

The mapping between an arbitrary rectangular element with vertices (x_i, y_i), $1 \leq i \leq 4$, and the reference element is given by

$$x = \sum_{j=1}^{4} x_j \hat{\psi}_j(\xi, \eta), y = \sum_{j=1}^{4} y_j \hat{\psi}_j(\xi, \eta). \tag{7.33}$$

From (7.12)-(7.13), we see that

$$\nabla \psi_j = (J^T)^{-1} \begin{bmatrix} \frac{\partial \hat{\psi}_j}{\partial \xi} \\ \frac{\partial \hat{\psi}_j}{\partial \eta} \end{bmatrix} = \frac{1}{det(J)} \begin{bmatrix} \frac{\partial y}{\partial \eta} & -\frac{\partial y}{\partial \xi} \\ -\frac{\partial x}{\partial \eta} & \frac{\partial x}{\partial \xi} \end{bmatrix} \begin{bmatrix} \frac{\partial \hat{\psi}_j}{\partial \xi} \\ \frac{\partial \hat{\psi}_j}{\partial \eta} \end{bmatrix} \tag{7.34}$$

$$= \frac{1}{det(J)} \begin{bmatrix} (\sum_{j=1}^{4} y_j \frac{\partial \hat{\psi}_j}{\partial \eta}) \frac{\partial \hat{\psi}_j}{\partial \xi} - (\sum_{j=1}^{4} y_j \frac{\partial \hat{\psi}_j}{\partial \xi}) \frac{\partial \hat{\psi}_j}{\partial \eta} \\ -(\sum_{j=1}^{4} x_j \frac{\partial \hat{\psi}_j}{\partial \eta}) \frac{\partial \hat{\psi}_j}{\partial \xi} + (\sum_{j=1}^{4} x_j \frac{\partial \hat{\psi}_j}{\partial \xi}) \frac{\partial \hat{\psi}_j}{\partial \eta} \end{bmatrix}. \tag{7.35}$$

Note that A_{ij} cannot be evaluated exactly as for the P_1 element. Hence we will use Gaussian quadrature over the reference rectangle. For example,

$$\int_E G(x, y) dx\, dy = \int_{\hat{E}} G(\hat{x}, \hat{y}) |J| d\xi d\eta$$

$$\approx \sum_{m=1}^{N} (\sum_{n=1}^{N} G(\xi_m, \eta_n) |J| \omega_n) \omega_m,$$

where ξ_m and ω_m (η_m and ω_n) are the one-dimensional quadrature points and weights, respectively. In our implementation, we use the Gaussian quadrature rule of order 2, in which case

$$\xi_1 = -\frac{1}{\sqrt{3}}, \quad \xi_2 = \frac{1}{\sqrt{3}}, \quad \omega_1 = \omega_2 = 1.$$

Below is the MATLAB code *elemA.m*, which is used to calculate the element matrix and right-hand side vector

$$(f, \phi_i) \approx (\frac{1}{4} \sum_{j=1}^{4} f_j, \phi_i), \quad i = 1, 2, 3, 4.$$

```
function [ke,rhse] = elemA(conn,x,y,gauss,rhs,e);

% 2d Q1 element stiffness matrix
ke = zeros(4,4);
rhse=zeros(4,1);
one = ones(1,4);
psiJ = [-1, +1, +1, -1]; etaJ = [-1, -1, +1, +1];

% get coordinates of element nodes
for j=1:4
   je = conn(e,j); xe(j) = x(je); ye(j) = y(je);
end

for i=1:2  % loop over gauss points in eta
  for j=1:2   % loop over gauss points in psi
    eta = gauss(i);  psi = gauss(j);
  % shape function: countcockwise starting at left-low corner
    NJ=0.25*(one + psi*psiJ).*(one + eta*etaJ);
    % derivatives of shape functions in reference coordinates
    NJpsi = 0.25*psiJ.*(one + eta*etaJ);   % 1x4 array
    NJeta = 0.25*etaJ.*(one + psi*psiJ);   % 1x4 array
    % derivatives of x and y wrt psi and eta
    xpsi = NJpsi*xe'; ypsi = NJpsi*ye';
    xeta = NJeta*xe'; yeta = NJeta*ye';
    Jinv = [yeta, -xeta; -ypsi, xpsi];     % 2x2 array
    jcob = xpsi*yeta - xeta*ypsi;
    % derivatives of shape functions in element coordinates
    NJdpsieta = [NJpsi; NJeta];            % 2x4 array
    NJdxy = Jinv*NJdpsieta;                % 2x4 array
    % assemble element stiffness matrix ke: 4x4 array
    ke = ke + (NJdxy(1,:))'*(NJdxy(1,:))/jcob ...
            + (NJdxy(2,:))'*(NJdxy(2,:))/jcob ...
            + NJ(1,:)'*NJ(1,:)*jcob;
    rhse = rhse + rhs*NJ'*jcob;
  end
end
```

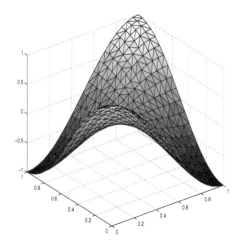

FIGURE 7.5

Numerical solutions obtained with $nx = ny = 20$.

The assembly procedure is all the same as the triangular element case. In our test, we solve the problem (7.27)-(7.28) with the exact solution

$$u = \cos \pi x \cos \pi y$$

which leads to $f = (2\pi^2 + 1)u$.

The problem is solved with different levels of refinement and convergence rate of $O(h^2)$ in the L^∞-norm is observed. More specifically, the maximum errors for $nx = 10, 20, 40$ are $0.0242, 0.0061, 0.0015$, respectively.

An examplary numerical solution with $nx = ny = 20$ is shown in Fig. 7.5. The MATLAB code *ellip_Q1.m* is shown below:

```
%-----------------------------------------------------------
% 2D Q1 FEM for solving
%      -Lap*u + u = f(x,y), on (0,length)x(0,height)
%      Neum BC=0
% Max err=0.0242,0.0061,0.0015 when nx=10,20,40
% so we did see O(h^2) convergence rate!
%-----------------------------------------------------------

clear all;
length = 1.0;
height = 1.0;
nx=20;
ny=20;
gauss = [-1/sqrt(3), 1/sqrt(3)];   % Gaussian quadrature point
```

```
% construct Q1 mesh
[x,y,conn,ne,np] = getQ1mesh(length,height,nx,ny);

Ag = zeros(np);
bg=zeros(np,1);

nloc = 4;    % number of nodes per element
for ie = 1:ne    % loop over all elements
   rhs= (feval(@SRC,x(conn(ie,1)),y(conn(ie,1))))...
       + feval(@SRC,x(conn(ie,2)),y(conn(ie,2)))...
       + feval(@SRC,x(conn(ie,3)),y(conn(ie,3)))...
       + feval(@SRC,x(conn(ie,4)),y(conn(ie,4))))/nloc;

   [A_loc,rhse] = elemA(conn,x,y,gauss,rhs,ie);
   % assemble local matrices into the global matrix
   for i=1:nloc;
      irow = conn(ie,i);    % global row index
      bg(irow)=bg(irow) + rhse(i);
      for j=1:nloc;
         icol = conn(ie,j);    %global column index
         Ag(irow, icol) = Ag(irow, icol) + A_loc(i,j);
      end;
   end;
end;

%solve the equation
u_fem = Ag\bg;

u_ex=cos(pi*x).*cos(pi*y);   % get the exact solution

disp('Max error='),
max(u_fem-u_ex'),                % find the max error

% plot the FEM solution
tri = delaunay(x,y);
trisurf(tri,x,y,u_fem);
```

7.7 Bibliographical remarks

In this chapter, we introduced the basic procedure for programming a finite element method for solving the second-order elliptic equation. Here we emphasize the simple and clear implementation instead of sophisticated

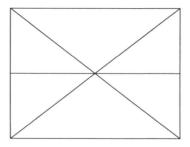

FIGURE 7.6
The initial grid.

usage of MATLAB functions. Finite element programming is a quite challenging task and can be written in a single book. Readers can find more advanced and sophisticated programming algorithms in other books (e.g., [1, 2, 3, 5, 6, 7, 8, 9, 10, 13, 14, 17, 18]). For example, [18] details how to program FEM for solving the Navier-Stokes equations in Fortran; [2] presents the package PLTMG, which solves elliptic problems using adaptive FEM in MATLAB; [10] introduces Diffpack, a sophisticated toolbox for solving PDEs based on C++; [15] elaborates on the adaptive finite element software ALBERTA written in ANSI-C; [6] introduces an open-source MATLAB package IFISS, which can be used to solve convection-diffusion, Stokes, and Navier-Stokes equations. Persson and Strang [12] developed a very nice and simple MATLAB mesh generator called DistMesh, which can generate unstructured triangular and tetrahedral meshes. Interested readers can download it from $http://www-math.mit.edu/\ persson/mesh/$. Readers also can download some nice finite element codes written in MATLAB at $http://matlabdb.mathematik.uni-stuttgart.de/$. Recently, the free package FEniCS [11, 16] aims to automate the solution of a range of PDEs by finite element methods. FEniCS provides a simple implementation based on the C++ and Python interfaces.

7.8   Exercises

1. Modify the function *gen_p1grid.m* to resolve the same problem as *ellip.m* on $\Omega = [0, 1]^2$ starting with the initial grid as shown in Fig. 7.6. Check the accuracy by solving the problem with different *nref*, say $nref = 2, 3, 4, 5$.

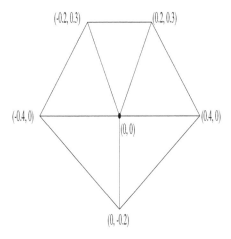

FIGURE 7.7
The initial grid for an irregular domain.

2. Let K be a triangle with vertices a_1, a_2, a_3, edge midpoints $a_{ij}, 1 \leq i < j \leq 3$, and $|K|$ be the area of K. Then the quadrature scheme

$$\int_K \phi(\boldsymbol{x})dx_1 dx_2 \approx \frac{|K|}{3} \sum_{1 \leq i < j \leq 3} \phi(a_{ij})$$

is exact for polynomials of degree ≤ 2. Furthermore, if we denote a_{123} for the barycenter of K, then the quadrature scheme

$$\int_K \phi(\boldsymbol{x})dx_1 dx_2 \approx |K|\phi(a_{123})$$

is exact for polynomials of degree ≤ 1. Moreover, the quadrature scheme

$$\int_K \phi(\boldsymbol{x})dx_1 dx_2 \approx \frac{|K|}{60}\left(3\sum_{i=1}^{3} \phi(a_i) + 8\sum_{1 \leq i < j \leq 3} \phi(a_{ij}) + 27\phi(a_{123})\right)$$

is exact for polynomials of degree ≤ 3.

3. Modify the codes *gen_p1grid.m* and *ellip.m* to solve the same problem as *ellip.m* on an irregular domain as shown in Fig. 7.7. Check the accuracy by solving the problem with $nref = 2, 3, 4, 5$.

4. Let Ω be a circle with radius r and Γ_h be an inscribed polygon approximating $\partial\Omega$. The polygon has side length h. Prove that the maximum distance between $\partial\Omega$ and Γ_h is about $\frac{h^2}{8r}$.

5. Let $\hat{K}$ be the reference triangle with vertices $\hat{a}_i, 1 \leq i \leq 3$, and mid-points $\hat{a}_{ij}$ of the side $\hat{a}_i\hat{a}_j$ of $\hat{K}$. For simplicity, we assume that

$$\hat{a}_1 = (0,0), \quad \hat{a}_2 = (1,0), \quad \hat{a}_3 = (0,1).$$

Suppose that a general triangle in the (x, y)-plane has vertices $a_i, 1 \leq i \leq 3$, and mid-points a_{ij} as follows

$$a_1 = (0,0), a_2 = (2,0), a_3 = (0,2), a_{12} = (1,0), a_{13} = (1,1), a_{23} = (0,1).$$

Let us define the isoparametric mapping F given by

$$F(\hat{x}) = \sum_{i=1}^{3} \lambda_i(\hat{x})(2\lambda_i(\hat{x}) - 1)a_i + \sum_{1 \leq i < j \leq 3} 4\lambda_i(\hat{x})\lambda_j(\hat{x})a_{ij}, \forall \, \hat{x} \in \hat{K}.$$

Prove that the mapping F is not invertible. Hint: Prove that the determinant of the Jacobi

$$J = \begin{bmatrix} \frac{\partial F_1}{\partial \hat{x}_1} & \frac{\partial F_1}{\partial \hat{x}_2} \\ \frac{\partial F_2}{\partial \hat{x}_1} & \frac{\partial F_2}{\partial \hat{x}_2} \end{bmatrix}$$

of F can be zero.

6. Let $\hat{K}$ be the unit square with vertices $\hat{a}_i, 1 \leq i \leq 4$, given by

$$\hat{a}_1 = (0,0), \quad \hat{a}_2 = (1,0), \quad \hat{a}_3 = (1,1), \quad \hat{a}_4 = (0,1).$$

A convex quadrilateral K is defined by its four vertices a_i as follows

$$a_1 = (0,0), \quad a_2 = (2,0), \quad a_3 = (2,3), \quad a_4 = (0,5).$$

Consider the mapping

$$F(\hat{x}) = (1 - \hat{x}_1)(1 - \hat{x}_2)a_1 + \hat{x}_1(1 - \hat{x}_2)a_2 + \hat{x}_1\hat{x}_2a_3 + (1 - \hat{x}_1)\hat{x}_2a_4.$$

Prove that the Jacobian of F is given by

$$J = 2(5 - 2\hat{x}_1).$$

Therefore, the mapping F is invertible.

7. Consider a non-convex quadrilateral K, whose four vertices are given by

$$a_1 = (2,0), \quad a_2 = (3,2), \quad a_3 = (5,3), \quad a_4 = (2,3).$$

Show that in this case, the above mapping F is not invertible.

8. Modify the code *ellip_Q1.m* to solve the problem (7.27) with Dirichlet boundary condition such that the exact solution is still $u = \cos \pi x \cos \pi y$. Hint: The Dirichlet boundary condition can be simply imposed as follows:

```
for m=1:np
    if(x(m)==0 | y(m)==0 | x(m)==1 | y(m)==1)
        for i=1:np
            bg(i) = bg(i) - Ag(i,m) * cos(pi*x(m))*cos(pi*y(m));
            Ag(i,m) = 0; Ag(m,i) = 0;
        end
        Ag(m,m) = 1.0; bg(m) = cos(pi*x(m))*cos(pi*y(m));
    end
end
```

References

[1] J.E. Akin. *Application and Implementation of Finite Element Methods.* Academic Press, London, 1982.

[2] R.E. Bank. *PLTMG: A Software Package for Solving Elliptic Partial Differential Equations: Users' Guide 8.0.* SIAM, Philadelphia, PA, 1998.

[3] G.F. Carey and J.T. Oden. *Finite Elements: Computational Aspects.* Prentice-Hall, Englewood Cliffs, NJ, 1983.

[4] P.G. Ciarlet. *The Finite Element Method for Elliptic Problems.* North-Holland, Amsterdam, 1978.

[5] L. Demkowicz. *Computing with hp-Adaptive Finite Elements.* Chapman & Hall/CRC, Boca Raton, FL, 2006.

[6] H.C. Elman, D.J. Silvester and A.J. Wathen. *Finite Elements and Fast Iterative Solvers with Applications in Incompressible Fluid Dynamics.* Oxford University Press, Oxford, UK, 2005.

[7] A. Ern and J.-L. Guermond. *Theory and Practice of Finite Elements.* Springer, New York, NY, 2004.

[8] J.C. Heinrich and D.W. Pepper. *The Finite Element Method: Advanced Concepts.* Taylor & Francis, Bristol, PA, 1996.

[9] T.J.R. Hughes. *Finite Element Method – Linear Static and Dynamic Finite Element Analysis.* Prentice-Hall, Englewood Cliffs, NJ, 1987.

[10] H.P. Langtangen. *Computational Partial Differential Equations: Numerical Methods and Diffpack Programming.* Springer-Verlag, Berlin, 2nd Edition, 2003.

[11] A. Logg and K.-A. Mardal and G.N. Wells (eds.). *Automated Solution of Differential Equations by the Finite Element Method.* Springer-Verlag, Berlin, 2012.

[12] P.-O. Persson and G. Strang. A simple mesh generator in MATLAB. *SIAM Review*, 46(2):329–345, 2004.

[13] C. Pozrikidis. *Introduction to Finite and Spectral Element Methods Using MATLAB.* Chapman & Hall/CRC, Boca Raton, FL, 2005.

[14] J.N. Reddy. *An Introduction to the Finite Element Method.* McGraw-Hill, Boston, 1993.

[15] A. Schmidt and K. G. Siebert. *Design of Adaptive Finite Element Software: The Finite Element Toolbox ALBERTA.* Springer, Berlin, 2005.

[16] L.R. Scott. *Introduction to Automated Modeling with FEniCS: Student Edition.* Computational Modeling Initiative, LLC, 2018.

[17] I.M. Smith and D.V. Griffiths. *Programming the Finite Element Method.* John Wiley & Sons, 4th Edition, Hoboken, NJ, 2004.

[18] C. Taylor and T.G. Hughes. *Finite Element Programming of the Navier-Stokes Equations.* Pineridge Press, Swansea, UK, 1981.

8

Mixed Finite Element Methods

In many applications, the governing PDEs include several variables (which are physically interesting) such as the velocities and pressure for fluid modeling using Navier-Stokes equations. The mixed finite element method is developed to approximate those variables simultaneously. One of the main issues in mixed methods is how to choose those finite element spaces such that the mixed methods will be more efficient and stable. In this chapter, we will introduce the basic theory and practical implementation for such methods.

In Sec. 8.1, we introduce an abstract framework for analyzing the mixed finite element method. Then in Sec. 8.2, we apply the framework to the second-order elliptic problem. Here we introduce some popular mixed finite element spaces and carry out the error estimate. By following similar ideas, we extend the discussion to the Stokes problem in Sec. 8.3, where some classic mixed spaces for the Stokes problem are introduced. Then in Sec. 8.4, we show the reader how to implement a mixed finite element method to solve the Stokes problem. Detailed implementation in MATLAB is discussed. Finally, in Sec. 8.5, we briefly introduce some mixed methods for both the steady and unsteady Navier-Stokes problem.

8.1 An abstract formulation

In this section we want to present an abstract theory for mixed finite element methods.

Let U and V be two Hilbert spaces, equipped with norms $||\cdot||_U$ and $||\cdot||_V$, respectively. Furthermore, let U' and V' be the corresponding dual spaces, i.e., the spaces of linear and continuous functional on U and V, respectively.

We consider the following problem: Given the linear functionals $F \in U'$ and $G \in V'$, find $(u, p) \in U \times V$ such that

$$a(u, v) + b(v, p) = F(v) \quad \forall \, v \in U, \tag{8.1}$$

$$b(u, q) = G(q) \quad \forall \, q \in V, \tag{8.2}$$

where $a(\cdot, \cdot) : U \times U \to R$, $b(\cdot, \cdot) : U \times V \to R$ are two bilinear forms, which satisfy the conditions:

$$|a(v_1, v_2)| \leq a^* ||v_1||_U ||v_2||_U \quad \forall \, v_1, v_2 \in U, \tag{8.3}$$

$$|b(v, q)| \leq b^* ||v||_U ||q||_V \quad \forall \, v \in U, \ q \in V, \tag{8.4}$$

where $a^* > 0$ and $b^* > 0$ are some given constants.

Also we define the linear space

$$U^G = \{v \in U : \ b(v, q) = G(q), \quad \forall \, q \in V\}, \tag{8.5}$$

and the compatibility condition: there exists a constant $\beta_* > 0$ such that

$$\forall \, q \in V, \ \exists 0 \neq v \in U \text{ such that } b(v, q) \geq \beta_* ||v||_U ||q||_V. \tag{8.6}$$

Note that the compatibility condition (8.6) is often called the inf-sup or Ladyzhenskaya-Babuska-Brezzi condition, since (8.6) is equivalent to

$$\inf_{q \in V} \sup_{v \in U} \frac{b(v, q)}{||v||_U ||q||_V} \geq \beta_*. \tag{8.7}$$

The following theorem assures the existence and uniqueness of the solution to the problem (8.1)-(8.2). Considering that the proof is too technical, we skip the proof here. Interested readers can consult books by Brezzi et al. [7] or Quarteroni and Valli [19, Ch.7].

THEOREM 8.1

Assume that the bilinear form $a(\cdot, \cdot)$ and $b(\cdot, \cdot)$ satisfy (8.3) and (8.4). Assume, moreover, that $a(\cdot, \cdot)$ is coercive on U^0 (the subspace of U^G with $G = 0$), i.e., there exists a constant $\alpha > 0$ such that

$$a(v, v) \geq \alpha ||v||_U^2 \quad \forall \, v \in U^0, \tag{8.8}$$

and $b(\cdot, \cdot)$ satisfies the inf-sup condition (8.7). Then, for each $F \in U', G \in V'$ there exists a unique solution $(u, p) \in U \times V$ to (8.1)-(8.2). Moreover, the map $(F, G) \rightarrow (u, p)$ is an isomorphism from $U' \times V'$ onto $U \times V$, and

$$||u||_U \leq \frac{1}{\alpha}[||F||_{U'} + \frac{\alpha + a^*}{\beta_*} ||G||_{V'}], \tag{8.9}$$

$$||p||_V \leq \frac{1}{\beta}[(1 + \frac{a^*}{\alpha})||F||_{U'} + \frac{a^*(\alpha + a^*)}{\alpha\beta_*} ||G||_{V'}]. \tag{8.10}$$

Recall that a linear mapping is an isomorphism if it is bijective and its inverse mapping is bounded. Furthermove, a mapping from one set X to another set Y is bijective if the mapping is both one-to-one (injective) and onto (surjective).

Now we can consider the mixed finite element method for approximating the problem (8.1)-(8.2). Let U_h and V_h be the respective finite dimensional subspaces of U and V. The discrete counterpart of (8.1)-(8.2) is: Find $u_h \in U_h$ and $p_h \in V_h$ such that

$$a(u_h, v) + b(v, p_h) = F(v) \quad \forall\, v \in U_h, \tag{8.11}$$
$$b(u_h, q) = G(q) \quad \forall\, q \in V_h. \tag{8.12}$$

The existence and uniqueness of the solution $(u_h, p_h) \in U_h \times V_h$ to the discrete problem (8.11) and (8.12) is guaranteed under the discrete inf-sup condition: there exists a constant $\beta_h > 0$ such that

$$\inf_{q_h \in V_h} \sup_{v_h \in U_h} \frac{b(v_h, q_h)}{||v_h||_U ||q_h||_V} \geq \beta_h \tag{8.13}$$

and the discrete coercivity assumption: there exists a constant $\alpha_h > 0$ such that

$$a(v_h, v_h) \geq \alpha_h ||v_h||_U^2 \quad \forall\, v \in U_h^0, \tag{8.14}$$

where U_h^0 is the subspace of U_h^G with $G = 0$, where the space

$$U_h^G = \{v_h \in U_h : \; b(v_h, q_h) = G(q_h), \quad \forall\, q_h \in V_h\}. \tag{8.15}$$

Furthermore, we have the following stability results:

$$||u_h||_U \leq \frac{1}{\alpha_h}[||F||_{U'} + \frac{\alpha_h + a^*}{\beta_h}||G||_{V'}], \tag{8.16}$$

$$||p_h||_V \leq \frac{1}{\beta_h}[(1 + \frac{a^*}{\alpha_h})||F||_{U'} + \frac{a^*(\alpha_h + a^*)}{\alpha_h \beta_h}||G||_{V'}]. \tag{8.17}$$

Now we can prove the main theorem of this section.

THEOREM 8.2

Let the assumptions of Theorem 8.1 be satisfied. Moreover, we assume that the discrete inf-sup condition (8.13) and the coercive condition (8.14) hold true. Hence the respective solutions (u, p) and (u_h, p_h) of (8.1)-(8.2) and (8.11)-(8.12) satisfy the error estimates

(i) $\quad ||u - u_h||_U$
$$\leq (1 + \frac{a^*}{\alpha_h}) \inf_{u_h^* \in U_h^G} ||u - u_h^*||_U + \frac{b^*}{\alpha_h} \inf_{p_h^* \in V_h} ||p - p_h^*||_V, \tag{8.18}$$

(ii) $\quad ||p - p_h||_V \leq \frac{a^*}{\beta_h}(1 + \frac{a^*}{\alpha_h}) \inf_{u_h^* \in U_h^G} ||u - u_h^*||_U$
$$+ (1 + \frac{b^*}{\beta_h} + \frac{a^* b^*}{\alpha_h \beta_h}) \inf_{p_h^* \in V_h} ||p - p_h^*||_V, \tag{8.19}$$

(iii) $\quad \inf_{u_h^* \in U_h^G} ||u - u_h^*||_U \leq (1 + \frac{b^*}{\beta_h}) \inf_{u_h \in U_h} ||u - u_h||_U. \tag{8.20}$

Proof. (i) Subtracting (8.11) from (8.1) with $v = v_h \in U_h$, we obtain

$$a(u_h - u_h^*, v_h) + b(v_h, p_h - p_h^*) = a(u - u_h^*, v_h) + b(v_h, p - p_h^*), \quad (8.21)$$

for any $u_h^* \in U_h^G$ and $p_h^* \in V_h$.

From (8.12) and (8.15), we see that

$$b(u_h - u_h^*, q_h) = 0 \quad \forall\, q \in V_h, \quad (8.22)$$

i.e., $u_h - u_h^* \in U_h^0$.

Applying (8.14) and (8.22) to (8.21) with $v_h = u_h - u_h^*$, and using the boundedness of (8.3)-(8.4), we have

$$\alpha_h \|u_h - u_h^*\|_U^2 \le a^* \|u - u_h^*\|_U \|u_h - u_h^*\|_U + b^* \|u_h - u_h^*\|_U \|p - p_h^*\|_V$$

i.e.,

$$\|u_h - u_h^*\|_U \le \frac{a^*}{\alpha_h} \|u - u_h^*\|_U + \frac{b^*}{\alpha_h} \|p - p_h^*\|_V,$$

which, along with the triangle inequality

$$\|u - u_h\|_U \le \|u - u_h^*\|_U + \|u_h - u_h^*\|_U,$$

completes the proof of (8.18).

(ii) Applying the discrete inf-sup condition (8.13), we obtain

$$\|p_h - p_h^*\|_V \le \frac{1}{\beta_h} \sup_{v_h \in U_h} \frac{b(v_h, p_h - p_h^*)}{\|v_h\|_U}, \quad \forall\, p_h^* \in V_h. \quad (8.23)$$

We can rewrite (8.21) as

$$b(v_h, p_h - p_h^*) = a(u - u_h, v_h) + b(v_h, p - p_h^*),$$

which along with (8.3)-(8.4) yields

$$b(v_h, p_h - p_h^*) \le a^* \|u - u_h\|_U \|v_h\|_U + b^* \|v_h\|_U \|p - p_h\|_V. \quad (8.24)$$

Substituting (8.24) into (8.23), we obtain

$$\|p_h - p_h^*\|_V \le \frac{1}{\beta_h} (a^* \|u - u_h\|_U + b^* \|p - p_h\|_V),$$

which, along with (8.18) and the triangle inequality, completes the proof of (8.19).

(iii) By the discrete inf-sup condition (8.13), for each $u_h \in U_h$, there exists a unique $r_h \in (U_h^0)^\perp$ such that (cf. [7, p. 55] or [19, p. 252])

$$b(r_h, q_h) = b(u - u_h, q_h) \quad \forall\, q_h \in V_h$$

and
$$||r_h||_U \le \frac{b^*}{\beta_h}||u - u_h||_U.$$

Let $u_h^* = r_h + u_h$, so $b(u_h^*, q_h) = b(u, q_h) = G(q_h)$, i.e., $u_h^* \in U_h^G$.
Hence we have

$$||u - u_h^*||_U \le ||u - u_h||_U + ||r_h||_U \le (1 + \frac{b^*}{\beta_h})||u - u_h||_U,$$

which concludes the proof of (8.20). □

8.2 Mixed methods for elliptic problems

Here we first show how to rewrite an elliptic model problem into a mixed variational form, upon which we introduced the mixed finite element method. Then we present three popular mixed finite element spaces for solving the elliptic problems. Finally, we show readers how to carry out the error estimates for the mixed finite element method.

8.2.1 The mixed variational formulation

We consider the model problem

$$-\nabla \cdot (\boldsymbol{a}\nabla p) = g \quad \text{in } \Omega \tag{8.25}$$
$$p = 0 \quad \text{on } \Gamma \tag{8.26}$$

where $\Omega \subset R^d$ ($d = 2$ *or* 3) is a bounded domain with boundary Γ, $\boldsymbol{a}$ is a $d \times d$ matrix and satisfies

$$0 < a_*|\xi|^2 \le \sum_{i,j=1}^{d} a_{ij}(\boldsymbol{x})\xi_i\xi_j \le a^*|\xi|^2 < \infty, \quad \forall \, \boldsymbol{x} \in \Omega, \xi \ne 0 \in R^d, \tag{8.27}$$

where we denote $|\xi|^2 = \sum_{i=1}^{d} \xi_i^2$.
   We define the space

$$L^2(\Omega) = \{v : v \text{ is defined on } \Omega \text{ and } \int_\Omega v^2 d\boldsymbol{x} < \infty\} \tag{8.28}$$

and

$$H(\text{div}; \Omega) = \{\boldsymbol{v} \in (L^2(\Omega))^d : \nabla \cdot \boldsymbol{v} \in L^2(\Omega)\}, \quad \nabla \cdot \boldsymbol{v} = \sum_{i=1}^{d} \frac{\partial v_i}{\partial x_i}, \tag{8.29}$$

with norm $||\boldsymbol{v}||_{H(\text{div};\Omega)} = (||\boldsymbol{v}||_{L^2(\Omega)}^2 + ||\nabla \cdot \boldsymbol{v}||_{L^2(\Omega)}^2)^{1/2}$. Furthermore, we define $U = H(\text{div}; \Omega), V = L^2(\Omega)$.

Introducing the new variable

$$u = a\nabla p, \tag{8.30}$$

we can rewrite the equation (8.25) as

$$\nabla \cdot u = -g \quad \text{in } \Omega. \tag{8.31}$$

Multiplying (8.30) by $v \in U$ and integrating over Ω, then applying Green's formula

$$(\nabla p, v) = <v \cdot n, p >_{\partial\Omega} -(p, \nabla \cdot v)$$

and boundary condition (8.26), we have

$$(a^{-1}u, v) + (p, \nabla \cdot v) = 0. \quad \forall\, v \in U. \tag{8.32}$$

Multiplying (8.31) by any $q \in V$, we obtain

$$(\nabla \cdot u, q) = -(g, q) \quad \forall\, q \in V. \tag{8.33}$$

Hence the problem (8.25)-(8.26) can be recast as a mixed variational problem: Find $u \in U$ and $p \in V$ such that

$$(a^{-1}u, v) + (\nabla \cdot v, p) = 0. \quad \forall\, v \in U, \tag{8.34}$$
$$(\nabla \cdot u, q) = -(g, q) \quad \forall\, q \in V. \tag{8.35}$$

Define

$$a(u, v) = (a^{-1}u, v), \quad \forall\, u, v \in U,$$
$$b(v, p) = (\nabla \cdot v, p), \quad \forall\, v \in U, p \in V.$$

Then (8.34)-(8.35) is in the form of (8.1)-(8.2) with

$$F(v) = 0, \quad G(q) = -(g, q).$$

It is easy to check that the bilinear forms $a(\cdot, \cdot)$ and $b(\cdot, \cdot)$ satisfy the continuity conditions (8.3) and (8.4) by using the assumption (8.27). Corresponding to (8.34)-(8.35), the subspace U^0 of U^G defined by (8.5) becomes

$$U^0 = \{v \in H(\text{div}; \Omega) : (\nabla \cdot v, q) = 0 \quad \forall\, q \in L^2(\Omega)\}. \tag{8.36}$$

Note that for any given $0 \neq q \in L^2(\Omega)$, we can find $v \in H(\text{div}; \Omega), v \neq 0$ such that $\text{div}\,v = q$. This can be assured by choosing $v = \nabla\phi$, where ϕ is the solution to

$$\triangle\phi = q \quad \text{in } \Omega, \tag{8.37}$$
$$\phi = 0 \quad \text{on } \Gamma. \tag{8.38}$$

Hence, for any $v \in U^0$ we see that $\nabla \cdot v = 0$. From (8.27), we have

$$a(v, v) = (a^{-1}v, v) \geq \frac{1}{a^*}||v||^2_{L^2(\Omega)} = \frac{1}{a^*}||v||^2_{H(\text{div};\Omega)}, \quad \forall \, v \in U^0,$$

i.e., $a(\cdot, \cdot)$ satisfies the condition (8.8).

Finally, we want to prove that the formulation (8.34)-(8.35) also satisfies the inf-sup condition (8.7). From above, for any $0 \neq q \in L^2(\Omega)$, there exists $0 \neq v \in H(\text{div}; \Omega)$. Hence

$$||v||_{L^2(\Omega)} = ||\nabla\phi||_0 \leq C_\Omega ||q||_0, \tag{8.39}$$

where in the last step we used ϕ multiplying (8.37) and the Poincaré-Friedrichs inequality [8, p. 12]

$$||\phi||_{L^2(\Omega)} \leq C_\Omega ||\nabla\phi||_0, \quad \forall \, \phi \in H_0^1(\Omega). \tag{8.40}$$

From (8.39) and $\nabla \cdot v = q$, we have

$$||v||_{H(\text{div};\Omega)} \equiv \sqrt{||v||_0^2 + ||\nabla \cdot v||_0^2} \leq \sqrt{1 + C_\Omega^2}||q||_0.$$

Consequently, we see that

$$\frac{b(v, q)}{||v||_U} = \frac{(q, q)}{||v||_{H(\text{div};\Omega)}} \geq \frac{1}{\sqrt{1 + C_\Omega^2}}||q||_0,$$

i.e., $b(\cdot, \cdot)$ satisfies the inf-sup condition (8.7). Thus all the conditions of Theorem 8.1 are satisfied, which guarantee the existence and uniqueness of the solution to the problem (8.34)-(8.35).

8.2.2 The mixed finite element spaces

From (8.34)-(8.35), we can consider the discrete problem: Find $u_h \in U_h$ and $p_h \in V_h$ such that

$$(a^{-1}u_h, v_h) + (\nabla \cdot v_h, p_h) = 0 \quad \forall \, v_h \in U_h, \tag{8.41}$$

$$(\nabla \cdot u_h, q_h) = -(g, q_h) \quad \forall \, q_h \in V_h, \tag{8.42}$$

where $U_h \subset H(\text{div}; \Omega)$ and $V_h \subset L^2(\Omega)$ are suitable finite dimensional spaces.

From Theorem 8.2, we need to choose U_h and V_h properly such that both the discrete inf-sup condition (8.13) and discrete coercivity condition (8.14) hold true, i.e., there exist constant $\beta_h > 0$ and $\alpha_h > 0$ such that

$$\inf_{q_h \in V_h} \sup_{v_h \in U_h} \frac{(\nabla \cdot v_h, q_h)}{||v_h||_{H(\text{div};\Omega)}||q||_0} \geq \beta_h, \tag{8.43}$$

and

$$a(v_h, v_h) \geq \alpha_h ||v_h||^2_{H(\text{div};\Omega)} \quad \forall \, v_h \in U_h^0, \tag{8.44}$$

where

$$U_h^0 = \{ \boldsymbol{v}_h \in U_h : (\nabla \cdot \boldsymbol{v}_h, q_h) = 0 \quad \forall \, q_h \in V_h \}. \tag{8.45}$$

Moreover, under the conditions (8.43) and (8.44), Theorem 8.2 leads to the following error estimate

$$\| \boldsymbol{u} - \boldsymbol{u}_h \|_{H(\mathrm{div};\Omega)} + \| p - p_h \|_0$$

$$\leq C[\inf_{\boldsymbol{v}_h \in U_h} \| \boldsymbol{u} - \boldsymbol{v}_h \|_{H(\mathrm{div};\Omega)} + \inf_{q_h \in V_h} \| p - q_h \|_0]. \tag{8.46}$$

For simplicity, here we only discuss the widely used Raviart-Thomas-Nédélec (RTN) spaces [17, 20]. Let us assume that $\Omega \subset R^d$ ($d = 2$ or 3) is a polygonal domain with Lipschitz boundary, and T_h is a regular family of triangulations of Ω.

Let $K \in T_h$ be a d-simplicial (i.e., triangular for $d = 2$ or tetrahedral for $d = 3$) element. Then we define

$$RT_r(K) = (P_r(K))^d \oplus \boldsymbol{x} \widetilde{P}_r(K) \quad \forall \, r \geq 0,$$

where $\widetilde{P}_r$ denotes the space of homogeneous polynomial of order r.

It is not difficult to check that the dimension of $RT_r(K)$ when $d = 2$ is

$$\dim(RT_r(K)) = d \cdot \dim(P_r(K)) + \dim(\widetilde{P}_r(K))$$

$$= 2 \cdot \frac{(r+2)(r+1)}{2} + (r+1) = (r+1)(r+3).$$

Similarly, we can show that the dimension of $RT_r(K)$ when $d = 3$ is

$$\dim(RT_r(K)) = 3 \cdot \frac{(r+3)(r+2)(r+1)}{6} + \frac{(r+2)(r+1)}{2}$$

$$= \frac{1}{2}(r+1)(r+2)(r+4).$$

The RTN spaces on the d-dimensional element are defined for each $r \geq 0$ by

$$U_h = \{ \boldsymbol{v} \in H(\mathrm{div}; \Omega) : \boldsymbol{v}|_K \in RT_r(K) \quad \forall \, K \in T_h \}, \tag{8.47}$$

$$V_h = \{ q \in L^2(\Omega) : q|_K \in P_r(K) \quad \forall \, K \in T_h \}. \tag{8.48}$$

Since the finite element spaces U_h and V_h are defined locally on each element $K \in K_h$, we simply denote

$$U_h(K) = U_h|_K, \quad V_h(K) = V_h|_K.$$

For a triangle or tetrahedron, the degrees of freedom for $\boldsymbol{v} \in U_h(K)$ are given by [20]:

$$< \boldsymbol{v} \cdot \boldsymbol{n}, w >_e, \quad \forall \, w \in P_r(e), \ e \in \partial K,$$

$$(\boldsymbol{v}, \boldsymbol{w})_K, \quad \forall \, \boldsymbol{w} \in (P_{r-1}(K))^d, \ d = 2 \text{ or } 3,$$

where $\boldsymbol{n}$ is the outward unit normal to $e \in \partial K$. It is remarked that a function $\boldsymbol{v} \in U_h$ is uniquely determined by these degrees of freedom. More specifically, we have

LEMMA 8.1
For $r \geq 0$, the following conditions imply $\boldsymbol{v} = 0$ on K:

$$< \boldsymbol{v} \cdot \boldsymbol{n}, w >_e = 0, \quad \forall\, w \in P_r(e), \ e \in \partial K, \tag{8.49}$$

$$(\boldsymbol{v}, \boldsymbol{w})_K = 0, \quad \forall\, \boldsymbol{w} \in (P_{r-1}(K))^d, \ d = 2 \text{ or } 3. \tag{8.50}$$

Proof. Due to the technicality, we leave the proof as an exercise for interested readers. □

Example 8.1
The lowest-order (i.e., $r = 0$) basis function of U_h on a triangle has the form

$$\boldsymbol{v}|_K = \begin{pmatrix} a_k \\ c_k \end{pmatrix} + b_k \begin{pmatrix} x_1 \\ x_2 \end{pmatrix}, \quad a_k, b_k, c_k \in R, \tag{8.51}$$

while the lowest-order U_h basis function on a tetrahedron has the form

$$\boldsymbol{v}|_K = \begin{pmatrix} a_k \\ c_k \\ d_k \end{pmatrix} + b_k \begin{pmatrix} x_1 \\ x_2 \\ x_3 \end{pmatrix}, \quad a_k, b_k, c_k, d_k \in R. \tag{8.52}$$

Here the unknowns a_k, b_k and c_k of (8.51) can be determined by the trace $< \boldsymbol{v} \cdot \boldsymbol{n}, 1 >_e$ along three edges of the triangle; while the unknowns a_k, b_k, c_k and d_k of (8.52) can be determined by the trace $< \boldsymbol{v} \cdot \boldsymbol{n}, 1 >_e$ along four faces of the tetrahedron.

For example, on a reference triangle $\hat{K}$ formed by vertices $\hat{A}_i$:

$$\hat{A}_1 = (0,0), \ \hat{A}_2 = (1,0), \ \hat{A}_3 = (0,1),$$

the three basis functions can be written as follows [16, p.66]:

$$\hat{\boldsymbol{N}}_1 = \begin{pmatrix} \hat{x}_1 \\ \hat{x}_2 \end{pmatrix}, \quad \hat{\boldsymbol{N}}_2 = \begin{pmatrix} -1 + \hat{x}_1 \\ \hat{x}_2 \end{pmatrix}, \quad \hat{\boldsymbol{N}}_3 = \begin{pmatrix} \hat{x}_1 \\ -1 + \hat{x}_2 \end{pmatrix}.$$

Note that $\hat{\boldsymbol{N}}_1, \hat{\boldsymbol{N}}_2$ and $\hat{\boldsymbol{N}}_3$ are associated with the three edges of $\hat{K}$ whose unit normal vectors are

$$\hat{\boldsymbol{n}}_1 = (\frac{1}{\sqrt{2}}, \frac{1}{\sqrt{2}})', \ \hat{\boldsymbol{n}}_2 = (-1,0)', \ \hat{\boldsymbol{n}}_3 = (0,-1)'.$$

It is easy to check that the basis functions $\hat{\boldsymbol{N}}_j$ satisfy the conditions

$$\int_{e_i} \hat{\boldsymbol{N}}_j \cdot \hat{\boldsymbol{n}}_i \, dl = \delta_{ij}, \quad i, j = 1, 2, 3,$$

where e_i denote the edges associating with the unit out normal vectors $\hat{n}_i, i = 1, 2, 3$. Moreover, the $H(div; \hat{K})$ interpolation can be written as

$$\Pi_{\hat{K}}^d v(\hat{x}_1, \hat{x}_2) = \sum_{j=1}^{3} (\int_{e_j} v \cdot \hat{n}_j dl) \hat{N}_j(\hat{x}_1, \hat{x}_2).$$

Similar technique can be used to derive the basis functions on a reference tetrahedron [16, p.65].   ☐

Example 8.2
The RTN spaces on rectangles are defined by

$$U_h(K) = Q_{r+1,r}(K) \times Q_{r,r+1}(K), \quad V_h(K) = Q_{r,r}(K) \quad \forall\, r \geq 0. \quad (8.53)$$

It is easy to check that the dimensions of $U_h(K)$ and $V_h(K)$ are

$$dimU_h(K) = 2(r+1)(r+2), \quad dimV_h(K) = (r+1)^2.$$

Furthermore, a function $v \in U_h(K)$ is uniquely defined by the degrees of freedom

$$< v \cdot n, w >_e, \quad \forall\, w \in P_r(e),\ e \in \partial K,$$
$$(v, w)_K, \quad \forall\, w \in Q_{r-1,r}(K) \times Q_{r,r-1}(K).$$

The lowest-order (i.e., $r = 0$) basis function v of U_h on a rectangle K takes the form

$$v|_K = \begin{pmatrix} a_1 + a_2 x_1 \\ a_3 + a_4 x_2 \end{pmatrix}, \quad a_i \in R, i = 1, 2, 3, 4. \quad (8.54)$$

The a_i's are determined by the traces $< v \cdot n, 1 >_e$ along four edges of the rectangle. More specifically, consider a rectangle $K = [x_1^c - h_x, x_1^c + h_x] \times [x_2^c - h_y, x_2^c + h_y]$ with the four edges e_j started from the bottom edge and oriented counterclockwise. After some algebra, we can obtain the basis functions N_j given as follows [16, p.59]:

$$N_1 = \begin{pmatrix} 0 \\ \frac{x_2 - (x_2^c + h_y)}{4h_x h_y} \end{pmatrix}, \quad N_2 = \begin{pmatrix} \frac{x_1 - (x_1^c - h_x)}{4h_x h_y} \\ 0 \end{pmatrix},$$

$$N_3 = \begin{pmatrix} 0 \\ \frac{x_2 - (x_2^c - h_y)}{4h_x h_y} \end{pmatrix}, \quad N_4 = \begin{pmatrix} \frac{x_1 - (x_1^c + h_x)}{4h_x h_y} \\ 0 \end{pmatrix}.$$

Furthermore, it is easy to check that the basis functions N_i satisfy the conditions:

$$\int_{e_j} N_i \cdot n_j dl = \delta_{ij}, \quad i, j = 1, \cdots, 4,$$

and the $H(div; K)$ interpolation function can be expressed as

$$\Pi_K^d \boldsymbol{v}(x_1, x_2) = \sum_{j=1}^{4} (\int_{e_j} \boldsymbol{v} \cdot \boldsymbol{n}_j dl) N_j(x_1, x_2),$$

and satisfies the interpolation conditions $\int_{e_j} (\Pi_K^d \boldsymbol{v} - \boldsymbol{v}) \cdot \boldsymbol{n}_j dl = 0.$ ☐

Example 8.3
The RTN spaces on cubes are defined for each $r \geq 0$ by

$$U_h(K) = Q_{r+1,r,r}(K) \times Q_{r,r+1,r}(K) \times Q_{r,r,r+1}(K), \quad V_h(K) = Q_{r,r,r}(K). \tag{8.55}$$

The dimensions of $U_h(K)$ and $V_h(K)$ are

$$dimU_h(K) = 3(r+1)^2(r+2), \quad dimV_h(K) = (r+1)^3.$$

Furthermore, a function $\boldsymbol{v} \in U_h(K)$ is uniquely defined by the degrees of freedom

$$< \boldsymbol{v} \cdot \boldsymbol{n}, w >_e, \quad \forall w \in Q_{r,r}(e), \ e \in \partial K,$$
$$(\boldsymbol{v}, \boldsymbol{w})_K, \quad \forall w \in Q_{r-1,r,r}(K) \times Q_{r,r-1,r}(K) \times Q_{r,r,r-1}(K).$$

Similarly, the lowest-order (i.e., $r = 0$) basis function $\boldsymbol{v}$ of U_h on a cube K takes the form

$$\boldsymbol{v}|_K = \begin{pmatrix} a_1 + a_2 x_1 \\ a_3 + a_4 x_2 \\ a_5 + a_6 x_3 \end{pmatrix}, \quad a_i \in R, i = 1, \cdots, 6, \tag{8.56}$$

and a_i's are uniquely determined by the traces $< \boldsymbol{v} \cdot \boldsymbol{n}, 1 >_e$ on six faces of the cube. ☐

8.2.3 The error estimates

To prove the error estimates for the mixed method (8.41)-(8.42), we need to prove the discrete inf-sup condition (8.43) and coercivity condition (8.44).
Note that all RTN spaces satisfy the property

$$\nabla \cdot U_h = V_h, \tag{8.57}$$

from which we see that for any $\boldsymbol{v}_h \in U_h^0$ (defined by (8.45)), $\nabla \cdot \boldsymbol{v}_h = 0$. This fact combining with the assumption (8.27) yields

$$a(\boldsymbol{v}_h, \boldsymbol{v}_h) \geq \frac{1}{a^*}||\boldsymbol{v}_h||_0^2 = \frac{1}{a^*}||\boldsymbol{v}_h||_{H(div;\Omega)}^2, \quad \forall \boldsymbol{v}_h \in U_h^0,$$

i.e., the condition (8.44) is true.

To prove the discrete inf-sup condition (8.43), we need a useful result due to Brezzi and Fortin [7].

LEMMA 8.2

Assume that the bilinear form $b(\cdot, \cdot)$ satisfies the continuous inf-sup condition (8.7). If there exists a projection operator $\Pi_h : U \to U_h$ such that

$$b(v - \Pi_h v, q) = 0 \quad \forall\, q \in V_h \tag{8.58}$$

and

$$||\Pi_h v||_U \le C^* ||v||_U \quad \forall\, v \in U, \tag{8.59}$$

where the constant $C^* > 0$ is independent of the mesh size h. Then the discrete inf-sup condition (8.43) holds true with $\beta_h = \beta_*/C^*$.

Proof. For any $q \in V_h$, using (8.7), (8.58) and (8.59), we have

$$\beta_* ||q||_V \le \sup_{v \in U} \frac{b(v, q)}{||v||_U} = \sup_{v \in U} \frac{b(\Pi_h v, q)}{||v||_U}$$

$$\le C^* \sup_{v \in U} \frac{b(\Pi_h v, q)}{||\Pi_h v||_U} \le C^* \sup_{w \in U_h} \frac{b(w, q)}{||w||_U},$$

which implies (8.13) with $\beta_h = \beta_*/C^*$. □

For the RTN spaces, the projection operator Π_h does exist [7]. First we can define the corresponding local interpolation operator $\Pi_K : (H^1(K))^d \to U_h \subset H(\text{div}; \Omega)$. For example, on a triangular or tetrahedral element, $\Pi_K : (H^1(K))^d \to RT_r(K)$ is defined by

$$< (v - \Pi_K v) \cdot n, w >_{\partial K} = 0, \quad \forall\, w \in P_r(\partial K),$$
$$(v - \Pi_K v, w)_K = 0, \quad \forall\, w \in (P_{r-1}(K))^d, \; d = 2 \text{ or } 3.$$

On a rectangular element K, $\Pi_K : (H^1(K))^2 \to Q_{r+1,r}(K) \times Q_{r,r+1}(K)$ is defined by

$$< (v - \Pi_K v) \cdot n, w >_{\partial K} = 0, \quad \forall\, w \in P_r(\partial K),$$
$$(v - \Pi_K v, w)_K = 0, \quad \forall\, w \in Q_{r-1,r}(K) \times Q_{r,r-1}(K),$$

while on a cubic element K, $\Pi_K : (H^1(K))^3 \to Q_{r+1,r,r}(K) \times Q_{r,r+1,r}(K) \times Q_{r,r,r+1}(K)$ is defined by

$$< (v - \Pi_K v) \cdot n, w >_{\partial K} = 0, \quad \forall\, w \in Q_{r,r}(\partial K),$$
$$(v - \Pi_K v, w)_K = 0, \quad \forall\, w \in Q_{r-1,r,r}(K) \times Q_{r,r-1,r}(K) \times Q_{r,r,r-1}(K).$$

LEMMA 8.3

Let $RTN(K)$ be the RTN spaces defined above, and Π_K be the interpolation operator: $(H^1(K))^d \to RTN(K)$. Moreover, let P_K be the L^2-projection onto the space $\text{div}(RTN(K))$, which is $P_r(K)$ or $Q_{r,r}(K)$ or $Q_{r,r,r}(K)$. Then we have

$$\text{div}(\Pi_K \boldsymbol{v}) = P_K(\text{div}\boldsymbol{v}) \quad \forall\, \boldsymbol{v} \in (H^1(\Omega))^d,\ d = 2\ \text{or}\ 3. \tag{8.60}$$

Proof. For any $w \in \text{div}(RTN(K))$, we have

$$(w, \nabla\cdot(\Pi_K \boldsymbol{v}-\boldsymbol{v}))_K = < w, (\Pi_K \boldsymbol{v}-\boldsymbol{v})\cdot\boldsymbol{n} >_{\partial K} -(\Pi_K \boldsymbol{v}-\boldsymbol{v}, \nabla\cdot w)_K = 0, \tag{8.61}$$

using the definition of Π_K.

By the definition of P_K, we have

$$(w, \nabla\cdot\boldsymbol{v})_K = (w, P_K(\nabla\cdot\boldsymbol{v}))_K, \quad \forall\, w \in \text{div}(RTN(K)). \tag{8.62}$$

Combining (8.61) and (8.62), we obtain

$$(\nabla\cdot(\Pi_K \boldsymbol{v}) - P_K(\nabla\cdot\boldsymbol{v}), w)_K = 0, \quad \forall\, w \in \text{div}(RTN(K)),$$

which leads to $\nabla\cdot(\Pi_K \boldsymbol{v}) - P_K(\nabla\cdot\boldsymbol{v}) = 0$, i.e., (8.60) is proved. $\square$

Considering that Π_K is defined in $(H^1(\Omega))^d$ and not in $H(\text{div};\Omega)$, we have to use a regularizing procedure introduced in [19, p. 237] to define the projection operator Π_h in $H(\text{div};\Omega)$. Let $\boldsymbol{v} \in H(\text{div};\Omega)$, define $\phi = \text{div}\boldsymbol{v}$ in Ω and extend ϕ to zero in $B \setminus \Omega$, where B is an open ball containing Ω.

Considering that for any $\phi \in L^2(\Omega)$, the problem

$$\triangle\psi = \phi \quad \text{in}\ B$$
$$\psi|_{\partial B} = 0 \quad \text{on}\ \partial B$$

has a solution $\psi \in H^2(B)$, which has the estimate

$$\|\psi\|_{H^2(B)} \le C\|\phi\|_{L^2(B)}.$$

Let us define $\boldsymbol{v}_* = (\nabla\psi)|_\Omega$. Hence $\boldsymbol{v}_* \in (H^1(\Omega))^d$, $\nabla\cdot\boldsymbol{v}_* = \triangle\psi = \text{div}\boldsymbol{v}$ in Ω, and

$$\|\boldsymbol{v}_*\|_{H^1(\Omega)} \le \|\psi\|_{H^2(B)} \le C\|\phi\|_{L^2(B)} = C\|\text{div}\boldsymbol{v}\|_{L^2(\Omega)}.$$

Now we can define the projection operator

$$\Pi_h(\boldsymbol{v})|_K = \Pi_K(\boldsymbol{v}_*),$$

from which we have

$$\|\Pi_h(\boldsymbol{v})\|_{H(\text{div};\Omega)} = \|\Pi_K(\boldsymbol{v}_*)\|_{H(\text{div};\Omega)} \le C\|\boldsymbol{v}_*\|_{H^1(\Omega)} \le C\|\text{div}\boldsymbol{v}\|_{L^2(\Omega)},$$

i.e., (8.59) holds true.

Note that (8.58) has proved to be true by (8.61), hence Lemma 8.2 implies that the discrete inf-sup condition (8.13) holds true. Thus the error estimate (8.46) is true, i.e.,

$$\|\boldsymbol{u} - \boldsymbol{u}_h\|_{H(\text{div};\Omega)} + \|p - p_h\|_0$$
$$\le C[\inf_{\boldsymbol{v}_h \in U_h} \|\boldsymbol{u} - \boldsymbol{v}_h\|_{H(\text{div};\Omega)} + \inf_{q_h \in V_h} \|p - q_h\|_0]. \tag{8.63}$$

Hence the error estimate becomes a problem of interpolation estimates for U_h and V_h, which is well known and stated in the following lemma.

LEMMA 8.4
[7, p. 132] Let T_h be a regular family of triangulation of Ω. There exists a constant C independent of h such that

$$||v - \Pi_h v||_0 \le Ch^m |v|_m, \quad ||\nabla \cdot (v - \Pi_h v)||_0 \le Ch^m |\nabla \cdot v|_m, \forall\, 1 \le m \le r+1.$$

Define the global L^2-projection P_h onto

$$V_h \equiv div(RTN(K)) : (P_h q)|_K = P_K(q|_K), \quad \forall\, q \in L^2(\Omega),$$

we have [19, p. 99]:

$$||q - P_h q||_0 \le Ch^m |q|_m \quad \forall\, q \in H^m(\Omega), \quad 0 \le m \le r+1. \tag{8.64}$$

Substituting Lemma 8.4 and (8.64) into (8.63), we finally obtain the error estimate

$$||u - u_h||_{H(div;\Omega)} + ||p - p_h||_0 \le Ch^m [|u|_m + |\nabla \cdot u|_m + |p|_m], \quad \forall\, 1 \le m \le r+1,$$

where (u, p) and (u_h, p_h) are the solutions to (8.34)-(8.35) and (8.41)-(8.42), respectively. The mixed finite element spaces are any RNT spaces of order $r \ge 0$.

8.3 Mixed methods for the Stokes problem

In this section, we first rewrite the Stokes problem into a mixed variational formulation, and its corresponding mixed finite element method. To guarantee the existence and stability for the mixed method, we introduce the inf-sup condition in both continuous and discrete forms. Finally, we listed many popular mixed finite element spaces, which satisfy the discrete inf-sup condition.

8.3.1   The mixed variational formulation

We now consider mixed finite element methods for the Stokes problem of incompressible flow of a viscous fluid. The model problem is: Find the pair (u, p) satisfying

$$-\mu \triangle u + \nabla p = f \quad \text{in } \Omega, \tag{8.65}$$

$$div u = 0 \quad \text{in } \Omega, \tag{8.66}$$

$$u = 0 \quad \text{on } \partial\Omega, \tag{8.67}$$

where Ω is a bounded Lipschitz domain in R^2, μ is the viscosity, u is the velocity field, f is the body force, and p is the pressure. The boundary condition (8.67) is often called the no-slip condition. Note that the pressure

p is unique up to an additive constant, which is usually fixed by enforcing the condition

$$\int_\Omega p\,dx = 0$$

or simply imposing the pressure to zero at one grid point.

We introduce the spaces

$$U = (H_0^1(\Omega))^2, \quad Q \equiv L_0^2(\Omega) = \{q \in L^2(\Omega) : \int_\Omega q\,dx = 0\}.$$

A mixed formulation corresponding to (8.65)-(8.67) is: Find $\boldsymbol{u} \in U$ and $p \in Q$ such that

$$\mu(\nabla\boldsymbol{u}, \nabla\boldsymbol{v}) - (\operatorname{div}\boldsymbol{v}, p) = (\boldsymbol{f}, \boldsymbol{v}), \quad \boldsymbol{v} \in U, \tag{8.68}$$
$$(\operatorname{div}\boldsymbol{u}, q) = 0, \quad q \in Q. \tag{8.69}$$

Let us introduce the bilinear forms $a(\cdot, \cdot) : U \times U \to R$ and $b(\cdot, \cdot) : U \times Q \to R$ as

$$a(\boldsymbol{u}, \boldsymbol{v}) = \mu(\nabla\boldsymbol{u}, \nabla\boldsymbol{v}), \quad b(\boldsymbol{v}, p) = -(\operatorname{div}\boldsymbol{v}, p).$$

Hence we can rewrite (8.68)-(8.69) as: Find $(\boldsymbol{u}, p) \in U \times Q$ such that

$$a(\boldsymbol{u}, \boldsymbol{v}) + b(\boldsymbol{v}, p) = (\boldsymbol{f}, \boldsymbol{v}), \quad \boldsymbol{v} \in U, \tag{8.70}$$
$$b(\boldsymbol{u}, q) = 0, \quad q \in Q. \tag{8.71}$$

The problem (8.70)-(8.71) satisfies an inf-sup condition as stated in the next lemma.

LEMMA 8.5
The bilinear form $b(\cdot, \cdot)$ satisfies

$$\inf_{q \in Q} \sup_{\boldsymbol{v} \in U} \frac{b(\boldsymbol{v}, q)}{\|\boldsymbol{v}\|_{H^1(\Omega)}\|q\|_{L^2(\Omega)}} \geq \beta_*, \tag{8.72}$$

where $\beta_* > 0$ is a constant.

Proof. From Arnold et al. [2], for any $q \in Q$, there exists $\boldsymbol{v} \in (H_0^1(\Omega))^2$ such that $\operatorname{div}\boldsymbol{v} = -q$ and

$$\|\boldsymbol{v}\|_{H^1(\Omega)} \leq C_\Omega\|q\|_{L^2(\Omega)},$$

where $C_\Omega > 0$ is a constant.

Hence we have

$$\frac{b(\boldsymbol{v}, q)}{\|\boldsymbol{v}\|_{H^1(\Omega)}} \geq \frac{1}{C_\Omega} \cdot \frac{b(\boldsymbol{v}, q)}{\|q\|_{L^2(\Omega)}} = \frac{1}{C_\Omega}\|q\|_{L^2(\Omega)},$$

which implies (8.72) with $\beta_* = 1/C_\Omega$. $\square$

Using (8.40), it is easy to see that for any $v \in (H_0^1(\Omega))^2$, we have

$$a(v,v) = \mu \|\nabla v\|_0^2 \geq \frac{\mu}{C_\Omega^2 + 1} (\|v\|_0^2 + \|\nabla v\|_0^2) \geq \frac{\mu}{C_\Omega^2 + 1} \|v\|_1^2,$$

which proves that $a(\cdot, \cdot)$ satisfies the coercivity over U. Thus $a(\cdot, \cdot)$ also satisfies the coercivity over the space

$$U^0 = \{v \in U : b(v,q) = 0 \quad \forall\, q \in Q\},$$

which is a subspace of U.

Hence by Theorem 8.1, the problem (8.68)-(8.69) has a unique solution.

Now we can consider mixed finite element approximations of the problem (8.68)-(8.69). Let T_h be a regular triangulation of Ω into triangles. Moreover, let U_h and Q_h be the corresponding finite dimensional subspaces of U and Q.

The mixed finite element method for the Stokes problem (8.65)-(8.67) is: Find $u_h \in U_h$ and $p_h \in Q_h$ such that

$$a(u_h, v_h) + b(v_h, p_h) = (f, v_h), \quad v_h \in U_h, \tag{8.73}$$

$$b(u_h, q_h) = 0, \quad q_h \in Q_h. \tag{8.74}$$

To guarantee the existence and stability for the mixed method (8.73)-(8.74), we have to construct the spaces U_h and Q_h very carefully. In the next section, we will state some popular choices of U_h and Q_h that satisfy the discrete inf-sup condition

$$\inf_{q_h \in Q_h} \sup_{v_h \in U_h} \frac{b(v_h, q_h)}{\|v_h\|_{H^1(\Omega)} \|q_h\|_{L^2(\Omega)}} \geq \beta_h > 0. \tag{8.75}$$

8.3.2   Mixed finite element spaces

Here we present many popular mixed finite element spaces, which satisfy the discrete inf-sup condition.

Example 8.4
The $P_2 - P_0$ element.

In this case, the spaces U_h and Q_h are defined as

$$U_h = \{v \in (H_0^1(\Omega))^2 : v|_K \in (P_2(K))^2, \ K \in T_h\}, \tag{8.76}$$

$$Q_h = \{q \in L_0^2(\Omega) : q|_K \in P_0(K), \ K \in T_h\}, \tag{8.77}$$

i.e., we use the continuous P_2 element for velocity field and the P_0 (piecewise constant) element for pressure on triangles. The degrees of freedom are indicated in Fig. 8.1. Here and below the symbol • denotes the values of the velocity components, while ○ refers to the value of the pressure.

The $P_2 - P_0$ element is proved to satisfy the discrete inf-sup condition [7, p. 221], thus Theorem 8.2 implies that

$$\|u - u_h\|_{H^1(\Omega)} + \|p - p_h\|_{L^2(\Omega)} \leq Ch(\|u\|_{H^2(\Omega)} + \|p\|_{H^1(\Omega)}). \tag{8.78}$$

▯

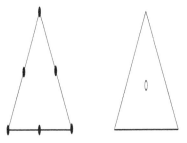

FIGURE 8.1
$P_2 - P_0$ element.

Example 8.5
The $Q_2 - Q_0$ element. Similarly, the $Q_2 - Q_0$ element can be defined as

$$U_h = \{v \in (H_0^1(\Omega))^2 : v|_K \in (Q_2(K))^2, K \in T_h\},$$
$$Q_h = \{q \in L_0^2(\Omega) : q|_K \in Q_0(K), K \in T_h\}.$$

It can be proved that this pair also satisfies the discrete inf-sup condition [15, p. 235] and the error estimate (8.78). ⬚

Example 8.6
Crouzeix-Raviart elements. The lowest Crouzeix-Raviart element is defined by

$$U_h = \{v \in (H_0^1(\Omega))^2 : v|_K = (P_2(K))^2 \oplus a_K \lambda_1 \lambda_2 \lambda_3, K \in T_h\}, \quad (8.79)$$
$$Q_h = \{q \in L_0^2(\Omega) : q|_K = P_1(K), K \in T_h\}, \quad (8.80)$$

where λ_i is the barycentric coordinates of the triangle defined in a previous chapter. Note that $b_K \equiv \lambda_1 \lambda_2 \lambda_3$ is cubic and vanishes on the edges of K, hence b_K is often called the cubic bubble function of K. Moreover, we want to remark that here we use the discontinuous P_1 element for pressure. We plot the degrees of freedom in Fig. 8.2.

This element is proved to satisfy the discrete inf-sup condition [11, p. 139] and has the convergence result

$$||u - u_h||_{H^1(\Omega)} + ||p - p_h||_{L^2(\Omega)} \leq Ch^2(||u||_{H^3(\Omega)} + ||p||_{H^2(\Omega)}). \quad (8.81)$$

Higher order Crouzeix-Raviart elements can be obtained using the discontinuous piecewise P_{k-1} function for pressure and piecewise $P_k^2 \oplus a_K p_{k-2} \lambda_1 \lambda_2 \lambda_3$ for velocities, where $p_{k-2} \in P_{k-2}$ for any $k \geq 3$. These elements satisfy the condition (8.75) and converge as $O(h^k)$. ⬚

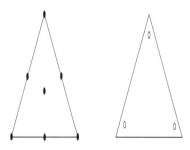

FIGURE 8.2
Crouzeix-Raviart element.

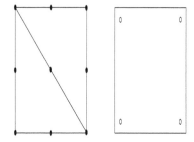

FIGURE 8.3
$P_k - Q_{k-1}$ elements ($k = 2$).

Example 8.7
Boland-Nicolaides elements. These elements were introduced by Boland and Nicolaides [6] for rectangles, with discontiunous piecewise Q_{k-1} ($k \geq 1$) pressure and piecewise P_k velocities on both triangles obtained by bisecting through one of the diagonals; see Fig. 8.3 for the $k = 2$ case.

The Boland-Nicolaides elements satisfy the discrete inf-sup condition and converge as $O(h^k)$ (see also [12, p. 36]). ☐

Example 8.8
Taylor-Hood elements. These elements were proposed by Taylor and Hood [22], where continuous P_k ($k \geq 2$) velocities and continuous P_{k-1} pressure are used on triangles, while on rectangles continuous Q_k ($k \geq 2$) velocities and continuous Q_{k-1} pressure are used. We show in Fig. 8.4 the corresponding degrees of freedom for the case $k = 2$ on a triangle.

The Taylor-Hood elements on both triangles and rectangles are proved to satisfy the discrete inf-sup condition and have the optimal error estimates

$$||\boldsymbol{u} - \boldsymbol{u}_h||_{H^1(\Omega)} + ||p - p_h||_{L^2(\Omega)} \leq Ch^k (||\boldsymbol{u}||_{H^{k+1}(\Omega)} + ||p||_{H^k(\Omega)}), \quad k \geq 2.$$

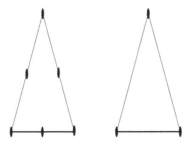

FIGURE 8.4
Taylor-Hood element ($k = 2$).

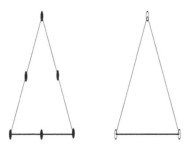

FIGURE 8.5
The MINI element.

For detailed proofs, see [7, 11, 22]. ☐

Example 8.9
MINI element. The MINI element was introduced by Arnold et al. [1], where we have

$$U_h = \{v \in (H_0^1(\Omega))^2 : v|_K = (P_1(K))^2 \oplus a_K \lambda_1 \lambda_2 \lambda_3, \ K \in T_h\}, \quad (8.82)$$
$$Q_h = \{q \in L_0^2(\Omega) \cap C^0(\overline{\Omega}) : q|_K = P_1(K), \ K \in T_h\}, \quad (8.83)$$

i.e., the velocities are approximated by piecewise P_1 augmented by a cubic bubble function. We plot the degrees of freedom for the MINI element in Fig. 8.5.

The MINI element satisfies the discrete inf-sup condition and have convergence result [1, (2.17)]:

$$||u - u_h||_{H^1(\Omega)} + ||p - p_h||_{L^2(\Omega)/R} \le Ch(||u||_{H^2(\Omega)} + ||p||_{H^1(\Omega)}) \le Ch||f||_{L^2(\Omega)}.$$

Before we conclude this section, we would like to remark that some simple choices such as $P_1 - P_0, Q_1 - Q_0, Q_2 - Q_1, P_2 - P_1$ (discontinuous), and equal interpolation methods (such as continuous $P_1 - P_1$ and $Q_1 - Q_1$) are unstable and have spurious pressure modes.   ▯

More detailed discussions on other mixed finite element spaces for the Stokes problem can be found in [7, 11].

8.4   An example MATLAB code for the Stokes problem

In this section, we will demonstrate how to solve the Stokes problem using the $P_2 - P_0$ element. For simplicity, we assume that the body force $\boldsymbol{f} = 0$, and Ω is a polygon discretized by a triangular mesh T_h with total ne elements and total ng nodes. Hence we can approximate the velocity and pressure by P_2 and P_0 functions as

$$\boldsymbol{u}_h = \sum_{j=1}^{ng} \begin{pmatrix} u_j^x \\ u_j^y \end{pmatrix} \phi_j^u(\boldsymbol{x}), \quad p_h = \sum_{j=1}^{ne} p_j \phi_j^p(\boldsymbol{x}), \tag{8.84}$$

where $\phi_j^u(\boldsymbol{x})$ is the continuous P_2 function, and $\phi_j^p(\boldsymbol{x}) = 1$ over each triangle E_j. Here u_j^x and u_j^y represent the x- and y-component of the approximate velocity field.

Substituting $\boldsymbol{v}_h = (\phi_i^u, 0)'$ into (8.73) gives the x-component equation

$$\mu \sum_j (\nabla \phi_j^u, \nabla \phi_i^u) u_j^x - \sum_j (\partial_x \phi_i^u, \phi_j^p) p_j = 0. \tag{8.85}$$

Similarly, substituting $\boldsymbol{v}_h = (0, \phi_i^u)'$ into (8.73) gives the y-component equation

$$\mu \sum_j (\nabla \phi_j^u, \nabla \phi_i^u) u_j^y - \sum_j (\partial_y \phi_i^u, \phi_j^p) p_j = 0. \tag{8.86}$$

Substituting $q_h = \phi_i^p$ into (8.74) leads to the equation

$$\sum_j [(\partial_x \phi_j^u, \phi_i^p) u_j^x + (\partial_y \phi_j^u, \phi_i^p) u_j^y] = 0. \tag{8.87}$$

Denote the element matrices

$$D_{u,ij} = \int \int_E \nabla \phi_j^u \cdot \nabla \phi_i^u \, dx dy,$$

$$D_{x,ij} = \int \int_E \partial_x \phi_j^u \phi_i^p \, dx dy, \quad D_{y,ij} = \int \int_E \partial_y \phi_j^u \phi_i^p \, dx dy, \tag{8.88}$$

and the vector solutions

$$\tilde{u}^x = \begin{bmatrix} u_1^x \\ \cdot \\ \cdot \\ \cdot \\ u_{ng}^x \end{bmatrix}, \quad \tilde{u}^y = \begin{bmatrix} u_1^y \\ \cdot \\ \cdot \\ \cdot \\ u_{ng}^y \end{bmatrix}, \quad \tilde{p} = \begin{bmatrix} p_1 \\ \cdot \\ \cdot \\ \cdot \\ p_{ne} \end{bmatrix}. \tag{8.89}$$

Assembling (8.85)-(8.87) leads to the global linear system

$$\begin{bmatrix} \mu D_u & 0 & -D_x^T \\ 0 & \mu D_u & -D_y^T \\ -D_x & -D_y & 0 \end{bmatrix} \begin{bmatrix} \tilde{u}^x \\ \tilde{u}^y \\ \tilde{p} \end{bmatrix} = 0, \tag{8.90}$$

where D_u is a $ng \times ng$ matrix, D_x and D_y are $ne \times ng$ matrices, and the superscript T denotes the matrix transpose.

Note that $\phi_i^p(\boldsymbol{x}) = 1$ over each element E_i, hence the evaluation of $D_{x,ij}$ and $D_{y,ij}$ can be simplified to

$$D_{x,ij} = \int\int_{E_i} \partial_x \phi_j^u \, dxdy, \quad D_{y,ij} = \int\int_{E_i} \partial_y \phi_j^u \, dxdy, \tag{8.91}$$

i.e., $D_{x,ij}$ and $D_{y,ij}$ are non-zero if the jth global node belongs to the ith element.

To evaluate the matrices D_u, D_x and D_y, we have to start with the P_2 function, which is defined on the 6-node triangular elements. The mesh generation of the 6-node triangles is similar to what we described in an earlier chapter. Here our implementation is inspired by Pozrikidis [18]. The basic idea is to divide each triangle into four subtriangles; the nodes and subtriangles are labeled as shown in Fig. 8.6.

The listed MATLAB code *gen_p2grid.m* can generate a 6-node triangular mesh from a coarse initial mesh by uniform refinement. An examplary initial grid and the grid after one refinement are shown in Fig. 8.7.

```
%---------------------------------------------------------
% Generate the 6-node triangular mesh
%     ne: total number of elements
%     np: total number of nodes
%     p(1:np,1:2):  (x,y) coordinates of all nodes
%     conn(1:ne,1:6): connectivity matrix
%     efl(1:ne,1:6): nodal type indicator
%     gfl(1:np,1): global nodal type indicator
%---------------------------------------------------------
function [ne,np,p,conn,efl,gfl] = gen_p2grid(nref)

% create an initial mesh with eight 6-node triangles
ne = 8;
```

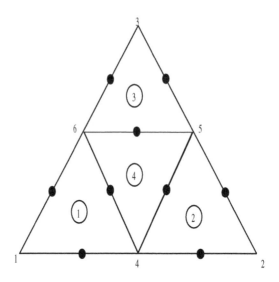

FIGURE 8.6

An examplary p_2 grid and its refinement.

```
x(1,1) = 0.0; y(1,1) = 0.0; efl(1,1)=0;  % 1st element
x(1,2) = 1.0; y(1,2) = 0.0; efl(1,2)=1;
x(1,3) = 1.0; y(1,3) = 1.0; efl(1,3)=1;

x(2,1) = 0.0; y(2,1) = 0.0; efl(2,1)=0;  % 2nd element
x(2,2) = 1.0; y(2,2) = 1.0; efl(2,2)=1;
x(2,3) = 0.0; y(2,3) = 1.0; efl(2,3)=1;

x(3,1) = 0.0; y(3,1) = 0.0; efl(3,1)=0;  % 3rd element
x(3,2) = 0.0; y(3,2) = 1.0; efl(3,2)=1;
x(3,3) =-1.0; y(3,3) = 1.0; efl(3,3)=1;

x(4,1) = 0.0; y(4,1) = 0.0; efl(4,1)=0;  % 4th element
x(4,2) =-1.0; y(4,2) = 1.0; efl(4,2)=1;
x(4,3) =-1.0; y(4,3) = 0.0; efl(4,3)=1;

x(5,1) = 0.0; y(5,1) = 0.0; efl(5,1)=0;  % 5th element
x(5,2) =-1.0; y(5,2) = 0.0; efl(5,2)=1;
x(5,3) =-1.0; y(5,3) =-1.0; efl(5,3)=1;

x(6,1) = 0.0; y(6,1) = 0.0; efl(6,1)=0;  % 6th element
x(6,2) =-1.0; y(6,2) =-1.0; efl(6,2)=1;
```

```
x(6,3) = 0.0; y(6,3) =-1.0; efl(6,3)=1;

x(7,1) = 0.0; y(7,1) = 0.0; efl(7,1)=0;   % 7th element
x(7,2) = 0.0; y(7,2) =-1.0; efl(7,2)=1;
x(7,3) = 1.0; y(7,3) =-1.0; efl(7,3)=1;

x(8,1) = 0.0; y(8,1) = 0.0; efl(8,1)=0;   % 8th element
x(8,2) = 1.0; y(8,2) =-1.0; efl(8,2)=1;
x(8,3) = 1.0; y(8,3) = 0.0; efl(8,3)=1;

for ie=1:8   % mid-edge nodes: other examples are different
   x(ie,4) = 0.5*(x(ie,1)+x(ie,2));
   y(ie,4) = 0.5*(y(ie,1)+y(ie,2)); efl(ie,4)=0;
   x(ie,5) = 0.5*(x(ie,2)+x(ie,3));
   y(ie,5) = 0.5*(y(ie,2)+y(ie,3)); efl(ie,5)=1;
   x(ie,6) = 0.5*(x(ie,3)+x(ie,1));
   y(ie,6) = 0.5*(y(ie,3)+y(ie,1)); efl(ie,6)=0;
end

%%%%%%%%%%%%%%%%%%%%%%%%%%%%%%%%%%%%%%%%%%%%%%%%%%%%%%%%%%%%%%%%%
% uniformly refine the grid
 for i=1:nref
   nm = 0; % count the new elements from each refinement
   for j=1:ne   % loop over current elements
      for k=1:4   % generate 4 sub-elements
         nm = nm+1;  % increase the element number by 1
         if (k==1) p1=1; p2=4; p3=6; end   % 1st sub-ele nodes
         if (k==2) p1=4; p2=2; p3=5; end   % 2nd sub-ele nodes
         if (k==3) p1=6; p2=5; p3=3; end   % 3rd sub-ele nodes
         if (k==4) p1=4; p2=5; p3=6; end   % 4th sub-ele nodes

         xn(nm,1)=x(j,p1); yn(nm,1)=y(j,p1); efln(nm,1)=efl(j,p1);
         xn(nm,2)=x(j,p2); yn(nm,2)=y(j,p2); efln(nm,2)=efl(j,p2);
         xn(nm,3)=x(j,p3); yn(nm,3)=y(j,p3); efln(nm,3)=efl(j,p3);
         xn(nm,4) = 0.5*(xn(nm,1)+xn(nm,2));   % mid-edge node
         yn(nm,4) = 0.5*(yn(nm,1)+yn(nm,2));
         xn(nm,5) = 0.5*(xn(nm,2)+xn(nm,3));
         yn(nm,5) = 0.5*(yn(nm,2)+yn(nm,3));
         xn(nm,6) = 0.5*(xn(nm,3)+xn(nm,1));
         yn(nm,6) = 0.5*(yn(nm,3)+yn(nm,1));

         if (efln(nm,1)==1 & efln(nm,2)==1)   % nodal type indicator
            efln(nm,4) = 1;
         else
            efln(nm,4) = 0;
```

```
        end
        if (efln(nm,2)==1 & efln(nm,3)==1)
            efln(nm,5) = 1;
        else
            efln(nm,5) = 0;
        end
        if (efln(nm,3)==1 & efln(nm,1)==1)
            efln(nm,6) = 1;
        else
            efln(nm,6) = 0;
        end
      end
    end % end of loop over current elements

    ne = 4*ne;  % number of elements increased by factor of four
    for k=1:ne     % relabel the new points
        for l=1:6
            x(k,l)=xn(k,l); y(k,l)=yn(k,l);  efl(k,l)=efln(k,l);
        end
    end
end % end of refinement loop

%%%%%%%%%%%%%%%%%%%%%%%%%%%%%%%%%%%%%%%%%%%%%%%%%%%%%%%%%%%%%%%%
% define the global nodes and the connectivity table
% we set the first element: nodes and connectivity
p(1,1)=x(1,1); p(1,2)=y(1,1); gfl(1,1)=efl(1,1);
p(2,1)=x(1,2); p(2,2)=y(1,2); gfl(2,1)=efl(1,2);
p(3,1)=x(1,3); p(3,2)=y(1,3); gfl(3,1)=efl(1,3);
p(4,1)=x(1,4); p(4,2)=y(1,4); gfl(4,1)=efl(1,4);
p(5,1)=x(1,5); p(5,2)=y(1,5); gfl(5,1)=efl(1,5);
p(6,1)=x(1,6); p(6,2)=y(1,6); gfl(6,1)=efl(1,6);

conn(1,1) = 1;  conn(1,2) = 2;  conn(1,3) = 3;
conn(1,4) = 4;  conn(1,5) = 5;  conn(1,6) = 6;

np = 6;            % we already have 6 nodes from 1st element
eps = 1.0e-8;

for i=2:ne         % loop over the rest elements
  for j=1:6        % loop over nodes of each element

  Iflag=0;
  for k=1:np
   if(abs(x(i,j)-p(k,1)) < eps & abs(y(i,j)-p(k,2)) < eps)
       Iflag = 1;   % the node has been recorded previously
```

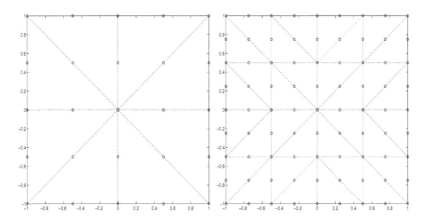

FIGURE 8.7
The initial mesh and the grid after one refinement.

```
        conn(i,j) = k;    % the jth local node of element i
    end
end

if(Iflag==0)   % record the node
    np = np+1;
    p(np,1)=x(i,j); p(np,2)=y(i,j); gfl(np,1) = efl(i,j);
    % the jth local node of element i becomes the new global node
    conn(i,j) = np;
end

end
end   % end of loop over elements

return;
```

As usual, the computation of elementary matrices is carried out on a reference element. We map an arbitrary triangle with vertices $a_i(x_i, y_i), 1 \leq i \leq 3$, and edge midpoints $a_{ij}, 1 \leq i < j \leq 3$ to the reference element with vertices

$$\hat{a}_1(0,0), \quad \hat{a}_2(1,0), \quad \hat{a}_3(0,1), \quad \hat{a}_{12}(\frac{1}{2},0), \quad \hat{a}_{23}(\frac{1}{2},\frac{1}{2}), \quad \hat{a}_{13}(0,\frac{1}{2}).$$

For clarity, we denote

$$\hat{a}_{12} = \hat{a}_4, \quad \hat{a}_{23} = \hat{a}_5, \quad \hat{a}_{13} = \hat{a}_6.$$

Hence the quadratic basis functions on the reference element can be obtained easily by satisfying the conditions

$$\hat{\psi}_i(\xi_j, \eta_j) = \delta_{ij},$$

which takes the forms

$$\hat{\psi}_1 = \zeta(2\zeta - 1), \quad \hat{\psi}_2 = \xi(2\xi - 1), \quad \hat{\psi}_3 = \eta(2\eta - 1),$$
$$\hat{\psi}_4 = 4\xi\zeta, \quad \hat{\psi}_5 = 4\xi\eta, \quad \hat{\psi}_6 = 4\eta\zeta,$$

where $(\xi, \eta, \zeta = 1 - \xi - \eta)$ are the triangle barycentric coordinates.

The mapping from an arbitrary triangle to the reference triangle is given by

$$x = \sum_{j=1}^{6} x_j \hat{\psi}_j(\xi, \eta), \quad y = \sum_{j=1}^{6} y_j \hat{\psi}_j(\xi, \eta). \tag{8.92}$$

We denote the Jacobi matrix of the mapping is denoted as

$$J \equiv \begin{bmatrix} \frac{\partial x}{\partial \xi} & \frac{\partial x}{\partial \eta} \\ \frac{\partial y}{\partial \xi} & \frac{\partial y}{\partial \eta} \end{bmatrix},$$

which determinant is denoted as $|J|$. The basis function on the physical triangle is defined as

$$\psi_j(x, y) = \hat{\psi}_j(\xi(x, y), \eta(x, y)),$$

from which we have

$$\frac{\partial \hat{\psi}_j}{\partial \xi} = \frac{\partial \psi_j}{\partial x} \frac{\partial x}{\partial \xi} + \frac{\partial \psi_j}{\partial y} \frac{\partial y}{\partial \xi}, \tag{8.93}$$

$$\frac{\partial \hat{\psi}_j}{\partial \eta} = \frac{\partial \psi_j}{\partial x} \frac{\partial x}{\partial \eta} + \frac{\partial \psi_j}{\partial y} \frac{\partial y}{\partial \eta}, \tag{8.94}$$

i.e.,

$$J^T \cdot \nabla \psi_j = \begin{bmatrix} \frac{\partial \hat{\psi}_j}{\partial \xi} \\ \frac{\partial \hat{\psi}_j}{\partial \eta} \end{bmatrix}, \quad j = 1, 2, 3,$$

where we denote the gradient

$$\nabla \psi_j = \begin{bmatrix} \frac{\partial \psi_j}{\partial x} \\ \frac{\partial \psi_j}{\partial y} \end{bmatrix}. \tag{8.95}$$

The evaluations of the P_2 basis functions and the gradient (8.95) are carried out by the function *p2basis.m* shown below.

```
%-------------------------------------------------------
% computation of the basis functions and their gradients
% over a 6-node triangle.
%-------------------------------------------------------
function [psi, gpsi, jac] = p2basis ...
    (x1,y1,x2,y2,x3,y3,x4,y4,x5,y5,x6,y6,xi,eta)
```

```
% compute the basis functions
psi(2) = xi*(2.0*xi-1.0);
psi(3) = eta*(2.0*eta-1.0);
psi(4) = 4.0*xi*(1.0-xi-eta);
psi(5) = 4.0*xi*eta;
psi(6) = 4.0*eta*(1.0-xi-eta);
psi(1) = 1.0-psi(2)-psi(3)-psi(4)-psi(5)-psi(6);

% compute xi derivatives of the basis functions
dps(2) =   4.0*xi-1.0;
dps(3) =   0.0;
dps(4) =   4.0*(1.0-xi-eta)-4.0*xi;
dps(5) =   4.0*eta;
dps(6) = -4.0*eta;
dps(1) = -dps(2)-dps(3)-dps(4)-dps(5)-dps(6);

% compute eta derivatives of the basis functions
pps(2) =   0.0;
pps(3) =   4.0*eta-1.0;
pps(4) = -4.0*xi;
pps(5) =   4.0*xi;
pps(6) =   4.0*(1.0-xi-eta)-4.0*eta;
pps(1) = -pps(2)-pps(3)-pps(4)-pps(5)-pps(6);

% compute the xi and eta derivatives of x
DxDxi = x1*dps(1) + x2*dps(2) + x3*dps(3) ...
      + x4*dps(4) + x5*dps(5) + x6*dps(6);
DyDxi = y1*dps(1) + y2*dps(2) + y3*dps(3) ...
      + y4*dps(4) + y5*dps(5) + y6*dps(6);

DxDeta = x1*pps(1) + x2*pps(2) + x3*pps(3) ...
      + x4*pps(4) + x5*pps(5) + x6*pps(6);
DyDeta = y1*pps(1) + y2*pps(2) + y3*pps(3) ...
      + y4*pps(4) + y5*pps(5) + y6*pps(6);

% compute the determinant of Jacobi matrix
jac = abs(DxDxi * DyDeta - DxDeta * DyDxi);

% compute the gradient of the basis functions
A11 = DxDxi;   A12 = DyDxi;
A21 = DxDeta; A22 = DyDeta;

Det = A11*A22-A21*A12;
```

```
for k=1:6
   B1 = dps(k); B2 = pps(k);
   Det1 = B1*A22 - B2*A12; Det2 = - B1*A21 + B2*A11;
   gpsi(k,1) = Det1/Det; gpsi(k,2) = Det2/Det;
end

return;
```

Using Lemma 7.1, we have

$$
\nabla \psi_j = (J^T)^{-1} \cdot \begin{bmatrix} \frac{\partial \hat{\psi}_j}{\partial \xi} \\ \frac{\partial \hat{\psi}_j}{\partial \eta} \end{bmatrix} = \frac{1}{|J|} \begin{bmatrix} \frac{\partial y}{\partial \eta} & -\frac{\partial y}{\partial \xi} \\ -\frac{\partial x}{\partial \eta} & \frac{\partial x}{\partial \xi} \end{bmatrix} \cdot \begin{bmatrix} \frac{\partial \hat{\psi}_j}{\partial \xi} \\ \frac{\partial \hat{\psi}_j}{\partial \eta} \end{bmatrix}
$$

$$
= \frac{1}{|J|} \begin{bmatrix} \sum_{i=1}^{6} y_i \frac{\partial \hat{\psi}_i}{\partial \eta} & -\sum_{i=1}^{6} y_i \frac{\partial \hat{\psi}_i}{\partial \xi} \\ -\sum_{i=1}^{6} x_i \frac{\partial \hat{\psi}_i}{\partial \eta} & \sum_{i=1}^{6} x_i \frac{\partial \hat{\psi}_i}{\partial \xi} \end{bmatrix} \cdot \begin{bmatrix} \frac{\partial \hat{\psi}_j}{\partial \xi} \\ \frac{\partial \hat{\psi}_j}{\partial \eta} \end{bmatrix} \tag{8.96}
$$

which being substituted into (8.88), we can evaluate the element matrix

$$
D_{u,ij} = \int \int_E \nabla \psi_i \cdot \nabla \psi_j dx dy = \int \int_{\hat{E}} \nabla \psi_i \cdot \nabla \psi_j |J| d\xi \eta, \tag{8.97}
$$

and similarly for $D_{x,ij}$ and $D_{y,ij}$.

Note that unlike the P_1 element, $D_{u,ij}$ cannot be evaluated exactly. However, we can approximate $D_{u,ij}$ by Gaussian quadrature over the triangle:

$$
\int \int_{\hat{E}} g(\xi, \eta) \approx \frac{1}{2} \sum_{k=1}^{nq} g(\xi_k, \eta_k) \omega_k,
$$

where ω_k is the weight, (ξ_k, η_k) are the integration point, nq is the number of quadrature points. Here we use the 7-point formula developed by [10], which is accurate for polynomials of degree ≤ 5. The details about the integration points and weights on our reference triangle can be seen in the code *p2quad.m*. Other quadrature formula can be implemented without much difficulty.

```
%---------------------------------------------------
% Abscissas (xi, eta) and weights (w) for Gaussian
% integration over our reference triangle
%
% m=7: order of the quadrature
%---------------------------------------------------
function [xi, eta, w] = p2quad(m)

al = 0.797426958353087;
be = 0.470142064105115;
ga = 0.059715871789770;
de = 0.101286507323456;
```

```
wt1 = 0.125939180544827;
wt2 = 0.132394152788506;

xi(1) = de;
xi(2) = al;
xi(3) = de;
xi(4) = be;
xi(5) = ga;
xi(6) = be;
xi(7) = 1.0/3.0;

eta(1) = de;
eta(2) = de;
eta(3) = al;
eta(4) = be;
eta(5) = be;
eta(6) = ga;
eta(7) = 1.0/3.0;

w(1) = wt1;   w(2) = wt1;   w(3) = wt1;
w(4) = wt2;   w(5) = wt2;   w(6) = wt2;
w(7) = 0.225;

return;
```

Using the Gaussian quadrature over the triangle and the gradient we just derived, we can calculate the matrices D_u, D_x and D_y, which are implemented in the functions *elmD.m* and *elmDxy.m*, respectively.

```
%------------------------------------------------
% Evaluation of element matrix D_u for a 6-node
% triangle using Gauss integration quadrature
%------------------------------------------------
function [edm, arel] = elmD ...
   (x1,y1, x2,y2, x3,y3, x4,y4, x5,y5, x6,y6, NQ)

% read the triangle quadrature
[xi, eta, w] = p2quad(NQ);

edm=zeros(6,6);   % initialization

% perform the quadrature
arel = 0.0;   % element area

for i=1:NQ
```

```
[psi, gpsi, jac] = p2basis ...
   (x1,y1,x2,y2,x3,y3,x4,y4,x5,y5,x6,y6,xi(i),eta(i));

cf = 0.5*jac*w(i);
for k=1:6
 for l=1:6
  edm(k,l) = edm(k,l) + (gpsi(k,1)*gpsi(l,1)    ...
                      +  gpsi(k,2)*gpsi(l,2) )*cf;
 end
end

 arel = arel + cf;
end

return;
```
_____
```
%-------------------------------------------------------
% Computation of the matrices D^x and D^y
%-------------------------------------------------------
function [Dx, Dy] = elmDxy ...
   (x1,y1, x2,y2, x3,y3, x4,y4, x5,y5, x6,y6, NQ)

% read the triangle quadrature
[xi, eta, w] = p2quad(NQ);

Dx = zeros(1,6); Dy = zeros(1,6);    %initialization

% perform the quadrature
for i=1:NQ
   [psi, gpsi, jac] = p2basis ...
      (x1,y1,x2,y2,x3,y3,x4,y4,x5,y5,x6,y6,xi(i),eta(i));

 cf = 0.5*jac*w(i);
 for k=1:6
   Dx(k) = Dx(k) + gpsi(k,1)*cf;
   Dy(k) = Dy(k) + gpsi(k,2)*cf;
 end
end

return;
```
_____

With all the preparation, now we demonstrate the main function *Stokes.m*, which is used to solve the flow in a square cavity driven by a translating lid. Here we assume that $\mu = 1$ and no-slip boundary conditions everywhere except at the lid, where $\boldsymbol{u} = (1,0)'$. An examplary velocity field and the corresponding mesh are presented in Fig. 8.8.

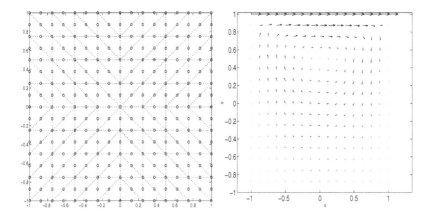

FIGURE 8.8
An example mesh and the corresponding velocity field obtained.

```
%----------------------------------------------------------
% Stokes.m: solve the Stokes flow in a cavity
%           using P2-P0 elements.
%
% The code needs:
%     1. gen_p2grid.m: generate 6-node triangle mesh
%     2. elmD.m: produce the element matrix D_u
%     3. elmDxy.m: produce the element matrix D_x, D_y
%     4. p2basis.m: calculate the P2 basis functions
%     5. p2quad.m: Gaussian quadrature on triangle
%----------------------------------------------------------
clear all;

vis = 1.0; % viscosity
Vel = 1.0;     % lid velocity
NQ = 7;      % gauss-triangle quadrature
nref = 3;    % discretization level

% generate p2 grid
[ne,np,p,conn,efl,gfl] = gen_p2grid(nref);
disp('Number of elements:'); ne

inodes = 0;
for j=1:np
 if(gfl(j,1)==0)    %interior nodes
   inodes = inodes+1;
 end
```

```
end
disp('Number of interior nodes:'); inodes

% specify the boundary velocity
for i=1:np
 if(gfl(i,1)==1)
   gfl(i,2) = 0.0; gfl(i,3) = 0.0; % x and y velocities
   if(p(i,2) > 0.999)
    gfl(i,2) = Vel;  % x velocity on the lid
   end
 end
end

% assemble the global diffusion matrix, Dx and Dy matrices
gdm = zeros(np,np); % initialization
gDx = zeros(ne,np);
gDy = zeros(ne,np);

for l=1:ne            % loop over the elements
   j=conn(l,1); x1=p(j,1); y1=p(j,2);
   j=conn(l,2); x2=p(j,1); y2=p(j,2);
   j=conn(l,3); x3=p(j,1); y3=p(j,2);
   j=conn(l,4); x4=p(j,1); y4=p(j,2);
   j=conn(l,5); x5=p(j,1); y5=p(j,2);
   j=conn(l,6); x6=p(j,1); y6=p(j,2);

   [edm_elm, arel] = elmD ...
    (x1,y1, x2,y2, x3,y3, x4,y4, x5,y5, x6,y6, NQ);

   [Dx, Dy] = elmDxy ...
    (x1,y1, x2,y2, x3,y3, x4,y4, x5,y5, x6,y6, NQ);

   for i=1:6
     i1 = conn(l,i);
     for j=1:6
       j1 = conn(l,j);
       gdm(i1,j1) = gdm(i1,j1) + edm_elm(i,j);
     end
     gDx(l,i1) = Dx(i);
     gDy(l,i1) = Dy(i);
   end
end

% form the final global coefficient matrix
nsys = 2*np+ne;   % total number of unknowns
```

```
Gm=zeros(nsys,nsys);
b=zeros(1,nsys);

for i=1:np      % first big block
  for j=1:np
    Gm(i,j) = vis*gdm(i,j);  Gm(np+i,np+j) = Gm(i,j);
  end
  for j=1:ne
    Gm(i,  2*np+j) = -gDx(j,i);
    Gm(np+i,2*np+j) = -gDy(j,i);
  end
end

% second big block
for i=1:ne
  for j=1:np
    Gm(2*np+i,j)    = -gDx(i,j);
    Gm(2*np+i,np+j) = -gDy(i,j);
  end
end

% compute RHS of the system and implement the Dirichlet BC
for j=1:np
 if(gfl(j,1)==1)
   for i=1:nsys
     b(i) = b(i) - Gm(i,j)*gfl(j,2) - Gm(i,np+j)*gfl(j,3);
     Gm(i,j) = 0; Gm(i,np+j) = 0;
     Gm(j,i) = 0; Gm(np+j,i) = 0;
   end
   Gm(j,j) = 1.0;
   Gm(np+j,np+j) = 1.0;
   b(j) = gfl(j,2);
   b(np+j) = gfl(j,3);
 end
end

% solve the linear system
Gm(:,nsys) = [];  % remove the last column
Gm(nsys,:) = [];  % remove the last row
b(nsys) = [];  % remove the last component

sol=Gm\b';

% recover the velocity
for i=1:np
```

```
ux(i) = sol(i);   uy(i) = sol(np+i);
end

% plot the velocity field
figure(1);
quiver(p(:,1)',p(:,2)',ux,uy);
hold on;
xlabel('x','fontsize',10)
ylabel('y','fontsize',10)
set(gca,'fontsize',15)
axis('equal')

% plot the mesh
figure(2);
for i=1:ne
  i1=conn(i,1); i2=conn(i,2); i3=conn(i,3);
  i4=conn(i,4); i5=conn(i,5); i6=conn(i,6);
  xp(1)=p(i1,1); yp(1)=p(i1,2); xp(2)=p(i4,1); yp(2)=p(i4,2);
  xp(3)=p(i2,1); yp(3)=p(i2,2); xp(4)=p(i5,1); yp(4)=p(i5,2);
  xp(5)=p(i3,1); yp(5)=p(i3,2); xp(6)=p(i6,1); yp(6)=p(i6,2);
  xp(7)=p(i1,1); yp(7)=p(i1,2);
  plot(xp, yp, ':');
  hold on;
  plot(xp, yp,'o','markersize',5);
end
```

8.5   Mixed methods for viscous incompressible flows

Here we give a brief discussion on developing and analyzing mixed finite element methods for solving both steady and unsteady Navier-Stokes problem.

8.5.1   The steady Navier-Stokes problem

The fundamental equations governing the steady flow of a homogeneous incompressible Newtonian fluid are the Navier-Stokes equations:

$$-\mu\triangle\boldsymbol{u} + (\boldsymbol{u}\cdot\nabla)\boldsymbol{u} + \nabla p = \boldsymbol{f} \quad \text{in } \Omega, \tag{8.98}$$

$$\nabla\cdot\boldsymbol{u} = 0 \quad \text{in } \Omega, \tag{8.99}$$

where we denote

$$(\boldsymbol{u}\cdot\nabla)\boldsymbol{u} = \sum_{j=1}^{d} u_j \frac{\partial\boldsymbol{u}}{\partial x_j}$$

and Ω for a bounded domain of R^d, $d = 2,3$. Furthermore, $\boldsymbol{u}$ denotes the velocity vector, $\mu > 0$ is the kinematic viscosity, $\boldsymbol{f}$ is the body force vector (per unit mass), $p = P/\rho_0$, P is the fluid pressure, and ρ_0 is the density.

There exist many different types of boundary conditions depending on different physical situations. The so-called no-slip condition

$$\boldsymbol{u} = 0 \quad \text{on } \partial\Omega, \tag{8.100}$$

is used for a viscous fluid confined in a fixed domain, in which case the particles of the fluid are adherent at the boundary due to the viscous effects.

The so-called non-friction condition is given by

$$\sum_{j=1}^{d} [\mu(\frac{\partial u_i}{\partial x_j} + \frac{\partial u_j}{\partial x_i}) - p\delta_{ij}]n_j = f_i^T \quad \text{on } \partial\Omega, \tag{8.101}$$

where $\boldsymbol{n} = (n_j)$ is the unit normal to the boundary, f_i^T is the boundary stress component, and δ_{ij} is the Kronecker notation (i.e., $\delta_{ij} = 0$ for $i \neq j$ and $\delta_{jj} = 1$). Note that $f_i^T = 0$ is often used on planes/lines of symmetry, and $f_i^T = $ constant is used as an inflow/outflow condition on a boundary. Other types of boundary conditions may be encountered in various modeling or simulation situations [19, p. 344].

Denote the two Hilbert spaces $U = (H_0^1(\Omega))^d$ and $Q = L_0^2(\Omega)$ as in Section 8.3, and introduce the trilinear form

$$c(\boldsymbol{w}; \boldsymbol{u}, \boldsymbol{v}) = (\boldsymbol{w} \cdot \nabla \boldsymbol{u}, \boldsymbol{v}) = \sum_{i=1}^{d}(\sum_{j=1}^{d} w_j \frac{\partial u_i}{\partial x_j}, v_i), \tag{8.102}$$

we can obtain the weak formulation of (8.98)-(8.100) as follows: Given $\boldsymbol{f} \in (L^2(\Omega))^d$, find $(\boldsymbol{u}, p) \in U \times Q$ such that

$$\mu(\nabla \boldsymbol{u}, \nabla \boldsymbol{v}) + c(\boldsymbol{u}; \boldsymbol{u}, \boldsymbol{v}) - (p, \nabla \cdot \boldsymbol{v}) = (\boldsymbol{f}, \boldsymbol{v}), \quad \forall \, \boldsymbol{v} \in U, \tag{8.103}$$
$$(\nabla \cdot \boldsymbol{u}, q) = 0, \quad \forall \, q \in Q. \tag{8.104}$$

With proper choice of the mixed finite element spaces U_h and Q_h as in Section 8.3, the mixed finite element method for the Navier-Stokes problem (8.98)-(8.100) can be obtained: Find $(\boldsymbol{u}_h, p_h) \in U_h \times Q_h$ such that

$$\mu(\nabla \boldsymbol{u}_h, \nabla \boldsymbol{v}) + c(\boldsymbol{u}_h; \boldsymbol{u}_h, \boldsymbol{v}) - (p_h, \nabla \cdot \boldsymbol{v}) = (\boldsymbol{f}, \boldsymbol{v}), \quad \forall \, \boldsymbol{v} \in U_h, \tag{8.105}$$
$$(\nabla \cdot \boldsymbol{u}_h, q) = 0, \quad \forall \, q \in Q_h. \tag{8.106}$$

Under suitable conditions, there exists a unique solution of (8.105)-(8.106) [11]. Note that (8.105)-(8.106) is a nonlinear system and can be solved by Newton methods. For an abstract nonlinear problem

$$F(z) = 0, \tag{8.107}$$

the Newton method is: Given an initial guess z^0, find z^{n+1} such that

$$\frac{\partial F}{\partial z}(z^n) \cdot (z^{n+1} - z^n) = -F(z^n), \quad \forall \, n \geq 0, \tag{8.108}$$

i.e., at each iteration step, we have to solve a linear problem for z^{n+1}.

Applying the Newton method (8.108) to (8.105)-(8.106), we obtain: Given $(\boldsymbol{u}_h^0, p_h^0)$, find $(\boldsymbol{u}_h^{n+1}, p_h^{n+1}) \in U_h \times Q_h$ such that

$$\mu(\nabla \boldsymbol{u}_h^{n+1}, \nabla \boldsymbol{v}) + c(\boldsymbol{u}_h^{n+1}; \boldsymbol{u}_h^n, \boldsymbol{v}) + c(\boldsymbol{u}_h^n; \boldsymbol{u}_h^{n+1}, \boldsymbol{v}) - (p_h^{n+1}, \nabla \cdot \boldsymbol{v})$$
$$= c(\boldsymbol{u}_h^n; \boldsymbol{u}_h^n, \boldsymbol{v}) + (\boldsymbol{f}, \boldsymbol{v}), \tag{8.109}$$
$$(\nabla \cdot \boldsymbol{u}_h^{n+1}, q) = 0, \tag{8.110}$$

for any $\boldsymbol{v} \in U_h$ and $q \in Q_h$.

It can be proved that for any $(\boldsymbol{u}_h^0, p_h^0)$ close to the true solution $(\boldsymbol{u}, p)$, the scheme (8.109)-(8.110) determines a unique sequence which converges quadratically to $(\boldsymbol{u}, p)$.

8.5.2   The unsteady Navier-Stokes problem

In this section we consider the unsteady viscous incompressible flows, which are described by the unsteady Navier-Stokes problem:

$$\frac{\partial \boldsymbol{u}}{\partial t} - \mu \triangle \boldsymbol{u} + (\boldsymbol{u} \cdot \nabla)\boldsymbol{u} + \nabla p = \boldsymbol{f} \quad \text{in } \Omega \times (0, T), \tag{8.111}$$
$$\nabla \cdot \boldsymbol{u} = 0 \quad \text{in } \Omega \times (0, T), \tag{8.112}$$
$$\boldsymbol{u} = 0 \quad \text{on } \partial\Omega \times (0, T), \tag{8.113}$$
$$\boldsymbol{u}|_{t=0} = \boldsymbol{u}_0 \quad \text{on } \Omega, \tag{8.114}$$

where $\boldsymbol{f} = \boldsymbol{f}(\boldsymbol{x}, t)$ and $\boldsymbol{u}_0 = \boldsymbol{u}_0(\boldsymbol{x}, t)$ are given functions.

The weak formulation of (8.111)-(8.114) is given as follows: For any $t \in (0, T)$, find $(\boldsymbol{u}(t), p(t)) \in U \times Q$ such that

$$\frac{d}{dt}(\boldsymbol{u}(t), \boldsymbol{v}) + \mu(\nabla \boldsymbol{u}(t), \nabla \boldsymbol{v}) + c(\boldsymbol{u}(t); \boldsymbol{u}(t), \boldsymbol{v}) - (p(t), \nabla \cdot \boldsymbol{v})$$
$$= (\boldsymbol{f}(t), \boldsymbol{v}), \quad \forall \, \boldsymbol{v} \in U, \tag{8.115}$$
$$(\nabla \cdot \boldsymbol{u}(t), q) = 0, \quad \forall \, q \in Q, \tag{8.116}$$
$$\boldsymbol{u}(0) = \boldsymbol{u}_0. \tag{8.117}$$

Similar to the steady problem, we can construct the semi-discrete scheme for approximating (8.115)-(8.117): For each $t \in [0, T]$, find $(\boldsymbol{u}_h(\cdot, t), p_h(\cdot, t)) \in U_h \times Q_h$ such that

$$\frac{d}{dt}(\boldsymbol{u}_h(t), \boldsymbol{v}) + \mu(\nabla \boldsymbol{u}_h(t), \nabla \boldsymbol{v}) + c(\boldsymbol{u}_h(t); \boldsymbol{u}_h(t), \boldsymbol{v}) - (p_h(t), \nabla \cdot \boldsymbol{v})$$
$$= (\boldsymbol{f}(t), \boldsymbol{v}), \quad \forall \, \boldsymbol{v} \in U_h, \tag{8.118}$$
$$(\nabla \cdot \boldsymbol{u}_h(t), q) = 0, \quad \forall \, q \in Q_h, \tag{8.119}$$
$$\boldsymbol{u}_h(0) = \boldsymbol{u}_{0,h}, \tag{8.120}$$

where $\boldsymbol{u}_{0,h}$ is an approximation to $\boldsymbol{u}_0$. A simple choice of $\boldsymbol{u}_{0,h}$ can be the L^2-projection of $\boldsymbol{u}_0$ onto U_h. Here the subspace $U_h \subset U$ and $Q_h \subset Q$ can be chosen as in Section 8.3.

The analysis of the scheme (8.118)-(8.120) has been conducted by many researchers [4, 13, 14]. An explicit matrix form of (8.118)-(8.120) can be easily derived. Assume that ϕ_j $(j = 1, \cdots, N_u)$ and ψ_k $(k = 1, \cdots, N_p)$ are the bases of U_h and Q_h, respectively, and denoting the finite element solution of (8.118)-(8.120) as follows

$$\boldsymbol{u}_h(\boldsymbol{x}, t) = \sum_{j=1}^{N_u} u_j(t)\phi_j(\boldsymbol{x}), \quad p_h(\boldsymbol{x}, t) = \sum_{j=1}^{N_p} p_j(t)\psi_j(\boldsymbol{x}),$$

we obtain the system of nonlinear equations:

$$M\frac{d\boldsymbol{u}}{dt}(t) + A\boldsymbol{u}(t) + C(\boldsymbol{u}(t))\boldsymbol{u}(t) + B p(t) = \boldsymbol{f}(t), \quad t \in (0, T), \quad (8.121)$$
$$B^T \boldsymbol{u}(t) = 0, \quad t \in (0, T), \quad (8.122)$$
$$\boldsymbol{u}(0) = \boldsymbol{u}_0, \quad (8.123)$$

where the matrices M, A, B, and C are given as follows:

$$M_{ij} = (\phi_i, \phi_j), \quad A_{ij} = \mu(\nabla\phi_j, \nabla\phi_i),$$
$$(C(\boldsymbol{u}))_{ij} = \sum_{l=1}^{N_u} u_l c(\phi_l; \phi_j, \phi_i), \quad B_{ij} = (\phi_j, \nabla \cdot \psi_i).$$

Furthermore, the vector $\boldsymbol{f} = (f_i(t)) = ((\boldsymbol{f}(t), \phi_i))$.

The problem (8.121)-(8.123) can be further discretized in time by schemes such as the backward Euler scheme and the Crank-Nicolson scheme.

8.6 Bibliographical remarks

In this chapter, we briefly introduced the fundamental theory for the mixed finite element method and its applications to elliptic problems and the Stokes problem. More detailed discussions on mixed methods can be found in other books [5, 7, 11, 12, 21]. Bahriawati and Carstensen [3] presented three short MATLAB programs for solving the Laplace equation with inhomogeneous mixed boundary conditions in 2-D with lowest-order Raviart-Thomas mixed finite elements; interested readers can download the program from http://www.math.hu-berlin.de/~cc/.

8.7   Exercises

1. Consider the problem

$$-\Delta u = f \quad \text{in } \Omega,$$
$$u = g_1 \quad \text{on } \Gamma_1,$$
$$\frac{\partial u}{\partial n} = g_2 \quad \text{on } \Gamma_2,$$

where Ω is a bounded domain of R^2 with boundary $\Gamma \equiv \Gamma_1 \cup \Gamma_2$, $\Gamma_1 \cap \Gamma_2 = \emptyset$, and f, g_1 and g_2 are given functions. Derive a mixed variational formulation for this problem and formulate a mixed finite element method using RTN spaces on triangles.

2. Consider the 1-D problem [9, Ch. 5]

$$-\frac{d^2 p}{dx^2} = f(x) \quad x \in I \equiv (0,1),$$
$$p(0) = p(1) = 0.$$

Let $u = -\frac{dp}{dx}, V = H^1(I), Q = L^2(I)$. Then prove that $(u,p) \in V \times Q$ satisfies the mixed variational form

$$(u,v) - (v_x, p) = 0 \quad v \text{ in } V, \tag{8.124}$$
$$(u_x, q) = (f, q) \quad q \in Q. \tag{8.125}$$

Furthermore, consider the Lagrangian functional

$$\mathcal{L}(v,q) = \frac{1}{2}(v,v) - (v_x, q) + (f, q), \quad v \in V, \ q \in Q.$$

Prove that the solution (u,p) of (8.124)-(8.125) is a saddle point of the functional $\mathcal{L}(v,q)$, i.e.,

$$\mathcal{L}(u,p) = \min_{v \in V} \max_{q \in Q} \mathcal{L}(v,q),$$

or

$$\mathcal{L}(u,q) \le \mathcal{L}(u,p) \le \mathcal{L}(v,p), \quad \forall \ (v,q) \in V \times Q.$$

3. Consider the Dirichlet problem

$$-\nabla \cdot (a(\boldsymbol{x})\nabla p + \boldsymbol{b}(\boldsymbol{x})p) + c(\boldsymbol{x})p = f(\boldsymbol{x}) \quad \text{in } \Omega \subset R^d \ (d = 2 \text{ or } 3),$$
$$p = -g(\boldsymbol{x}) \quad \text{on } \partial\Omega.$$

Let $\boldsymbol{u} = -(a\nabla p + \boldsymbol{b}p)$. Prove that $\boldsymbol{u} \in H(\text{div}; \Omega)$ and $p \in L^2(\Omega)$ satisfy the mixed weak formulation

$$(a^{-1}\boldsymbol{u}, \boldsymbol{v}) - (\nabla \cdot \boldsymbol{v}, p) + (a^{-1}\boldsymbol{b}p, \boldsymbol{v}) = < g, \boldsymbol{v} \cdot \boldsymbol{n} >_{\partial\Omega} \quad \forall \ \boldsymbol{v} \in H(\text{div}; \Omega),$$

$$(\nabla \cdot \boldsymbol{u}, q) + (cp, q) = (f, q) \quad \forall \, q \in L^2(\Omega).$$

4. Let $RT_r(K)$ $(r \geq 0)$ be the Raviart-Thomas space on the triangle K. Show that a function $\boldsymbol{v} \in RT_r(K)$ is uniquely defined by the degrees of freedom [20]

$$< \boldsymbol{v} \cdot \boldsymbol{n}, w >_e, \quad \forall \, w \in P_r(e), \ e \in \partial K,$$
$$(\boldsymbol{v}, \boldsymbol{w})_K, \quad \forall \, \boldsymbol{w} \in (P_{r-1}(K))^2.$$

5. Modify the code *Stokes.m* to solve the Stokes flow in a rectangular cavity $[-1, 1.5] \times [-0.8, 0.8]$ with $\mu = 1$ and lid velocity $\boldsymbol{u} = (1, 0)'$. Plot the velocity field and the pressure.

6. For a general triangle K with vertices $a_i, 1 \leq i \leq 3$, we can define the unique affine invertible mapping

$$x \equiv F_K(\hat{x}) = B_K \hat{x} + c_K$$

such that

$$F_K(\hat{a}_i) = a_i, \quad 1 \leq i \leq 3,$$

where $\hat{a}_i, 1 \leq i \leq 3$, are the vertices of a reference triangle $\hat{K}$. For any vector function $\hat{q} = (\hat{q}_1, \hat{q}_2)$ defined on $\hat{K}$, we can associate a function q defined on K by

$$q(x) = \frac{1}{J_K} B_K \hat{q}(F_K^{-1}(x)), \tag{8.126}$$

where $J_K = det(B_K)$. Prove that the divergences q and $\hat{q}$ are related by the following

$$\nabla \cdot \boldsymbol{q} = \frac{1}{J_K} \hat{\nabla} \cdot \hat{\boldsymbol{q}},$$

where we denote

$$\nabla \cdot \boldsymbol{q} = \frac{\partial q_1}{\partial x_1} + \frac{\partial q_2}{\partial x_2}, \quad \hat{\nabla} \cdot \hat{\boldsymbol{q}} = \frac{\partial \hat{q}_1}{\partial \hat{x}_1} + \frac{\partial \hat{q}_2}{\partial \hat{x}_2}.$$

Hint: Let $B_K = \begin{pmatrix} b_{11} & b_{12} \\ b_{21} & b_{22} \end{pmatrix}$, then use the chain rule

$$\nabla \cdot \boldsymbol{q} = \left(\frac{\partial q_1}{\partial \hat{x}_1} \frac{\partial \hat{x}_1}{\partial x_1} + \frac{\partial q_1}{\partial \hat{x}_2} \frac{\partial \hat{x}_2}{\partial x_1} \right) + \left(\frac{\partial q_2}{\partial \hat{x}_1} \frac{\partial \hat{x}_1}{\partial x_2} + \frac{\partial q_2}{\partial \hat{x}_2} \frac{\partial \hat{x}_2}{\partial x_2} \right)$$

and the following

$$\frac{\partial \hat{x}_1}{\partial x_1} = \frac{b_{22}}{J_K}, \quad \frac{\partial \hat{x}_2}{\partial x_1} = -\frac{b_{21}}{J_K}, \quad \frac{\partial \hat{x}_1}{\partial x_2} = -\frac{b_{12}}{J_K}, \quad \frac{\partial \hat{x}_2}{\partial x_2} = \frac{b_{11}}{J_K}.$$

7. For any function $\hat{q} \in (H^l(\hat{K})^2$ (integer $l \geq 0$) defined by (8.126), prove that

$$|\hat{q}|_{H^l(\hat{K})} \leq C \|B_K\|^l \|B_K^{-1}\| \, |J_K|^{\frac{1}{2}} |q|_{H^l(K)},$$

where $||B_K||$ and $||B_K^{-1}||$ denote the spectral norms of B_K and B_K^{-1}, respectively.

8. Consider the parabolic problem

$$\frac{\partial p}{\partial t} - (\frac{\partial^2 p}{\partial x^2} + \frac{\partial^2 p}{\partial y^2}) = f(x, y), \quad \text{in } \Omega \times (0, t_F),$$

$$p(x, y, 0) = p_0(x, y), \quad \text{in } \Omega,$$

$$p = 0 \quad \text{on } \partial\Omega \times (0, t_F).$$

Develop a mixed variational formulation for this problem and formulate a mixed finite element method using some mixed spaces defined in this chapter. Also formulate a fully discrete scheme using the Crank-Nicolson method and prove a corresponding stability result.

9. Consider the biharmonic problem

$$\triangle^2 u = f, \quad \text{in } \Omega,$$

$$u = \frac{\partial u}{\partial n} = 0, \quad \text{on } \partial\Omega.$$

Let $p = \triangle u \in H^1(\Omega)$. Prove that $(u, p) \in H_0^1(\Omega) \times H^1(\Omega)$ satisfies the mixed weak formulation

$$(p, q) + (\nabla q, \nabla u) = 0, \quad \forall q \in H^1(\Omega),$$

$$(\nabla p, \nabla v) = -(f, v), \quad \forall v \in H_0^1(\Omega).$$

10. Prove that for the trilinear form $c(\boldsymbol{w}; \boldsymbol{u}, \boldsymbol{v})$, there exists a constant $c > 0$ such that

$$|c(\boldsymbol{w}; \boldsymbol{u}, \boldsymbol{v})| \le c|\boldsymbol{w}|_1|\boldsymbol{u}|_1|\boldsymbol{v}|_1 \quad \forall \, \boldsymbol{w}, \boldsymbol{u}, \boldsymbol{v} \in (H_0^1(\Omega))^d.$$

[Hint: Use the embedding theorem: $H^1(\Omega) \hookrightarrow L^4(\Omega)$ for both 2D and 3D domain Ω.]

11. Assume that

$$|\frac{\partial F}{\partial z}(z_1) - \frac{\partial F}{\partial z}(z_2)| \le L|z_1 - z_2|, \quad \forall \, z_1, z_2 \in B(z; \epsilon),$$

where $B(z; \epsilon)$ denotes a ball centered at z with radius ϵ. Then there exists a constant δ with $0 < \delta < \epsilon$ such that for any initial guess $z^0 \in B(z; \delta)$, the Newton method (8.108) converges quadratically to z, i.e.,

$$|z^{n+1} - z| \le C|z^n - z|^2,$$

holds true for some constant $C > 0$.

12. Modify the code *Stokes.m* by implementing $Q_2 - Q_0$ element.

References

[1] D.N. Arnold, F. Brezzi and M. Fortin. A stable finite element for the Stokes equations. *Calcolo*, 21:337–344, 1984.

[2] D.N. Arnold, L.R. Scott and M. Vogelius. Regular inversion of the divergence operator with Dirichlet boundary conditions on a polygon. *Annali della Scuola Normale Superiore di Pisa* 15(2):169–192, 1988.

[3] C. Bahriawati and C. Carstensen. Three MATLAB implementations of the lowest-order Raviart-Thomas MFEM with a posteriori error control. *Comput. Methods Appl. Math.*, 5(4):333–361, 2005.

[4] C. Bernardi and G. Raugel. A conforming finite element method for the time-dependent Navier-Stokes equations. *SIAM J. Numer. Anal.*, 22(3):455–473, 1985.

[5] D. Boffi, F. Brezzi and M. Fortin. *Mixed Finite Element Methods and Applications*. Springer-Verlag, Berlin, 2013.

[6] J.M. Boland and R.A. Nicolaides. Stability of finite elements under divergence constraints. *SIAM J. Numer. Anal.*, 20:722–731, 1983.

[7] F. Brezzi and M. Fortin. *Mixed and Hybrid Finite Element Methods*. Springer-Verlag, Berlin/Heidelberg, 1991.

[8] P.G. Ciarlet. *The Finite Element Method for Elliptic Problems*. North-Holland, Amsterdam, 1978.

[9] Z. Chen. *Finite Element Methods and Their Applications*. Springer, New York, NY, 2005.

[10] G.R. Cowper. Gaussian quadrature formulas for triangles. *Internat. J. Numer. Methods Engrg.*, 7:405–408, 1973.

[11] V. Girault and P.-A. Raviart. *Finite Element Methods for Navier-Stokes Equations*. Springer-Verlag, Berlin, 1986.

[12] Max D. Gunzburger. *Finite Element Methods for Viscous Incompressible Flows*. Academic Press, Boston, MA, 1989.

[13] J.G. Heywood and R. Rannacher. Finite element approximation of the nonstationary Navier-Stokes problem. I. Regularity of solutions and second-order error estimates for spatial discretization. *SIAM J. Numer. Anal.*, 19(2):275–311, 1982.

[14] J.G. Heywood and R. Rannacher. Finite-element approximation of the nonstationary Navier-Stokes problem. IV. Error analysis for second-order time discretization. *SIAM J. Numer. Anal.*, 27(2):353–384, 1990.

[15] C. Johnson. *Numerical Solution of Partial Differential Equations by the Finite Element Method.* Cambridge University Press, New York, 1988.

[16] J. Li and Y. Huang. *Time-Domain Finite Element Methods for Maxwell's Equations in Metamaterials.* Springer-Verlag, Berlin, 2013.

[17] J.-C. Nédélec. Mixed finite elements in R^3. *Numer. Math.*, 35:315–341, 1980.

[18] C. Pozrikidis. *Introduction to Finite and Spectral Element Methods Using MATLAB.* Chapman & Hall/CRC, Boca Raton, FL, 2005.

[19] A. Quarteroni and A. Valli. *Numerical Approximation of Partial Differential Equations.* Springer-Verlag, Berlin, 1994.

[20] P.-A. Raviart and J.M. Thomas. A mixed finite element method for 2nd order elliptic problems. In *Mathematical Aspects of Finite Element Methods* (Lecture Notes in Math., Vol. 606), Springer, Berlin, 292–315, 1977.

[21] J.E. Roberts and J.-M. Thomas. Mixed and hybrid methods. *Handbook of Numerical Analysis* (eds. by P. G. Ciarlet and J.-L. Lions), Vol. II, 523–639, North-Holland, Amsterdam, 1991.

[22] C. Taylor and P. Hood. A numerical solution of the Navier-Stokes equations using the finite element technique. *Internat. J. Comput. & Fluids*, 1:73–100, 1973.

9

Finite Element Methods for Electromagnetics

In our daily life, we have to interact with all sorts of electromagnetic waves. Examples include heating our food using a microwave oven, talking to friends through a wireless communication network, and taking a medical exam with magnetic resonance imaging (MRI). Modeling such wave interactions becomes a very important research area often called computational electromagnetics (CEM).

In this chapter, we will briefly introduce how the finite element method is used in CEM. In Sec. 9.1, we introduce Maxwell's equations in both the time domain and frequency domain. In Sec. 9.2, we show how to develop and analyze the time-domain finite difference method for solving Maxwell's equation in free space. Then we discuss some basic time-domain Galerkin methods and mixed methods in Sec. 9.3. In Sec. 9.4, we extend the discussion to frequency-domain methods such as the standard Galerkin method and the discontinuous Galerkin method for the vector wave equation. Finally, we introduce some recent work in modeling Maxwell's equations in dispersive media including the negative index metamaterials in Sec. 9.5.

9.1 Introduction to Maxwell's equations

The fundamental equations governing all macroscopic electromagnetic phenomena are Maxwell's equations:

$$\text{Faraday's law (1831): } \nabla \times \boldsymbol{E} = -\frac{\partial \boldsymbol{B}}{\partial t}, \tag{9.1}$$

$$\text{Maxwell-Ampere law (1820): } \nabla \times \boldsymbol{H} = \frac{\partial \boldsymbol{D}}{\partial t} + \boldsymbol{J}, \tag{9.2}$$

$$\text{Gauss's law: } \nabla \cdot \boldsymbol{D} = \rho, \tag{9.3}$$

$$\text{Gauss's law -- magnetic: } \nabla \cdot \boldsymbol{B} = 0, \tag{9.4}$$

where $\boldsymbol{E}(\boldsymbol{x}, t)$ and $\boldsymbol{H}(\boldsymbol{x}, t)$ are the electric and magnetic fields, $\boldsymbol{D}(\boldsymbol{x}, t)$ and $\boldsymbol{B}(\boldsymbol{x}, t)$ are the corresponding electric and magnetic flux densities, $\boldsymbol{J}$ is the electric current density, and ρ is the electric charge density.

Note that among the four equations above, only three are independent. Usually the first three equations are chosen as such independent equations. Furthermore, constitutive relations are needed for describing the macroscopic properties of the underlying medium. For a simple medium, we have

$$D = \epsilon E, \quad B = \mu H, \quad J = \sigma E + J_s, \tag{9.5}$$

where J_s denotes a specified current, ϵ, μ and σ denote the permittivity, permeability, and conductivity of the medium, respectively. These parameters become tensors for anisotropic media. For inhomogeneous media, these parameters are functions of position too. For more complex media, the constitutive relations can become more complicated.

Frequently, the electromagnetic fields are assumed to be time-harmonic (i.e., harmonically oscillating with a single frequency ω), in which case, we may write

$$E(x,t) = e^{-i\omega t}\widehat{E}(x), \quad H(x,t) = e^{-i\omega t}\widehat{H}(x), \quad J_s(x,t) = e^{-i\omega t}\widehat{J}_s(x), \tag{9.6}$$

where the so-called phasors $\widehat{E}, \widehat{H}$ and $\widehat{J}_s$ are independent of time, but complex-valued.

Substituting (9.6) and (9.5) into (9.1) and (9.2), we obtain the time-harmonic Maxwell's equations

$$\nabla \times \widehat{E} = i\omega\mu\widehat{H}, \tag{9.7}$$

$$\nabla \times \widehat{H} = -i\omega(\epsilon + i\frac{\sigma}{\omega})\widehat{E} + \widehat{J}_s. \tag{9.8}$$

We can eliminate $\widehat{H}$ from (9.7) and (9.8) to obtain the equation

$$\nabla \times (\mu^{-1}\nabla \times \widehat{E}) - \omega^2(\epsilon + i\frac{\sigma}{\omega})\widehat{E} = i\omega\widehat{J}_s, \tag{9.9}$$

which is often called the vector wave equation. Similarly, we can eliminate $\widehat{E}$ to obtain an equation for $\widehat{H}$ only.

The corresponding time-domain vector wave equation of (9.9) can be written as

$$\nabla \times (\frac{1}{\mu}\nabla \times E) + \epsilon\frac{\partial^2 E}{\partial t^2} + \sigma\frac{\partial E}{\partial t} = -\frac{\partial J_s}{\partial t}, \tag{9.10}$$

which can be obtained from (9.1) and (9.2) by eliminating H and using (9.5).

Boundary conditions are needed for a complete description of an electromagnetic problem. Between two media, say medium 1 and medium 2, the fields must satisfy the interface conditions

$$n \times (E_2 - E_1) = -M_s, \tag{9.11}$$

$$n \times (H_2 - H_1) = J_s, \tag{9.12}$$

$$n \cdot (D_2 - D_1) = \rho_s, \tag{9.13}$$

$$n \cdot (B_2 - B_1) = \mu_s, \tag{9.14}$$

where (M_s, J_s) and (ρ_s, μ_s) are the magnetic and electric current and surface charge densities, respectively, where n is unit vector normal to the interface pointing from medium 1 to medium 2.

The two most common cases are:

(i) $M_s = J_s = \rho_s = \mu_s = 0$, i.e., the interface is free of source and charge;

(ii) $M_s = \mu_s = 0, \rho_s, J_s \neq 0$, i.e., there exists a surface electric current density J_s and a surface charge density ρ_s.

When medium 2 becomes a perfect conductor (hence $E_2 = H_2 = 0$), we obtain the so-called perfect conducting boundary condition:

$$n \times E_1 = 0, \quad n \cdot B_1 = 0. \tag{9.15}$$

In this case, the boundary can always support a surface current and a surface charge, i.e., we have

$$n \times H_1 = -J_s, \quad n \cdot D_1 = -\rho_s. \tag{9.16}$$

When medium 2 is not a perfect conductor, we have the impedance or imperfectly conducting boundary condition:

$$n \times H_1 - \lambda(n \times E_1) \times n = 0, \tag{9.17}$$

where the impedance λ is a positive function of position.

9.2 The time-domain finite difference method

Due to its simplicity in implementation and robustness, the time-domain finite difference (FDTD) method, originally proposed by Yee [73] in 1966, is one of the most popular numerical methods for solving time-domain Maxwell's equations. The FDTD method is different from the standard finite difference method in that both spatial and time grids are staggered. In this section, we illustrate the basic techniques for developing and analyzing the FDTD method through a 2D model problem. Here we adopt the notation and analysis method developed in our previous work on the FDTD method for Maxwell's equations in metamaterials [51]. The same techniques can be easily extended to the 3D Maxwell's equations in complex media.

9.2.1 The semi-discrete scheme

In free space, the constitute equations (9.5) become as follows:

$$D = \epsilon_0 E, \quad B = \mu_0 H, \quad J = J_s, \tag{9.18}$$

where $\epsilon_0 = 8.854 \times 10^{-12}$ farads/meter, and $\mu_0 = 4\pi \times 10^{-7}$ henrys/meter are the free space permittivity and permeability, respectively. Substituting (9.18) into (9.1)-(9.2) and still denoting $\boldsymbol{J}_s$ by $\boldsymbol{J}$, we obtain Maxwell's equations in free space:

$$\epsilon_0 \frac{\partial \boldsymbol{E}}{\partial t} = \nabla \times \boldsymbol{H} - \boldsymbol{J}, \quad \forall\, (\boldsymbol{x}, t) \in \Omega \times (0, T), \tag{9.19}$$

$$\mu_0 \frac{\partial \boldsymbol{H}}{\partial t} = -\nabla \times \boldsymbol{E}, \quad \forall\, (\boldsymbol{x}, t) \in \Omega \times (0, T). \tag{9.20}$$

To complete the problem, we assume that (9.19)-(9.20) are supplemented with the perfect conduct (PEC) boundary condition

$$\boldsymbol{n} \times \boldsymbol{E} = \boldsymbol{0} \quad \text{on } \partial\Omega, \tag{9.21}$$

and the initial conditions

$$\boldsymbol{E}(\boldsymbol{x}, 0) = \boldsymbol{E}_0(\boldsymbol{x}), \quad \boldsymbol{H}(\boldsymbol{x}, 0) = \boldsymbol{H}_0(\boldsymbol{x}), \tag{9.22}$$

where $\boldsymbol{n}$ denotes the outward unit normal vector, $\boldsymbol{E}_0(\boldsymbol{x})$ and $\boldsymbol{H}_0(\boldsymbol{x})$ are some given proper functions.

To avoid the technicality caused by the 3D problems, below we only consider the 2D transverse electric (TE) case of (9.19)-(9.22), in which $\boldsymbol{E} = (E_x, E_y), \boldsymbol{H} = H_z := H, \boldsymbol{J} = (J_x, J_y)$, and the curls $\nabla \times \boldsymbol{E} = \frac{\partial E_y}{\partial x} - \frac{\partial E_x}{\partial y}$ and $\nabla \times \boldsymbol{H} = (\frac{\partial H}{\partial y}, -\frac{\partial H}{\partial x})'$. Here the subindices x, y and z denote the components in the x, y and z directions, respectively. For simplicity, we consider the physical domain Ω is a rectangle $[a, b] \times [c, d]$, which is partitioned by a non-uniform grid

$$a = x_0 < x_1 < \cdots < x_{N_x} = b, \quad c = y_0 < y_1 < \cdots < y_{N_y} = d.$$

Following the classic FDTD method, we place the unknowns E_x, E_y and H at the mid-points of the horizontal edges, at the mid-points of the vertical edges, and at the element centers (cf. Fig. 9.1), respectively. Also we denote the corresponding approximate solutions as follows (by suppressing the explicit dependence on time t):

$$E_{x, i+\frac{1}{2}, j}, \quad i = 0, \cdots, N_x - 1, \; j = 0, \cdots, N_y,$$
$$E_{y, i, j+\frac{1}{2}}, \quad j = 0, \cdots, N_y - 1, \; i = 0, \cdots, N_x,$$
$$H_{i+\frac{1}{2}, j+\frac{1}{2}}, \quad i = 0, \cdots, N_x - 1, \; j = 0, \cdots, N_y - 1.$$

For convenience, we denote the following three types of rectangles:

$$T_{ij} = (x_i, x_{i+1}) \times (y_j, y_{j+1}),$$
$$T_{i-\frac{1}{2}, j} = (x_{i-\frac{1}{2}}, x_{i+\frac{1}{2}}) \times (y_j, y_{j+1}),$$
$$T_{i, j-\frac{1}{2}} = (x_i, x_{i+1}) \times (y_{j-\frac{1}{2}}, y_{j+\frac{1}{2}}),$$

and their corresponding areas $|T_{ij}|, |T_{i-\frac{1}{2},j}|$ and $|T_{i,j-\frac{1}{2}}|$. To distinguish the role of non-uniform mesh, we denote $h_x = \max_{0 \le i \le N_x - 1}(x_{i+1} - x_i)$ and $h_y = \max_{0 \le j \le N_y - 1}(y_{j+1} - y_j)$ for the maximal mesh sizes in the x and y directions, respectively. The global mesh size h is defined as $h := \max(h_x, h_y)$.

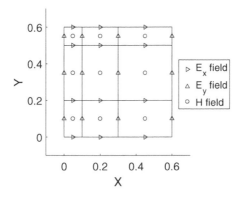

FIGURE 9.1
The staggered grid for solving 2D Maxwell's equations.

Integrating the x-component of (9.19), i.e., $\epsilon_0 \frac{\partial E_x}{\partial t} = \frac{\partial H}{\partial y} - J_x$, on $T_{i,j-\frac{1}{2}}$ (for any $0 \le i \le N_x - 1, 1 \le j \le N_y - 1$), we obtain

$$\int_{x_i}^{x_{i+1}} \int_{y_{j-\frac{1}{2}}}^{y_{j+\frac{1}{2}}} \epsilon_0 \frac{\partial E_x}{\partial t} = \int_{x_i}^{x_{i+1}} [H(x, y_{j+\frac{1}{2}}, t) - H(x, y_{j-\frac{1}{2}}, t)] dx - \int_{T_{i,j-\frac{1}{2}}} J_x. \tag{9.23}$$

Approximating the first two integrals in (9.23) by the mid-point quadrature rule, we have

$$\epsilon_0 |T_{i,j-\frac{1}{2}}| \cdot \frac{\partial E_x}{\partial t}\big|_{i+\frac{1}{2},j} = (x_{i+1} - x_i)(H_{i+\frac{1}{2},j+\frac{1}{2}} - H_{i+\frac{1}{2},j-\frac{1}{2}}) - \int_{T_{i,j-\frac{1}{2}}} J_x. \tag{9.24}$$

Similarly, integrating the y-component of (9.19), i.e., $\epsilon_0 \frac{\partial E_y}{\partial t} = -\frac{\partial H}{\partial x} - J_y$, on $T_{i-\frac{1}{2},j}$ (for any $1 \le i \le N_x - 1, 0 \le j \le N_y - 1$) yields

$$\int_{x_{i-\frac{1}{2}}}^{x_{i+\frac{1}{2}}} \int_{y_j}^{y_{j+1}} \epsilon_0 \frac{\partial E_y}{\partial t} = -\int_{y_j}^{y_{j+1}} [H(x_{i+\frac{1}{2}}, y, t) - H(x_{i-\frac{1}{2}}, y, t)] dy - \int_{T_{i-\frac{1}{2},j}} J_y. \tag{9.25}$$

Approximating the first two integrals in (9.25) by the mid-point quadrature rule, we have

$$\epsilon_0 |T_{i-\frac{1}{2},j}| \cdot \frac{\partial E_y}{\partial t}\big|_{i,j+\frac{1}{2}} = -(y_{j+1} - y_j)(H_{i+\frac{1}{2},j+\frac{1}{2}} - H_{i-\frac{1}{2},j+\frac{1}{2}}) - \int_{T_{i-\frac{1}{2},j}} J_y. \tag{9.26}$$

By the same technique, integrating the 2D form of (9.20), i.e., $\mu_0\frac{\partial H}{\partial t} = -(\frac{\partial E_y}{\partial x} - \frac{\partial E_x}{\partial y})$, on T_{ij} (for any $0 \le i \le N_x - 1, 0 \le j \le N_y - 1$) yields

$$\int_{x_i}^{x_{i+1}} \int_{y_j}^{y_{j+1}} \mu_0\frac{\partial H}{\partial t} = -\int_{x_i}^{x_{i+1}} \int_{y_j}^{y_{j+1}} (\frac{\partial E_y}{\partial x} - \frac{\partial E_x}{\partial y}) \tag{9.27}$$

$$= -\int_{y_j}^{y_{j+1}} (E_y(x_{i+1}, y, t) - E_y(x_i, y, t))dy + \int_{x_i}^{x_{i+1}} (E_x(x, y_{j+1}, t) - E_x(x, y_j, t))dx.$$

Further application of the mid-point quadrature rule leads to

$$\mu_0|T_{ij}| \cdot \frac{\partial H}{\partial t}|_{i+\frac{1}{2},j+\frac{1}{2}} = -(y_{j+1} - y_j)(E_{y,i+1,j+\frac{1}{2}} - E_{y,i,j+\frac{1}{2}})$$
$$+(x_{i+1} - x_i)(E_{x,i+\frac{1}{2},j+1} - E_{x,i+\frac{1}{2},j}). \tag{9.28}$$

In summary, we obtain the semi-discrete scheme (9.24), (9.26) and (9.28) for solving the 2D Maxwell's equations.

9.2.1.1   The stability analysis

To simplify the notation, we introduce the following mesh-dependent energy norms:

$$||E_x||_E^2 = \sum_{\substack{0\le i\le N_x-1 \\ 1\le j\le N_y-1}} |T_{i,j-\frac{1}{2}}| \cdot |E_{x,i+\frac{1}{2},j}|^2,$$

$$||E_y||_E^2 = \sum_{\substack{1\le i\le N_x-1 \\ 0\le j\le N_y-1}} |T_{i-\frac{1}{2},j}| \cdot |E_{y,i,j+\frac{1}{2}}|^2,$$

$$||H||_H^2 = \sum_{\substack{0\le i\le N_x-1 \\ 0\le j\le N_y-1}} |T_{ij}| \cdot |H_{i+\frac{1}{2},j+\frac{1}{2}}|^2.$$

In the rest of this subsection, we show how to use the energy method to prove the stability for the above derived semi-discrete scheme.

THEOREM 9.1
The solution of the semi-discrete scheme (9.24), (9.26) and (9.28) satisfies the stability:

$$\max_{0\le t\le T} [\epsilon_0(||E_x||_E^2 + ||E_y||_E^2) + \mu_0||H||_H^2](t) \tag{9.29}$$

$$\le 2[\epsilon_0(||E_x||_E^2 + ||E_y||_E^2) + \mu_0||H||_H^2](0) + \frac{4T}{\epsilon_0}\int_0^T (||J_x||^2 + ||J_y||^2)(s)ds.$$

Proof. Multiplying (9.24), (9.26) and (9.28) by $E_{x,i+\frac{1}{2},j}$, $E_{y,i,j+\frac{1}{2}}$, and $H_{i+\frac{1}{2},j+\frac{1}{2}}$, respectively, then summing up each result over its corresponding rectangular elements, and adding all results together, we obtain the sum of the left-hand side:

$$LHS = \frac{1}{2}\frac{d}{dt}[\epsilon_0(||E_x||_E^2 + ||E_y||_E^2) + \mu_0||H||_H^2] \tag{9.30}$$

and the sum of the right-hand side:

$$
\begin{aligned}
RHS = &\sum_{\substack{0 \le i \le N_x-1 \\ 1 \le j \le N_y-1}} (x_{i+1} - x_i)(H_{i+\frac{1}{2},j+\frac{1}{2}} - H_{i+\frac{1}{2},j-\frac{1}{2}})E_{x,i+\frac{1}{2},j} \\
& - \sum_{\substack{1 \le i \le N_x-1 \\ 0 \le j \le N_y-1}} (y_{j+1} - y_j)(H_{i+\frac{1}{2},j+\frac{1}{2}} - H_{i-\frac{1}{2},j+\frac{1}{2}})E_{y,i,j+\frac{1}{2}} \\
& - \sum_{\substack{0 \le i \le N_x-1 \\ 0 \le j \le N_y-1}} (y_{j+1} - y_j)(E_{y,i+1,j+\frac{1}{2}} - E_{y,i,j+\frac{1}{2}})H_{i+\frac{1}{2},j+\frac{1}{2}} \\
& + \sum_{\substack{0 \le i \le N_x-1 \\ 0 \le j \le N_y-1}} (x_{i+1} - x_i)(E_{x,i+\frac{1}{2},j+1} - E_{x,i+\frac{1}{2},j})H_{i+\frac{1}{2},j+\frac{1}{2}} \\
& - \sum_{\substack{0 \le i \le N_x-1 \\ 1 \le j \le N_y-1}} (\int_{T_{i,j-\frac{1}{2}}} J_x)E_{x,i+\frac{1}{2},j} - \sum_{\substack{1 \le i \le N_x-1 \\ 0 \le j \le N_y-1}} (\int_{T_{i-\frac{1}{2},j}} J_y)E_{y,i,j+\frac{1}{2}} \\
:= &\sum_{k=1}^{6} RHS_k. \tag{9.31}
\end{aligned}
$$

Note that the sum of the first four terms in (9.31) actually equals zero:

$$
\begin{aligned}
&\sum_{k=1}^{4} RHS_k \\
= &\sum_{0 \le i \le N_x-1} (x_{i+1} - x_i) \sum_{0 \le j \le N_y-1} [H_{i+\frac{1}{2},j+\frac{1}{2}}E_{x,i+\frac{1}{2},j+1} - H_{i+\frac{1}{2},j-\frac{1}{2}}E_{x,i+\frac{1}{2},j}] \\
& - \sum_{0 \le j \le N_y-1} (y_{j+1} - y_j) \sum_{0 \le i \le N_x-1} [H_{i+\frac{1}{2},j+\frac{1}{2}}E_{y,i+1,j+\frac{1}{2}} - H_{i-\frac{1}{2},j+\frac{1}{2}}E_{y,i,j+\frac{1}{2}}] \\
= &\sum_{0 \le i \le N_x-1} (x_{i+1} - x_i)[H_{i+\frac{1}{2},N_y-\frac{1}{2}}E_{x,i+\frac{1}{2},N_y} - H_{i+\frac{1}{2},-\frac{1}{2}}E_{x,i+\frac{1}{2},0}] \tag{9.32} \\
& - \sum_{0 \le j \le N_y-1} (y_{j+1} - y_j)[H_{N_x-\frac{1}{2},j+\frac{1}{2}}E_{y,N_x,j+\frac{1}{2}} - H_{-\frac{1}{2},j+\frac{1}{2}}E_{y,0,j+\frac{1}{2}}] = 0,
\end{aligned}
$$

where we used the PEC boundary condition (9.21), which in the 2D case is equivalent to

$$E_{x,i+\frac{1}{2},N_y} = E_{x,i+\frac{1}{2},0} = 0, \quad E_{y,N_x,j+\frac{1}{2}} = E_{y,0,j+\frac{1}{2}} = 0, \tag{9.33}$$

for all $i \in [0, N_x - 1]$ and $j \in [0, N_y - 1]$.

By the Cauchy-Schwarz inequality, we can bound the sum of the last two terms in (9.31) as follows:

$$\sum_{k=5}^{6} RHS_k$$

$$\leq \left(\sum_{\substack{0 \leq i \leq N_x-1 \\ 1 \leq j \leq N_y-1}} \frac{1}{|T_{i,j-\frac{1}{2}}|} \left(\int_{T_{i,j-\frac{1}{2}}} J_x \right)^2 \right)^{1/2} \left(\sum_{\substack{0 \leq i \leq N_x-1 \\ 1 \leq j \leq N_y-1}} |T_{i,j-\frac{1}{2}}| \cdot |E_{x,i+\frac{1}{2},j}|^2 \right)^{1/2}$$

$$+ \left(\sum_{\substack{1 \leq i \leq N_x-1 \\ 0 \leq j \leq N_y-1}} \frac{1}{|T_{i-\frac{1}{2},j}|} \left(\int_{T_{i-\frac{1}{2},j}} J_y \right)^2 \right)^{1/2} \left(\sum_{\substack{1 \leq i \leq N_x-1 \\ 0 \leq j \leq N_y-1}} |T_{i-\frac{1}{2},j}| \cdot |E_{y,i,j+\frac{1}{2}}|^2 \right)^{1/2}$$

$$\leq \left(\sum_{\substack{0 \leq i \leq N_x-1 \\ 1 \leq j \leq N_y-1}} \int_{T_{i,j-\frac{1}{2}}} |J_x|^2 \right)^{1/2} ||E_x||_E + \left(\sum_{\substack{1 \leq i \leq N_x-1 \\ 0 \leq j \leq N_y-1}} \int_{T_{i-\frac{1}{2},j}} |J_y|^2 \right)^{1/2} ||E_y||_E$$

$$\leq \frac{\delta \epsilon_0}{2} (||E_x||_E^2 + ||E_y||_E^2) + \frac{1}{2\delta\epsilon_0}(||J_x||^2 + ||J_y||^2), \tag{9.34}$$

where $\delta > 0$ is an arbitrary constant to be determined later.

Equating (9.31) and (9.30), substituting the estimates (9.32)-(9.34) into (9.31), then integrating the result with respect to time from $t = 0$ to any $t \in (0, T]$, we have

$$[\epsilon_0(||E_x||_E^2 + ||E_y||_E^2) + \mu_0||H||_H^2](t) - [\epsilon_0(||E_x||_E^2 + ||E_y||_E^2) + \mu_0||H||_H^2](0)$$

$$\leq \delta \int_0^t \epsilon_0(||E_x||_E^2 + ||E_y||_E^2)(s)ds + \frac{1}{\delta\epsilon_0} \int_0^t (||J_x||^2 + ||J_y||^2)(s)ds$$

$$\leq \delta T \max_{0 \leq t \leq T} [\epsilon_0(||E_x||_E^2 + ||E_y||_E^2)(t)] + \frac{1}{\delta\epsilon_0} \int_0^T (||J_x||^2 + ||J_y||^2)(s)ds. \tag{9.35}$$

Now taking the maximum of the left-hand side of (9.35) for t over $[0, T]$, then choosing $\delta = \frac{1}{2T}$, we obtain

$$\max_{0 \leq t \leq T} [\epsilon_0(||E_x||_E^2 + ||E_y||_E^2) + \mu_0||H||_H^2](t)$$

$$\leq 2[\epsilon_0(||E_x||_E^2 + ||E_y||_E^2) + \mu_0||H||_H^2](0) + \frac{4T}{\epsilon_0} \int_0^T (||J_x||^2 + ||J_y||^2)(s)ds,$$

which completes the proof. □

9.2.1.2   The error estimate

To make the error analysis easy to follow, we denote the errors by their corresponding script letters. For example, the error of E_x at point $(x_{i+\frac{1}{2}}, y_j, t)$

is denoted by $\mathcal{E}_{x,i+\frac{1}{2},j} = E_x(x_{i+\frac{1}{2}}, y_j, t) - E_{x,i+\frac{1}{2},j}$, where $E_x(x_{i+\frac{1}{2}}, y_j, t)$ and $E_{x,i+\frac{1}{2},j}$ denote the exact and numerical solutions of E_x at point $(x_{i+\frac{1}{2}}, y_j, t)$, respectively. Similarly, we introduce the errors of E_y and H:

$$\mathcal{E}_{y,i,j+\frac{1}{2}} = E_y(x_i, y_{j+\frac{1}{2}}, t) - E_{y,i,j+\frac{1}{2}},$$
$$\mathcal{H}_{i+\frac{1}{2},j+\frac{1}{2}} = H(x_{i+\frac{1}{2}}, y_{j+\frac{1}{2}}, t) - H_{i+\frac{1}{2},j+\frac{1}{2}}.$$

By the definition of errors, and from (9.23) and (9.24), we obtain

$$\epsilon_0 |T_{i,j-\frac{1}{2}}| \cdot \frac{\partial \mathcal{E}_x}{\partial t}\Big|_{i+\frac{1}{2},j} = \epsilon_0 \left(\int_{T_{i,j-\frac{1}{2}}} \frac{\partial E_x}{\partial t}(x_{i+\frac{1}{2}}, y_j, t) - |T_{i,j-\frac{1}{2}}| \cdot \frac{\partial E_x}{\partial t}\Big|_{i+\frac{1}{2},j} \right)$$

$$= \epsilon_0 \left(\int_{T_{i,j-\frac{1}{2}}} \frac{\partial E_x}{\partial t}(x_{i+\frac{1}{2}}, y_j, t) - \int_{T_{i,j-\frac{1}{2}}} \frac{\partial E_x}{\partial t}(x, y, t) \right)$$

$$+ \left[\int_{x_i}^{x_{i+1}} (H(x, y_{j+\frac{1}{2}}, t) - H(x, y_{j-\frac{1}{2}}, t)) dx - \int_{T_{i,j-\frac{1}{2}}} J_x(x, y, t) \right]$$

$$- \left[(x_{i+1} - x_i)(H_{i+\frac{1}{2},j+\frac{1}{2}} - H_{i+\frac{1}{2},j-\frac{1}{2}}) - \int_{T_{i,j-\frac{1}{2}}} J_x(x, y, t) \right]$$

$$= \epsilon_0 \left(\int_{T_{i,j-\frac{1}{2}}} \frac{\partial E_x}{\partial t}(x_{i+\frac{1}{2}}, y_j, t) - \int_{T_{i,j-\frac{1}{2}}} \frac{\partial E_x}{\partial t}(x, y, t) \right)$$

$$+ (x_{i+1} - x_i)(\mathcal{H}_{i+\frac{1}{2},j+\frac{1}{2}} - \mathcal{H}_{i+\frac{1}{2},j-\frac{1}{2}}) + \int_{x_i}^{x_{i+1}} (H(x, y_{j+\frac{1}{2}}, t) - H(x, y_{j-\frac{1}{2}}, t)) dx$$

$$- (x_{i+1} - x_i)(H(x_{i+\frac{1}{2}}, y_{j+\frac{1}{2}}, t) - H(x_{i+\frac{1}{2}}, y_{j-\frac{1}{2}}, t)),$$

which leads to the error equation for E_x:

$$\epsilon_0 |T_{i,j-\frac{1}{2}}| \cdot \frac{\partial \mathcal{E}_x}{\partial t}\Big|_{i+\frac{1}{2},j} = (x_{i+1} - x_i)(\mathcal{H}_{i+\frac{1}{2},j+\frac{1}{2}} - \mathcal{H}_{i+\frac{1}{2},j-\frac{1}{2}})$$

$$+ \epsilon_0 \int_{T_{i,j-\frac{1}{2}}} \left(\frac{\partial E_x}{\partial t}(x_{i+\frac{1}{2}}, y_j, t) - \frac{\partial E_x}{\partial t}(x, y, t) \right)$$

$$+ \left[\int_{x_i}^{x_{i+1}} (H(x, y_{j+\frac{1}{2}}, t) - H(x, y_{j-\frac{1}{2}}, t)) dx \right.$$

$$\left. - \int_{x_i}^{x_{i+1}} (H(x_{i+\frac{1}{2}}, y_{j+\frac{1}{2}}, t) - H(x_{i+\frac{1}{2}}, y_{j-\frac{1}{2}}, t)) dx \right]$$

$$:= (x_{i+1} - x_i)(\mathcal{H}_{i+\frac{1}{2},j+\frac{1}{2}} - \mathcal{H}_{i+\frac{1}{2},j-\frac{1}{2}}) + r_{1,ij} + r_{2,ij}. \tag{9.36}$$

Similarly, we can obtain the error equation for E_y:

$$\epsilon_0 |T_{i-\frac{1}{2},j}| \cdot \frac{\partial \mathcal{E}_y}{\partial t}\Big|_{i,j+\frac{1}{2}} = -(y_{j+1} - y_j)(\mathcal{H}_{i+\frac{1}{2},j+\frac{1}{2}} - \mathcal{H}_{i-\frac{1}{2},j+\frac{1}{2}})$$

$$+ \epsilon_0 \int_{T_{i-\frac{1}{2},j}} \left(\frac{\partial E_y}{\partial t}(x_i, y_{j+\frac{1}{2}}, t) - \frac{\partial E_y}{\partial t}(x, y, t) \right)$$

$$-[\int_{y_j}^{y_{j+1}} (H(x_{i+\frac{1}{2}},y,t) - H(x_{i-\frac{1}{2}},y,t))dy$$

$$- \int_{y_j}^{y_{j+1}} (H(x_{i+\frac{1}{2}},y_{j+\frac{1}{2}},t) - H(x_{i-\frac{1}{2}},y_{j+\frac{1}{2}},t))dy]$$

$$:= -(y_{j+1} - y_j)(\mathcal{H}_{i+\frac{1}{2},j+\frac{1}{2}} - \mathcal{H}_{i-\frac{1}{2},j+\frac{1}{2}}) + r_{3,ij} + r_{4,ij}. \tag{9.37}$$

By the same technique, we can obtain the error equation for H:

$$\mu_0|T_{ij}| \cdot \frac{\partial \mathcal{H}}{\partial t}|_{i+\frac{1}{2},j+\frac{1}{2}} = -(y_{j+1} - y_j)(\mathcal{E}_{y,i+1,j+\frac{1}{2}} - \mathcal{E}_{y,i,j+\frac{1}{2}})$$

$$+(x_{i+1} - x_i)(\mathcal{E}_{x,i+\frac{1}{2},j+1} - \mathcal{E}_{x,i+\frac{1}{2},j})$$

$$+\mu_0 \int_{T_{ij}} (\frac{\partial H}{\partial t}(x_{i+\frac{1}{2}},y_{j+\frac{1}{2}},t) - \frac{\partial H}{\partial t}(x,y,t))$$

$$-[\int_{y_j}^{y_{j+1}} (E_y(x_{i+1},y,t) - E_y(x_i,y,t))dy$$

$$- \int_{y_j}^{y_{j+1}} (E_y(x_{i+1},y_{j+\frac{1}{2}},t) - E_y(x_i,y_{j+\frac{1}{2}},t))dy]$$

$$:= -(y_{j+1} - y_j)(\mathcal{E}_{y,i+1,j+\frac{1}{2}} - \mathcal{E}_{y,i,j+\frac{1}{2}})$$

$$+(x_{i+1} - x_i)(\mathcal{E}_{x,i+\frac{1}{2},j+1} - \mathcal{E}_{x,i+\frac{1}{2},j}) + r_{5,ij} + r_{6,ij}. \tag{9.38}$$

With the above preparations, we can prove the following convergence result.

THEOREM 9.2
Suppose that the solution of the model problem (9.19)-(9.22) possesses the following regularity property:

$$E_x, E_y, H \in C([0,T]; C^3(\overline{\Omega})) \cap C^1([0,T]; C^2(\overline{\Omega})).$$

If the initial error

$$[\epsilon_0(||\mathcal{E}_x||_E^2 + ||\mathcal{E}_y||_E^2) + \mu_0||\mathcal{H}||_H^2]^{\frac{1}{2}}(0) \le C(h_x^2 + h_y^2), \tag{9.39}$$

holds true, then we have

$$\max_{0\le t\le T}[\epsilon_0(||\mathcal{E}_x||_E^2 + ||\mathcal{E}_y||_E^2) + \mu_0||\mathcal{H}||_H^2]^{\frac{1}{2}}(t) \le CT(h_x^2 + h_y^2),$$

where the constant $C > 0$ is independent of h_x, h_y and T.

Proof. Note that the error equations (9.36)-(9.38) have exactly the same forms as the scheme (9.24), (9.26) and (9.28) with extra terms $r_{k,ij}, k = 1, \cdots, 6$. Hence the error analysis can be proved by following the same technique as that used for the stability analysis. Now the major issue becomes estimating those $r_{k,ij}, k = 1, \cdots, 6$.

By the Taylor expansion, for any function f we can easily prove that

$$\int_{T_{i,j-\frac{1}{2}}} (f(x,y,t) - f(x_{i+\frac{1}{2}}, y_j, t)) dx dy$$

$$= \int_{T_{i,j-\frac{1}{2}}} [(x - x_{i+\frac{1}{2}})\frac{\partial f}{\partial x}(p_*) + (y - y_j)\frac{\partial f}{\partial y}(p_*) + \frac{1}{2}(x - x_{i+\frac{1}{2}})^2 \frac{\partial^2 f}{\partial x^2}(p_1)$$

$$+ (x - x_{i+\frac{1}{2}})(y - y_j)\frac{\partial^2 f}{\partial x \partial y}(p_2) + \frac{1}{2}(y - y_j)^2 \frac{\partial^2 f}{\partial y^2}(p_3)]$$

$$\leq \int_{T_{i,j-\frac{1}{2}}} C[h_x^2 |\frac{\partial^2 f}{\partial x^2}|_\infty + h_y^2|\frac{\partial^2 f}{\partial y^2}|_\infty], \tag{9.40}$$

where we denote $p_* = (x_{i+\frac{1}{2}}, y_j, t)$, and p_1, p_2 and p_3 for some midpoints between p_* and (x, y, t). Here and in the rest of this section, we denote the essential supremum $|u|_\infty = \inf\{M : |u(\boldsymbol{x}, t)| \leq M \text{ a.e. in } \overline{\Omega} \times [0, T]\}$.

Applying (9.40) to $f = \frac{\partial E_x}{\partial t}$, we obtain

$$r_{1,ij} = (O(h_x^2)|\frac{\partial^3 E_x}{\partial t \partial x^2}|_\infty + O(h_y^2)|\frac{\partial^3 E_x}{\partial t \partial y^2}|_\infty) \cdot |T_{i,j-\frac{1}{2}}|.$$

To bound $r_{2,ij}$, we note that

$$\int_{x_i}^{x_{i+1}} (H(x, y_{j+\frac{1}{2}}, t) - H(x, y_{j-\frac{1}{2}}, t)) dx$$

$$- \int_{x_i}^{x_{i+1}} (H(x_{i+\frac{1}{2}}, y_{j+\frac{1}{2}}, t) - H(x_{i+\frac{1}{2}}, y_{j-\frac{1}{2}}, t)) dx$$

$$= \int_{x_i}^{x_{i+1}} [\int_{y_{j-\frac{1}{2}}}^{y_{j+\frac{1}{2}}} (\frac{\partial H}{\partial y}(x, y, t) - \frac{\partial H}{\partial y}(x_{i+\frac{1}{2}}, y, t)) dy] dx = O(h_x^2)|\frac{\partial^3 H}{\partial y \partial x^2}|_\infty |T_{i,j-\frac{1}{2}}|,$$

which leads to

$$r_{2,ij} = O(h_x^2)|\frac{\partial^3 H}{\partial y \partial x^2}|_\infty \cdot |T_{i,j-\frac{1}{2}}|.$$

By carrying out the above technique to the E_y error equation, we have

$$r_{3,ij} = (O(h_x^2)|\frac{\partial^3 E_y}{\partial t \partial x^2}|_\infty + O(h_y^2)|\frac{\partial^3 E_y}{\partial t \partial y^2}|_\infty) \cdot |T_{i-\frac{1}{2},j}|,$$

$$r_{4,ij} = -\int_{T_{i-\frac{1}{2},j}} (\frac{\partial H}{\partial x}(x, y, t) - \frac{\partial H}{\partial x}(x, y_{j+\frac{1}{2}}, t)) = O(h_y^2)|\frac{\partial^3 H}{\partial x \partial y^2}|_\infty |T_{i-\frac{1}{2},j}|.$$

Using the same technique to the H error equation, we have

$$r_{5,ij} = (O(h_x^2)|\frac{\partial^3 H}{\partial t \partial x^2}|_\infty + O(h_y^2)|\frac{\partial^3 H}{\partial t \partial y^2}|_\infty) \cdot |T_{ij}|,$$

$$r_{6,ij} = -\int_{T_{ij}} \left(\frac{\partial E_y}{\partial x}(x,y,t) - \frac{\partial E_y}{\partial x}(x,y_{j+\frac{1}{2}},t)\right) = O(h_y^2)\left|\frac{\partial^3 E_y}{\partial x \partial y^2}\right|_{\infty}|T_{ij}|.$$

Let us denote the error energy

$$Q(t) = [\epsilon_0(||\mathcal{E}_x||_E^2 + ||\mathcal{E}_y||_E^2) + \mu_0||\mathcal{H}||_H^2](t).$$

Multiplying $\mathcal{E}_{x,i+\frac{1}{2},j}$ to (9.36), $\mathcal{E}_{y,i,j+\frac{1}{2}}$ to (9.37), $\mathcal{H}_{i+\frac{1}{2},j+\frac{1}{2}}$ to (9.38), summing up the results for all i and j, then using estimates such as the following:

$$\sum_{\substack{0 \le i \le N_x - 1 \\ 1 \le j \le N_y - 1}} r_{1,ij}\mathcal{E}_{x,i+\frac{1}{2},j}$$

$$\le \sum_{\substack{0 \le i \le N_x - 1 \\ 1 \le j \le N_y - 1}} [\delta|T_{i,j-\frac{1}{2}}| \cdot |\mathcal{E}_{x,i+\frac{1}{2},j}|^2 + \frac{1}{4\delta}(O(h_x^2) + O(h_y^2))^2|T_{i,j-\frac{1}{2}}|]$$

$$\le \delta\epsilon_0||\mathcal{E}_x||_E^2 + \frac{1}{4\delta}(O(h_x^2) + O(h_y^2))^2,$$

and the estimate (9.31) with E and H replaced by $\mathcal{E}$ and $\mathcal{H}$, we obtain

$$\frac{1}{2}\frac{d}{dt}Q(t) \le \frac{C}{\delta}(h_x^2 + h_y^2)^2 + 2\delta Q(t),$$

where $\delta > 0$ is a small constant to be determined.

Integrating the above inequality with respect to time from $t = 0$ to any $t \in (0,T]$, we have

$$Q(t) \le Q(0) + \frac{Ct}{\delta}(h_x^2 + h_y^2)^2 + 4\delta \int_0^t Q(s)ds$$

$$\le Q(0) + \frac{CT}{\delta}(h_x^2 + h_y^2)^2 + 4\delta T \max_{0 \le t \le T} Q(t). \qquad (9.41)$$

Now choosing $\delta = \frac{1}{8T}$, then taking the maximum of (9.41) over $[0,T]$, and using the initial error assumption (9.39), we obtain

$$\max_{0 \le t \le T} Q(t) \le Q(0) + CT^2(h_x^2 + h_y^2)^2 + \frac{1}{2}\max_{0 \le t \le T} Q(t), \qquad (9.42)$$

which completes the proof. □

9.2.2 The fully discrete scheme

To construct a fully discrete scheme, we divide the time interval $[0,T]$ into $N_t + 2$ uniform intervals, i.e., we have discrete times $t_i = i\tau$, where $i = 0, 1, \cdots, N_t + 2$ and the time step size $\tau = T/(N_t + 2)$.

Approximating those time directives in the semi-discrete schemes (9.24), (9.26) and (9.28) by the central difference at time levels $n + \frac{1}{2}$, $n + \frac{1}{2}$ and

$n + 1$, respectively, we can obtain the following fully discrete scheme: Given initial approximations $E^0_{x,i+\frac{1}{2},j}$, $E^0_{y,i,j+\frac{1}{2}}$, $H^{\frac{1}{2}}_{i+\frac{1}{2},j+\frac{1}{2}}$, for any $0 \le n \le N_t$, solve $E^{n+1}_{x,i+\frac{1}{2},j}$, $E^{n+1}_{y,i,j+\frac{1}{2}}$, $H^{n+\frac{3}{2}}_{i+\frac{1}{2},j+\frac{1}{2}}$ from:

$$\epsilon_0 \frac{E^{n+1}_{x,i+\frac{1}{2},j} - E^n_{x,i+\frac{1}{2},j}}{\tau}$$
$$= \frac{H^{n+\frac{1}{2}}_{i+\frac{1}{2},j+\frac{1}{2}} - H^{n+\frac{1}{2}}_{i+\frac{1}{2},j-\frac{1}{2}}}{y_{j+\frac{1}{2}} - y_{j-\frac{1}{2}}} - \frac{1}{|T_{i,j-\frac{1}{2}}|} \int_{T_{i,j-\frac{1}{2}}} J^{n+\frac{1}{2}}_x, \qquad (9.43)$$

$$\epsilon_0 \frac{E^{n+1}_{y,i,j+\frac{1}{2}} - E^n_{y,i,j+\frac{1}{2}}}{\tau}$$
$$= -\frac{H^{n+\frac{1}{2}}_{i+\frac{1}{2},j+\frac{1}{2}} - H^{n+\frac{1}{2}}_{i-\frac{1}{2},j+\frac{1}{2}}}{x_{i+\frac{1}{2}} - x_{i-\frac{1}{2}}} - \frac{1}{|T_{i-\frac{1}{2},j}|} \int_{T_{i-\frac{1}{2},j}} J^{n+\frac{1}{2}}_y, \qquad (9.44)$$

$$\mu_0 \frac{H^{n+\frac{3}{2}}_{i+\frac{1}{2},j+\frac{1}{2}} - H^{n+\frac{1}{2}}_{i+\frac{1}{2},j+\frac{1}{2}}}{\tau}$$
$$= -\frac{E^{n+1}_{y,i+1,j+\frac{1}{2}} - E^{n+1}_{y,i,j+\frac{1}{2}}}{x_{i+1} - x_i} + \frac{E^{n+1}_{x,i+\frac{1}{2},j+1} - E^{n+1}_{x,i+\frac{1}{2},j}}{y_{j+1} - y_j}. \qquad (9.45)$$

Note that this scheme is very easy to be implemented. At each time step, we can first solve (9.43) and (9.44) in parallel for E^{n+1}_x and E^{n+1}_y, then solve (9.45) for $H^{n+\frac{3}{2}}$.

9.2.2.1 The stability analysis

To analyze the fully discrete scheme, the following discrete Gronwall inequality is often needed.

LEMMA 9.1

[66, p.14] Assume that k_n is a non-negative sequence, and that the sequence u_n satisfies

$$u_0 \le g_0, \quad u_n \le g_0 + \sum_{s=0}^{n-1} k_s u_s, \quad n \ge 1.$$

If $g_0 \ge 0$, we have

$$u_n \le g_0 \exp(\sum_{s=0}^{n-1} k_s), \quad n \ge 1.$$

Before we start the stability analysis, let us introduce some notation. Let $C_v = 1/\sqrt{\epsilon_0 \mu_0}$ be the wave propagation speed in free space. For any grid function $u_{i,j}$, we denote the backward difference operators ∇_x and ∇_y:

$$\nabla_x u_{i+1,j} = \frac{u_{i+1,j} - u_{i,j}}{x_{i+1} - x_i}, \quad \nabla_y u_{i,j+1} = \frac{u_{i,j+1} - u_{i,j}}{y_{j+1} - y_j},$$

and the constant $C_{inv} > 0$ satisfying the inverse inequality

$$||\nabla_x u|| \leq C_{inv} h_x^{-1}||u||, \quad ||\nabla_y u|| \leq C_{inv} h_y^{-1}||u||, \tag{9.46}$$

for those discrete energy norms introduced in last section.

The rest of this section is devoted to the proof of the following discrete stability for the scheme (9.43)-(9.45).

THEOREM 9.3
Assume that the time step size τ satisfies the constraint

$$\tau \leq \min(\frac{h_y}{2C_{inv}C_v}, \frac{h_x}{2C_{inv}C_v}, 1), \tag{9.47}$$

then the solution of the fully discrete scheme (9.43)-(9.45) satisfies the following stability:

$$\epsilon_0(||E_x^{N_t+1}||_E^2 + ||E_y^{N_t+1}||_E^2) + \mu_0||H^{N_t+\frac{3}{2}}||_H^2$$

$$\leq 3\exp(\tau N_t)[\epsilon_0(||E_x^0||_E^2 + ||E_y^0||_E^2) + \mu_0||H^{\frac{1}{2}}||_H^2], \tag{9.48}$$

where the constant $C > 0$ is independent of τ, h_x and h_y.

Proof. Multiplying (9.43), (9.44) and (9.45) by $\tau|T_{i,j-\frac{1}{2}}|(E_{x,i+\frac{1}{2},j}^{n+1} + E_{x,i+\frac{1}{2},j}^n)$, $\tau|T_{i-\frac{1}{2},j}|(E_{y,i,j+\frac{1}{2}}^{n+1} + E_{y,i,j+\frac{1}{2}}^n)$, and $\tau|T_{ij}|(H_{i+\frac{1}{2},j+\frac{1}{2}}^{n+\frac{3}{2}} + H_{i+\frac{1}{2},j+\frac{1}{2}}^{n+\frac{1}{2}})$, respectively, then adding all results together, we can obtain the sum of the left-hand side as

$$LHS = \epsilon_0(||E_x^{n+1}||_E^2 - ||E_x^n||_E^2) + \epsilon_0(||E_y^{n+1}||_E^2 - ||E_y^n||_E^2)$$

$$+\mu_0(||H^{n+\frac{3}{2}}||_H^2 - ||H^{n+\frac{1}{2}}||_H^2),$$

and the sum of the right-hand side as

$$RHS = \tau \sum_{\substack{0 \leq i \leq N_x-1 \\ 1 \leq j \leq N_y-1}} [(x_{i+1} - x_i)(H_{i+\frac{1}{2},j+\frac{1}{2}}^{n+\frac{1}{2}} - H_{i+\frac{1}{2},j-\frac{1}{2}}^{n+\frac{1}{2}})$$

$$- \int_{T_{i,j-\frac{1}{2}}} J_x^{n+\frac{1}{2}}](E_{x,i+\frac{1}{2},j}^{n+1} + E_{x,i+\frac{1}{2},j}^n)$$

$$+\tau \sum_{\substack{1 \leq i \leq N_x-1 \\ 0 \leq j \leq N_y-1}} [-(y_{j+1} - y_j)(H_{i+\frac{1}{2},j+\frac{1}{2}}^{n+\frac{1}{2}} - H_{i-\frac{1}{2},j+\frac{1}{2}}^{n+\frac{1}{2}})$$

$$-\int_{T_{i-\frac{1}{2},j}} J_y^{n+\frac{1}{2}}](E_{y,i,j+\frac{1}{2}}^{n+1} + E_{y,i,j+\frac{1}{2}}^n) \tag{9.49}$$

$$+\tau \sum_{\substack{0\leq i\leq N_x-1 \\ 0\leq j\leq N_y-1}} [-(y_{j+1}-y_j)(E_{y,i+1,j+\frac{1}{2}}^{n+1} - E_{y,i,j+\frac{1}{2}}^{n+1})$$

$$+(x_{i+1}-x_i)(E_{x,i+\frac{1}{2},j+1}^{n+1} - E_{x,i+\frac{1}{2},j}^{n+1})](H_{i+\frac{1}{2},j+\frac{1}{2}}^{n+\frac{3}{2}} + H_{i+\frac{1}{2},j+\frac{1}{2}}^{n+\frac{1}{2}}).$$

Using the PEC boundary condition (9.33), we can extend the first sum in (9.49) to $j = 0$ and the second sum in (9.49) to $i = 0$. Then regrouping those terms in (9.49), we can rewrite RHS as follows:

$$RHS$$

$$= \tau \sum_{0\leq i\leq N_x-1} (x_{i+1}-x_i) \sum_{0\leq j\leq N_y-1} [(H_{i+\frac{1}{2},j+\frac{1}{2}}^{n+\frac{1}{2}} - H_{i+\frac{1}{2},j-\frac{1}{2}}^{n+\frac{1}{2}})(E_{x,i+\frac{1}{2},j}^{n+1} + E_{x,i+\frac{1}{2},j}^n)$$

$$+(E_{x,i+\frac{1}{2},j+1}^{n+1} - E_{x,i+\frac{1}{2},j}^{n+1})(H_{i+\frac{1}{2},j+\frac{1}{2}}^{n+\frac{3}{2}} + H_{i+\frac{1}{2},j+\frac{1}{2}}^{n+\frac{1}{2}})]$$

$$+\tau \sum_{0\leq j\leq N_y-1} (y_{j+1}-y_j) \sum_{0\leq i\leq N_x-1} [(H_{i-\frac{1}{2},j+\frac{1}{2}}^{n+\frac{1}{2}} - H_{i+\frac{1}{2},j+\frac{1}{2}}^{n+\frac{1}{2}})(E_{y,i,j+\frac{1}{2}}^{n+1} + E_{y,i,j+\frac{1}{2}}^n)$$

$$+(E_{y,i,j+\frac{1}{2}}^{n+1} - E_{y,i+1,j+\frac{1}{2}}^{n+1})(H_{i+\frac{1}{2},j+\frac{1}{2}}^{n+\frac{3}{2}} + H_{i+\frac{1}{2},j+\frac{1}{2}}^{n+\frac{1}{2}})]$$

$$-\tau \sum_{\substack{0\leq i\leq N_x-1 \\ 1\leq j\leq N_y-1}} (\int_{T_{i,j-\frac{1}{2}}} J_x^{n+\frac{1}{2}})(E_{x,i+\frac{1}{2},j}^{n+1} + E_{x,i+\frac{1}{2},j}^n) \tag{9.50}$$

$$-\tau \sum_{\substack{1\leq i\leq N_x-1 \\ 0\leq j\leq N_y-1}} (\int_{T_{i-\frac{1}{2},j}} J_y^{n+\frac{1}{2}})(E_{y,i,j+\frac{1}{2}}^{n+1} + E_{y,i,j+\frac{1}{2}}^n)$$

$$:= \tau \left[\sum_{0\leq i\leq N_x-1} (x_{i+1}-x_i)R_1 + \sum_{0\leq j\leq N_y-1} (y_{j+1}-y_j)R_2 + R_3 + R_4 \right].$$

To evaluate the above RHS, below we evaluate each term separately. First, note that

$$\sum_{n=0}^{N_t} R_1 = \sum_{n=0}^{N_t} \sum_{0\leq j\leq N_y-1} [(H_{i+\frac{1}{2},j+\frac{1}{2}}^{n+\frac{1}{2}} - H_{i+\frac{1}{2},j-\frac{1}{2}}^{n+\frac{1}{2}})(E_{x,i+\frac{1}{2},j}^{n+1} + E_{x,i+\frac{1}{2},j}^n)$$

$$+(E_{x,i+\frac{1}{2},j+1}^{n+1} - E_{x,i+\frac{1}{2},j}^{n+1})(H_{i+\frac{1}{2},j+\frac{1}{2}}^{n+\frac{3}{2}} + H_{i+\frac{1}{2},j+\frac{1}{2}}^{n+\frac{1}{2}})]$$

$$= \sum_{n=0}^{N_t} \sum_{0\leq j\leq N_y-1} [(H_{i+\frac{1}{2},j+\frac{1}{2}}^{n+\frac{1}{2}} E_{x,i+\frac{1}{2},j}^n - H_{i+\frac{1}{2},j+\frac{1}{2}}^{n+\frac{3}{2}} E_{x,i+\frac{1}{2},j}^{n+1})$$

$$+(H_{i+\frac{1}{2},j+\frac{1}{2}}^{n+\frac{1}{2}} E_{x,i+\frac{1}{2},j+1}^{n+1} - H_{i+\frac{1}{2},j-\frac{1}{2}}^{n+\frac{1}{2}} E_{x,i+\frac{1}{2},j}^{n+1})]$$

$$+\sum_{n=0}^{N_t} \sum_{0\leq j\leq N_y-1} [(H_{i+\frac{1}{2},j+\frac{1}{2}}^{n+\frac{3}{2}} E_{x,i+\frac{1}{2},j+1}^{n+1} - H_{i+\frac{1}{2},j+\frac{1}{2}}^{n+\frac{1}{2}} E_{x,i+\frac{1}{2},j+1}^n)$$

$$+(H^{n+\frac{1}{2}}_{i+\frac{1}{2},j+\frac{1}{2}}E^n_{x,i+\frac{1}{2},j+1} - H^{n+\frac{1}{2}}_{i+\frac{1}{2},j-\frac{1}{2}}E^n_{x,i+\frac{1}{2},j})]$$

$$= \sum_{0\le j\le N_y-1} (H^{\frac{1}{2}}_{i+\frac{1}{2},j+\frac{1}{2}}E^0_{x,i+\frac{1}{2},j} - H^{N_t+\frac{3}{2}}_{i+\frac{1}{2},j+\frac{1}{2}}E^{N_t+1}_{x,i+\frac{1}{2},j})$$

$$+\sum_{n=0}^{N_t}(H^{n+\frac{1}{2}}_{i+\frac{1}{2},N_y+\frac{1}{2}}E^{n+1}_{x,i+\frac{1}{2},N_y} - H^{n+\frac{1}{2}}_{i+\frac{1}{2},-\frac{1}{2}}E^{n+1}_{x,i+\frac{1}{2},0})$$

$$+\sum_{0\le j\le N_y-1}(H^{N_t+\frac{3}{2}}_{i+\frac{1}{2},j+\frac{1}{2}}E^{N_t+1}_{x,i+\frac{1}{2},j+1} - H^{\frac{1}{2}}_{i+\frac{1}{2},j+\frac{1}{2}}E^0_{x,i+\frac{1}{2},j+1})$$

$$+\sum_{n=0}^{N_t}(H^{n+\frac{1}{2}}_{i+\frac{1}{2},N_y+\frac{1}{2}}E^n_{x,i+\frac{1}{2},N_y}) - H^{n+\frac{1}{2}}_{i+\frac{1}{2},-\frac{1}{2}}E^n_{x,i+\frac{1}{2},0})$$

$$= \sum_{0\le j\le N_y-1}(H^{\frac{1}{2}}_{i+\frac{1}{2},j+\frac{1}{2}}E^0_{x,i+\frac{1}{2},j} - H^{N_t+\frac{3}{2}}_{i+\frac{1}{2},j+\frac{1}{2}}E^{N_t+1}_{x,i+\frac{1}{2},j}) \qquad (9.51)$$

$$+\sum_{0\le j\le N_y-1}(H^{N_t+\frac{3}{2}}_{i+\frac{1}{2},j+\frac{1}{2}}E^{N_t+1}_{x,i+\frac{1}{2},j+1} - H^{\frac{1}{2}}_{i+\frac{1}{2},j+\frac{1}{2}}E^0_{x,i+\frac{1}{2},j+1})$$

$$= \sum_{0\le j\le N_y-1}(y_{j+1}-y_j)(H^{N_t+\frac{3}{2}}_{i+\frac{1}{2},j+\frac{1}{2}}\nabla_y E^{N_t+1}_{x,i+\frac{1}{2},j+1} - H^{\frac{1}{2}}_{i+\frac{1}{2},j+\frac{1}{2}}\nabla_y E^0_{x,i+\frac{1}{2},j+1}),$$

where we used the PEC boundary condition (9.33) in the second last step, and the backward difference operator ∇_y in the last step.

Similary, we can evaluate the R_2 term as follows:

$$\sum_{n=0}^{N_t} R_2 = \sum_{n=0}^{N_t}\sum_{0\le i\le N_x-1}[(H^{n+\frac{1}{2}}_{i-\frac{1}{2},j+\frac{1}{2}} - H^{n+\frac{1}{2}}_{i+\frac{1}{2},j+\frac{1}{2}})(E^{n+1}_{y,i,j+\frac{1}{2}} + E^n_{y,i,j+\frac{1}{2}})$$

$$+(E^{n+1}_{y,i,j+\frac{1}{2}} - E^{n+1}_{y,i+1,j+\frac{1}{2}})(H^{n+\frac{3}{2}}_{i+\frac{1}{2},j+\frac{1}{2}} + H^{n+\frac{1}{2}}_{i+\frac{1}{2},j+\frac{1}{2}})]$$

$$= \sum_{n=0}^{N_t}\sum_{0\le i\le N_x-1}[(H^{n+\frac{1}{2}}_{i-\frac{1}{2},j+\frac{1}{2}}E^{n+1}_{y,i,j+\frac{1}{2}} - H^{n+\frac{1}{2}}_{i+\frac{1}{2},j+\frac{1}{2}}E^{n+1}_{y,i+1,j+\frac{1}{2}})$$

$$+(-H^{n+\frac{1}{2}}_{i+\frac{1}{2},j+\frac{1}{2}}E^n_{y,i,j+\frac{1}{2}} + H^{n+\frac{3}{2}}_{i+\frac{1}{2},j+\frac{1}{2}}E^{n+1}_{y,i,j+\frac{1}{2}})]$$

$$+\sum_{n=0}^{N_t}\sum_{0\le i\le N_x-1}[(H^{n+\frac{1}{2}}_{i-\frac{1}{2},j+\frac{1}{2}}E^n_{y,i,j+\frac{1}{2}} - H^{n+\frac{1}{2}}_{i+\frac{1}{2},j+\frac{1}{2}}E^n_{y,i+1,j+\frac{1}{2}})$$

$$+(H^{n+\frac{1}{2}}_{i+\frac{1}{2},j+\frac{1}{2}}E^n_{y,i+1,j+\frac{1}{2}} - H^{n+\frac{3}{2}}_{i+\frac{1}{2},j+\frac{1}{2}}E^{n+1}_{y,i+1,j+\frac{1}{2}})]$$

$$= \sum_{n=0}^{N_t}(H^{n+\frac{1}{2}}_{-\frac{1}{2},j+\frac{1}{2}}E^{n+1}_{y,0,j+\frac{1}{2}} - H^{n+\frac{1}{2}}_{N_x-\frac{1}{2},j+\frac{1}{2}}E^{n+1}_{y,N_x,j+\frac{1}{2}})$$

$$+\sum_{0\le i\le N_x-1}(-H^{\frac{1}{2}}_{i+\frac{1}{2},j+\frac{1}{2}}E^0_{y,i,j+\frac{1}{2}} + H^{N_t+\frac{3}{2}}_{i+\frac{1}{2},j+\frac{1}{2}}E^{N_t+1}_{y,i,j+\frac{1}{2}})$$

$$+\sum_{n=0}^{N_t}(H^{n+\frac{1}{2}}_{-\frac{1}{2},j+\frac{1}{2}}E^n_{y,0,j+\frac{1}{2}} - H^{n+\frac{1}{2}}_{N_x+\frac{1}{2},j+\frac{1}{2}}E^n_{y,N_x,j+\frac{1}{2}})$$

$$+ \sum_{0 \le i \le N_x - 1} (H^{\frac{1}{2}}_{i+\frac{1}{2},j+\frac{1}{2}} E^0_{y,i+1,j+\frac{1}{2}} - H^{N_t+\frac{3}{2}}_{i+\frac{1}{2},j+\frac{1}{2}} E^{N_t+1}_{y,i+1,j+\frac{1}{2}})$$

$$= \sum_{0 \le i \le N_x - 1} (-H^{\frac{1}{2}}_{i+\frac{1}{2},j+\frac{1}{2}} E^0_{y,i,j+\frac{1}{2}} + H^{N_t+\frac{3}{2}}_{i+\frac{1}{2},j+\frac{1}{2}} E^{N_t+1}_{y,i,j+\frac{1}{2}}) \tag{9.52}$$

$$+ \sum_{0 \le i \le N_x - 1} (H^{\frac{1}{2}}_{i+\frac{1}{2},j+\frac{1}{2}} E^0_{y,i+1,j+\frac{1}{2}} - H^{N_t+\frac{3}{2}}_{i+\frac{1}{2},j+\frac{1}{2}} E^{N_t+1}_{y,i+1,j+\frac{1}{2}})$$

$$= \sum_{0 \le i \le N_x - 1} (x_{i+1} - x_i)(-H^{N_t+\frac{3}{2}}_{i+\frac{1}{2},j+\frac{1}{2}} \nabla_x E^{N_t+1}_{y,i+1,j+\frac{1}{2}} + H^{\frac{1}{2}}_{i+\frac{1}{2},j+\frac{1}{2}} \nabla_x E^0_{y,i+1,j+\frac{1}{2}}),$$

where the PEC boundary condition (9.33) was used in the second last step, and the backward difference operator ∇_x was used in the last step.

By using the Cauchy-Schwarz inequality and the discrete energy norm notation, we have

$$\sum_{n=0}^{N_t} R_3 \le \sum_{n=0}^{N_t} [\frac{\delta \epsilon_0}{2} ||E_x^{n+1} + E_x^n||_E^2 + \frac{1}{2\delta\epsilon_0} ||J_x^{n+\frac{1}{2}}||^2]$$

$$\le \sum_{n=0}^{N_t} [\delta\epsilon_0(||E_x^{n+1}||_E^2 + ||E_x^n||_E^2) + \frac{1}{2\delta\epsilon_0} ||J_x^{n+\frac{1}{2}}||^2]. \tag{9.53}$$

Similarly, we can obtain

$$\sum_{n=0}^{N_t} R_4 \le \sum_{n=0}^{N_t} [\delta\epsilon_0(||E_y^{n+1}||_E^2 + ||E_y^n||_E^2) + \frac{1}{2\delta\epsilon_0} ||J_y^{n+\frac{1}{2}}||^2]. \tag{9.54}$$

Equating LHS and RHS, then summing up (9.50) from $n = 0$ to N_t, and substituting the estimates (9.51)-(9.54), we have

$$\epsilon_0(||E_x^{N_t+1}||_E^2 - ||E_x^0||_E^2) + \epsilon_0(||E_y^{N_t+1}||_E^2 - ||E_y^0||_E^2) + \mu_0(||H^{N_t+\frac{3}{2}}||_H^2 - ||H^{\frac{1}{2}}||_H^2)$$

$$\le \tau \sum_{\substack{0 \le i \le N_x - 1 \\ 0 \le j \le N_y - 1}} |T_{ij}|(H^{N_t+\frac{3}{2}}_{i+\frac{1}{2},j+\frac{1}{2}} \nabla_y E^{N_t+1}_{x,i+\frac{1}{2},j+1} - H^{\frac{1}{2}}_{i+\frac{1}{2},j+\frac{1}{2}} \nabla_y E^0_{x,i+\frac{1}{2},j+1})$$

$$+ \tau \sum_{\substack{0 \le i \le N_x - 1 \\ 0 \le j \le N_y - 1}} |T_{ij}|(-H^{N_t+\frac{3}{2}}_{i+\frac{1}{2},j+\frac{1}{2}} \nabla_x E^{N_t+1}_{y,i+1,j+\frac{1}{2}} + H^{\frac{1}{2}}_{i+\frac{1}{2},j+\frac{1}{2}} \nabla_x E^0_{y,i+1,j+\frac{1}{2}})$$

$$+ \tau \sum_{n=0}^{N_t} [\delta\epsilon_0(||E_x^{n+1}||_E^2 + ||E_x^n||_E^2) + \frac{1}{2\delta\epsilon_0} ||J_x^{n+\frac{1}{2}}||^2]$$

$$+ \tau \sum_{n=0}^{N_t} [\delta\epsilon_0(||E_y^{n+1}||_E^2 + ||E_y^n||_E^2) + \frac{1}{2\delta\epsilon_0} ||J_y^{n+\frac{1}{2}}||^2]. \tag{9.55}$$

Now we just need to bound those right-hand side terms of (9.55). Using the Cauchy-Schwarz inequality and the inverse estimate (9.46), we have

$$\tau \sum_{\substack{0 \le i \le N_x - 1 \\ 0 \le j \le N_y - 1}} |T_{ij}| \cdot H^{N_t+\frac{3}{2}}_{i+\frac{1}{2},j+\frac{1}{2}} \nabla_y E^{N_t+1}_{x,i+\frac{1}{2},j+1}$$

$$\leq \tau \Big(\sum_{\substack{0\leq i\leq N_x-1 \\ 0\leq j\leq N_y-1}} |T_{ij}|\cdot|H_{i+\frac{1}{2},j+\frac{1}{2}}^{N_t+\frac{3}{2}}|^2 \Big)^{1/2} \Big(\sum_{\substack{0\leq i\leq N_x-1 \\ 0\leq j\leq N_y-1}} |T_{ij}|\cdot|\nabla_y E_{x,i+\frac{1}{2},j+1}^{N_t+1}|^2 \Big)^{1/2}$$

$$= \tau \|H^{N_t+\frac{3}{2}}\|_H \|\nabla_y E_x^{N_t+1}\|_E$$

$$\leq \delta\mu_0 \|H^{N_t+\frac{3}{2}}\|_H^2 + \frac{1}{4\delta}\cdot\frac{(\tau C_{inv}h_y^{-1})^2}{\mu_0\epsilon_0}\cdot\epsilon_0\|E_x^{N_t+1}\|_E^2. \tag{9.56}$$

Similarly, we can obtain

$$\tau \sum_{\substack{0\leq i\leq N_x-1 \\ 0\leq j\leq N_y-1}} |T_{ij}|\cdot H_{i+\frac{1}{2},j+\frac{1}{2}}^{N_t+\frac{3}{2}} \nabla_x E_{y,i+1,j+\frac{1}{2}}^{N_t+1}$$

$$\leq \delta\mu_0\|H^{N_t+\frac{3}{2}}\|_H^2 + \frac{1}{4\delta}\cdot\frac{(\tau C_{inv}h_x^{-1})^2}{\mu_0\epsilon_0}\cdot\epsilon_0\|E_y^{N_t+1}\|_E^2. \tag{9.57}$$

Substituting the estimates (9.56)-(9.57) into (9.55), then using the notation $C_v = \frac{1}{\sqrt{\epsilon_0\mu_0}}$ and combining like terms, we obtain

$$[\epsilon_0(\|E_x^{N_t+1}\|_E^2 + \|E_y^{N_t+1}\|_E^2) + \mu_0\|H^{N_t+\frac{3}{2}}\|_H^2] - [\epsilon_0(\|E_x^0\|_E^2 + \|E_y^0\|_E^2) + \mu_0\|H^{\frac{1}{2}}\|_H^2]$$

$$\leq [\delta\tau + \frac{(\tau C_{inv}C_v h_y^{-1})^2}{4\delta}]\cdot\epsilon_0\|E_x^{N_t+1}\|_E^2 + [\delta\tau + \frac{(\tau C_{inv}C_v h_x^{-1})^2}{4\delta}]\cdot\epsilon_0\|E_y^{N_t+1}\|_E^2$$

$$+ 2\delta\mu_0\|H^{N_t+\frac{3}{2}}\|_H^2 + [\frac{(\tau C_{inv}C_v h_y^{-1})^2}{4\delta}\cdot\epsilon_0\|E_x^0\|_E^2 + \frac{(\tau C_{inv}C_v h_x^{-1})^2}{4\delta}\cdot\epsilon_0\|E_y^0\|_E^2]$$

$$+ 2\delta\mu_0\|H^{\frac{1}{2}}\|_H^2 + 2\delta\tau\sum_{n=0}^{N_t}\epsilon_0(\|E_x^n\|_E^2 + \|E_y^n\|_E^2). \tag{9.58}$$

Denote

$$Q^n := \epsilon_0(\|E_x^n\|_E^2 + \|E_y^n\|_E^2) + \mu_0\|H^{n+\frac{1}{2}}\|_H^2.$$

Now choosing δ and τ small enough so that the left-hand side terms of (9.58) can control those corresponding terms on the right-hand side. A specific choice can be

$$\delta = \frac{1}{4},\ \tau\leq 1,\ \tau C_{inv}C_v h_y^{-1}\leq \frac{1}{2},\ \tau C_{inv}C_v h_x^{-1}\leq \frac{1}{2},$$

i.e., the time constraint stated in (9.47), we have

$$Q^{N_t+1} - Q^0 \leq \frac{1}{2}Q^{N_t+1} + \frac{1}{2}Q^0 + \frac{\tau}{2}\sum_{n=0}^{N_t}\epsilon_0(\|E_x^n\|_E^2 + \|E_y^n\|_E^2). \tag{9.59}$$

Application of the discrete Gronwall's inequality to (9.59) completes the proof. □

9.2.2.2 The error estimate

To make the error analysis easy to follow, we denote the errors by their corresponding script letters. For example, the error of E_x at point $(x_{i+\frac{1}{2}}, y_j, t_n)$ is denoted by $\mathcal{E}^n_{x,i+\frac{1}{2},j} = E_x(x_{i+\frac{1}{2}}, y_j, t_n) - E^n_{x,i+\frac{1}{2},j}$, where $E_x(x_{i+\frac{1}{2}}, y_j, t_n)$ and $E^n_{x,i+\frac{1}{2},j}$ denote the exact and numerical solutions of E_x at point $(x_{i+\frac{1}{2}}, y_j, t_n)$, respectively. Similar error notations $\mathcal{E}^n_{y,i,j+\frac{1}{2}}$ and $\mathcal{H}^{n+\frac{1}{2}}_{i+\frac{1}{2},j+\frac{1}{2}}$ are used for solutions E_y and H, respectively.

Below we first derive the error equations for E_x, E_y and H one by one.

The error equation for E_x

Multiplying (9.43) by $|T_{i,j-\frac{1}{2}}|$, we can rewrite (9.43) as

$$\epsilon_0 |T_{i,j-\frac{1}{2}}| \frac{(E^{n+1}_{x,i+\frac{1}{2},j} - E^n_{x,i+\frac{1}{2},j})}{\tau}$$

$$= (x_{i+1} - x_i)(H^{n+\frac{1}{2}}_{i+\frac{1}{2},j+\frac{1}{2}} - H^{n+\frac{1}{2}}_{i+\frac{1}{2},j-\frac{1}{2}}) - \int_{T_{i,j-\frac{1}{2}}} J_x(x, y, t_{n+\frac{1}{2}}),$$

from which we can easily obtain the error equation for E_x:

$$\epsilon_0 |T_{i,j-\frac{1}{2}}| \frac{(\mathcal{E}^{n+1}_{x,i+\frac{1}{2},j} - \mathcal{E}^n_{x,i+\frac{1}{2},j})}{\tau} = (x_{i+1} - x_i)(\mathcal{H}^{n+\frac{1}{2}}_{i+\frac{1}{2},j+\frac{1}{2}} - \mathcal{H}^{n+\frac{1}{2}}_{i+\frac{1}{2},j-\frac{1}{2}}) + R_1,$$

$$(9.60)$$

where the local truncation error term R_1 is given by

$$R_1 = \epsilon_0 |T_{i,j-\frac{1}{2}}| \frac{(E_x(x_{i+\frac{1}{2}}, y_j, t_{n+1}) - E_x(x_{i+\frac{1}{2}}, y_j, t_n))}{\tau}$$

$$- (x_{i+1} - x_i)(H(x_{i+\frac{1}{2}}, y_{j+\frac{1}{2}}, t_{n+\frac{1}{2}}) - H(x_{i+\frac{1}{2}}, y_{j-\frac{1}{2}}, t_{n+\frac{1}{2}}))$$

$$+ \int_{T_{i,j-\frac{1}{2}}} J_x(x, y, t_{n+\frac{1}{2}}). \qquad (9.61)$$

Integrating (9.23) from $t = t_n$ to t_{n+1} and dividing the result by τ, we have

$$\frac{\epsilon_0}{\tau} \int_{T_{i,j-\frac{1}{2}}} (E_x(x, y, t_{n+1}) - E_x(x, y, t_n)) dx dy \qquad (9.62)$$

$$= \frac{1}{\tau} \int_{t_n}^{t_{n+1}} \int_{x_i}^{x_{i+1}} (H(x, y_{j+\frac{1}{2}}, t) - H(x, y_{j-\frac{1}{2}}, t)) dx dt - \frac{1}{\tau} \int_{t_n}^{t_{n+1}} \int_{T_{i,j-\frac{1}{2}}} J_x(x, y, t).$$

Subtracting (9.62) from (9.61), we can rewrite R_1 as follows:

$$R_1 = \frac{\epsilon_0}{\tau} \int_{T_{i,j-\frac{1}{2}}} \Big[(E_x(x_{i+\frac{1}{2}}, y_j, t_{n+1}) - E_x(x, y, t_{n+1}))$$

$$- (E_x(x_{i+\frac{1}{2}}, y_j, t_n) - E_x(x, y, t_n)) \Big] dx dy$$

$$- \left\{ \int_{x_i}^{x_{i+1}} (H(x_{i+\frac{1}{2}}, y_{j+\frac{1}{2}}, t_{n+\frac{1}{2}}) - H(x_{i+\frac{1}{2}}, y_{j-\frac{1}{2}}, t_{n+\frac{1}{2}})) dx \right.$$

$$- \frac{1}{\tau} \int_{t_n}^{t_{n+1}} \int_{x_i}^{x_{i+1}} (H(x, y_{j+\frac{1}{2}}, t) - H(x, y_{j-\frac{1}{2}}, t)) dx dt \Big\}$$

$$+ \left[\int_{T_{i,j-\frac{1}{2}}} J_x(x, y, t_{n+\frac{1}{2}}) dx dy - \frac{1}{\tau} \int_{t_n}^{t_{n+1}} \int_{T_{i,j-\frac{1}{2}}} J_x(x, y, t) dx dy dt \right]$$

$$= R_{11} + R_{12} + R_{13}. \tag{9.63}$$

Following the same technique used for deriving (9.40), for any function f we can prove that

$$\int_{T_{i,j-\frac{1}{2}}} (f(x, y, t_{n+1}) - f(x_{i+\frac{1}{2}}, y_j, t_{n+1})) dx dy$$

$$- \int_{T_{i,j-\frac{1}{2}}} (f(x, y, t_n) - f(x_{i+\frac{1}{2}}, y_j, t_n)) dx dy$$

$$= \int_{T_{i,j-\frac{1}{2}}} [\frac{1}{2}(x - x_{i+\frac{1}{2}})^2 (\frac{\partial^2 f}{\partial x^2}(q_1, t_{n+1}) - \frac{\partial^2 f}{\partial x^2}(q_1, t_n))$$

$$+ \frac{1}{2}(y - y_j)^2 (\frac{\partial^2 f}{\partial y^2}(q_2, t_{n+1}) - \frac{\partial^2 f}{\partial y^2}(q_2, t_n))] dx dy \tag{9.64}$$

$$= \tau \int_{T_{i,j-\frac{1}{2}}} \left[\frac{1}{2}(x - x_{i+\frac{1}{2}})^2 \frac{\partial^3 f}{\partial t \partial x^2}(q_1, t_*) + \frac{1}{2}(y - y_j)^2 \frac{\partial^3 f}{\partial t \partial y^2}(q_2, t_*) \right] dx dy,$$

where we denote q_1 and q_2 for some points between $(x_{i+\frac{1}{2}}, y_j)$ and (x, y), and t_* for some point between t_n and t_{n+1}. In the last step, we used the following Taylor expansion

$$g(t_{n+1}) - g(t_n) = \tau \frac{\partial g}{\partial t}(t_*)$$

with $g = \frac{\partial^2 f}{\partial x^2}$ and $g = \frac{\partial^2 f}{\partial y^2}$, respectively.

Applying (9.64) with $f = E_x$, we can bound R_{11} as follows:

$$R_{11} = \frac{\epsilon_0}{\tau} \int_{T_{i,j-\frac{1}{2}}} \left[\frac{1}{2}(x - x_{i+\frac{1}{2}})^2 \tau \frac{\partial^3 E_x}{\partial t \partial x^2}(q_1, t_*) + \frac{1}{2}(y - y_j)^2 \tau \frac{\partial^3 E_x}{\partial t \partial y^2}(q_2, t_*) \right]$$

$$= (O(h_x^2)|\frac{\partial^3 E_x}{\partial t \partial x^2}|_\infty + O(h_y^2)|\frac{\partial^3 E_x}{\partial t \partial y^2}|_\infty)|T_{i,j-\frac{1}{2}}|.$$

Similarly, by the Taylor expansion, we can estimate R_{12} as follows:

$$R_{12} = - \int_{x_i}^{x_{i+1}} \int_{y_{j-\frac{1}{2}}}^{y_{j+\frac{1}{2}}} \frac{\partial H}{\partial y}(x_{i+\frac{1}{2}}, y, t_{n+\frac{1}{2}}) + \frac{1}{\tau} \int_{t_n}^{t_{n+1}} \int_{x_i}^{x_{i+1}} \int_{y_{j-\frac{1}{2}}}^{y_{j+\frac{1}{2}}} \frac{\partial H}{\partial y}(x, y, t)$$

$$= -\int_{x_i}^{x_{i+1}} \int_{y_{j-\frac{1}{2}}}^{y_{j+\frac{1}{2}}} \left[\frac{\partial H}{\partial y}(x_{i+\frac{1}{2}}, y, t_{n+\frac{1}{2}}) - \frac{\partial H}{\partial y}(x, y, t_{n+\frac{1}{2}}) \right] dydx$$

$$+ \int_{x_i}^{x_{i+1}} \int_{y_{j-\frac{1}{2}}}^{y_{j+\frac{1}{2}}} \frac{1}{\tau} \int_{t_n}^{t_{n+1}} \left[\frac{\partial H}{\partial y}(x, y, t) - \frac{\partial H}{\partial y}(x, y, t_{n+\frac{1}{2}}) \right] dtdydx$$

$$= \int_{T_{i,j-\frac{1}{2}}} \frac{1}{2}(x - x_{i+\frac{1}{2}})^2 \frac{\partial^3 H}{\partial x^2 \partial y}(x_*, y, t_{n+\frac{1}{2}}) dxdy$$

$$+ \int_{T_{i,j-\frac{1}{2}}} \frac{1}{\tau} \int_{t_n}^{t_{n+1}} \frac{1}{2}(t - t_{n+\frac{1}{2}})^2 \frac{\partial^3 H}{\partial t^2 \partial y}(x, y, t_*) dtdydx$$

$$= (O(h_x^2)|\frac{\partial^3 H}{\partial x^2 \partial y}|_\infty + O(\tau^2)|\frac{\partial^3 H}{\partial t^2 \partial y}|_\infty)|T_{i,j-\frac{1}{2}}|,$$

where x_* is some number between $x_{i+\frac{1}{2}}$ and x, and t_* is some number between $t_{n+\frac{1}{2}}$ and t.

Using exactly the same argument, we can estimate R_{13} as follows:

$$R_{13} = \frac{1}{\tau} \int_{t_n}^{t_{n+1}} \int_{T_{i,j-\frac{1}{2}}} (J_x(x, y, t_{n+\frac{1}{2}}) - J_x(x, y, t)) dxdydt$$

$$= O(\tau^2)|\frac{\partial^2 J_x}{\partial t^2}|_\infty |T_{i,j-\frac{1}{2}}|.$$

The error equation for E_y

Multiplying (9.44) by $|T_{i-\frac{1}{2},j}|$, we can easily derive the error equation for E_y:

$$\epsilon_0 |T_{i-\frac{1}{2},j}| \frac{(\mathcal{E}_{y,i,j+\frac{1}{2}}^{n+1} - \mathcal{E}_{y,i,j+\frac{1}{2}}^n)}{\tau} = -(y_{j+1} - y_j)(\mathcal{H}_{i+\frac{1}{2},j+\frac{1}{2}}^{n+\frac{1}{2}} - \mathcal{H}_{i-\frac{1}{2},j+\frac{1}{2}}^{n+\frac{1}{2}}) + R_2,$$
(9.65)

where the local truncation error R_2 is given by

$$R_2 = \epsilon_0 |T_{i-\frac{1}{2},j}| \frac{(E_y(x_i, y_{j+\frac{1}{2}}, t_{n+1}) - E_y(x_i, y_{j+\frac{1}{2}}, t_n))}{\tau}$$

$$+ (y_{j+1} - y_j)(H(x_{i+\frac{1}{2}}, y_{j+\frac{1}{2}}, t_{n+\frac{1}{2}}) - H(x_{i-\frac{1}{2}}, y_{j+\frac{1}{2}}, t_{n+\frac{1}{2}}))$$

$$+ \int_{T_{i-\frac{1}{2},j}} J_y(x, y, t_{n+\frac{1}{2}}).$$
(9.66)

Integrating (9.25) from $t = t_n$ to t_{n+1} and dividing the result by τ, we have

$$\frac{\epsilon_0}{\tau} \int_{T_{i-\frac{1}{2},j}} (E_y(x, y, t_{n+1}) - E_y(x, y, t_n)) dxdy$$
(9.67)

$$= -\frac{1}{\tau} \int_{t_n}^{t_{n+1}} \int_{y_j}^{y_{j+1}} (H(x_{i+\frac{1}{2}}, y, t) - H(x_{i-\frac{1}{2}}, y, t)) - \frac{1}{\tau} \int_{t_n}^{t_{n+1}} \int_{T_{i-\frac{1}{2},j}} J_y(x, y, t).$$

Subtracting (9.67) from (9.66), we can rewrite R_2 as follows:

$$
R_2 = \frac{\epsilon_0}{\tau} \int_{T_{i-\frac{1}{2},j}} \left[(E_y(x_i, y_{j+\frac{1}{2}}, t_{n+1}) - E_y(x, y, t_{n+1})) \right.
$$

$$
\left. - (E_y(x_i, y_{j+\frac{1}{2}}, t_n) - E_y(x, y, t_n)) \right] dxdy
$$

$$
- \{ \int_{y_j}^{y_{j+1}} (H(x_{i+\frac{1}{2}}, y_{j+\frac{1}{2}}, t_{n+\frac{1}{2}}) - H(x_{i-\frac{1}{2}}, y_{j+\frac{1}{2}}, t_{n+\frac{1}{2}}))dy
$$

$$
- \frac{1}{\tau} \int_{t_n}^{t_{n+1}} \int_{y_j}^{y_{j+1}} (H(x_{i+\frac{1}{2}}, y, t) - H(x_{i-\frac{1}{2}}, y, t))dydt \}
$$

$$
+ \left[\int_{T_{i-\frac{1}{2},j}} J_y(x, y, t_{n+\frac{1}{2}})dxdy - \frac{1}{\tau} \int_{t_n}^{t_{n+1}} \int_{T_{i-\frac{1}{2},j}} J_y(x, y, t)dxdydt \right]
$$

$$
= R_{21} + R_{22} + R_{23}. \tag{9.68}
$$

Following the same technique developed above for R_1, we can show that

$$
R_{21} = (O(h_x^2)|\frac{\partial^3 E_y}{\partial t \partial x^2}|_\infty + O(h_y^2)|\frac{\partial^3 E_y}{\partial t \partial y^2}|_\infty)|T_{i-\frac{1}{2},j}|,
$$

$$
R_{22} = (O(h_y^2)|\frac{\partial^3 H}{\partial y^2 \partial x}|_\infty + O(\tau^2)|\frac{\partial^3 H}{\partial t^2 \partial x}|_\infty)|T_{i-\frac{1}{2},j}|,
$$

$$
R_{23} = (O(\tau^2)|\frac{\partial^2 J_y}{\partial t^2}|_\infty)|T_{i-\frac{1}{2},j}|.
$$

The error equation for H

Multiplying (9.45) by $|T_{i,j}|$, we can easily obtain the error equation for H:

$$
\mu_0 |T_{i,j}| \frac{(\mathcal{H}_{i+\frac{1}{2},j+\frac{1}{2}}^{n+\frac{3}{2}} - \mathcal{H}_{i+\frac{1}{2},j+\frac{1}{2}}^{n+\frac{1}{2}})}{\tau} = -(y_{j+1} - y_j)(\mathcal{E}_{y,i+1,j+\frac{1}{2}}^{n+1} - \mathcal{E}_{y,i,j+\frac{1}{2}}^{n+1})
$$

$$
+ (x_{i+1} - x_i)(\mathcal{E}_{x,i+\frac{1}{2},j+1}^{n+1} - \mathcal{E}_{x,i+\frac{1}{2},j}^{n+1}) + R_3, \tag{9.69}
$$

where the local truncation error R_3 is given by

$$
R_3 = \frac{\mu_0 |T_{i,j}|}{\tau} (H(x_{i+\frac{1}{2}}, y_{j+\frac{1}{2}}, t_{n+\frac{3}{2}}) - H(x_{i+\frac{1}{2}}, y_{j+\frac{1}{2}}, t_{n+\frac{1}{2}}))
$$

$$
+ (y_{j+1} - y_j)(E_y(x_{i+1}, y_{j+\frac{1}{2}}, t_{n+1}) - E_y(x_i, y_{j+\frac{1}{2}}, t_{n+1})) \tag{9.70}
$$

$$
- (x_{i+1} - x_i)(E_x(x_{i+\frac{1}{2}}, y_{j+1}, t_{n+1}) - E_x(x_{i+\frac{1}{2}}, y_j, t_{n+1})).
$$

Integrating (9.27) from $t = t_{n+\frac{1}{2}}$ to $t_{n+\frac{3}{2}}$ and dividing the result by τ, we obtain

$$
\frac{\mu_0}{\tau} \int_{T_{i,j}} (H(x, y, t_{n+\frac{3}{2}}) - H(x, y, t_{n+\frac{1}{2}}))dxdy
$$

$$= -\frac{1}{\tau} \int_{t_{n+\frac{1}{2}}}^{t_{n+\frac{3}{2}}} \int_{T_{i,j}} (\frac{\partial E_y}{\partial x} - \frac{\partial E_x}{\partial y})dxdydt. \tag{9.71}$$

Subtracting (9.71) from (9.70), we can rewrite R_3 as follows:

$$R_3 = \frac{\mu_0}{\tau} \int_{T_{i,j}} \{(H(x_{i+\frac{1}{2}},y_{j+\frac{1}{2}},t_{n+\frac{3}{2}}) - H(x,y,t_{n+\frac{3}{2}}))$$

$$-(H(x_{i+\frac{1}{2}},y_{j+\frac{1}{2}},t_{n+\frac{1}{2}}) - H(x,y,t_{n+\frac{1}{2}}))\}dxdy \tag{9.72}$$

$$+\{\int_{T_{i,j}} (\frac{\partial E_y}{\partial x}(x,y_{j+\frac{1}{2}},t_{n+1}) - \frac{\partial E_x}{\partial y}(x_{i+\frac{1}{2}},y,t_{n+1}))dxdy$$

$$-\frac{1}{\tau} \int_{t_{n+\frac{1}{2}}}^{t_{n+\frac{3}{2}}} \iint_{T_{i,j}} (\frac{\partial E_y}{\partial x}(x,y,t) - \frac{\partial E_x}{\partial y}(x,y,t))dxdydt\} := R_{31} + R_{32}.$$

By the Taylor expansion, we can obtain

$$R_{31} = (O(h_x^2)|\frac{\partial^3 H}{\partial t\partial x^2}|_\infty + O(h_y^2)|\frac{\partial^3 H}{\partial t\partial y^2}|_\infty)|T_{i,j}|,$$

$$R_{32} = (O(h_y^2)|\frac{\partial^3 E_y}{\partial y^2\partial x}|_\infty + O(\tau^2)|\frac{\partial^3 E_y}{\partial t^2\partial x}|_\infty$$

$$+O(h_x^2)|\frac{\partial^3 E_x}{\partial x^2\partial y}|_\infty + O(\tau^2)|\frac{\partial^3 E_x}{\partial t^2\partial y}|_\infty)|T_{i,j}|.$$

The final error estimate

With the above preparations, we can now prove the major error estimate result.

THEOREM 9.4

Suppose that the solution of (9.19)-(9.22) possesses the following regularity property:

$$E_x, E_y, H \in C([0,T];C^3(\overline{\Omega})) \cap C^1([0,T];C^2(\overline{\Omega})) \cap C^2([0,T];C^1(\overline{\Omega})).$$

If the initial error

$$||\mathcal{E}_x^0||_E + ||\mathcal{E}_y^0||_E + ||\mathcal{H}^{\frac{1}{2}}||_H \le C(h_x^2 + h_y^2 + \tau^2), \tag{9.73}$$

holds true, then for any $1 \le n \le N_t$ we have

$$\left[\epsilon_0(||\mathcal{E}_x^{n+1}||_E^2 + ||\mathcal{E}_y^{n+1}||_E^2) + \mu_0||\mathcal{H}^{n+\frac{3}{2}}||_H^2\right]^{1/2}$$
$$\le C(h_x^2 + h_y^2 + \tau^2), \tag{9.74}$$

where the constant $C > 0$ is independent of τ, h_x and h_y.

Proof. Note that the error equations (9.60), (9.65) and (9.69) have exactly the same form as (9.43)-(9.45) with extra right-hand side terms representing the errors introduced by time discretization and space discretization. Hence, we can follow exactly the same technique developed in the proof of Theorem 9.3 to obtain (cf. (9.55)):

$$
\epsilon_0(||\mathcal{E}_x^{N_t+1}||_E^2 - ||\mathcal{E}_x^0||_E^2) + \epsilon_0(||\mathcal{E}_y^{N_t+1}||_E^2 - ||\mathcal{E}_y^0||_E^2)
$$

$$
+\mu_0(||\mathcal{H}^{N_t+\frac{3}{2}}||_H^2 - ||\mathcal{H}^{\frac{1}{2}}||_H^2)
$$

$$
\leq \tau \sum_{\substack{0\leq i\leq N_x-1 \\ 0\leq j\leq N_y-1}} |T_{ij}|(\mathcal{H}_{i+\frac{1}{2},j+\frac{1}{2}}^{N_t+\frac{3}{2}} \nabla_y \mathcal{E}_{x,i+\frac{1}{2},j+1}^{N_t+1} - \mathcal{H}_{i+\frac{1}{2},j+\frac{1}{2}}^{\frac{1}{2}} \nabla_y \mathcal{E}_{x,i+\frac{1}{2},j+1}^0)
$$

$$
+\tau \sum_{\substack{0\leq i\leq N_x-1 \\ 0\leq j\leq N_y-1}} |T_{ij}|(-\mathcal{H}_{i+\frac{1}{2},j+\frac{1}{2}}^{N_t+\frac{3}{2}} \nabla_x \mathcal{E}_{y,i+1,j+\frac{1}{2}}^{N_t+1} + \mathcal{H}_{i+\frac{1}{2},j+\frac{1}{2}}^{\frac{1}{2}} \nabla_x \mathcal{E}_{y,i+1,j+\frac{1}{2}}^0)
$$

$$
+\tau \sum_{n=0}^{N_t} \sum_{\substack{0\leq i\leq N_x-1 \\ 0\leq j\leq N_y-1}} (\mathcal{E}_{x,i+\frac{1}{2},j}^{n+1} + \mathcal{E}_{x,i+\frac{1}{2},j}^n)R_1
$$

$$
+\tau \sum_{n=0}^{N_t} \sum_{\substack{0\leq i\leq N_x-1 \\ 0\leq j\leq N_y-1}} (\mathcal{E}_{y,i,j+\frac{1}{2}}^{n+1} + \mathcal{E}_{y,i,j+\frac{1}{2}}^n)R_2
$$

$$
+\tau \sum_{n=0}^{N_t} \sum_{\substack{0\leq i\leq N_x-1 \\ 0\leq j\leq N_y-1}} (\mathcal{H}_{i+\frac{1}{2},j+\frac{1}{2}}^{n+\frac{3}{2}} + \mathcal{H}_{i+\frac{1}{2},j+\frac{1}{2}}^{n+\frac{1}{2}})R_3. \tag{9.75}
$$

All terms except those containing R_i on the RHS of (9.75) can be bounded as in the proof of Theorem 9.3. The R_i terms can be easily bounded by the Cauchy-Schwarz inequality. For example, we have

$$
\tau \sum_{n=0}^{N_t} \sum_{\substack{0\leq i\leq N_x-1 \\ 0\leq j\leq N_y-1}} (\mathcal{E}_{x,i+\frac{1}{2},j}^{n+1} + \mathcal{E}_{x,i+\frac{1}{2},j}^n)R_1
$$

$$
\leq \tau \sum_{n=0}^{N_t} \sum_{\substack{0\leq i\leq N_x-1 \\ 0\leq j\leq N_y-1}} |T_{i,j-\frac{1}{2}}|C(h_x^2 + h_y^2 + \tau^2)(|\mathcal{E}_{x,i+\frac{1}{2},j}^{n+1}| + |\mathcal{E}_{x,i+\frac{1}{2},j}^n|)
$$

$$
\leq \tau \sum_{n=0}^{N_t} \sum_{\substack{0\leq i\leq N_x-1 \\ 0\leq j\leq N_y-1}} |T_{i,j-\frac{1}{2}}| \left[\frac{C}{\delta}(h_x^2 + h_y^2 + \tau^2)^2 + \frac{\delta}{2}(|\mathcal{E}_{x,i+\frac{1}{2},j}^{n+1}|^2 + |\mathcal{E}_{x,i+\frac{1}{2},j}^n|^2) \right]
$$

$$
\leq \frac{CT}{\delta}(h_x^2 + h_y^2 + \tau^2)^2 + \tau \sum_{n=0}^{N_t} \frac{\delta}{2}(||\mathcal{E}_{x,i+\frac{1}{2},j}^{n+1}||_E^2 + ||\mathcal{E}_{x,i+\frac{1}{2},j}^n||_E^2),
$$

where we used the inequality $ab \leq \frac{1}{\delta}a^2 + \frac{\delta}{4}b^2$, where the constant $\delta > 0$.

Choosing δ small enough so that $\|\mathcal{E}^{N_t+1}_{x,i+\frac{1}{2},j}\|^2_E$ etc. can be bounded by the corresponding terms on the left-hand side of (9.75). The proof is completed by using the discrete Gronwall inequality. $\Box$

9.3 The time-domain finite element method

In this section, we present and analyze three types of time-domain finite element methods, i.e., the mixed method, the standard Galerkin method, and the discontinuous Galerkin method.

9.3.1 The mixed method

Substituting (9.5) into (9.1) and (9.2), we can obtain Maxwell's equations in a simple medium:

$$\epsilon \boldsymbol{E}_t + \sigma \boldsymbol{E} - \nabla \times \boldsymbol{H} = -\boldsymbol{J}_s, \quad \text{in } \Omega \times (0, T), \tag{9.76}$$

$$\mu \boldsymbol{H}_t + \nabla \times \boldsymbol{E} = 0, \quad \text{in } \Omega \times (0, T), \tag{9.77}$$

where Ω denotes a bounded Lipschitz continuous polyhedral domain in R^3. To make the problem more general, we assume that ϵ, σ, and μ vary with the position, and these coefficients are piecewise smooth, real, bounded, and positive, i.e., there exist constants $\epsilon_{min}, \epsilon_{max}, \sigma_{max}, \mu_{min}$, and μ_{max} such that, for all $\boldsymbol{x} \in \bar{\Omega}$,

$$0 < \epsilon_{min} \leq \epsilon(\boldsymbol{x}) \leq \epsilon_{max} < \infty,$$
$$0 < \mu_{min} \leq \mu(\boldsymbol{x}) \leq \mu_{max} < \infty,$$
$$0 \leq \sigma(\boldsymbol{x}) \leq \sigma_{max} < \infty.$$

For simplicity, we assume that the boundary $\partial\Omega$ of domain Ω is a perfect conductor, i.e.,

$$\boldsymbol{n} \times \boldsymbol{E} = 0 \quad \text{on } \partial\Omega \times (0, T). \tag{9.78}$$

In addition, initial conditions are assumed to be

$$\boldsymbol{E}(\boldsymbol{x}, 0) = \boldsymbol{E}_0(\boldsymbol{x}) \quad \text{and} \quad \boldsymbol{H}(\boldsymbol{x}, 0) = \boldsymbol{H}_0(\boldsymbol{x}), \quad \forall \, \boldsymbol{x} \in \Omega, \tag{9.79}$$

where $\boldsymbol{E}_0$ and $\boldsymbol{H}_0$ are given functions, and $\boldsymbol{H}_0$ satisfies

$$\nabla \cdot (\mu \boldsymbol{H}_0) = 0 \quad \text{in } \Omega, \quad \boldsymbol{H}_0 \cdot \boldsymbol{n} = 0 \quad \text{on } \partial\Omega. \tag{9.80}$$

Before we move on, we need to introduce some functional spaces. Let $C^m(0, T; X)$ be the space of m times continuously differentiable functions from $[0, T]$ into the Hilbert space X, and

$$H(\text{curl}; \Omega) = \{\boldsymbol{v} \in (L^2(\Omega))^3; \nabla \times \boldsymbol{v} \in (L^2(\Omega))^3\},$$

with equipped norm

$$||v||_{0,\text{curl}} = (||v||_0^2 + ||\nabla \times v||_0^2)^{1/2}.$$

Furthermore, for any Hilbert space X and integer $p \in [1, \infty)$, we define

$$L^p(0, T; X) = \{v \in X; \int_0^T ||v(s)||_X^p ds < \infty\}$$

with the norm

$$||v||_{L^p(0,T;X)} = (\int_0^T ||v(s)||_X^p ds)^{1/p}.$$

The weak formulation of (9.76)-(9.77) can be obtained as follows. Multiplying (9.76) by a test function $\phi \in (L^2(\Omega))^3$ and integrating over Ω, then multiplying (9.77) by a test function $\psi \in H(\text{curl}; \Omega)$ and integrating over Ω, and using the Stokes formula

$$\int_\Omega \nabla \times E \cdot \psi = \int_{\partial\Omega} n \times E \cdot \psi + \int_\Omega E \cdot \nabla \times \psi, \qquad (9.81)$$

we obtain the weak formulation of (9.76)-(9.77): Find
$(E, H) \in [C^1(0, T; (L^2(\Omega))^3) \cap C^0(0, T; H(\text{curl}; \Omega))]^2$ such that

$$(\epsilon E_t, \phi) + (\sigma E, \phi) - (\nabla \times H, \phi) = -(J_s, \phi), \quad \forall \phi \in (L^2(\Omega))^3, \ (9.82)$$
$$(\mu H_t, \psi) + (E, \nabla \times \psi) = 0, \quad \forall \psi \in H(\text{curl}; \Omega), \qquad (9.83)$$

for $0 < t \leq T$ with the initial conditions

$$E(x, 0) = E_0(x), \quad H(x, 0) = H_0(x).$$

Notice that the boundary condition (9.78) is used in deriving (9.83).

To define a finite element method, we need to construct finite-dimensional spaces

$$U_h \subset (L^2(\Omega))^3 \quad \text{and} \quad V_h \subset H(\text{curl}; \Omega).$$

We assume that Ω is discretized by a regular finite element mesh T_h, of tetrahedra with maximum diameter h. Following [56, p. 1617], let P_k denote the standard space of polynomials of total degree less than or equal to k, and let $\tilde{P}_k$ denote the space of homogeneous polynomials of order k. Define $S_k \subset (\tilde{P}_k)^3$ and $R_k \subset (P_k)^3$ by

$$S_k = \{p \in (\tilde{P}_k)^3 \mid p(\mathbf{x}) \cdot \mathbf{x} = 0, \quad \mathbf{x} \in R^3\},$$
$$R_k = (P_{k-1})^3 \oplus S_k.$$

Note that the dimension of S_k can be calculated as follows:

$$\dim(S_k) = 3 \cdot \dim(\widetilde{P}_k) - \dim(\widetilde{P}_{k+1})$$
$$= 3\left(\dim(P_k) - \dim(P_{k-1})\right) - \left(\dim(P_{k+1}) - \dim(P_k)\right)$$
$$= 3\left[\frac{(k+3)(k+2)(k+1)}{3!} - \frac{(k+2)(k+1)k}{3!}\right]$$

$$-\left[\frac{(k+4)(k+3)(k+2)}{3!} - \frac{(k+3)(k+2)(k+1)}{3!}\right]$$

$$= 3\frac{(k+2)(k+1)}{2} - \frac{(k+3)(k+2)}{2} = k(k+2).$$

Hence the dimension of R_k is

$$\dim(R_k) = 3 \cdot \dim(P_{k-1}) + \dim(S_k)$$

$$= 3 \cdot \frac{(k+2)(k+1)k}{3!} + k(k+2) = \frac{(k+3)(k+2)k}{2}.$$

Following [62, 56], let K be a tetrahedron in Ω with general edge e and face f. Let τ_e be a unique vector parallel to e. Let $v \in (W^{1,l}(K))^3$ for some $l > 2$. We define the following three sets of moments of v on K:

$$M_e(v) = \{\int_e (v \cdot \tau_e)qds : \ \forall \ q \in P_{k-1}(e) \text{ for 6 edges } e \text{ of } K\}, \qquad (9.84)$$

$$M_f(v) = \{\int_f v \times \mathbf{n} \cdot \mathbf{q}dA : \ \forall \ \mathbf{q} \in (P_{k-2}(f))^2 \text{ for 4 faces } f \text{ of } K\}, (9.85)$$

$$M_K(v) = \{\int_K v \cdot \mathbf{q}dx : \ \forall \ \mathbf{q} \in (P_{k-3}(K))^3\}. \qquad (9.86)$$

Hence for any $v \in (W^{1,l}(\Omega))^3, l > 2$, an interpolation operator $\Pi_h v \in V_h$ can be defined such that $\Pi_h v|_K \in R_k$ has the same moments (9.84)-(9.86) as v on K for each $K \in T_h$. Furthermore, these three sets of degrees of freedom are proved to be R_k-unisolvent and $H(curl)$ conforming [62]. Note that a finite element is said to be conforming in $H(curl)$ if the tangential components of $\Pi_h v|_{K_1}$ and $\Pi_h v|_{K_2}$ are the same on the common face of K_1 and K_2.

Furthermore, for any integer $k \geq 1$, we define [56, Eqs. (3.2) and (3.8)]

$$U_h = \{u_h \in (L^2(\Omega))^3 : \ u_h|_K \in (P_{k-1})^3 \ \forall \ K \in T_h\}, \qquad (9.87)$$

$$V_h = \{v_h \in H(curl; \Omega) : \ v_h|_K \in R_k \ \forall \ K \in T_h\}, \qquad (9.88)$$

then any function in V_h can be uniquely defined by the degrees of freedom (9.84)-(9.86) on each $K \in T_h$.

The following interpolation properties for the Nédélec spaces have been proved (see [62, Theorem 2] and [56, Theorem 3.2]).

LEMMA 9.2
If $v \in (H^{l+1}(\Omega))^3, 1 \leq l \leq k$, then

$$||v - \Pi_h v||_0 + ||\nabla \times (v - \Pi_h v)||_0 \leq Ch^l||v||_{l+1}.$$

We also need the standard $(L^2(\Omega))^3$ projection operator $P_h : (L^2(\Omega))^3 \rightarrow U_h$ defined as:

$$(P_h u - u, \phi_h) = 0 \quad \forall \ \phi_h \in U_h.$$

Furthermore, we have the error estimate

$$||\boldsymbol{u} - P_h\boldsymbol{u}||_0 \leq Ch^l||\boldsymbol{u}||_l, \qquad 0 \leq l \leq k. \tag{9.89}$$

Now we can consider the semi-discrete mixed method for solving (9.82)-(9.83): Find $(\boldsymbol{E}_h, \boldsymbol{H}_h) \in C^1(0, T; U_h) \times C^1(0, T; V_h)$ such that

$$(\epsilon\frac{\partial \boldsymbol{E}_h}{\partial t}, \boldsymbol{\phi}) + (\sigma\boldsymbol{E}_h, \boldsymbol{\phi}) - (\nabla \times \boldsymbol{H}_h, \boldsymbol{\phi}) = -(\boldsymbol{J}_s, \boldsymbol{\phi}), \quad \forall \boldsymbol{\phi} \in U_h, \tag{9.90}$$

$$(\mu\frac{\boldsymbol{H}_h}{\partial t}, \boldsymbol{\psi}) + (\boldsymbol{E}_h, \nabla \times \boldsymbol{\psi}) = 0, \quad \forall \boldsymbol{\psi} \in V_h, \tag{9.91}$$

for $0 < t \leq T$, and with initial conditions

$$\boldsymbol{E}_h(0) = P_h\boldsymbol{E}_0 \quad \text{and} \quad \boldsymbol{H}_h(0) = \Pi_h\boldsymbol{H}_0. \tag{9.92}$$

The scheme (9.90)-(9.92) was first proposed and analyzed by Monk [56], who proved the following optimal convergence result.

THEOREM 9.5
Suppose that the solution $(\boldsymbol{E}, \boldsymbol{H})$ of (9.76)-(9.77) satisfies $\boldsymbol{E} \in C^1(0, T; (H^k(\Omega))^3)$ and $\boldsymbol{H} \in C^1(0, T; (H^{k+1}(\Omega))^3)$. Let U_h and V_h be given, respectively, by (9.87)-(9.88), and $(\boldsymbol{E}_h(t), \boldsymbol{H}_h(t)) \in U_h \times V_h$ be the solution of (9.90)-(9.92). Then there exists a constant C independent of h and t such that for any $k \geq 1$,

$$||(\boldsymbol{H} - \boldsymbol{H}_h)(t)||_0 + ||(\boldsymbol{E} - \boldsymbol{E}_h)(t)||_0 + ||(\boldsymbol{E} - \boldsymbol{E}_h)(t)||_{L^2(0,t;(L^2(\Omega))^3)}$$
$$\leq Ch^k\{||\boldsymbol{H}_0||_{k+1} + ||\boldsymbol{E}_0||_k + ||\boldsymbol{H}_t||_{L^1(0,t;(H^{k+1}(\Omega))^3)}$$
$$+ ||\boldsymbol{H}||_{L^1(0,t;(H^{k+1}(\Omega))^3)} + ||\boldsymbol{E}_t||_{L^1(0,t;(H^k(\Omega))^3)} + ||\boldsymbol{E}||_{L^2(0,t;(H^k(\Omega))^3)}\}.$$

The proof of the theorem is quite technical and lengthy; interested readers can consult the original paper [56].

There exist many possible fully discrete schemes for solving (9.90)-(9.91). One popular method used in computational electromagnetics is the leapfrog scheme. Let us divide the time interval $[0, T]$ into M uniform subintervals by points

$$0 = t_0 < t_1 < \cdots < t_M = T, \quad \text{where} \ t_n = n \cdot \Delta t, \ \Delta t = T/M.$$

We let $\boldsymbol{E}_h^n$ be the approximation of $\boldsymbol{E}_h(t)$ at t_n, $\boldsymbol{H}_h^{n+\frac{1}{2}}$ be the approximation of $\boldsymbol{H}_h(t)$ at $t_{n+\frac{1}{2}} = (n + \frac{1}{2})\Delta t$. The leapfrog scheme can be formulated as follows: Given $(\boldsymbol{E}_h^n, \boldsymbol{H}_h^{n+\frac{1}{2}})$ the new time approximation $(\boldsymbol{E}_h^{n+1}, \boldsymbol{H}_h^{n+\frac{3}{2}})$ is obtained by successively solving the equations

$$(\epsilon\frac{\boldsymbol{E}_h^{n+1} - \boldsymbol{E}_h^n}{\Delta t}, \boldsymbol{\phi}) + (\sigma\frac{\boldsymbol{E}_h^{n+1} + \boldsymbol{E}_h^n}{2}, \boldsymbol{\phi}) - (\nabla \times \boldsymbol{H}_h^{n+\frac{1}{2}}, \boldsymbol{\phi}) = -(\boldsymbol{J}_s^{n+\frac{1}{2}}, \boldsymbol{\phi}),$$

$$\forall \, \phi \in U_h, \tag{9.93}$$

$$(\mu \frac{H_h^{n+\frac{3}{2}} - H_h^{n+\frac{1}{2}}}{\triangle t}, \psi) + (E_h^{n+1}, \nabla \times \psi) = 0, \quad \forall \, \psi \in V_h. \tag{9.94}$$

The initial value $H_h^{\frac{1}{2}}$ can be computed using a Taylor expansion, i.e.,

$$H_h^{\frac{1}{2}} \approx H^{\frac{1}{2}} = H(0) + \frac{1}{2} \triangle t \frac{\partial H}{\partial t}(0) = H_0 - \frac{\triangle t}{2\mu} \nabla \times E_0. \tag{9.95}$$

Notice that the leapfrog scheme (9.93)-(9.94) is an explicit scheme, which can be written in matrix form [56]:

$$M_\epsilon \frac{E_h^{n+1} - E_h^n}{\triangle t} + M_\sigma \frac{E_h^{n+1} + E_h^n}{2} - M_c H_h^{n+\frac{1}{2}} = -J_s^{n+\frac{1}{2}}, \tag{9.96}$$

$$M_\mu \frac{H_h^{n+\frac{3}{2}} - H_h^{n+\frac{1}{2}}}{\triangle t} + M_c^T E_h^{n+1} = 0, \tag{9.97}$$

where both M_ϵ and M_σ are symmetric positive definite block diagonal matrices of size $dim U_h \times dim U_h$, M_μ is symmetric positive definite and of size $dim V_h \times dim V_h$, M_c is a matrix of $dim U_h \times dim V_h$, and M_c^T is the transpose of M_c. In the leapfrog scheme, at each time step, we first solve E_h^{n+1} from (9.96), then $H_h^{n+\frac{3}{2}}$ from (9.97).

To eliminate the time step constraints for explicit schemes, fully implicit schemes can be used. For example, we can develop the Crank-Nicolson scheme for solving (9.90)-(9.91): Find $(E_h^{n+1}, H_h^{n+1}), n \geq 1$, such that

$$(\epsilon \frac{E_h^{n+1} - E_h^n}{\triangle t}, \phi) + (\sigma \frac{E_h^{n+1} + E_h^n}{2}, \phi) - (\nabla \times \frac{H_h^{n+1} + H_h^n}{2}, \phi)$$

$$= -(J_s^{n+\frac{1}{2}}, \phi), \tag{9.98}$$

$$(\mu \frac{H_h^{n+1} - H_h^n}{\triangle t}, \psi) + (\frac{E_h^{n+1} + E_h^n}{2}, \nabla \times \psi) = 0, \tag{9.99}$$

for any $\phi \in U_h$ and $\psi \in V_h$. Notice that (9.98)-(9.99) can be written in matrix form as

$$M_\epsilon \frac{E_h^{n+1} - E_h^n}{\triangle t} + M_\sigma \frac{E_h^{n+1} + E_h^n}{2} - M_c \frac{H_h^{n+1} + H_h^n}{2} = -J_s^{n+\frac{1}{2}},$$

$$M_\mu \frac{H_h^{n+1} - H_h^n}{\triangle t} + M_c^T \frac{E_h^{n+1} + E_h^n}{2} = 0.$$

The unconditional stability of the Crank-Nicolson scheme can be proved easily by choosing $\phi = E_h^{n+1} + E_h^n$ in (9.98) and $\psi = H_h^{n+1} + H_h^n$ in (9.99) and adding the results together.

It is claimed that [56, p. 1632] the leapfrog and the Crank-Nicolson schemes can be proved to be convergent with error $O((\triangle t)^2 + h^k)$, where k is the degree of the basis functions in subspaces U_h and V_h defined in (9.87) and (9.88), respectively.

9.3.2   The standard Galerkin method

Here we will discuss some methods for solving the vector wave equation in time domain

$$\epsilon E_{tt} + \sigma E_t + \nabla \times (\frac{1}{\mu}\nabla \times E) = -\frac{\partial J_s}{\partial t} \quad \text{in } \Omega \times (0, T), \tag{9.100}$$

with initial conditions

$$E(x, 0) = E_0(x)$$

and

$$E_t(x, 0) = E_1(x) \equiv \frac{1}{\epsilon(x)}(-J_s(x, 0) + \nabla \times H_0(x) - \sigma(x)E_0(x)). \tag{9.101}$$

As in the last subsection, we still assume that Ω is a bounded Lipschitz continuous polyhedral domain in R^3. The parameters ϵ, μ, and σ satisfy the same conditions as mentioned in the last subsection.

In order to obtain a weak formulation of (9.100), we define the space

$$H_0(\text{curl}; \Omega) = \{v \in (L^2(\Omega))^3; \nabla \times v \in (L^2(\Omega))^3, n \times v = 0 \text{ on } \partial\Omega\}.$$

Multiplying (9.100) by a test function $\phi \in H_0(\text{curl}; \Omega)$ and integrating by parts, we obtain the weak formulation for (9.100): Find $E(t) \in H_0(\text{curl}; \Omega)$ such that

$$(\epsilon E_{tt}, \phi) + (\sigma E_t, \phi) + (\frac{1}{\mu}\nabla \times E, \nabla \times \phi) = -(\frac{\partial J_s}{\partial t}, \phi) \quad \forall\, \phi \in H_0(\text{curl}; \Omega), \tag{9.102}$$

subject to the initial conditions

$$E(0) = E_0 \quad \text{and} \quad E_t(0) = E_1. \tag{9.103}$$

Now we need to construct a finite element space $V_h \subset H_0(\text{curl}; \Omega)$. For simplicity, let us consider the second type Nédélec spaces on tetrahedra [63]. Denote

$$D_k = (P_{k-1})^3 \oplus \{p(x)x : p \in \tilde{P}_{k-1}\}, \quad k \geq 1, \tag{9.104}$$

where P_{k-1} denotes the set of polynomials of total degree at most $k - 1$, $\tilde{P}_{k-1}$ denotes the set of homogeneous polynomials of degree $k - 1$.

Let K be a nondegenerate tetrahedron with faces f and edges e. Following Nédélec [63], we can define the finite element space as follows:

$$V_h = \{v \in H(\text{curl}; \Omega); v|_K \in (P_k)^3, \forall\, K \in T_h, n \times v = 0 \text{ on } \partial\Omega\}, \tag{9.105}$$

where each function v is uniquely defined by the following three sets of degrees of freedom:

$$M_e(v) = \{\int_e (v \cdot \tau)q ds : \quad q \in P_k(e) \text{ on each edge } e \text{ of } K\},$$

$$M_f(\boldsymbol{v}) = \{\int_f \boldsymbol{v} \cdot \boldsymbol{q} dA : \quad \boldsymbol{q} \in D_{k-1}(f) \text{ tangent to each face } f \text{ of } K\},$$

$$M_K(\boldsymbol{v}) = \{\int_K \boldsymbol{v} \cdot \boldsymbol{q} dV : \quad \boldsymbol{q} \in D_{k-2}(K)\}.$$

Here τ denotes the tangent vector along the direction of e. An interpolation operator Π_h can be defined such that $\Pi_h \boldsymbol{v} \in V_h$ and $\Pi_h \boldsymbol{v}$ has the same degrees of freedom as $\boldsymbol{v}$ on each tetrahedron $K \in T_h$. Nédélec [63] proved that these degrees of freedom are $(P_k)^3$ unisolvent and $H(curl)$ conforming.

Now we can formulate the semi-discrete scheme for solving (9.100)-(9.101): Find $\boldsymbol{E}_h(t) \in V_h$ such that

$$(\epsilon \frac{\partial^2 \boldsymbol{E}_h}{\partial t^2}, \phi) + (\sigma \frac{\partial \boldsymbol{E}_h}{\partial t}, \phi) + (\frac{1}{\mu} \nabla \times \boldsymbol{E}_h, \nabla \times \phi) = -(\frac{\partial \boldsymbol{J}_s}{\partial t}, \phi) \quad \forall \phi \in V_h, \quad (9.106)$$

subject to the initial conditions

$$\boldsymbol{E}_h(0) = \Pi_h \boldsymbol{E}_0 \quad \text{and} \quad \frac{\partial \boldsymbol{E}_h(0)}{\partial t} = \Pi_h \boldsymbol{E}_1. \tag{9.107}$$

The error analysis for the time-dependent vector wave equation (9.100) in R^3 was first carried out by Monk [58], who proved the following optimal energy-norm error estimates on general polyhedral domains.

THEOREM 9.6
Let $\boldsymbol{E}$ and $\boldsymbol{E}_h$ be the solutions of (9.100)-(9.101) and (9.106)-(9.107), respectively. Under the regularity assumptions that $\boldsymbol{E}_t(t), \boldsymbol{E}_{tt}(t) \in H^{k+1}(\Omega), 0 \le t \le T$, then we have

$$||(\boldsymbol{E} - \boldsymbol{E}_h)(t)||_{0,curl} + ||(\boldsymbol{E} - \boldsymbol{E}_h)_t(t)||_0$$

$$\le Ch^k \{||\boldsymbol{E}_0||_{k+1} + \max_{0 \le s \le t} ||\boldsymbol{E}_t(s)||_{k+1} + \int_0^t ||\boldsymbol{E}_{tt}(s)||_{k+1} ds\}, \quad (9.108)$$

where the constant $C = C(T)$ is independent of the mesh size h.

Due to its technicality, we skipped the proof, which can be found in Monk's paper [58]. Actually, the above estimates hold true for other elements such as: the first type Nédélec elements on tetrahedra and cubes [62], and $H^1(\Omega)$ conforming elements with continuous piecewise polynomials of degree k, since the proof only uses the following interpolation estimate:

$$||\boldsymbol{v} - \Pi_h \boldsymbol{v}||_0 + h||\boldsymbol{v} - \Pi_h \boldsymbol{v}||_{0,curl} \le Ch^{k+1}||\boldsymbol{v}||_{k+1}, \quad \text{for any } \boldsymbol{v} \in (H^{k+1}(\Omega))^3. \tag{9.109}$$

Notice that the L^2 error estimate from (9.108) is not optimal. Under the assumptions that Ω is a convex polyhedral domain, $\sigma \equiv 0$, and ϵ and μ are constants, Monk [58] proved the following almost optimal error estimate

$$||(\boldsymbol{E} - \boldsymbol{E}_h)(t)||_0 \le Ch^{k+1-\delta}, \tag{9.110}$$

where $\delta > 0$ is an arbitrary small constant.

Fully discrete finite element methods were considered later by Ciarlet and Zou [15]. For example, the backward Euler scheme can be formulated as follows: For $n = 1, 2, \cdots, M$, find $\boldsymbol{E}_h^n \in V_h$ such that

$$(\epsilon \frac{\boldsymbol{E}_h^n - 2\boldsymbol{E}_h^{n-1} + \boldsymbol{E}_h^{n-2}}{(\triangle t)^2}, \phi) + (\sigma \frac{\boldsymbol{E}_h^n - \boldsymbol{E}_h^{n-1}}{\triangle t}, \phi) + (\frac{1}{\mu}\nabla \times \boldsymbol{E}_h^n, \nabla \times \phi)$$

$$= -(\frac{\partial \boldsymbol{J}_s}{\partial t}(t_n), \phi) \quad \forall \, \phi \in V_h, \tag{9.111}$$

subject to the initial approximations

$$\boldsymbol{E}_h^0 = \Pi_h \boldsymbol{E}_0, \quad \boldsymbol{E}_h^0 - \boldsymbol{E}_h^{-1} = \triangle t \Pi_h \boldsymbol{E}_1. \tag{9.112}$$

Similarly, the Crank-Nicolson scheme can be developed as follows: For $n = 0, 1, \cdots, M - 1$ and any $\phi \in V_h$, find $\boldsymbol{E}_h^{n+1} \in V_h$ such that

$$(\epsilon \frac{\boldsymbol{E}_h^{n+1} - 2\boldsymbol{E}_h^n + \boldsymbol{E}_h^{n-1}}{(\triangle t)^2}, \phi) + (\sigma \frac{\boldsymbol{E}_h^{n+1} - \boldsymbol{E}_h^{n-1}}{2\triangle t}, \phi)$$

$$+(\frac{1}{\mu}\nabla \times \frac{\boldsymbol{E}_h^{n+1} + \boldsymbol{E}_h^{n-1}}{2}, \nabla \times \phi) = -(\frac{\partial \boldsymbol{J}_s}{\partial t}(t_n), \phi), \tag{9.113}$$

subject to the initial approximations

$$\boldsymbol{E}_h^0 = \Pi_h \boldsymbol{E}_0, \quad \boldsymbol{E}_h^1 - \boldsymbol{E}_h^{-1} = 2\triangle t \Pi_h \boldsymbol{E}_1. \tag{9.114}$$

Ciarlet and Zou [15] proved the optimal error estimates for the backward Euler scheme in both the energy norm and the L^2 norm for the lowest-order first and second type of $H(\mathrm{curl}; \Omega)$-conforming Nédélec elements using less regularity requirements than Monk assumed [58]. More specifically, the finite element space V_h can be chosen either as

$$V_h = \{\boldsymbol{v} \in H(\mathrm{curl}; \Omega) : \, \boldsymbol{v}|_K \in (P_1)^3, \forall \, K \in T_h, \boldsymbol{n} \times \boldsymbol{v}_h = 0 \text{ on } \partial\Omega\},$$

or

$$V_h = \{\boldsymbol{v} \in H(\mathrm{curl}; \Omega) : \, \boldsymbol{v}|_K = a_K + b_K \times \boldsymbol{x}, a_K \text{ and } b_K \in R^3, \forall \, K \in T_h,$$
$$\boldsymbol{n} \times \boldsymbol{v}_h = 0 \text{ on } \partial\Omega\}.$$

Then for general polyhedral domains, Ciarlet and Zou [15] proved that

$$\max_{1 \le n \le M} (||\frac{\boldsymbol{E}_h^n - \boldsymbol{E}_h^{n-1}}{\triangle t} - \frac{\partial \boldsymbol{E}^n}{\partial t}||_0 + ||\nabla \times (\boldsymbol{E}_h^n - \boldsymbol{E}^n)||_0) \le C(\triangle t + h),$$

where C is a constant independent of both the time step $\triangle t$ and the mesh size h. When Ω is a convex polyhedral domain, $\sigma \equiv 0$, and ϵ and μ are constants, they proved the optimal error estimate

$$||\boldsymbol{E}_h^n - \boldsymbol{E}^n||_0 \le C(\triangle t + h^2), \quad n = 1, 2, \cdots, M,$$

where C is independent of $\triangle t$ and h.

9.3.3 The discontinuous Galerkin method

The discontinuous Galerkin (DG) finite element method was originally introduced for the neutron transport equation in the 1970s. Later it was used for the weak enforcement of continuity of the Galerkin method for diffusion problems. In recent years, the DG methods have been extended to broader applications [16]. For mathematical analysis, see recent survey articles [4, 18]. The main advantages of DG methods are the local mass conservation property, great flexibility in handling nonmatching grids, elements of various types and shapes, and parallelization. In this subsection, we will introduce several DG methods developed for solving time-dependent Maxwell's equations.

First, we want to introduce the nodal DG method developed by Hesthaven and Warburton [26]. For clarity, we keep similar notation of [26] and rewrite Maxwell's equations (9.76)-(9.77) with $\sigma = 0$ in conservation form

$$\lambda \frac{\partial q}{\partial t} + \nabla \cdot F(q) = s, \tag{9.115}$$

where we denote

$$\lambda = \begin{bmatrix} \epsilon(x) & 0 \\ 0 & \mu(x) \end{bmatrix}, \quad q = \begin{bmatrix} E \\ H \end{bmatrix}, \quad F_i(q) = \begin{bmatrix} -e_i \times H \\ e_i \times E \end{bmatrix}. \tag{9.116}$$

Here e_i $(i = 1, 2, 3)$ are the Cartesian unit vectors in x, y, and z directions, vector $F(q)$ is represented as $F(q) = [F_1(q), F_2(q), F_3(q)]$, and vector $s = [s^E, s^H]^T$ is the general source term.

Before we move on, let us first elaborate on how it (9.115) is derived. Note that

$$e_1 \times H = \begin{vmatrix} \vec{i} & \vec{j} & \vec{k} \\ 1 & 0 & 0 \\ H_x & H_y & H_z \end{vmatrix} = -H_z \vec{j} + H_y \vec{k}.$$

Similarly, we can show that

$$e_2 \times H = H_z \vec{i} - H_x \vec{k}, \quad e_3 \times H = -H_y \vec{i} + H_x \vec{j},$$

where subscripts x, y, and z denote the components of H.

Hence the first component of $\nabla \cdot F$ equals

$$\frac{\partial F_1^{(1)}}{\partial x} + \frac{\partial F_2^{(1)}}{\partial y} + \frac{\partial F_3^{(1)}}{\partial z}$$

$$= \frac{\partial}{\partial x}(-e_1 \times H) + \frac{\partial}{\partial y}(-e_2 \times H) + \frac{\partial}{\partial z}(-e_3 \times H)$$

$$= \frac{\partial}{\partial x}(H_z \vec{j} - H_y \vec{k}) + \frac{\partial}{\partial y}(-H_z \vec{i} + H_x \vec{k}) + \frac{\partial}{\partial z}(H_y \vec{i} - H_x \vec{j})$$

$$= (\frac{\partial H_y}{\partial z} - \frac{\partial H_z}{\partial y})\vec{i} + (\frac{\partial H_z}{\partial x} - \frac{\partial H_x}{\partial z})\vec{j} + (\frac{\partial H_x}{\partial y} - \frac{\partial H_y}{\partial x})\vec{k}$$

$$= \nabla \times H.$$

By the same technique, we can see that the second component of $\nabla \cdot \boldsymbol{F}$ equals $-\nabla \times \boldsymbol{E}$, from which we can conclude that (9.115) is indeed equivalent to (9.76)-(9.77) with $\sigma = 0$ and a general source s on the right-hand side.

To derive the DG method, we assume that Ω is decomposed into tetrahedral mesh T_h, i.e., $\Omega = \cup_{K \in T_h} K$. We require that (9.115) satisfy elementwise the weak form

$$\int_K (\lambda \frac{\partial \boldsymbol{q}_h}{\partial t} + \nabla \cdot \boldsymbol{F}_h - \boldsymbol{s}_h) \phi(\boldsymbol{x}) d\boldsymbol{x} = \oint_{\partial K} \psi(\boldsymbol{x}) \boldsymbol{n} \cdot (\boldsymbol{F}_h - \boldsymbol{F}_h^*) d\boldsymbol{x}, \quad (9.117)$$

where ϕ and ψ are test functions, $\boldsymbol{F}_h^*$ is a numerical flux, and the DG solution $\boldsymbol{q}_h$ on each element K is represented as

$$\boldsymbol{q}_h(\boldsymbol{x}, t) = \sum_{j=0}^{N} q(\boldsymbol{x}_j, t) L_j(\boldsymbol{x}) = \sum_{j=0}^{N} q_j(t) L_j(\boldsymbol{x}). \quad (9.118)$$

Here $L_j(\boldsymbol{x})$ are the 3-D Lagrange interpolation polynomials based on the $N+1$ nodal points. To make the basis a complete polynomial of degree n, we need $\frac{1}{6}(n+1)(n+2)(n+3)$ nodal points, which number is the total degrees of freedom for n-th order polynomial $(P_n(x))^3$.

Note that (9.117) imposes the boundary or interface conditions rather weakly by penalizing the interface integral. Also it gives a tremendous freedom in designing different schemes by choosing various ϕ, ψ, and $\boldsymbol{F}_h^*$. One popular choice [26] is to choose the test functions

$$\phi(\boldsymbol{x}) = \psi(\boldsymbol{x}) = L_i(\boldsymbol{x}) \quad (9.119)$$

and upwind flux

$$\boldsymbol{n} \cdot (\boldsymbol{F}_h - \boldsymbol{F}_h^*) = \begin{pmatrix} \boldsymbol{n} \times (Z^+[\boldsymbol{H}_h] - \boldsymbol{n} \times [\boldsymbol{E}_h])/(Z^+ + Z^-) \\ \boldsymbol{n} \times (-\boldsymbol{n} \times [\boldsymbol{H}_h] - Y^+[\boldsymbol{E}_h])/(Y^+ + Y^-) \end{pmatrix} \quad (9.120)$$

where

$$[\boldsymbol{E}_h] = \boldsymbol{E}_h^+ - \boldsymbol{E}_h^-, \quad [\boldsymbol{H}_h] = \boldsymbol{H}_h^+ - \boldsymbol{H}_h^-$$

denote the jumps across an interface. Here $Z^\pm$ and $Y^\pm$ are the local impedance and conductance defined as

$$Z^\pm = 1/Y^\pm = \sqrt{\frac{\mu^\pm}{\epsilon^\pm}},$$

where the superscript "+" refers to the value from the neighboring element while "-" refers to the value inside the element.

Now substituting (9.118)-(9.120) into (9.117), we obtain the nodal DG semi-discrete scheme

$$\sum_{j=0}^{N} (M_{ij}^\epsilon \frac{dE_j}{dt} - S_{ij} \times H_j - M_{ij} s_j^E)$$

$$= \sum_l F_{il}(\boldsymbol{n}_l \times \frac{Z_l^+[H_l] - \boldsymbol{n}_l \times [E_l]}{Z_l^+ + Z_l^-}), \tag{9.121}$$

$$\sum_{j=0}^{N}(M_{ij}^\mu \frac{dH_j}{dt} + S_{ij} \times E_j - M_{ij}s_j^H)$$

$$= \sum_l F_{il}(\boldsymbol{n}_l \times \frac{-\boldsymbol{n}_l \times [H_l] - Y_l^+[E_l]}{Y_l^+ + Y_l^-}), \tag{9.122}$$

where the local material mass matrices

$$M_{ij}^\epsilon = \int_K L_i(\boldsymbol{x})\epsilon(\boldsymbol{x})L_j(\boldsymbol{x})d\boldsymbol{x}, \quad M_{ij}^\mu = \int_K L_i(\boldsymbol{x})\mu(\boldsymbol{x})L_j(\boldsymbol{x})d\boldsymbol{x},$$

$$M_{ij} = \int_K L_i(\boldsymbol{x})L_j(\boldsymbol{x})d\boldsymbol{x}, \quad S_{ij} = \int_K L_i(\boldsymbol{x})\nabla L_j(\boldsymbol{x})d\boldsymbol{x},$$

and the face mass matrix

$$F_{il} = \oint_{\partial K} L_i(\boldsymbol{x})L_j(\boldsymbol{x})ds.$$

It is proved [26] that if the boundary of Ω is either periodic or terminated with a perfectly conducting boundary, then the semi-discrete approximation (9.121)-(9.122) is stable in the sense that

$$\frac{d}{dt}(EGY) \leq C(EGY + ||s^E||^2_{L^2(\Omega)} + ||s^H||^2_{L^2(\Omega)}),$$

where EGY denotes the global energy

$$EGY = \sum_{K \in T_h} \frac{1}{2} \int_K (\mu|\boldsymbol{H}_h|^2 + \epsilon|\boldsymbol{E}_h|^2)d\boldsymbol{x}.$$

The following convergence result has been obtained in [26].

THEOREM 9.7
Assume that the solution to Maxwell's equations $q \in W^{p,2}(K), p \geq 2$, for any $K \in T_h$. Then the DG solution q_h of (9.121)-(9.122) converges to the exact solution, and the following error estimate holds true

$$\sum_{K \in T_h} (\int_K |q - q_h|^2 d\boldsymbol{x})^{1/2}$$

$$\leq C \sum_{K \in T_h} (\frac{h^\sigma}{n^p}||q(0)||_{W^{p,2}(K)} + t\frac{h^{\sigma-1}}{n^{p-2}} \max_{s \in [0,t]} ||q(s)||_{W^{p,2}(K)}),$$

where n is the degree of the basis polynomial, $\sigma = \min(p, n+1)$, and $h = \max_{K \in T_h}(diam(K))$. The constant C depends upon the material properties, but is independent of h and n.

Another slightly different DG method was proposed in [13] as below:

$$\int_K \lambda \frac{\partial q_h}{\partial t} \psi d\boldsymbol{x} - \int_K \boldsymbol{F}(\boldsymbol{q}_h) \cdot \nabla \psi d\boldsymbol{x} + \oint_{\partial K} \hat{\boldsymbol{F}}_h \cdot \boldsymbol{n} \psi = \int_K s \psi d\boldsymbol{x}, \qquad (9.123)$$

for any test function $\psi \in (P_k(K))^3$, the space of polynomials of degree k in the element K. Here $\hat{\boldsymbol{F}}_h \cdot \boldsymbol{n}$ is the numerical flux on ∂K, which can be chosen as the same upwind flux defined by (9.120).

Substituting the approximate solution

$$\boldsymbol{q}_h|_K = \sum_{j=1}^{N_k} q_j(t) \psi_j(\boldsymbol{x}) \qquad (9.124)$$

into (9.123), we obtain a system of ordinary differential equations

$$\frac{d\boldsymbol{q}}{dt} = L\boldsymbol{q} + S(t), \qquad (9.125)$$

which can be solved by any time difference scheme. In [13], the mth-order, mth-stage strong stability preserving Runge-Kutta scheme with low storage was used. Interested readers can find more details about the implementation in the original paper [13].

Finally, we would like to introduce the locally divergence-free DG method developed by Cockburn et al. [17]. Rewrite the 2-D Maxwell equations

$$\frac{\partial H_x}{\partial t} = -\frac{\partial E_z}{\partial y}, \quad \frac{\partial H_y}{\partial t} = \frac{\partial E_z}{\partial x}, \quad \frac{\partial E_z}{\partial t} = \frac{\partial H_y}{\partial x} - \frac{\partial H_x}{\partial y} \qquad (9.126)$$

into the conservation form

$$\frac{\partial \boldsymbol{q}}{\partial t} + \nabla \cdot \boldsymbol{F}(\boldsymbol{q}) = 0, \qquad (9.127)$$

where $\boldsymbol{q} = (H_x, H_y, E_z)^T$, $\boldsymbol{F}(\boldsymbol{q}) = (\boldsymbol{F}_1, \boldsymbol{F}_2, \boldsymbol{F}_3)$,

$$\boldsymbol{F}_1 = (0, -E_z, -H_y), \quad \boldsymbol{F}_2 = (E_z, 0, H_x), \quad \boldsymbol{F}_3 = (0, 0, 0).$$

Note that the system (9.126) has a divergence-free solution (H_x, H_y) at all times if initially it is divergence free. To construct a locally divergence-free DG method, we need to develop the locally divergence-free finite element space

$$\boldsymbol{V}_h^k = \{\boldsymbol{v} : \boldsymbol{v}|_K \in (P_k(K))^3, K \in T_h, (\frac{\partial v_1}{\partial x} + \frac{\partial v_2}{\partial y})|_K = 0\}$$

$$= \boldsymbol{V}_{h,0}^k \oplus \{v_3 : v_3|_K \in P_k(K), K \in T_h\}. \qquad (9.128)$$

We can obtain the bases for $\boldsymbol{V}_{h,0}^k|_K$ by taking the curl of bases of $P_{k+1}(K)$. Note that the 2-D curl operator is defined as

$$\nabla \times \boldsymbol{f} = \begin{pmatrix} \frac{\partial \boldsymbol{f}}{\partial y} \\ -\frac{\partial \boldsymbol{f}}{\partial x} \end{pmatrix}. \qquad (9.129)$$

Hence if K is a rectangular element, the bases of $V_{h,0}^1|_K$ are $\nabla \times P_2(K)$, which leads to

$$\begin{pmatrix} 0 \\ -1 \end{pmatrix}, \begin{pmatrix} 1 \\ 0 \end{pmatrix}, \begin{pmatrix} 0 \\ -2x \end{pmatrix}, \begin{pmatrix} x \\ -y \end{pmatrix}, \begin{pmatrix} 2y \\ 0 \end{pmatrix}, \qquad (9.130)$$

from bases x, y, x^2, xy, y^2, respectively.

Similarly, for $k = 2$, the bases of $V_{h,0}^2|_K$ can be obtained from $\nabla \times P_3(K)$, which includes (9.130) and the extra terms below

$$\begin{pmatrix} 0 \\ -3x^2 \end{pmatrix}, \begin{pmatrix} x^2 \\ -2xy \end{pmatrix}, \begin{pmatrix} 2xy \\ -y^2 \end{pmatrix}, \begin{pmatrix} 3y^2 \\ 0 \end{pmatrix}, \qquad (9.131)$$

obtained from bases x^3, x^2y, xy^2, y^3, respectively.

Now we can form the DG method for (9.127): find $q_h \in V_h^k$ such that

$$\int_K \frac{\partial q_h}{\partial t} v d\boldsymbol{x} - \int_K \boldsymbol{F}(q_h) \cdot \nabla v d\boldsymbol{x} + \sum_{e \in \partial K} \oint_e \hat{\boldsymbol{F}}_h \cdot \boldsymbol{n} v = 0, \ \forall \, v \in V_h^k, \quad (9.132)$$

where the numerical flux can be the upwind flux defined by (9.120). In particular, for the 2-D problem (9.126), the upwind flux can be written as

$$\hat{\boldsymbol{F}}_h \cdot \boldsymbol{n} = \begin{pmatrix} n_2(\overline{E_z} - \frac{n_2}{2}[H_x] + \frac{n_1}{2}[H_y]) \\ -n_1(\overline{E_z} - \frac{n_2}{2}[H_x] + \frac{n_1}{2}[H_y]) \\ n_2\overline{H_x} - n_1\overline{H_y} - \frac{1}{2}[E_z] \end{pmatrix}, \qquad (9.133)$$

where we denote the average and jump as

$$\bar{v} = 0.5(v^- + v^+), \quad [v] = v^+ - v^-.$$

The following error estimate has been proved in [17].

THEOREM 9.8
Let $q = (H_x, H_y, E_z)$ be the smooth exact solution of (9.126), and $q_h = (H_{x,h}, H_{y,h}, E_{z,h})$ the solution of (9.132)-(9.133) in V_h^k. Then

$$\|q - q_h\|_0 \leq Ch^{k+\frac{1}{2}}\|q\|_{k+1}, \quad \forall \, k \geq 1. \qquad (9.134)$$

In [17], the Lax-Friedrichs flux

$$\hat{\boldsymbol{F}}_h \cdot \boldsymbol{n} = \begin{pmatrix} n_2(\overline{E_z} - \frac{n_2}{2}[H_x] + \frac{n_1}{2}[H_y]) - \frac{1}{2}(n_1^2[H_x] + n_1n_2[H_y]) \\ -n_1(\overline{E_z} - \frac{n_2}{2}[H_x] + \frac{n_1}{2}[H_y]) - \frac{1}{2}(n_1n_2[H_x] + n_2^2[H_y]) \\ n_2\overline{H_x} - n_1\overline{H_y} - \frac{1}{2}[E_z] \end{pmatrix} \qquad (9.135)$$

is proved to be another good choice, which achieves the same error estimate as (9.134) for a globally divergence-free finite element space.

9.4   The frequency-domain finite element method

In this section, we present and analyze three types of finite element methods for solving frequency-domain Maxwell's equations, i.e., the standard Galerkin method, the discontinuous Galerkin method, and the mixed discontinuous Galerkin method.

9.4.1   The standard Galerkin method

Let us consider the vector wave equation

$$\nabla \times (\frac{1}{\mu} \nabla \times \boldsymbol{E}) - \omega^2 (\epsilon + i\frac{\sigma}{\omega}) \boldsymbol{E} = i\omega \boldsymbol{J}_s, \quad \text{in } \Omega, \tag{9.136}$$

with a perfectly conducting boundary

$$\boldsymbol{n} \times \boldsymbol{E} = 0 \quad \text{on } \partial\Omega, \tag{9.137}$$

where Ω is a bounded polygonal domain in R^3 with boundary surface $\partial\Omega$ and unit outward normal $\boldsymbol{n}$.

For simplicity, we assume that Ω is discretized by a family of tetrahedral meshes T_h. The meshes are regular and quasi-uniform. Then we can construct a family of finite dimensional subspaces $V_0^h \subset H_0(\text{curl}; \Omega)$ by using Nédélec's first type spaces, i.e.,

$$V_0^h = \{ v_h \in V_h : \ \boldsymbol{n} \times v_h = 0 \quad \text{on } \partial\Omega \}, \tag{9.138}$$

where V_h is defined by (9.88).

The standard Galerkin method for solving (9.136)-(9.137) is: Find $\boldsymbol{E}_h \in V_0^h$ such that

$$(\mu^{-1} \nabla \times \boldsymbol{E}_h, \nabla \times \phi) - \omega^2 ((\epsilon + i\frac{\sigma}{\omega}) \boldsymbol{E}_h, \phi) = (i\omega \boldsymbol{J}_s, \phi) \quad \forall \phi \in V_0^h. \tag{9.139}$$

If σ is strictly positive on $\overline{\Omega}$, the bilinear form in (9.139) is coercive on $H_0(\text{curl}; \Omega)$, i.e., there exists a constant $C > 0$ such that [45]:

$$|(\mu^{-1} \nabla \times \boldsymbol{v}, \nabla \times \boldsymbol{v}) - \omega^2 ((\epsilon + i\frac{\sigma}{\omega}) \boldsymbol{v}, \boldsymbol{v})| \geq C ||\boldsymbol{v}||_{0,\text{curl}}^2, \quad \forall \boldsymbol{v} \in H_0(\text{curl}; \Omega). \tag{9.140}$$

Hence by Ceá's lemma and the interpolation estimate (Lemma 9.2), we easily obtain [59, Theorem 3.1]:

THEOREM 9.9
Suppose that σ is continuous and strictly positive on $\overline{\Omega}$ and $\boldsymbol{E} \in (H^{k+1}(\Omega))^3$. Then there exists a unique solution $\boldsymbol{E}_h$ to (9.139). Furthermore, the following error estimate holds true:

$$||\boldsymbol{E} - \boldsymbol{E}_h||_{0,curl} \leq Ch^k ||\boldsymbol{E}||_{k+1},$$

where the constant $C > 0$ is independent of h.

When $\sigma \equiv 0$, the bilinear form in (9.139) is not coercive on $H_0(\text{curl}; \Omega)$. In this case, (9.136) may fail to have a solution at a countably infinite discrete set of values of w which are termed the interior Maxwell eigenvalues. To guarantee convergence, we need more restrictive assumptions. By applying a discrete Helmholtz decomposition, Monk [59] proved the following optimal error estimate:

THEOREM 9.10

Assume that Ω is a convex polygonal domain in R^3, $\sigma \equiv 0$, ϵ and μ are constants, and $\boldsymbol{E} \in (H^{k+1}(\Omega))^3$. Then provided that ω is not an interior Maxwell eigenvalue, and h is sufficiently small, there exists a unique solution $\boldsymbol{E}_h$ of (9.139). Furthermore, there is a constant C independent of h such that

$$||\boldsymbol{E} - \boldsymbol{E}_h||_{0,curl} \leq Ch^k ||\boldsymbol{E}||_{k+1}.$$

Notice that both Theorems 9.9 and 9.10 only provide sub-optimal error estimates in the L^2-norm. Using Nédélec's second type elements (9.105), Monk [59] proved the almost optimal error estimate in L^2-norm:

THEOREM 9.11

Let the same assumptions as Theorem 9.10 hold true, and $\boldsymbol{E}_h$ is the solution of (9.139) with V_0^h as Nédélec's second type elements. Then for any $\delta > 0$ sufficiently small, there exists a constant $C = C(\delta)$ such that h is small enough, we have

$$||\boldsymbol{E} - \boldsymbol{E}_h||_{0,curl} \leq Ch^k ||\boldsymbol{E}||_{k+1}, \quad ||\boldsymbol{E} - \boldsymbol{E}_h||_0 \leq Ch^{k+1-\delta} ||\boldsymbol{E}||_{k+1}.$$

The same convergence results as Theorem 9.9 holds true for Nédélec's second type elements.

9.4.2   The discontinuous Galerkin method

Recently, DG methods have begun to find their applications in solving Maxwell's equations in the frequency domain [32, 33, 34, 65]. Here we will present some related DG methods.

For simplicity, we consider the regular tetrahedra mesh T_h of Ω defined in the last subsection. We denote h_K the diameter of the tetrahedral element $K \in T_h$, h the mesh size of T_h given by $h \equiv \max_{K \in T_h} h_K$, F_h^I the set of all interior faces of elements in T_h, F_h^B the set of all boundary faces, and the set $F_h \equiv F_h^I \cup F_h^B$.

Next we define some trace operators for piecewise smooth vector function $\boldsymbol{v}$ and scalar function q. For an interior face $f \in F_h^I$ shared by two neighboring elements K^+ and K^- with unit outward normal vectors $\boldsymbol{n}^+$ and $\boldsymbol{n}^-$, respectively, we denote by $\boldsymbol{v}^\pm$ and $q^\pm$ the traces of $\boldsymbol{v}$ and q taken from $K^\pm$, respectively. Furthermore, we define the jumps across f by

$$[[v]]_T \equiv n^+ \times v^+ + n^- \times v^-, \quad [[v]]_N \equiv v^+ \cdot n^+ + v^- \cdot n^-, \quad [[q]]_N \equiv q^+ n^+ + q^- n^-,$$

and the averages across f by

$$\{\{v\}\} \equiv \frac{1}{2}(v^+ + v^-), \quad \{\{q\}\} \equiv \frac{1}{2}(q^+ + q^-).$$

On a boundary face $f \in F_h^B$, we define

$$[[v]]_T \equiv n \times v, \quad \{\{v\}\} \equiv v, \quad [[q]]_N \equiv qn.$$

For two neighboring faces ∂K_1 and ∂K_2, we have

$$\int_{\partial K_1} v_1 \cdot n_1 \times u_1 + \int_{\partial K_2} v_2 \cdot n_2 \times u_2$$

$$= \int_{\partial K_1} \frac{v_1 + v_2}{2} \cdot n_1 \times u_1 + \int_{\partial K_2} \frac{v_1 + v_2}{2} \cdot n_2 \times u_2$$

$$+ \int_{\partial K_1} \frac{v_1 - v_2}{2} \cdot n_1 \times u_1 - \int_{\partial K_2} \frac{v_1 - v_2}{2} \cdot n_2 \times u_2$$

$$= \int_{\partial K_1 \cap \partial K_2} (v_1 + v_2) \cdot \frac{1}{2}(n_1 \times u_1 + n_2 \times u_2)$$

$$- \int_{\partial K_1} n_1 \times (v_1 - v_2) \cdot \frac{u_1}{2} + \int_{\partial K_2} n_2 \times (v_1 - v_2) \cdot \frac{u_2}{2}, \quad (9.141)$$

where we used the identity

$$a \cdot (b \times c) = (a \times b) \cdot c = -(b \times a) \cdot c,$$

which holds true for any vectors $a, b,$ and c.

Using the fact $n_2 = -n_1$, we obtain

$$-n_1 \times (v_1 - v_2) \cdot \frac{u_1}{2} + n_2 \times (v_1 - v_2) \cdot \frac{u_2}{2}$$

$$= -(n_1 \times v_1 + n_2 \times v_2) \cdot \frac{u_1}{2} - (n_1 \times v_1 + n_2 \times v_2) \cdot \frac{u_2}{2}$$

$$= -(n_1 \times v_1 + n_2 \times v_2) \cdot \frac{u_1 + u_2}{2}. \quad (9.142)$$

Substituting (9.142) into (9.141) and using the trace operators defined earlier, we have

LEMMA 9.3

$$\int_{\partial K_1} v_1 \cdot n_1 \times u_1 + \int_{\partial K_2} v_2 \cdot n_2 \times u_2$$

$$= \int_{\partial K_1 \cap \partial K_2} \{\{v\}\} \cdot [[u]]_T - \int_{\partial K_1 \cap \partial K_2} [[v]]_T \cdot \{\{u\}\}. \qquad (9.143)$$

Using Stokes' formula (9.81) and (9.143), we obtain

$$\int_\Omega \nabla \times (\mu^{-1}\nabla \times E) \cdot v$$

$$= \sum_{K \in T_h} [\int_{\partial K} n \times (\mu^{-1}\nabla \times E) \cdot v + \int_K \mu^{-1}\nabla \times E \cdot \nabla \times v]$$

$$= \sum_{f \in F_h} \int_f \{\{v\}\} \cdot [[\mu^{-1}\nabla \times E]]_T - \sum_{f \in F_h} \int_f [[v]]_T \cdot \{\{\mu^{-1}\nabla \times E\}\}$$

$$+ \int_\Omega (\mu^{-1}\nabla \times E) \cdot (\nabla \times v). \qquad (9.144)$$

For a given partition T_h of Ω and approximation order $l \geq 1$, we define the following discontinuous finite element space

$$V_h = \{v \in (L^2(\Omega))^3, \ v|_K \in (P^l(K))^3, \ \forall \, K \in T_h\}, \qquad (9.145)$$

where $P^l(K)$ denotes the space of polynomial of total degree l on K.

Now we can construct a direct interior penalty DG method for solving (9.136)-(9.137): Find $E_h \in V_h$ such that

$$a_h(E_h, \overline{\phi}) - \omega^2((\epsilon + i\frac{\sigma}{\omega})E_h, \overline{\phi}) = (i\omega J_s, \overline{\phi}), \quad \forall \, \phi \in V_h, \qquad (9.146)$$

where the discrete (complex valued) sesquilinear form

$$a_h(u, v) = (\mu^{-1}\nabla_h \times u, \nabla_h \times \overline{v}) - \sum_{f \in F_h} \int_f [[u]]_T \cdot \{\{\mu^{-1}\nabla_h \times \overline{v}\}\}ds$$

$$- \sum_{f \in F_h} \int_f [[\overline{v}]]_T \cdot \{\{\mu^{-1}\nabla_h \times u\}\}ds + \sum_{f \in F_h} \int_f a[[u]]_T \cdot [[\overline{v}]]_T ds. \qquad (9.147)$$

Here $\nabla_h \times$ denotes the elementwise curl, a denotes the jumps across faces f by

$$a = \alpha m^{-1} \hbar^{-1},$$

where $\hbar$ and m are respectively defined by

$$\hbar|_f = \begin{cases} \min\{h_K, h_{K'}\}, & f \in F_h^I, \ f = \partial K \cap \partial K', \\ h_K, & f \in F_h^B, \ f = \partial K \cap \partial \Omega, \end{cases}$$

and

$$m|_f = \begin{cases} \min\{\mu_K, \mu_{K'}\}, & f \in F_h^I, \ f = \partial K \cap \partial K', \\ \mu_K, & f \in F_h^B, \ f = \partial K \cap \partial \Omega. \end{cases}$$

Note that the sesquilinear form a_h of (9.147) is obtained from (9.144) with the facts that the exact solution $\boldsymbol{E} \in H_0(\text{curl}; \Omega)$ and $\mu^{-1}\nabla \times \boldsymbol{E} \in H(\text{curl}; \Omega)$, which lead to

$$[[\mu^{-1}\nabla \times \boldsymbol{E}]]_T = 0 \quad \text{and} \quad [[\boldsymbol{E}]]_T = 0.$$

The last term in a_h of (9.147) is a penalty term added in order to make the sesquilinear form of (9.146) be coercive. Notice that

$$\sum_{f \in F_h} \int_f [[\boldsymbol{u}]]_T \cdot \{\{\mu^{-1}\nabla_h \times \overline{\boldsymbol{v}}\}\} ds$$

$$= \sum_{f \in F_h} \int_f a^{1/2}[[\boldsymbol{u}]]_T \cdot a^{-1/2}m^{1/2}\hbar^{1/2}\{\{\mu^{-1}\nabla_h \times \overline{\boldsymbol{v}}\}\} ds$$

$$\leq \alpha^{-\frac{1}{2}}\left(\sum_{f \in F_h} \|a^{\frac{1}{2}}[[\boldsymbol{u}]]_T\|^2_{0,f}\right)^{\frac{1}{2}}\left(\sum_{f \in F_h} \|m^{\frac{1}{2}}\hbar^{\frac{1}{2}}\{\{\mu^{-1}\nabla_h \times \overline{\boldsymbol{v}}\}\}\|^2_{0,f}\right)^{\frac{1}{2}}, \quad (9.148)$$

and

$$a_h(\boldsymbol{v}, \boldsymbol{v}) = \sum_{K \in T_h} \|\mu^{-1/2}\nabla \times \boldsymbol{v}\|^2_{0,K} + \sum_{f \in F_h} \|a^{1/2}[[\boldsymbol{v}]]_T\|^2_{0,f}$$

$$- 2\sum_{f \in F_h} \int_f [[\boldsymbol{v}]]_T \cdot \{\{\mu^{-1}\nabla_h \times \boldsymbol{v}\}\} ds.$$

Using the definition of m and $\hbar$ and the assumption that μ is a piecewise constant, we have

$$\sum_{f \in F_h} \|m^{1/2}\hbar^{1/2}\{\{\mu^{-1}\nabla_h \times \boldsymbol{v}\}\}\|^2_{0,f}$$

$$\leq \sum_{K \in T_h} h_K \mu_K \|\mu_K^{-1}\nabla_h \times \boldsymbol{v}\|^2_{0,\partial K} = \sum_{K \in T_h} h_K \|\mu_K^{-1/2}\nabla_h \times \boldsymbol{v}\|^2_{0,\partial K}$$

$$\leq \sum_{K \in T_h} C^2_{inv}\|\mu_K^{-1/2}\nabla_h \times \boldsymbol{v}\|^2_{0,K}, \quad (9.149)$$

where we used the inverse inequality

$$\|w\|_{0,\partial K} \leq C_{inv}h_K^{-1/2}\|w\|_{0,K}, \quad \forall\, w \in (P^l(K))^3,$$

with constant C_{inv} depending only on the shape-regularity of the mesh and the approximation order l.

Finally, using the definition a_h of (9.147) and substituting (9.149) into (9.148), we obtain

$$a_h(\boldsymbol{v}, \boldsymbol{v}) = \sum_{K \in T_h} \|\mu_K^{-1/2}\nabla \times \boldsymbol{v}\|^2_{0,K} + \sum_{f \in F_h} \|a^{1/2}[[\boldsymbol{v}]]_T\|^2_{0,f}$$

$$- 2\sum_{f \in F_h} \int_f [[\boldsymbol{v}]]_T\{\{\mu^{-1}\nabla_h \times \boldsymbol{v}\}\} ds$$

$$\geq (1 - \alpha^{-1/2} C_{inv}) [\sum_{K \in T_h} ||\mu^{-1/2} \nabla \times v||^2_{0,K} + \sum_{f \in F_h} ||a^{1/2} [[v]]_T||^2_{0,f}],$$

which leads to

$$a_h(v,v) \geq \frac{1}{2} [\sum_{K \in T_h} ||\mu^{-1/2} \nabla \times v||^2_{0,K} + \sum_{f \in F_h} ||a^{1/2} [[v]]_T||^2_{0,f}], \qquad (9.150)$$

provided that $\alpha \geq \alpha_{min} = 4C^2_{inv}$.

The following optimal error estimates in the energy norm have been proved in [32].

THEOREM 9.12

Assume that the solution E of (9.136)-(9.137) satisfies the regularity assumptions $E, \nabla \times E \in (H^s(T_h))^3, s > \frac{1}{2}$, and let E_h be the DG solution to (9.146). Furthermore, we set $V(h) = H_0(curl; \Omega) + V_h$ and its corresponding norm

$$||v||^2_{V(h)} = ||\epsilon^{1/2} v||^2_{0,\Omega} + ||\mu^{-1/2} \nabla_h \times v||^2_{0,\Omega} + ||\mu^{-1/2} \hbar^{-1/2} [[v]]_T||^2_{0,F_h}. \quad (9.151)$$

Then for any $\alpha \geq \alpha_{min}$ and $0 < h \leq h_0$ (i.e., mesh size is small enough), the following optimal error estimate holds true

$$||E - E_h||_{V(h)} \leq C h^{\min(s,l)} [||\epsilon E||_{s,T_h} + ||\mu^{-1} \nabla \times E||_{s,T_h}], \qquad (9.152)$$

where we denote $l \geq 1$ the polynomial approximation degree defined in the space V_h of (9.145), and the norm $||v||_{s,T_h} = (\sum_{K \in T_h} ||v||^2_{s,K})^{1/2}$.

9.4.3 The mixed DG method

A mixed formulation for (9.136)-(9.137) was first proposed by Demkowicz et al. [21] and later was modified by Houston et al. [33, 34] for some mixed DG methods. To make the presentation clear, we assume that $\sigma = 0$ and denote $j = i\omega J_s$ in (9.136).

For any function $E \in H_0(curl; \Omega)$, we have the $L^2_\epsilon(\Omega)^3$-orthogonal Helmholtz decomposition [12]:

$$E = u + \nabla p, \quad u \in H_0(curl; \Omega) \cap H(div^0_\epsilon; \Omega), \quad p \in H^1_0(\Omega), \qquad (9.153)$$

where we let $L^2_\epsilon(\Omega)^3$ denote the space of square integrable functions on Ω equipped with the inner product $(v,w)_\epsilon = \int_\Omega \epsilon v \cdot w dx$, and we let

$$H(div^0_\epsilon; \Omega) = \{v \in L^2(\Omega)^3 : \nabla \cdot (\epsilon v) = 0 \text{ in } \Omega\}.$$

Using the decomposition (9.153), we can reformulate the problem (9.136)-(9.137) as follows [65]: Find $u \in H(curl; \Omega)$ and $p \in H^1(\Omega)$ such that

$$\nabla \times (\mu^{-1} \nabla \times u) - \omega^2 \epsilon u - \omega^2 \epsilon \nabla p = j, \quad \text{in } \Omega, \qquad (9.154)$$

$$\nabla \cdot (\epsilon \boldsymbol{u}) = 0, \quad \text{in } \Omega, \tag{9.155}$$

$$\boldsymbol{n} \times \boldsymbol{u} = 0, \quad p = 0 \quad \text{on } \partial\Omega. \tag{9.156}$$

A mixed DG formulation was presented in [65], and optimal error estimates were proved for smooth material coefficients by employing a duality approach. Subsequent work [33, 34] has shown that most of the stabilization terms introduced in [65] are unnecessary. Hence we shall present the most recent work of Houston et al. [34].

Let $V = H_0(\text{curl}; \Omega), Q = H_0^1(\Omega)$. The standard variational formulation of (9.154)-(9.156) is [21]: Find $(\boldsymbol{u}, p) \in V \times Q$ such that

$$a(\boldsymbol{u}, \boldsymbol{v}) + b(\boldsymbol{v}, p) = \int_\Omega \boldsymbol{j} \cdot \boldsymbol{v} d\boldsymbol{x}, \quad \boldsymbol{v} \in V, \tag{9.157}$$

$$b(\boldsymbol{u}, q) = 0, \quad q \in Q, \tag{9.158}$$

where the bilinear forms a and b are defined by

$$a(\boldsymbol{u}, \boldsymbol{v}) = (\mu^{-1}\nabla \times \boldsymbol{u}, \nabla \times \boldsymbol{v}) - \omega^2(\epsilon \boldsymbol{u}, \boldsymbol{v}), \tag{9.159}$$

$$b(\boldsymbol{v}, p) = -\omega^2(\epsilon \boldsymbol{v}, \nabla p). \tag{9.160}$$

To construct a mixed DG method, we define the finite element spaces:

$$V_h = P^l(T_h)^3, \quad Q_h = P^{l+1}(T_h), \quad l \geq 1. \tag{9.161}$$

A DG mixed method for (9.154)-(9.156) can be formulated as follows: Find $(\boldsymbol{u}_h.p_h) \in V_h \times Q_h$ such that

$$a_h(\boldsymbol{u}_h, \boldsymbol{v}) + b_h(\boldsymbol{v}, p_h) = \int_\Omega \boldsymbol{j} \cdot \boldsymbol{v} d\boldsymbol{x}, \quad \boldsymbol{v} \in V_h, \tag{9.162}$$

$$b_h(\boldsymbol{u}_h, q) - c_h(p_h, q) = 0, \quad q \in Q_h, \tag{9.163}$$

where we define

$$a_h(\boldsymbol{u}, \boldsymbol{v}) = (\mu^{-1}\nabla_h \times \boldsymbol{u}, \nabla_h \times \boldsymbol{v}) - \int_{F_h} [[\boldsymbol{u}]]_T \cdot \{\{\mu^{-1}\nabla_h \times \boldsymbol{v}\}\} ds$$

$$- \int_{F_h} [[\boldsymbol{v}]]_T \{\{\mu^{-1}\nabla_h \times \boldsymbol{u}\}\} ds + \int_{F_h} a[[\boldsymbol{u}]]_T [[\boldsymbol{v}]]_T ds - \omega^2(\epsilon \boldsymbol{u}, \boldsymbol{v}),$$

$$b_h(\boldsymbol{v}, p) = -\omega^2(\epsilon \boldsymbol{v}, \nabla_h p) + \omega^2 \int_{F_h} \{\{\epsilon \boldsymbol{v}\}\} \cdot [[p]]_N ds, \tag{9.164}$$

$$c_h(p, q) = \int_{F_h} c[[p]]_N \cdot [[q]]_N ds. \tag{9.165}$$

Here for simplicity, we denote $\int_{F_h} \phi(s)ds = \sum_{f \in F_h} \int_f \phi(s)ds$.

Note that a_h corresponds to the interior penalty discretization of the curl-curl operator, b_h corresponds to the usual DG formulation for the divergence operator, and c_h is the interior penalty form that enforces the continuity of p_h weakly.

We can choose the interior penalty function a as we did in the last subsection, i.e.,

$$a = \alpha m^{-1} \hbar^{-1}.$$

The penalty function c can be chosen similarly, i.e.,

$$c = \gamma e(\boldsymbol{x}) \hbar^{-1},$$

where

$$e(\boldsymbol{x}) = \begin{cases} \max\{\epsilon_K, \epsilon_{K'}\}, & \text{if } \boldsymbol{x} \in \partial K \cap \partial K', \\ \epsilon_K \ (\text{or } \epsilon_{K'}), & \text{if } \boldsymbol{x} \in \partial K \cap \partial\Omega \ (\text{or } \partial K' \cap \partial\Omega). \end{cases}$$

With delicate analysis, the following optimal energy-norm error estimates have been obtained by Houston et al. [34].

THEOREM 9.13
Assume that the solution $(\boldsymbol{u}, p)$ of (9.154)-(9.156) satisfies the regularity:

$$\epsilon \boldsymbol{u} \in H^s(T_h)^3, \quad \mu^{-1}\nabla \times \boldsymbol{u} \in H^s(T_h)^3, \quad p \in H^{s+1}(T_h), \quad s > \frac{1}{2}.$$

Then for $\alpha \geq \alpha_{min}$ and $\gamma > 0$, the mixed DG solution $(\boldsymbol{u}_h, p_h)$ of (9.162)-(9.163) satisfies the following optimal error estimates

$$|||(\boldsymbol{u} - \boldsymbol{u}_h, p - p_h)|||_{DG} \leq C h^{\min(s,l)} [|||\epsilon \boldsymbol{u}||_{s,T_h} + ||\mu^{-1}\nabla \times \boldsymbol{u}||_{s,T_h} + ||p||_{s+1,T_h}],$$

where the positive constant C is independent of the mesh size h. Here we denote the DG-norm

$$|||(\boldsymbol{v}, q)|||_{DG} = (||\boldsymbol{v}||^2_{V(h)} + ||q||^2_{Q(h)})^{1/2}, \quad V(h) = V + V_h, \quad Q(h) = Q + Q_h,$$

where

$$||\boldsymbol{v}||^2_{V(h)} = ||\mu^{-1/2}\nabla_h \times \boldsymbol{v}||^2_{0,\Omega} + ||m^{-1/2}\hbar^{-1/2}[[\boldsymbol{v}]]_T||^2_{0,F_h} + \omega^2||\epsilon^{1/2}\boldsymbol{v}||^2_{0,\Omega},$$

$$||q||^2_{Q(h)} = \omega^2||\epsilon^{1/2}\nabla_h q||^2_{0,\Omega} + ||e^{1/2}\hbar^{-1/2}[[q]]_N||^2_{0,F_h}.$$

9.5 Maxwell's equations in dispersive media

In previous sections, we discussed Maxwell's equations in a simple medium such as air in the free space. While in real life, the properties of most electromagnetic materials are wavelength dependent. Such materials are generally called dispersive media, and examples include human tissue, water, soil, snow, ice, plasma, optical fibers, and radar-absorbing materials. Some concrete applications of wave interactions with dispersive media include geophysical probing and subsurface studies of the moon and other planets; high power and ultra-wide-band (UWB) radar systems, in which it is necessary to model UWB electromagnetic pulse propagation through plasmas (i.e., the ionosphere); ground penetrating radar (GPR) detection of buried objects (such as buried waste drums, metallic or dielectric pipes, pollution plumes) in soil media; and electromagnetic wave interactions with biological and water-based substances.

In early 1990, researchers started modeling of wave propagation in dispersive media. But most work is exclusively restricted to the finite-difference time-domain (FDTD) method due to its much simpler implementation compared to the finite element methods. In 2001, Jiao and Jin [39] initiated the application of the time-domain finite element method (TDFEM) for dispersive media. Readers can find some early work on modeling of dispersive media in book chapters ([42, Ch. 8], [71, Ch. 9], and [40, Ch. 11]). In this section, we will introduce some finite element methods for solving Maxwell's equations in dispersive media based on our recent work in this area [46, 47, 48, 52, 54].

9.5.1 Isotropic cold plasma

Electromagnetic wave propagation in isotropic nonmagnetized cold electron plasma is governed by the following equations:

$$\epsilon_0 \frac{\partial E}{\partial t} = \nabla \times H - J \tag{9.166}$$

$$\mu_0 \frac{\partial H}{\partial t} = -\nabla \times E \tag{9.167}$$

$$\frac{\partial J}{\partial t} + \nu J = \epsilon_0 \omega_p^2 E \tag{9.168}$$

where E is the electric field, H is the magnetic field, ϵ_0 is the permittivity of free space, μ_0 is the permeability of free space, J is the polarization current density, ω_p is the plasma frequency, $\nu \geq 0$ is the electron-neutral collision frequency. Solving (9.168) with the assumption that the initial electron velocity is zero, we obtain

$$J(E) \equiv J(x, t; E)$$
$$= \epsilon_0 \omega_p^2 e^{-\nu t} \int_0^t e^{\nu s} E(x, s) ds = \epsilon_0 \omega_p^2 \int_0^t e^{-\nu(t-s)} E(x, s) ds. \tag{9.169}$$

Substituting (9.169) into (9.166), we obtain the following system of equations depending only on E and H:

$$\epsilon_0 E_t - \nabla \times H + J(E) = 0 \quad \text{in } \Omega \times (0, T), \tag{9.170}$$
$$\mu_0 H_t + \nabla \times E = 0 \quad \text{in } \Omega \times (0, T), \tag{9.171}$$

where $\boldsymbol{J}(\boldsymbol{E})$ is defined by (9.169).

To complete the problem, we assume a perfect conducting boundary condition of

$$\boldsymbol{n} \times \boldsymbol{E} = 0 \quad \text{on} \quad \partial\Omega \times (0,T), \tag{9.172}$$

and the simple initial conditions

$$\boldsymbol{E}(\boldsymbol{x},0) = \boldsymbol{E}_0(\boldsymbol{x}) \quad \text{and} \quad \boldsymbol{H}(\boldsymbol{x},0) = \boldsymbol{H}_0(\boldsymbol{x}) \quad \boldsymbol{x} \in \Omega, \tag{9.173}$$

where $\boldsymbol{E}_0$ and $\boldsymbol{H}_0$ are some given functions and $\boldsymbol{H}_0$ satisfies

$$\nabla \cdot (\mu_0 \boldsymbol{H}_0) = 0 \quad \text{in} \ \Omega, \quad \boldsymbol{H}_0 \cdot \boldsymbol{n} = 0 \quad \text{on} \ \partial\Omega. \tag{9.174}$$

We can solve the problem (9.170)-(9.173) by a mixed finite element method. Multiplying (9.170) by a test function ϕ and integrating over domain Ω, and then multiplying (9.171) by test function ψ and integrating over domain Ω, we obtain the weak formulation: Find the solution $(\boldsymbol{E}, \boldsymbol{H}) \in C^1(0,T;(L^2(\Omega))^3) \times C^1(0,T;H(\mathrm{curl};\Omega))$ of (9.170)-(9.171) such that

$$\epsilon_0(\boldsymbol{E}_t, \phi) - (\nabla \times \boldsymbol{H}, \phi) + (\boldsymbol{J}(\boldsymbol{E}), \phi) = 0 \quad \forall \, \phi \in (L^2(\Omega))^3, \tag{9.175}$$
$$\mu_0(\boldsymbol{H}_t, \psi) + (\boldsymbol{E}, \nabla \times \psi) = 0 \quad \forall \, \psi \in H(\mathrm{curl};\Omega) \tag{9.176}$$

for $0 < t \leq T$ with the initial conditions

$$\boldsymbol{E}(0) = \boldsymbol{E}_0 \quad \text{and} \quad \boldsymbol{H}(0) = \boldsymbol{H}_0. \tag{9.177}$$

Note that the boundary condition (9.172) is imposed weakly in (9.176).

For simplicity, we assume that Ω is a bounded and convex Lipschitz polyhedral domain in $\mathcal{R}^3$. Furthermore, we assume that Ω is discretized by a regular finite element mesh T_h, of tetrahedra of maximum diameter h.

We shall use the same mixed finite element spaces (9.84)-(9.86) developed previously for the simple medium, i.e., we can define

$$\mathbf{U}_h = \{\boldsymbol{u}_h \in (L^2(\Omega))^3 \ : \ \boldsymbol{u}_h|_K \in (P_{k-1})^3 \quad \forall \, K \in T_h\},$$
$$\mathbf{V}_h = \{\boldsymbol{v}_h \in H(\mathrm{curl};\Omega) \ : \ \boldsymbol{v}_h|_K \in R_k \quad \forall \, K \in T_h\}.$$

To construct a fully discrete scheme for (9.175)-(9.177), we divide the time interval $(0,T)$ into M uniform subintervals by points $0 = t^0 < t^1 < \cdots < t^M = T$, where $t^k = k\tau$, and denote the k-th subinterval by $I^k = (t^{k-1}, t^k]$. Moreover, we define $\boldsymbol{u}^k = \boldsymbol{u}(\cdot, k\tau)$ for $0 \leq k \leq M$, and denote the first-order backward finite difference:

$$\partial_\tau \boldsymbol{u}^k = (\boldsymbol{u}^k - \boldsymbol{u}^{k-1})/\tau.$$

Now we can formulate our fully discrete mixed finite element scheme for (9.175)-(9.177) as follows: for $k = 1, 2, \cdots, M$, find $\boldsymbol{E}_h^k \in \mathbf{U}_h, \boldsymbol{H}_h^k \in \mathbf{V}_h$ such that

$$\epsilon_0(\partial_\tau \boldsymbol{E}_h^k, \phi_h) - (\nabla \times \boldsymbol{H}_h^k, \phi_h) + (\boldsymbol{J}_h^k, \phi_h) = 0 \quad \forall \, \phi_h \in \mathbf{U}_h, \tag{9.178}$$

$$\mu_0(\partial_\tau \boldsymbol{H}_h^k, \boldsymbol{\psi}_h) + (\boldsymbol{E}_h^k, \nabla \times \boldsymbol{\psi}_h) = 0 \quad \forall \, \boldsymbol{\psi}_h \in \mathbf{V}_h \quad (9.179)$$

for $0 < t \leq T$, subject to the initial conditions

$$\boldsymbol{E}^h(0) = P_h \boldsymbol{E}_0 \quad \text{and} \quad \boldsymbol{H}^h(0) = \Pi_h \boldsymbol{H}_0. \qquad (9.180)$$

Here P_h is the standard $(L^2(\Omega))^3$ projection operator P_h onto $\mathbf{U}_h$, and Π_h is the Nédélec interpolation operator. Furthermore, $\boldsymbol{J}_h^k$ is defined by the recursive formula:

$$\boldsymbol{J}_h^0 = 0, \quad \boldsymbol{J}_h^k = e^{-\nu\tau}(\boldsymbol{J}_h^{k-1} + \epsilon_0 \omega_p^2 \tau \boldsymbol{E}_h^{k-1}), \quad k \geq 1. \qquad (9.181)$$

We can estimate the difference between $\boldsymbol{J}^k$ and $\boldsymbol{J}_h^k$ by the following.

LEMMA 9.4
Let $\boldsymbol{J}^k \equiv \boldsymbol{J}(\boldsymbol{E}(\cdot, t^k))$ defined by (9.169), $\boldsymbol{J}_h^k$ defined by (9.181). Then for any $1 \leq n \leq M$, we have

$$|\boldsymbol{J}_h^n - \boldsymbol{J}^n| \leq \epsilon_0 \omega_p^2 \tau \sum_{k=0}^{n-1} |\boldsymbol{E}_h^k - \boldsymbol{E}^k| + \epsilon_0 \omega_p^2 \tau \int_0^{t^n} |\nu \boldsymbol{E}(t) + \boldsymbol{E}_t(t)| dt.$$

Proof. [*] By definition (9.169), we have

$$\boldsymbol{J}^k \equiv \boldsymbol{J}(\boldsymbol{E}(\cdot, t^k)) = \epsilon_0 \omega_p^2 e^{-\nu t^k} \int_0^{t^k} e^{\nu s} \boldsymbol{E}(x, s) ds$$

$$= \epsilon_0 \omega_p^2 e^{-\nu t^k} \Big[\int_0^{t^{k-1}} e^{\nu s} \boldsymbol{E}(x, s) ds + \int_{t^{k-1}}^{t^k} e^{\nu s} \boldsymbol{E}(x, s) ds \Big]$$

$$= e^{-\nu\tau} \boldsymbol{J}^{k-1} + \epsilon_0 \omega_p^2 e^{-\nu t^k} \int_{I^k} e^{\nu s} \boldsymbol{E}(x, s) ds. \qquad (9.182)$$

Subtracting (9.182) from (9.181), we obtain

$$\boldsymbol{J}_h^k - \boldsymbol{J}^k$$

$$= e^{-\nu\tau}(\boldsymbol{J}_h^{k-1} - \boldsymbol{J}^{k-1}) + \epsilon_0 \omega_p^2 [\tau e^{-\nu\tau} \boldsymbol{E}_h^{k-1} - \int_{I^k} e^{-\nu(t^k - s)} \boldsymbol{E}(x, s) ds]$$

$$= e^{-\nu\tau}(\boldsymbol{J}_h^{k-1} - \boldsymbol{J}^{k-1}) + \epsilon_0 \omega_p^2 [\tau e^{-\nu\tau}(\boldsymbol{E}_h^{k-1} - \boldsymbol{E}^{k-1})$$

$$+ e^{-\nu t^k} \int_{I^k} (e^{\nu t^{k-1}} \boldsymbol{E}^{k-1} - e^{\nu s} \boldsymbol{E}(s)) ds]$$

[*]Reprinted from Lemma 2.3 of *Computer Methods in Applied Mechanics and Engineering*, Vol. 196, Jichun Li, Error analysis of fully discrete mixed finite element schemes for 3-D Maxwell's equations in dispersive media, p. 3081–3094. Copyright (2007), with permission from Elsevier.

$$= \sum_{i=1}^{3} R_i. \tag{9.183}$$

It is easy to see that

$$R_1 \le |\boldsymbol{J}_h^{k-1} - \boldsymbol{J}^{k-1}|,$$
$$R_2 = \epsilon_0 \omega_p^2 \tau e^{-\nu \tau} (\boldsymbol{E}_h^{k-1} - \boldsymbol{E}^{k-1}) \le \epsilon_0 \omega_p^2 \tau |\boldsymbol{E}_h^{k-1} - \boldsymbol{E}^{k-1}|.$$

Using the following identity

$$\int_{I^k} |f(t^{k-1}) - f(t)| dt = \int_{I^k} |\int_{t^{k-1}}^{t} f_s(s) ds| dt$$
$$\le \int_{I^k} (\int_{I^k} |f_s(s)| ds) dt = \tau \int_{I^k} |f_t(t)| dt \tag{9.184}$$

with $f(t) = e^{\nu t} \boldsymbol{E}(x,t)$, we obtain

$$R_3 \le \epsilon_0 \omega_p^2 e^{-\nu t^k} \tau \int_{I^k} |(e^{\nu t} \boldsymbol{E}(x,t))_t| dt$$
$$= \epsilon_0 \omega_p^2 \tau \int_{I^k} e^{-\nu(t^k - t)} |\nu \boldsymbol{E}(t) + \boldsymbol{E}_t(t)| dt$$
$$\le \epsilon_0 \omega_p^2 \tau \int_{I^k} |\nu \boldsymbol{E}(t) + \boldsymbol{E}_t(t)| dt.$$

Combining the above estimates, we have

$$|\boldsymbol{J}_h^k - \boldsymbol{J}^k| \le |\boldsymbol{J}_h^{k-1} - \boldsymbol{J}^{k-1}| + \epsilon_0 \omega_p^2 \tau [|\boldsymbol{E}_h^{k-1} - \boldsymbol{E}^{k-1}| + \int_{I^k} |\nu \boldsymbol{E}(t) + \boldsymbol{E}_t(t)| dt].$$

Summing both sides over $k = 1, 2, \cdots, n$, and using the fact $\boldsymbol{J}_h^0 = \boldsymbol{J}^0 = 0$ lead to

$$|\boldsymbol{J}_h^n - \boldsymbol{J}^n| \le \epsilon_0 \omega_p^2 \tau \sum_{k=1}^{n} [|\boldsymbol{E}_h^{k-1} - \boldsymbol{E}^{k-1}| + \int_{I^k} |\nu \boldsymbol{E}(t) + \boldsymbol{E}_t(t)| dt]$$
$$= \epsilon_0 \omega_p^2 \tau \sum_{k=0}^{n-1} |\boldsymbol{E}_h^k - \boldsymbol{E}^k| + \epsilon_0 \omega_p^2 \tau \int_0^{t^n} |\nu \boldsymbol{E}(t) + \boldsymbol{E}_t(t)| dt,$$

which concludes our proof. $\square$

The following optimal error estimates have been obtained in [47].

THEOREM 9.14
Let $(\boldsymbol{E}^n, \boldsymbol{H}^n)$ and $(\boldsymbol{E}_h^n, \boldsymbol{H}_h^n)$ be the solutions of (9.175)-(9.177) and (9.178)-(9.181) at time $t = t^n$, respectively. Assume that

$$\boldsymbol{E}(t), \boldsymbol{E}_t(t), \boldsymbol{H}_t(t) \in (H^l(\Omega))^3, \nabla \times \boldsymbol{E}_t(t), \nabla \times \boldsymbol{H}_t(t) \in (L^2(\Omega))^3,$$

$$\boldsymbol{H}(t), \nabla \times \boldsymbol{H}(t) \in (H^{l+1}(\Omega))^3,$$

for all $0 \leq t \leq T$. *Then there is a constant* $C = C(T, \epsilon_0, \mu_0, \omega_p, \nu, \boldsymbol{E}, \boldsymbol{H})$, *independent of both the time step* τ *and the finite element mesh size* h, *such that*

$$\max_{1 \leq n \leq M} (\|\boldsymbol{E}^n - \boldsymbol{E}_h^n\|_0 + \|\boldsymbol{H}^n - \boldsymbol{H}_h^n\|_0) \leq C(\tau + h^l), \tag{9.185}$$

where $1 \leq l \leq k$, *and* k *is the degree of basis functions in* $\mathbf{U}_h$ *and* $\mathbf{V}_h$.

REMARK 9.1   If the solutions are not smooth enough, we can obtain the following estimate

$$\max_{1 \leq n \leq M} (\|\boldsymbol{E}^n - \boldsymbol{E}_h^n\|_0 + \|\boldsymbol{H}^n - \boldsymbol{H}_h^n\|_0) \leq C(\tau + h^\alpha), \quad \frac{1}{2} < \alpha \leq 1, \tag{9.186}$$

for the lowest order Nédélec element (i.e., $k = 1$ in our definitions of $\mathbf{U}_h$ and $\mathbf{V}_h$) under the reasonable regularity assumptions [3, 15]: for any $t \in [0, T]$ and $\frac{1}{2} < \alpha \leq 1$,

$$\boldsymbol{E}(t), \boldsymbol{E}_t(t) \in (H^\alpha(\Omega))^3, \quad \boldsymbol{H}(t), \boldsymbol{H}_t(t) \in (H^\alpha(\mathrm{curl}; \Omega))^3,$$
$$\boldsymbol{E}(t), \boldsymbol{E}_t(t), \nabla \times \boldsymbol{E}_t(t), \nabla \times \boldsymbol{H}_t(t) \in (L^2(\Omega))^3.$$

Finally, the same error estimates (9.185) hold true for Nédélec $H(\mathrm{curl})$ conforming cubic elements [62] defined by

$$\mathbf{U}_h|_K = (u_1, u_2, u_3) \in Q_{k-1,k,k} \times Q_{k,k-1,k} \times Q_{k,k,k-1}, \quad k \geq 1,$$
$$\mathbf{V}_h|_K = (v_1, v_2, v_3) \in (Q_{k-1,k-1,k-1})^3, \quad k \geq 1.$$

■

9.5.2   Debye medium

For the single-pole model of Debye, the governing equations are [47]:

$$\epsilon_0 \epsilon_\infty \frac{\partial \boldsymbol{E}}{\partial t} = \nabla \times \boldsymbol{H} - \frac{1}{t_0}[(\epsilon_s - \epsilon_\infty)\epsilon_0 \boldsymbol{E} - \mathbf{P}] \tag{9.187}$$

$$\mu_0 \frac{\partial \boldsymbol{H}}{\partial t} = -\nabla \times \boldsymbol{E} \tag{9.188}$$

$$\frac{\partial \mathbf{P}}{\partial t} + \frac{1}{t_0}\mathbf{P} = \frac{(\epsilon_s - \epsilon_\infty)\epsilon_0}{t_0} \boldsymbol{E} \tag{9.189}$$

where $\mathbf{P}$ is the polarization vector, ϵ_∞ is the permittivity at infinite frequency, ϵ_s is the permittivity at zero frequency, t_0 is the relaxation time, and the rest have the same meaning as those stated previously for plasma.

We introduce the polarization current

$$\boldsymbol{J}(\mathbf{x}, t) \equiv \frac{1}{t_0}[(\epsilon_s - \epsilon_\infty)\epsilon_0 \boldsymbol{E} - \mathbf{P}],$$

from which and solving (9.189) for $\mathbf{P}$ with initial condition $\mathbf{P} = 0$, we obtain

$$\mathbf{J}(\mathbf{x}, t) = \frac{(\epsilon_s - \epsilon_\infty)\epsilon_0}{t_0} \mathbf{E}(\mathbf{x}, t) - \frac{(\epsilon_s - \epsilon_\infty)\epsilon_0}{t_0^2} \int_0^t e^{-(t-s)/t_0} \mathbf{E}(\mathbf{x}, s) ds.$$

Introducing the pseudo-polarization current

$$\tilde{\mathbf{J}}(\mathbf{E}) \equiv \tilde{\mathbf{J}}(\mathbf{x}, t; \mathbf{E}) = \frac{(\epsilon_s - \epsilon_\infty)\epsilon_0}{t_0^2} \int_0^t e^{-\frac{(t-s)}{t_0}} \mathbf{E}(\mathbf{x}, s) ds, \tag{9.190}$$

we can rewrite the governing equations for a Debye medium: Find $\mathbf{E}$ and $\mathbf{H}$, which satisfy

$$\epsilon_0\epsilon_\infty \mathbf{E}_t - \nabla \times \mathbf{H} + \frac{(\epsilon_s - \epsilon_\infty)\epsilon_0}{t_0} \mathbf{E} - \tilde{\mathbf{J}}(\mathbf{E}) = 0 \quad \text{in } \Omega \times (0, T), \tag{9.191}$$

$$\mu_0 \mathbf{H}_t + \nabla \times \mathbf{E} = 0 \quad \text{in } \Omega \times (0, T), \tag{9.192}$$

and the same boundary and initial conditions as those stated previously for plasma.

From (9.191)-(9.192), we can easily obtain the weak formulation: For any $\phi \in (L^2(\Omega))^3$ and $\psi \in H(\text{curl}; \Omega)$, find the solution $(\mathbf{E}, \mathbf{H}) \in C^1(0, T; (L^2(\Omega))^3) \times C^1(0, T; H(\text{curl}; \Omega))$ such that

$$\epsilon_0\epsilon_\infty (\mathbf{E}_t, \phi) - (\nabla \times \mathbf{H}, \phi) + \frac{(\epsilon_s - \epsilon_\infty)\epsilon_0}{t_0} (\mathbf{E}, \phi)$$

$$-(\tilde{\mathbf{J}}(\mathbf{E}), \phi) = 0, \tag{9.193}$$

$$\mu_0(\mathbf{H}_t, \psi) + (\mathbf{E}, \nabla \times \psi) = 0 \tag{9.194}$$

for $0 < t \leq T$ with the initial conditions

$$\mathbf{E}(0) = \mathbf{E}_0 \quad \text{and} \quad \mathbf{H}(0) = \mathbf{H}_0. \tag{9.195}$$

Using the same mixed finite element spaces and notation as those introduced for plasma, we can formulate the fully discrete mixed finite element scheme for our Debye model as follows: for $k = 1, 2, \cdots, M$, find $\mathbf{E}_h^k \in \mathbf{U}_h, \mathbf{H}_h^k \in \mathbf{V}_h$ such that

$$\epsilon_0\epsilon_\infty (\partial_\tau \mathbf{E}_h^k, \phi_h) - (\nabla \times \mathbf{H}_h^k, \phi_h) + \frac{(\epsilon_s - \epsilon_\infty)\epsilon_0}{t_0} (\mathbf{E}_h^k, \phi_h)$$

$$-(\tilde{\mathbf{J}}_h^k, \phi_h) = 0 \quad \forall \phi_h \in \mathbf{U}_h, \tag{9.196}$$

$$\mu_0(\partial_\tau \mathbf{H}_h^k, \psi_h) + (\mathbf{E}_h^k, \nabla \times \psi_h) = 0 \quad \forall \psi_h \in \mathbf{V}_h \tag{9.197}$$

for $0 < t \leq T$, subject to the initial conditions

$$\mathbf{E}^h(0) = P_h \mathbf{E}_0 \quad \text{and} \quad \mathbf{H}^h(0) = \Pi_h \mathbf{H}_0. \tag{9.198}$$

Here $\tilde{\boldsymbol{J}}_h^k$ is defined by the recursive formula

$$\tilde{\boldsymbol{J}}_h^0 = 0, \quad \tilde{\boldsymbol{J}}_h^k = e^{-\frac{\tau}{t_0}}(\tilde{\boldsymbol{J}}_h^{k-1} + \frac{(\epsilon_s - \epsilon_\infty)\epsilon_0}{t_0^2}\tau\boldsymbol{E}_h^{k-1}), \quad k \geq 1. \qquad (9.199)$$

Note that τ denotes the time step size as in the earlier section for plasma. The existence and uniqueness of the solution for the linear system (9.196)-(9.199) is assured with the same arguments as discussed for plasma.

First, we have the following estimate for the difference between $\tilde{\boldsymbol{J}}_h^n$ and $\tilde{\boldsymbol{J}}^n$ [47].

LEMMA 9.5
Let $\tilde{\boldsymbol{J}}^k \equiv \tilde{\boldsymbol{J}}(\boldsymbol{E}(\cdot, t^k))$ be defined by (9.190), $\tilde{\boldsymbol{J}}_h^k$ defined by (9.199). Then for any $1 \leq n \leq M$, we have

$$|\tilde{\boldsymbol{J}}_h^n - \tilde{\boldsymbol{J}}^n| \leq \frac{(\epsilon_s - \epsilon_\infty)\epsilon_0}{t_0^2}\tau[\sum_{k=0}^{n-1} |\boldsymbol{E}_h^k - \boldsymbol{E}^k| + \int_0^{t^n} |\frac{1}{t_0}\boldsymbol{E}(t) + \boldsymbol{E}_t(t)|dt].$$

For a Debye medium, the following optimal error estimates have been proved in [47].

THEOREM 9.15
Let $(\boldsymbol{E}^n, \boldsymbol{H}^n)$ and $(\boldsymbol{E}_h^n, \boldsymbol{H}_h^n)$ be the solutions of (9.193)-(9.195) and (9.196)-(9.199) at time $t = t^n$, respectively. Assume that

$$\boldsymbol{E}(t), \boldsymbol{E}_t(t), \boldsymbol{H}_t(t) \in (H^l(\Omega))^3, \nabla \times \boldsymbol{E}_t(t), \nabla \times \boldsymbol{H}_t(t) \in (L^2(\Omega))^3,$$
$$\boldsymbol{H}(t), \nabla \times \boldsymbol{H}(t) \in (H^{l+1}(\Omega))^3,$$

for all $0 \leq t \leq T$. Then there is a constant $C = C(T, \epsilon_0, \epsilon_s, \epsilon_\infty, t_0, \mu_0, \boldsymbol{E}, \boldsymbol{H})$, independent of both the time step τ and the finite element mesh size h, such that

$$\max_{1 \leq n \leq M} (\|\boldsymbol{E}^n - \boldsymbol{E}_h^n\|_0 + \|\boldsymbol{H}^n - \boldsymbol{H}_h^n\|_0) \leq C(\tau + h^l),$$

where $1 \leq l \leq k$, and k is the degree of basis functions in $\mathbf{U}_h$ and $\mathbf{V}_h$.

Finally, we like to mention that a unified DG method has been developed for Maxwell's equations in linear dispersive media in [55]. More specifically, in [55], the 2-D Maxwell equations in the coupled media with a single-pole Debye medium and the perfectly matched layer are written in conservation form and nodal basis functions are used in the DG method. But no error analysis is provided there.

9.5.3 Lorentz medium

The Lorentzian two-pole model is described by the following equations [47]:

$$\epsilon_0 \epsilon_\infty \frac{\partial \boldsymbol{E}}{\partial t} = \nabla \times \boldsymbol{H} - \boldsymbol{J}, \tag{9.200}$$

$$\mu_0 \frac{\partial \boldsymbol{H}}{\partial t} = -\nabla \times \boldsymbol{E} \tag{9.201}$$

$$\frac{\partial \boldsymbol{J}}{\partial t} = -\nu \boldsymbol{J} + (\epsilon_s - \epsilon_\infty)\epsilon_0 \omega_1^2 \boldsymbol{E} - \omega_1^2 \mathbf{P}, \tag{9.202}$$

$$\frac{\partial \mathbf{P}}{\partial t} = \boldsymbol{J} \tag{9.203}$$

where, in addition to the notation defined earlier, ω_1 is the resonant frequency, ν is the damping coefficient, $\mathbf{P}$ is the polarization vector, and $\boldsymbol{J}$ is the polarization current.

Differentiating both sides of (9.202) with respect to t and using (9.203), we obtain

$$\frac{d^2 \boldsymbol{J}}{dt^2} + \nu \frac{d\boldsymbol{J}}{dt} + \omega_1^2 \boldsymbol{J} = (\epsilon_s - \epsilon_\infty)\epsilon_0 \omega_1^2 \frac{d\boldsymbol{E}}{dt}. \tag{9.204}$$

Solving (9.204) with the fact that $\boldsymbol{E}(\mathbf{x}, 0) = 0$, we have

$$\boldsymbol{J}(\mathbf{x}, t) = \frac{(\epsilon_s - \epsilon_\infty)\epsilon_0 \omega_1^3}{\alpha} \int_0^t \boldsymbol{E}(\mathbf{x}, s) \cdot e^{-\delta(t-s)} \cdot \sin(\gamma - \alpha(t-s)) ds, \tag{9.205}$$

where we introduced the notation

$$\cos\gamma = \delta/\omega_1, \quad \sin\gamma = \alpha/\omega_1, \quad \alpha = \sqrt{\omega_1^2 - \nu^2/4}.$$

Here we assume that $\nu/2 < \omega_1$, which is the interesting case in real applications.

Substituting (9.205) into (9.200), we obtain the governing equations for $\boldsymbol{E}$ and $\boldsymbol{H}$ only:

$$\epsilon_0 \epsilon_\infty \boldsymbol{E}_t - \nabla \times \boldsymbol{H} + \hat{\boldsymbol{J}}(\boldsymbol{E}) = 0 \quad \text{in } \Omega \times (0, T), \tag{9.206}$$

$$\mu_0 \boldsymbol{H}_t + \nabla \times \boldsymbol{E} = 0 \quad \text{in } \Omega \times (0, T), \tag{9.207}$$

with the same boundary and initial conditions as those stated for plasma. For clarity, we use $\hat{\boldsymbol{J}}$ to represent the polarization current $\boldsymbol{J}$ for a Lorentz medium and rewrite (9.205) as

$$\hat{\boldsymbol{J}}(\boldsymbol{E}) = \tilde{\beta} \int_0^t e^{-\delta(t-s)} \cdot \sin(\gamma - \alpha(t-s)) \cdot \boldsymbol{E}(\mathbf{x}, s) ds, \tag{9.208}$$

$$= Im(\tilde{\beta} e^{j\gamma} \int_0^t e^{-(\delta + j\alpha)(t-s)} \boldsymbol{E}(\mathbf{x}, s) ds) \equiv Im(\overline{\boldsymbol{J}}(\boldsymbol{E})), \tag{9.209}$$

where $j = \sqrt{-1}, \tilde{\beta} = (\epsilon_s - \epsilon_\infty)\epsilon_0\omega_1^3/\sqrt{\omega_1^2 - \frac{\nu^2}{4}}$, and $Im(A)$ means the imaginary part of the complex number A.

From (9.206)-(9.207), we can easily obtain the weak formulation: find the solution
$(\boldsymbol{E}, \boldsymbol{H}) \in C^1(0, T; (L^2(\Omega))^3) \times C^1(0, T; H(\text{curl}; \Omega))$ such that

$$\epsilon_0\epsilon_\infty(\boldsymbol{E}_t, \boldsymbol{\phi}) - (\nabla \times \boldsymbol{H}, \boldsymbol{\phi}) + (\hat{\boldsymbol{J}}(\boldsymbol{E}), \boldsymbol{\phi}) = 0 \quad \forall\, \boldsymbol{\phi} \in (L^2(\Omega))^3, (9.210)$$
$$\mu_0(\boldsymbol{H}_t, \boldsymbol{\psi}) + (\boldsymbol{E}, \nabla \times \boldsymbol{\psi}) = 0 \quad \forall\, \boldsymbol{\psi} \in H(\text{curl}; \Omega) \qquad (9.211)$$

for $0 < t \leq T$ with the initial conditions

$$\boldsymbol{E}(0) = \boldsymbol{E}_0 \quad \text{and} \quad \boldsymbol{H}(0) = \boldsymbol{H}_0. \qquad (9.212)$$

Using the same mixed finite element spaces and notation as those introduced for plasma, we can formulate our fully discrete mixed finite element scheme for a Lorentz medium as follows: for $k = 1, 2, \cdots, M$, find $\boldsymbol{E}_h^k \in \boldsymbol{U}_h, \boldsymbol{H}_h^k \in \boldsymbol{V}_h$ such that

$$\epsilon_0\epsilon_\infty(\partial_\tau \boldsymbol{E}_h^k, \boldsymbol{\phi}_h) - (\nabla \times \boldsymbol{H}_h^k, \boldsymbol{\phi}_h) + (\hat{\boldsymbol{J}}_h^k, \boldsymbol{\phi}_h) = 0 \quad \forall\, \boldsymbol{\phi}_h \in \boldsymbol{U}_h, (9.213)$$
$$\mu_0(\partial_\tau \boldsymbol{H}_h^k, \boldsymbol{\psi}_h) + (\boldsymbol{E}_h^k, \nabla \times \boldsymbol{\psi}_h) = 0 \quad \forall\, \boldsymbol{\psi}_h \in \boldsymbol{V}_h \qquad (9.214)$$

for $0 < t \leq T$, subject to the initial conditions

$$\boldsymbol{E}^h(0) = P_h\boldsymbol{E}_0 \quad \text{and} \quad \boldsymbol{H}^h(0) = \Pi_h\boldsymbol{H}_0. \qquad (9.215)$$

Here we denote

$$\hat{\boldsymbol{J}}_h^k \equiv Im(\overline{\boldsymbol{J}}_h^k), \qquad (9.216)$$

where $\overline{\boldsymbol{J}}_h^k$ satisfies the following recursive identities:

$$\overline{\boldsymbol{J}}_h^0 = 0, \quad \overline{\boldsymbol{J}}_h^k = e^{-(\delta+j\alpha)\tau}(\overline{\boldsymbol{J}}_h^{k-1} + \tilde{\beta}e^{j\gamma}\tau\boldsymbol{E}_h^{k-1}), \quad j = \sqrt{-1}, \quad k \geq 1. \quad (9.217)$$

Following the same notation used for plasma, τ still denotes the time step size. The existence and uniqueness of the solution for the linear system (9.213)-(9.216) is assured with the same arguments as discussed earlier for plasma.

To carry out the error analysis for a Lorentz medium, we can prove the following lemma [47].

LEMMA 9.6
Let $\hat{\boldsymbol{J}}^k \equiv \hat{\boldsymbol{J}}(\boldsymbol{E}(\cdot, t^k))$ defined by (9.208), $\hat{\boldsymbol{J}}_h^k$ defined by (9.216). Then for any $1 \leq n \leq M$, we have

$$|\hat{\boldsymbol{J}}_h^n - \hat{\boldsymbol{J}}^n| \leq \tilde{\beta}\tau[\sum_{k=0}^{n-1} |\boldsymbol{E}_h^k - \boldsymbol{E}^k| + \int_0^{t^n} (\omega_1|\boldsymbol{E}(t)| + |\boldsymbol{E}_t(t)|)dt].$$

For a Lorentz medium, we proved the following optimal error estimates in [47].

THEOREM 9.16
Let $(\boldsymbol{E}^n, \boldsymbol{H}^n)$ and $(\boldsymbol{E}_h^n, \boldsymbol{H}_h^n)$ be the solutions of (9.206)-(9.207) and (9.213)-(9.216) at time $t = t^n$, respectively. Assume that

$$\boldsymbol{E}(t), \boldsymbol{E}_t(t), \boldsymbol{H}_t(t) \in (H^l(\Omega))^3, \nabla \times \boldsymbol{E}_t(t), \nabla \times \boldsymbol{H}_t(t) \in (L^2(\Omega))^3,$$
$$\boldsymbol{H}(t), \nabla \times \boldsymbol{H}(t) \in (H^{l+1}(\Omega))^3,$$

for all $0 \leq t \leq T$. Then there is a constant $C = C(T, \epsilon_0, \epsilon_s, \epsilon_\infty, \nu, \mu_0, \omega_1, \boldsymbol{E}, \boldsymbol{H})$, independent of both the time step τ and the finite element mesh size h, such that

$$\max_{1 \leq n \leq M} (\|\boldsymbol{E}^n - \boldsymbol{E}_h^n\|_0 + \|\boldsymbol{H}^n - \boldsymbol{H}_h^n\|_0) \leq C(\tau + h^l),$$

where $1 \leq l \leq k$, and k is the degree of basis functions in $\mathbf{U}_h$ and $\mathbf{V}_h$.

9.5.4 Double-negative metamaterials

In 1968, Veselago postulated the existence of electromagnetic material in which both permittivity and permeability were negative real values. Until 2000, such materials were realized in practice by arranging periodic arrays of small metallic wires and split-ring resonators [68, 70]. These artificially structured periodic media (often termed as metamaterials) have very unusual electromagnetic properties and open great potential applications in diverse areas such as interconnects for wireless telecommunications, radar and defense, nanolithography with light, medical imaging with super-resolution, and so on [22].

Here we will focus on the modeling of double-negative (DNG) metamaterials, in which both permittivity and permeability are negative. Note that DNG materials are also referred to as left-hand (LH) media, negative-index materials (NIM), and backward-wave media (BW). Furthermore, DNG metamaterials are dispersive and are often described by lossy Drude polarization and magnetization models.

The governing equations for modeling the wave propagation in a DNG medium are given by:

$$\epsilon_0 \frac{\partial \boldsymbol{E}}{\partial t} = \nabla \times \boldsymbol{H} - \boldsymbol{J}, \tag{9.218}$$

$$\mu_0 \frac{\partial \boldsymbol{H}}{\partial t} = -\nabla \times \boldsymbol{E} - \boldsymbol{K}, \tag{9.219}$$

$$\frac{\partial \boldsymbol{J}}{\partial t} + \Gamma_e \boldsymbol{J} = \epsilon_0 \omega_{pe}^2 \boldsymbol{E}, \tag{9.220}$$

$$\frac{\partial \boldsymbol{K}}{\partial t} + \Gamma_m \boldsymbol{K} = \mu_0 \omega_{pm}^2 \boldsymbol{H}, \tag{9.221}$$

where $\boldsymbol{J}$ and $\boldsymbol{K}$ are the induced electric and magnetic currents, respectively. Furthermore, ω_{pe} and ω_{pm} are the electric and magnetic plasma frequencies, and Γ_e and Γ_m are the electric and magnetic damping frequencies.

Solving (9.220) and (9.221) with initial electric and magnetic currents $\boldsymbol{J}_0(\boldsymbol{x})$ and $\boldsymbol{K}_0(\boldsymbol{x})$, respectively, we obtain

$$\boldsymbol{J}(\boldsymbol{x},t;\boldsymbol{E}) = \boldsymbol{J}_0(\boldsymbol{x}) + \epsilon_0\omega_{pe}^2 \int_0^t e^{-\Gamma_e(t-s)} \boldsymbol{E}(\boldsymbol{x},s)ds$$
$$\equiv \boldsymbol{J}_0(\boldsymbol{x}) + \boldsymbol{J}(\boldsymbol{E}), \tag{9.222}$$

$$\boldsymbol{K}(\boldsymbol{x},t;\boldsymbol{H}) = \boldsymbol{K}_0(\boldsymbol{x}) + \mu_0\omega_{pm}^2 \int_0^t e^{-\Gamma_m(t-s)} \boldsymbol{H}(\boldsymbol{x},s)ds$$
$$\equiv \boldsymbol{K}_0(\boldsymbol{x}) + \boldsymbol{K}(\boldsymbol{H}). \tag{9.223}$$

In summary, the model equations for wave propagation in a DNG medium become: find $(\boldsymbol{E},\boldsymbol{H})$ such that

$$\epsilon_0\boldsymbol{E}_t - \nabla \times \boldsymbol{H} + \boldsymbol{J}(\boldsymbol{E}) = -\boldsymbol{J}_0(\boldsymbol{x}) \quad \text{in } \Omega \times (0,T), \tag{9.224}$$
$$\mu_0\boldsymbol{H}_t + \nabla \times \boldsymbol{E} + \boldsymbol{K}(\boldsymbol{H}) = -\boldsymbol{K}_0(\boldsymbol{x}) \quad \text{in } \Omega \times (0,T). \tag{9.225}$$

subject to the same boundary and initial conditions as those stated earlier for plasma.

Multiplying (9.224) by a test function $\boldsymbol{\phi} \in (L^2(\Omega))^3$, Eq. (9.225) by $\boldsymbol{\psi} \in H(\text{curl};\Omega)$, and integrating over Ω, we can obtain the weak formulation for (9.222)-(9.223): For any $\boldsymbol{\phi} \in (L^2(\Omega))^3$ and $\boldsymbol{\psi} \in H(\text{curl};\Omega)$, find $(\boldsymbol{E},\boldsymbol{H}) \in [C^1(0,T;(L^2(\Omega))^3) \cap C^0(0,T;H(\text{curl};\Omega))]^2$ of (9.224)-(9.225) such that

$$\epsilon_0(\boldsymbol{E}_t,\boldsymbol{\phi}) - (\nabla \times \boldsymbol{H},\boldsymbol{\phi}) + (\boldsymbol{J}(\boldsymbol{E}),\boldsymbol{\phi}) = (-\boldsymbol{J}_0,\boldsymbol{\phi}), \tag{9.226}$$
$$\mu_0(\boldsymbol{H}_t,\boldsymbol{\psi}) + (\boldsymbol{E},\nabla \times \boldsymbol{\psi}) + (\boldsymbol{K}(\boldsymbol{H}),\boldsymbol{\psi}) = (-\boldsymbol{K}_0,\boldsymbol{\psi}). \tag{9.227}$$

A fully discrete mixed finite element scheme for (9.224)-(9.225) can be formulated as follows: For $k = 1,2,\cdots,M$, and any $\boldsymbol{\phi}_h \in \boldsymbol{U}_h, \boldsymbol{\psi}_h \in \boldsymbol{V}_h$, find $\boldsymbol{E}_h^k \in \boldsymbol{U}_h, \boldsymbol{H}_h^k \in \boldsymbol{V}_h$ such that

$$\epsilon_0(\partial_\tau \boldsymbol{E}_h^k,\boldsymbol{\phi}_h) - (\nabla \times \boldsymbol{H}_h^k,\boldsymbol{\phi}_h) + (\boldsymbol{J}_h^{k-1},\boldsymbol{\phi}_h) = (-\boldsymbol{J}_0,\boldsymbol{\phi}_h), \tag{9.228}$$
$$\mu_0(\partial_\tau \boldsymbol{H}_h^k,\boldsymbol{\psi}_h) + (\boldsymbol{E}_h^k,\nabla \times \boldsymbol{\psi}_h) + (\boldsymbol{K}_h^{k-1},\boldsymbol{\psi}_h) = (-\boldsymbol{K}_0,\boldsymbol{\psi}_h) \tag{9.229}$$

for $0 < t \le T$, subject to the initial conditions

$$\boldsymbol{E}^h(0) = P_h\boldsymbol{E}_0 \quad \text{and} \quad \boldsymbol{H}^h(0) = \Pi_h\boldsymbol{H}_0. \tag{9.230}$$

Here the spaces $\boldsymbol{U}_h$ and $\boldsymbol{V}_h$, and the operators P_h and Π_h have the same meaning as we defined for plasma. Furthermore, $\boldsymbol{J}_h^k$ and $\boldsymbol{K}_h^k$ are recursively defined as

$$\boldsymbol{J}_h^0 = 0, \quad \boldsymbol{J}_h^k = e^{-\Gamma_e\tau}\boldsymbol{J}_h^{k-1} + \epsilon_0\omega_{pe}^2\tau\boldsymbol{E}_h^k, \quad k \ge 1, \tag{9.231}$$
$$\boldsymbol{K}_h^0 = 0, \quad \boldsymbol{K}_h^k = e^{-\Gamma_m\tau}\boldsymbol{K}_h^{k-1} + \mu_0\omega_{pm}^2\tau\boldsymbol{H}_h^k, \quad k \ge 1. \tag{9.232}$$

Note that (9.231) is a little different from (9.181). Using the same technique as Lemmas 9.4, we can still prove the following lemma [46].

LEMMA 9.7
Let $\boldsymbol{J}^k \equiv \boldsymbol{J}(\boldsymbol{E}(\cdot, t^k))$ and $\boldsymbol{K}^k \equiv \boldsymbol{K}(\boldsymbol{H}(\cdot, t^k))$ defined by (9.222) and (9.223), $\boldsymbol{J}_h^k$ and $\boldsymbol{K}_h^k$ defined by (9.231) and (9.232), respectively. Then for any $1 \leq n \leq M$, we have

(i) $|\boldsymbol{J}_h^n - \boldsymbol{J}^n| \leq \epsilon_0 \omega_{pe}^2 \tau \sum_{k=1}^{n} |\boldsymbol{E}_h^k - \boldsymbol{E}^k| + \epsilon_0 \omega_{pe}^2 \tau \int_0^{t^n} |\Gamma_e \boldsymbol{E}(t) + \boldsymbol{E}_t(t)| dt,$

(ii) $|\boldsymbol{K}_h^n - \boldsymbol{K}^n| \leq \epsilon_0 \omega_{pm}^2 \tau \sum_{k=1}^{n} |\boldsymbol{H}_h^k - \boldsymbol{H}^k| + \epsilon_0 \omega_{pm}^2 \tau \int_0^{t^n} |\Gamma_m \boldsymbol{H}(t) + \boldsymbol{H}_t(t)| dt.$

To prove our main result, we need another lemma [46] below.

LEMMA 9.8
Let $B = H^1(curl; \Omega)$ or $B = (H^\alpha(\Omega))^3$ with $\alpha \geq 0$. For any $\boldsymbol{u} \in H^1(0, T; B)$, we have the following estimates

$$(i) \quad \|\partial_\tau \boldsymbol{u}^k\|_B^2 \leq \frac{1}{\tau} \int_{t^{k-1}}^{t^k} \|\boldsymbol{u}_t(t)\|_B^2 dt, \qquad (9.233)$$

$$(ii) \quad \|\boldsymbol{u}^k - \frac{1}{\tau} \int_{I^k} \boldsymbol{u}(t) dt\|_B^2 \leq \tau \int_{I^k} \|\boldsymbol{u}_t(t)\|_B^2 dt, \qquad (9.234)$$

$$(iii) \quad \|\boldsymbol{u}^{k-1} - \frac{1}{\tau} \int_{I^k} \boldsymbol{u}(t) dt\|_B^2 \leq \tau \int_{I^k} \|\boldsymbol{u}_t(t)\|_B^2 dt. \qquad (9.235)$$

Furthermore, we have the optimal error estimate.

THEOREM 9.17
Let $(\boldsymbol{E}^n, \boldsymbol{H}^n)$ and $(\boldsymbol{E}_h^n, \boldsymbol{H}_h^n)$ be the solutions of (9.226)-(9.227) and (9.228)-(9.232) at time $t = t^n$, respectively. Assume that

$$\boldsymbol{E}(t), \boldsymbol{E}_t(t), \boldsymbol{H}_t(t) \in (H^l(\Omega))^3, \nabla \times \boldsymbol{E}_t(t), \nabla \times \boldsymbol{H}_t(t) \in (L^2(\Omega))^3,$$
$$\boldsymbol{H}(t), \nabla \times \boldsymbol{H}(t) \in (H^{l+1}(\Omega))^3,$$

for all $0 \leq t \leq T$. Then there is a positive constant

$$C = C(T, \epsilon_0, \mu_0, \omega_{pe}, \omega_{pm}, \Gamma_e, \Gamma_m, \boldsymbol{E}, \boldsymbol{H}),$$

independent of both the time step τ and the finite element mesh size h, such that

$$\max_{1 \leq n \leq M} (\|\boldsymbol{E}^n - \boldsymbol{E}_h^n\|_0 + \|\boldsymbol{H}^n - \boldsymbol{H}_h^n\|_0) \leq C(\tau + h^l),$$

where $1 \leq l \leq k$, and k is the degree of basis functions in $\mathbf{U}_h$ and $\mathbf{V}_h$.

Proof. [†] Integrating the weak formulation (9.226)-(9.227) in time over I^k and choosing $\phi = \frac{1}{\tau}\phi_h, \psi = \frac{1}{\tau}\psi_h$ leads to

$$\frac{\epsilon_0}{\tau}(\mathbf{E}^k - \mathbf{E}^{k-1}, \phi_h) - (\nabla \times \frac{1}{\tau}\int_{I^k} \mathbf{H}(s)ds, \phi_h) + (\frac{1}{\tau}\int_{I^k} \mathbf{J}(\mathbf{E}(s))ds, \phi_h) = 0,$$

$$\frac{\mu_0}{\tau}(\mathbf{H}^k - \mathbf{H}^{k-1}, \psi_h) + (\frac{1}{\tau}\int_{I^k} \mathbf{E}(s)ds, \nabla \times \psi_h) + (\frac{1}{\tau}\int_{I^k} \mathbf{K}(\mathbf{H}(s))ds, \psi_h) = 0,$$

for any $\phi_h \in \mathbf{U}_h$ and $\psi_h \in \mathbf{V}_h$. Then from each subtracting, (9.228) and (9.229), respectively, we obtain the error equations

$$\epsilon_0(\partial_\tau(\mathbf{E}^k - \mathbf{E}_h^k), \phi_h) - (\nabla \times (\frac{1}{\tau}\int_{I^k} \mathbf{H}(s)ds - \mathbf{H}_h^k), \phi_h)$$

$$+(\frac{1}{\tau}\int_{I^k} \mathbf{J}(\mathbf{E}(s))ds - \mathbf{J}_h^{k-1}, \phi_h) = 0 \quad \forall \phi_h \in \mathbf{U}_h, \qquad (9.236)$$

$$\mu_0(\partial_\tau(\mathbf{H}^k - \mathbf{H}_h^k), \psi_h) + (\frac{1}{\tau}\int_{I^k} \mathbf{E}(s)ds - \mathbf{E}_h^k, \nabla \times \psi_h)$$

$$+(\frac{1}{\tau}\int_{I^k} \mathbf{K}(\mathbf{H}(s))ds - \mathbf{K}_h^{k-1}, \psi_h) = 0 \quad \forall \psi_h \in \mathbf{V}_h. \qquad (9.237)$$

Denote $\xi_h^k = P_h\mathbf{E}^k - \mathbf{E}_h^k, \eta_h^k = \Pi_h\mathbf{H}^k - \mathbf{H}_h^k$. Then choosing $\phi_h = \xi_h^k, \psi_h = \eta_h^k$ in (9.236)-(9.237), the above error equations can be rewritten as

$$\epsilon_0(\partial_\tau\xi_h^k, \xi_h^k) - (\nabla \times \eta_h^k, \xi_h^k)$$

$$= \epsilon_0(\partial_\tau(P_h\mathbf{E}^k - \mathbf{E}^k), \xi_h^k) - (\nabla \times (\Pi_h\mathbf{H}^k - \frac{1}{\tau}\int_{I^k} \mathbf{H}(s)ds), \xi_h^k)$$

$$+(\mathbf{J}_h^{k-1} - \mathbf{J}^{k-1}, \xi_h^k) + (\mathbf{J}^{k-1} - \frac{1}{\tau}\int_{I^k} \mathbf{J}(\mathbf{E}(s))ds, \xi_h^k),$$

$$\mu_0(\partial_\tau\eta_h^k, \eta_h^k) + (\xi_h^k, \nabla \times \eta_h^k)$$

$$= \mu_0(\partial_\tau(\Pi_h\mathbf{H}^k - \mathbf{H}^k), \eta_h^k) + (P_h\mathbf{E}^k - \frac{1}{\tau}\int_{I^k} \mathbf{E}(s)ds, \nabla \times \eta_h^k)$$

$$+(\mathbf{K}_h^{k-1} - \mathbf{K}^{k-1}, \eta_h^k) + (\mathbf{K}^{k-1} - \frac{1}{\tau}\int_{I^k} \mathbf{K}(\mathbf{H}(s))ds, \eta_h^k).$$

Adding the above two equations together, then multiplying both sides of the resultant by τ, and using the inequality

$$a(a - b) \geq \frac{1}{2}(a^2 - b^2),$$

[†]Reprinted from Sec. 4.2 of *Journal of Computational and Applied Mathematics*, Vol. 209, Jichun Li, Error analysis of mixed finite element methods for wave propagation in double negative metamaterials, p. 81–96. Copyright (2006), with permission from Elsevier.

we obtain

$$\frac{\epsilon_0}{2}(\|\xi_h^k\|_0^2 - \|\xi_h^{k-1}\|_0^2) + \frac{\mu_0}{2}(\|\eta_h^k\|_0^2 - \|\eta_h^{k-1}\|_0^2)$$

$$\leq \epsilon_0\tau(\partial_\tau(P_h\boldsymbol{E}^k - \boldsymbol{E}^k), \xi_h^k) - \tau(\nabla \times (\Pi_h\boldsymbol{H}^k - \boldsymbol{H}^k), \xi_h^k)$$

$$-\tau(\nabla \times (\boldsymbol{H}^k - \frac{1}{\tau}\int_{I^k}\boldsymbol{H}(s)ds), \xi_h^k) + \tau(\boldsymbol{J}_h^{k-1} - \boldsymbol{J}^{k-1}, \xi_h^k)$$

$$+\tau(\boldsymbol{J}^{k-1} - \frac{1}{\tau}\int_{I^k}\boldsymbol{J}(\boldsymbol{E}(s))ds, \xi_h^k) + \mu_0\tau(\partial_\tau(\Pi_h\boldsymbol{H}^k - \boldsymbol{H}^k), \eta_h^k)$$

$$+\tau(P_h\boldsymbol{E}^k - \boldsymbol{E}^k, \nabla \times \eta_h^k) + \tau(\boldsymbol{E}^k - \frac{1}{\tau}\int_{I^k}\boldsymbol{E}(s)ds, \nabla \times \eta_h^k)$$

$$+\tau(\boldsymbol{K}_h^{k-1} - \boldsymbol{K}^{k-1}, \eta_h^k) + \tau(\boldsymbol{K}^{k-1} - \frac{1}{\tau}\int_{I^k}\boldsymbol{K}(\boldsymbol{H}(s))ds, \eta_h^k)$$

$$= \sum_{i=1}^{10}(I)_i. \tag{9.238}$$

In the rest we shall estimate $(I)_i$ one by one for $i = 1, 2, \cdots, 10$.

Using the Cauchy-Schwarz inequality, Lemma 9.8 and the projection error estimate, we have

$$(I)_1 \leq \frac{1}{2}\epsilon_0\tau\|\xi_h^k\|_0^2 + \frac{1}{2}\epsilon_0\tau\|\partial_\tau(P_h\boldsymbol{E}^k - \boldsymbol{E}^k)\|_0^2$$

$$\leq \frac{1}{2}\epsilon_0\tau\|\xi_h^k\|_0^2 + \frac{1}{2}\epsilon_0\int_{I^k}\|(P_h\boldsymbol{E} - \boldsymbol{E})_t(t)\|_0^2dt$$

$$\leq \frac{1}{2}\epsilon_0\tau\|\xi_h^k\|_0^2 + C\epsilon_0 h^{2l}\int_{I^k}\|\boldsymbol{E}_t(t)\|_l^2dt.$$

Similarly, with the interpolation error estimate, we can easily obtain

$$(I)_2 \leq \frac{1}{2}\tau\|\xi_h^k\|_0^2 + \frac{1}{2}\tau\|\nabla \times (\Pi_h\boldsymbol{H}^k - \boldsymbol{H}^k)\|_0^2$$

$$\leq \frac{1}{2}\tau\|\xi_h^k\|_0^2 + C\tau h^{2l}\|\nabla \times \boldsymbol{H}^k\|_{l+1}^2.$$

By the Cauchy-Schwarz inequality and Lemma 9.8, we have

$$(I)_3 \leq \frac{1}{2}\tau\|\xi_h^k\|_0^2 + \frac{1}{2}\tau\|\nabla \times (\boldsymbol{H}^k - \frac{1}{\tau}\int_{I^k}\boldsymbol{H}(s)ds)\|_0^2$$

$$= \frac{1}{2}\tau\|\xi_h^k\|_0^2 + \frac{1}{2}\tau\|(\nabla \times \boldsymbol{H})^k - \frac{1}{\tau}\int_{I^k}\nabla \times \boldsymbol{H}(s)ds\|_0^2$$

$$\leq \frac{1}{2}\tau\|\xi_h^k\|_0^2 + \frac{1}{2}\tau^2\int_{I^k}\|\nabla \times \boldsymbol{H}_t(t)\|_0^2dt.$$

Using the Cauchy-Schwarz inequality and Lemma 9.7, we have

$$(I)_4 \leq \frac{1}{2}\tau\|\xi_h^k\|_0^2 + \frac{1}{2}\tau\|\boldsymbol{J}_h^{k-1} - \boldsymbol{J}^{k-1}\|_0^2$$

$$\leq \frac{1}{2}\tau||\xi_h^k||_0^2 + C\tau^3||\sum_{j=1}^{k-1}|\boldsymbol{E}_h^j - \boldsymbol{E}^j| + \int_0^{t^{k-1}}|\boldsymbol{\Gamma}_e\boldsymbol{E}(t) + \boldsymbol{E}_t(t)|dt||_0^2.$$

Using the Cauchy-Schwarz inequality and inequalities

$$|a+b|^2 \leq 2(a^2+b^2), \quad |\sum_{j=1}^k a_jb_j|^2 \leq (\sum_{j=1}^k |a_j|^2)(\sum_{j=1}^k |b_j|^2),$$

we can obtain

$$\tau^3|\sum_{j=1}^{k-1}|\boldsymbol{E}_h^j - \boldsymbol{E}^j| + \int_0^{t^{k-1}}|\boldsymbol{\Gamma}_e\boldsymbol{E}(t) + \boldsymbol{E}_t(t)|dt|^2$$

$$\leq 2\tau^3[(\sum_{j=1}^{k-1}1^2)(\sum_{j=1}^{k-1}|\boldsymbol{E}_h^j - \boldsymbol{E}^j|^2) + (\int_0^{t^{k-1}}1^2dt)(\int_0^{t^{k-1}}|\boldsymbol{\Gamma}_e\boldsymbol{E}(t) + \boldsymbol{E}_t(t)|^2dt]$$

$$\leq 2\tau^3[(k-1)\sum_{j=1}^{k-1}|\boldsymbol{E}_h^j - \boldsymbol{E}^j|^2 + T\int_0^T|\boldsymbol{\Gamma}_e\boldsymbol{E}(t) + \boldsymbol{E}_t(t)|^2dt]$$

$$\leq C\tau^2[\sum_{j=1}^{k-1}(|\xi_h^j|^2 + |P_h\boldsymbol{E}^j - \boldsymbol{E}^j|^2) + \tau\int_0^T|\boldsymbol{\Gamma}_e\boldsymbol{E}(t) + \boldsymbol{E}_t(t)|^2dt],$$

where in the last step we used the triangle inequality, the fact that $k\tau \leq T$, and absorbed the dependence of T into the generic constant C.

Hence, integrating the above inequality in Ω gives

$$\tau^3||\sum_{j=1}^{k-1}|\boldsymbol{E}_h^j - \boldsymbol{E}^j| + \int_0^{t^{k-1}}|\boldsymbol{\Gamma}_e\boldsymbol{E}(t) + \boldsymbol{E}_t(t)|dt||_0^2$$

$$\leq C\tau^2[\sum_{j=1}^{k-1}(||\xi_h^j||_0^2 + h^{2l}||\boldsymbol{E}^j||_l^2) + \tau\int_0^T||\boldsymbol{\Gamma}_e\boldsymbol{E}(t) + \boldsymbol{E}_t(t)||_0^2dt]$$

$$\leq C\tau^2\sum_{j=1}^{k-1}||\xi_h^j||_0^2 + C\tau h^{2l}\max_{0\leq t\leq T}||\boldsymbol{E}(t)||_l^2 + C\tau^3\int_0^T||\boldsymbol{\Gamma}_e\boldsymbol{E}(t) + \boldsymbol{E}_t(t)||_0^2dt,$$

where we used the fact that $k\tau \leq T$ for any $1 \leq k \leq M$.

Therefore, we have

$$(I)_4 \leq \frac{1}{2}\tau||\xi_h^k||_0^2 + C\tau^2\sum_{j=1}^{k-1}||\xi_h^j||_0^2 + C\tau h^{2l}\max_{0\leq t\leq T}||\boldsymbol{E}(t)||_l^2$$

$$+C\tau^3\int_0^T||\boldsymbol{\Gamma}_e\boldsymbol{E}(t) + \boldsymbol{E}_t(t)||_0^2dt.$$

Using the Cauchy-Schwarz inequality and Lemma 9.8, we have

$$(I)_5 \le \frac{1}{2}\tau||\xi_h^k||_0^2 + \frac{1}{2}\tau||\boldsymbol{J}^{k-1} - \frac{1}{\tau}\int_{I^k}\boldsymbol{J}(\boldsymbol{E}(s))ds||_0^2$$

$$\le \frac{1}{2}\tau||\xi_h^k||_0^2 + \frac{1}{2}\tau^2\int_{I^k}||\boldsymbol{J}_t(\boldsymbol{E}(t))||_0^2 dt$$

From the definition of $\boldsymbol{J}$, we have

$$\boldsymbol{J}_t(\boldsymbol{E}) = -\Gamma_e\boldsymbol{J} + \epsilon_0\omega_{pe}^2\boldsymbol{E} = -\Gamma_e\epsilon_0\omega_{pe}^2\int_0^t e^{-\Gamma_e(t-s)}\boldsymbol{E}(s)ds + \epsilon_0\omega_{pe}^2\boldsymbol{E},$$

integrating which in Ω and using the Cauchy-Schwarz inequality, we can obtain

$$||\boldsymbol{J}_t(\boldsymbol{E}(t))||_0^2 \le C||\int_0^t e^{-\Gamma_e(t-s)}\boldsymbol{E}(s)ds||_0^2 + C||\boldsymbol{E}(t)||_0^2$$

$$\le C(\int_\Omega\int_0^t |e^{-\Gamma_e(t-s)}|^2 ds d\Omega)(\int_0^t ||\boldsymbol{E}(s)||_0^2 ds) + C||\boldsymbol{E}(t)||_0^2$$

$$\le C(\int_0^T ||\boldsymbol{E}(s)||_0^2 ds + ||\boldsymbol{E}(t)||_0^2),$$

where we have absorbed the dependence of $\Gamma_e, \epsilon_0, \omega_{pe}$ and T into the generic constant C.

Hence, we have

$$(I)_5 \le \frac{1}{2}\tau||\xi_h^k||_0^2 + C\tau^3(\int_0^T ||\boldsymbol{E}(s)||_0^2 ds + ||\boldsymbol{E}(t)||_0^2).$$

Similarly, using Lemma 9.8, we obtain

$$(I)_6 \le \frac{1}{2}\mu_0\tau||\eta_h^k||_0^2 + \frac{1}{2}\mu_0\tau||\partial_\tau(\Pi_h\boldsymbol{H}^k - \boldsymbol{H}^k)||_0^2$$

$$\le \frac{1}{2}\mu_0\tau||\eta_h^k||_0^2 + \frac{1}{2}\mu_0\int_{I^k}||(\Pi_h\boldsymbol{H} - \boldsymbol{H})_t(t)||_0^2 dt$$

$$\le \frac{1}{2}\mu_0\tau||\eta_h^k||_0^2 + Ch^{2l}\int_{I^k}||\boldsymbol{H}_t(t)||_l^2 dt.$$

Using the fact $\nabla \times \eta_h^k \subset \boldsymbol{U}_h^k$, and the definition of P_h, we have

$$(I)_7 = \tau(P_h\boldsymbol{E}^k - \boldsymbol{E}^k, \nabla \times \eta_h^k) = 0.$$

Using the perfect conducting boundary condition and integration by parts, we have

$$(I)_8 = \tau(\nabla \times (\boldsymbol{E}^k - \frac{1}{\tau}\int_{I^k}\boldsymbol{E}(s)ds), \eta_h^k)$$

$$\leq \frac{1}{2}\tau||\eta_h^k||_0^2 + \frac{1}{2}\tau||(\nabla \times \boldsymbol{E})^k - \frac{1}{\tau}\int_{I^k}(\nabla \times \boldsymbol{E})(s)ds||_0^2$$

$$\leq \frac{1}{2}\tau||\eta_h^k||_0^2 + \frac{1}{2}\tau^2\int_{I^k}||\nabla \times \boldsymbol{E}_t(t)||_0^2dt.$$

The estimates of $(I)_9$ and $(I)_{10}$ can be carried out with the same procedures as that of $(I)_4$ and $(I)_5$, i.e., we shall have

$$(I)_9 \leq \frac{1}{2}\tau||\eta_h^k||_0^2 + C\tau^2\sum_{j=1}^{k-1}||\eta_h^j||_0^2 + C\tau h^{2l}\max_{0\leq t\leq T}||\boldsymbol{H}(t)||_l^2$$

$$+C\tau^3\int_0^T ||\Gamma_m\boldsymbol{H}(t) + \boldsymbol{H}_t(t)||_0^2dt,$$

$$(I)_{10} \leq \frac{1}{2}\tau||\eta_h^k||_0^2 + C\tau^3(\int_0^T ||\boldsymbol{H}(s)||_0^2ds + ||\boldsymbol{H}(t)||_0^2).$$

Summing both sides of (9.238) over $k = 1, 2, \cdots, n$, using the estimates obtained above for $(I)_i, 1 \leq i \leq 10$, and the fact $\sum_{k=1}^n \int_{I^k} a\,dt = \int_0^{t^n} a\,dt \leq \int_0^T |a|dt$, we obtain

$$\frac{\epsilon_0}{2}||\xi_h^n||_0^2 + \frac{\mu_0}{2}||\eta_h^n||_0^2$$

$$\leq \frac{\epsilon_0}{2}||\xi_h^0||_0^2 + \frac{\mu_0}{2}||\eta_h^0||_0^2 + C_1\tau\sum_{k=1}^n(||\xi_h^k||_0^2 + ||\eta_h^k||_0^2)$$

$$+Ch^{2l}\int_0^T ||\boldsymbol{E}_t(t)||_l^2dt + Ch^{2l}\max_{0\leq t\leq T}||(\nabla \times \boldsymbol{H})(t)||_{l+1}^2$$

$$+C\tau^2\int_0^T ||\nabla \times \boldsymbol{H}_t(t)||_0^2dt + C\sum_{k=1}^n[\tau^2\sum_{j=1}^{k-1}||\xi_h^j||_0^2 + \tau h^{2l}\max_{0\leq t\leq T}||\boldsymbol{E}(t)||_l^2$$

$$+\tau^3\int_0^T ||\Gamma_e\boldsymbol{E}(t) + \boldsymbol{E}_t(t)||_0^2dt] + C\sum_{k=1}^n\tau^3(\int_0^T ||\boldsymbol{E}(s)||_0^2ds + ||\boldsymbol{E}(t)||_0^2)$$

$$+Ch^{2l}\int_0^T ||\boldsymbol{H}_t(t)||_l^2dt + C\tau^2\int_0^T ||\nabla \times \boldsymbol{E}_t(t)||_0^2dt$$

$$+C\sum_{k=1}^n[\tau^2\sum_{j=1}^{k-1}||\eta_h^j||_0^2 + \tau h^{2l}\max_{0\leq t\leq T}||\boldsymbol{H}(t)||_l^2$$

$$+\tau^3\int_0^T ||\Gamma_m\boldsymbol{H}(t) + \boldsymbol{H}_t(t)||_0^2dt]$$

$$+C\sum_{k=1}^n\tau^3(\int_0^T ||\boldsymbol{H}(s)||_0^2ds + ||\boldsymbol{H}(t)||_0^2). \tag{9.239}$$

Absorbing the dependence of $\boldsymbol{E}$ and $\boldsymbol{H}$ into the generic constant C, and using the fact that $\xi_h^0 = \eta_h^0 = 0$ due to our initial assumptions and $n\tau \leq T$, we can simply rewrite (9.239) as

$$||\xi_h^n||_0^2 + ||\eta_h^n||_0^2 \le C(\tau^2 + h^{2l}) + C_2\tau \sum_{k=1}^{n}(||\xi_h^k||_0^2 + ||\eta_h^k||_0^2), \qquad (9.240)$$

which is equivalent to

$$||\xi_h^n||_0^2 + ||\eta_h^n||_0^2 \le C(\tau^2 + h^{2l}) + C_3\tau \sum_{k=1}^{n-1}(||\xi_h^k||_0^2 + ||\eta_h^k||_0^2), \qquad (9.241)$$

where we used the Cauchy-Schwarz inequality for the $k = n$ term in the summation and absorbed $||\xi_h^n||_0^2$ and $||\eta_h^n||_0^2$ into the left side.

Using the Gronwall inequality, we finally obtain

$$||\xi_h^n||_0^2 + ||\eta_h^n||_0^2 \le C(\tau^2 + h^{2l})e^{C_3 n\tau} \le C(\tau^2 + h^{2l})e^{C_3 T},$$

which along with the triangle inequality and the interpolation estimates, we have

$$\begin{aligned}
&||\boldsymbol{E}^n - \boldsymbol{E}_h^n||_0^2 + ||\boldsymbol{H}^n - \boldsymbol{H}_h^n||_0^2 \\
&\le 2(||\boldsymbol{E}^n - P_h\boldsymbol{E}^n||_0^2 + ||\xi_h^n||_0^2) + 2(||\boldsymbol{H}^n - \Pi_h\boldsymbol{H}^n||_0^2 + ||\eta_h^n||_0^2) \\
&\le Ch^{2l}||\boldsymbol{E}^n||_l^2 + Ch^{2l}||\boldsymbol{H}^n||_{l+1}^2 + 2(||\xi_h^n||_0^2 + ||\eta_h^n||_0^2) \\
&\le C(h^{2l} + \tau^2),
\end{aligned}$$

which concludes the proof. □

9.6 Bibliographical remarks

In this chapter, we first introduce the popular FDTD method. Basic stability and error analysis techniques for the FDTD method are illustrated through a 2-D model problem. Due to its simple implementation and resonable accuracy, the FDTD method remains one of the dominant computational electrodynamics techniques. Interested readers can find various interesting applications and further developments of the FDTD method in the book by Taflove and Hagness [71] and the book by Inan and Marshall [38]. Clearly documented MATLAB codes for 1-D, 2-D and 3-D are provided in [71]. Mathematical analysis of the FDTD method can be found in papers [9, 49, 53, 61] and references therein. Recently, numerical analysis has also been performed for the ADI-FDTD method [24, 31, 35, 44].

For more advanced analysis of FEM for Maxwell's equations, readers should consult the book by Monk [57]. An overview of finite element analysis for

electromagnetic problems as of 2001 can be found in the review paper by
Hiptmair [29]. Readers interested in hp FEM can consult papers [1, 43] and
books [19, 20]. Some posteriori error estimators have been developed for
Maxwell's equations [8, 60, 64]. Various domain decomposition methods and
preconditioners have been studied recently for solving Maxwell's equations
[2, 5, 25, 28, 30, 37, 72]. As for analysis of the perfectly matched layer problems, readers can consult papers [6, 7, 10, 14, 36] and references therein.
For practical implementation of finite element methods for solving Maxwell's
equations, readers should refer to more specialized books such as the book
by Silvester and Ferrari [69] (where many classical FEMs are described and
Fortran source codes are provided), the book by Demkowicz [19] (where hp
FEMs for Maxwell's equations are discussed and 1-D and 2-D Fortran codes
are provided in CD-ROM), and the book by Hesthaven and Warburton [27]
(where discontinuous Galerkin finite element MATLAB codes are described
and can be downloaded online). For more rigorous mathematical treatment
of Maxwell's equations and various numerical methods developed in recent
years, readers can consult some recent books [11, 23, 41, 50, 67] and references therein.

9.7   Exercises

1. Prove that at each time step, the coefficient matrix for the system (9.178)-
(9.181) can be written as

$$Q \equiv \begin{pmatrix} A & -B \\ B' & C \end{pmatrix},$$

where the stiffness matrices $A = \frac{\epsilon_0}{\tau}(\mathbf{U}_h, \mathbf{U}_h)$ and $C = \frac{\mu_0}{\tau}(\mathbf{V}_h, \mathbf{V}_h)$ are symmetric positive definite (SPD), and the matrix $B = (\nabla \times \mathbf{V}_h, \mathbf{U}_h)$. Here B'
denotes the transpose of B.

2. Prove that the determinant of Q equals

$$det(Q) = det(A) \cdot det(C + B'A^{-1}B),$$

which is nonzero. This fact guarantees the existence and uniqueness of solution for the system (9.178)-(9.181) at each time step.

3. Prove Lemma 9.5.

4. Prove Lemma 9.6.

5. Prove that (9.205) is a solution of (9.204).

6. Prove Lemma 9.7.

7. For the isotropic cold plasma model (9.166)-(9.168), prove the following
stability

$$\epsilon_0 \|\boldsymbol{E}(t)\|_0^2 + \mu_0 \|\boldsymbol{H}(t)\|_0^2 + \frac{1}{\epsilon_0 \omega_p^2} \|\boldsymbol{J}(t)\|_0^2$$

$$\leq \epsilon_0||\boldsymbol{E}_0||_0^2 + \mu_0||\boldsymbol{H}_0||_0^2 + \frac{1}{\epsilon_0\omega_1^2}||\boldsymbol{J}_0||_0^2, \quad \forall\, t \in [0,T],$$

where $(\boldsymbol{E}(t), \boldsymbol{H}(t), \boldsymbol{J}(t))$ is the solution of (9.166)-(9.168), and $(\boldsymbol{E}_0, \boldsymbol{H}_0, \boldsymbol{J}_0)$ is the corresponding initial function.

8. For the Debye medium model (9.187)-(9.189), prove the following stability

$$\epsilon_0\epsilon_\infty||\boldsymbol{E}(t)||_0^2 + \mu_0||\boldsymbol{H}(t)||_0^2 + \frac{1}{(\epsilon_s - \epsilon_\infty)\epsilon_0}||\boldsymbol{P}(t)||_0^2$$

$$\leq \epsilon_0\epsilon_\infty||\boldsymbol{E}_0||_0^2 + \mu_0||\boldsymbol{H}_0||_0^2 + \frac{1}{(\epsilon_s - \epsilon_\infty)\epsilon_0}||\boldsymbol{P}_0||_0^2, \quad \forall\, t \in [0,T],$$

where $(\boldsymbol{E}(t), \boldsymbol{H}(t), \boldsymbol{P}(t))$ is the solution of (9.187)-(9.189), and $(\boldsymbol{E}_0, \boldsymbol{H}_0, \boldsymbol{P}_0)$ is the corresponding initial function.

9. Prove that

$$|u(t^k) - \frac{1}{\tau}\int_{t^{k-1}}^{t^k} u(t)dt|^2 \leq \tau \int_{t^{k-1}}^{t^k} |u_t(t)|^2 dt,$$

for any $0 < t^{k-1} < t^k$. Here we denote $\tau = t^k - t^{k-1}$.

10. Prove that

$$|\frac{1}{2}(u(t^k) + u(t^{k-1})) - \frac{1}{\tau}\int_{t^{k-1}}^{t^k} u(t)dt|^2 \leq \frac{\tau^3}{4} \int_{t^{k-1}}^{t^k} |u_{tt}(t)|^2 dt,$$

for any $0 < t^{k-1} < t^k$. Here we denote $\tau = t^k - t^{k-1}$.

11. Prove that the total number of degrees of freedom from (9.84)-(9.86) is $N_k = \frac{k(k+2)(k+3)}{2}$.

12. Verify that the total number of degrees of freedom for the second type Nédélec spaces on tetrahedra is

$$\dim((P_k)^3) = \frac{(k+1)(k+2)(k+3)}{2}.$$

References

[1] M. Ainsworth and J. Coyle. Hierarchic *hp*-edge element families for Maxwell's equations on hybrid quadrilateral/triangular meshes. *Comput. Methods Appl. Mech. Engrg.*, 190(49–50):6709–6733, 2001.

[2] A. Alonso and A. Valli. A domain decomposition approach for heterogeneous time-harmonic Maxwell equations. *Comput. Methods Appl. Mech. Engrg.*, 143(1–2):97–112, 1997.

[3] C. Amrouche, C. Bernardi, M. Dauge and V. Girault. Vector potentials in three-dimensional non-smooth domains. *Math. Methods Appl. Sci.*, 21:823–864, 1998.

[4] D.N. Arnold, F. Brezzi, B. Cockburn and L.D. Marini. Unified analysis of discontinuous Galerkin methods for elliptic problems. *SIAM J. Numer. Anal.*, 39(5):1749–1779, 2002.

[5] D.N. Arnold, R.S. Falk and R. Winther. Multigrid in $H(\text{div})$ and $H(\text{curl})$. *Numer. Math.*, 85(2):197–217, 2000.

[6] G. Bao, P. Li and H. Wu. An adaptive edge element method with perfectly matched absorbing layers for wave scattering by biperiodic structures. *Math. Comp.*, 79:1–34, 2010.

[7] E. Bécache, P. Joly, M. Kachanovska and V. Vinoles. Perfectly matched layers in negative index metamaterials and plasmas. *ESAIM: Proc. Surv.*, 50:113–132, 2015

[8] R. Beck, R. Hiptmair, R.H.W. Hoppe and B. Wohlmuth. Residual based a posteriori error estimators for eddy current computation. *M2AN Math. Model. Numer. Anal.*, 34(1):159–182, 2000.

[9] V.A. Bokil and N.L. Gibson. Convergence analysis of Yee schemes for Maxwell's equations in Debye and Lorentz dispersive media. *Int. J. Numer. Anal. Model.*, 11(4):657–687, 2014.

[10] J.H. Bramble and J.E. Pasciak. Analysis of a finite element PML approximation for the three dimensional time-harmonic Maxwell problem. *Math. Comp.*, 77(261):1–10, 2008.

[11] W. Cai. *Computational Methods for Electromagnetic Phenomena.* Cambridge University Press, New York, 2013.

[12] S. Caorsi, P. Fernandes and M. Raffetto. On the convergence of Galerkin finite element approximations of electromagnetic eigenproblems. *SIAM J. Numer. Anal.*, 38(2):580–607, 2000.

[13] M.-H. Chen, B. Cockburn and F. Reitich. High-order RKDG methods for computational electromagnetics. *J. Sci. Comput.*, 22/23:205–226, 2005.

[14] Z. Chen and H. Wu. An adaptive finite element method with perfectly matched absorbing layers for the wave scattering by periodic structures. *SIAM J. Numer. Anal.*, 41(3):799–826, 2003.

[15] P. Ciarlet, Jr. and J. Zou. Fully discrete finite element approaches for time-dependent Maxwell's equations. *Numer. Math.*, 82:193–219, 1999.

[16] B. Cockburn, G.E. Karniadakis and C.-W. Shu (eds.). *Discontinuous Galerkin Methods: Theory, Computation and Applications.* Springer, Berlin, 2000.

[17] B. Cockburn, F. Li and C.-W. Shu. Locally divergence-free discontinuous Galerkin methods for the Maxwell equations. *J. Comput. Phys.*, 194(2):588–610, 2004.

[18] B. Cockburn and C.-W. Shu. Runge-Kutta discontinuous Galerkin methods for convection-dominated problems. *J. Sci. Comput.*, 16(3):173–261, 2001.

[19] L. Demkowicz. *Computing with hp-Adaptive Finite Elements, Vol. 1: One and Two Dimensional Elliptic and Maxwell Problems.* Chapman & Hall/CRC, Boca Raton, FL, 2006.

[20] L. Demkowicz, J. Kurtz, D. Pardo, M. Paszynski, W. Rachowicz and A. Zdunek. *Computing with hp-Adaptive Finite Elements, Vol. 2: Frontiers: Three Dimensional Elliptic and Maxwell Problems with Applications.* Chapman & Hall/CRC, Boca Raton, FL, 2008.

[21] L. Demkowicz and L. Vardapetyan. Modeling of electromagnetic absorption/scattering problems using *hp*-adaptive finite elements. *Comput. Methods Appl. Mech. Engrg.*, 152(1-2):103–124, 1998.

[22] G.V. Eleftheriades and K.G. Balmain (eds.). *Negative Refraction Metamaterials: Fundamental Properties and Applications.* Wiley-Interscience, New York, NY, 2005.

[23] M. Fabrizio and A. Morro. *Electromagnetism of Continuous Media.* Oxford University Press, Oxford, UK, 2003.

[24] L. Gao, X. Li and W. Chen. New energy identities and super convergence analysis of the energy conserved splitting FDTD methods for 3D Maxwell's Equations. *Math. Meth. Appl. Sci.*, 36:440–455, 2013.

[25] J. Gopalakrishnan, J.E. Pasciak and L.F. Demkowicz. Analysis of a multigrid algorithm for time harmonic Maxwell equations. *SIAM J. Numer. Anal.*, 42(1):90–108, 2004.

[26] J.S. Hesthaven and T. Warburton. Nodal high-order methods on unstructured grids. I. Time-domain solution of Maxwell's equations. *J. Comput. Phys.*, 181(1):186–221, 2002.

[27] J.S. Hesthaven and T. Warburton. *Nodal Discontinuous Galerkin Methods: Algorithms, Analysis, and Applications.* Springer, New York, NY, 2007.

[28] R. Hiptmair. Multigrid method for Maxwell's equations. *SIAM J. Numer. Anal.*, 36(1):204–225, 1999.

[29] R. Hiptmair. Finite elements in computational electromagnetism. *Acta Numer.*, 11:237–339, 2002.

[30] R. Hiptmair and J. Xu. Nodal auxiliary space preconditioning in H(curl) and H(div) spaces. *SIAM J. Numer. Anal.*, 45(6):2483–2509, 2007.

[31] M. Hochbruck, T. Jahnke and R. Schnaubelt. Convergence of an ADI splitting for Maxwell's equations. *Numer. Math.*, 129:535–561, 2015.

[32] P. Houston, I. Perugia, A. Schneebeli and D. Schötzau. Interior penalty method for the indefinite time-harmonic Maxwell equations. *Numer. Math.*, 100(3):485–518, 2005.

[33] P. Houston, I. Perugia and D. Schötzau. Mixed discontinuous Galerkin approximation of the Maxwell operator. *SIAM J. Numer. Anal.*, 42(1):434–459, 2004.

[34] P. Houston, I. Perugia and D. Schötzau. Mixed discontinuous Galerkin approximation of the Maxwell operator: non-stabilized formulation. *J. Sci. Comput.*, 22/23:315–346, 2005.

[35] Y. Huang, M. Chen, J. Li and Y. Lin. Numerical analysis of a leapfrog ADI-FDTD method for Maxwell's equations in lossy media. *Comput. Math. Appl.*, 76:938–956, 2018.

[36] Y. Huang, H. Jia and J. Li. Analysis and application of an equivalent Berenger's PML model. *J. Comput. Appl. Math.*, 333:157–169, 2018.

[37] Q. Hu and J. Zou. Substructuring preconditioners for saddle-point problems arising from Maxwell's equations in three dimensions. *Math. Comp.*, 73(245):35–61, 2004.

[38] U.S. Inan and R.A. Marshall. *Numerical Electromagnetics: The FDTD Method.* Cambridge Univ Press, Cambridge, UK, 2011.

[39] D. Jiao and J.-M. Jin. Time-domain finite-element modeling of dispersive media. *IEEE Microwave and Wireless Components Letters*, 11:220–223, 2001.

[40] J. Jin. *The Finite Element Method in Electromagnetics.* John Wiley & Sons, New York, NY, 2nd Edition, 2002.

[41] A. Kirsch and F. Hettlich. *The Mathematical Theory of Time-Harmonic Maxwell's Equations: Expansion-, Integral-, and Variational Methods.* Springer International Publishing Switzerland, 2015.

[42] K. Kunz and R.J. Luebbers. *The Finite-Difference Time-Domain Method for Electromagnetics.* CRC Press, Boca Raton, FL, 1993.

[43] P.D. Ledger and K. Morgan. The application of the *hp*-finite element method to electromagnetic problems. *Arch. Comput. Methods Engrg.*, 12(3):235–302, 2005.

[44] J. Lee and B. Fornberg. Some unconditionally stable time stepping methods for the 3D Maxwell's equations. *J. Comp. Appl. Math.*, 166:497–523, 2004.

[45] R. Leis. Exterior boundary-value problems in mathematical physics. *Trends in Applications of Pure Mathematics to Mechanics.* Vol. II, pp. 187–203, Monographs Stud. Math., 5, Pitman, Boston, MA, London, UK, 1979.

[46] J. Li. Error analysis of mixed finite element methods for wave propagation in double negative metamaterials. *J. Comp. Appl. Math.*, 209:81–96, 2007.

[47] J. Li. Error analysis of fully discrete mixed finite element schemes for 3-D Maxwell's equations in dispersive media. *Comput. Methods Appl. Mech. Engrg.*, 196:3081–3094, 2007.

[48] J. Li and Y. Chen. Analysis of a time-domain finite element method for 3-D Maxwell's equations in dispersive media. *Comput. Methods Appl. Mech. Engrg.*, 195:4220–4229, 2006.

[49] J. Li, M. Chen and M. Chen. Developing and analyzing fourth-order difference methods for metamaterials Maxwell's equations. *Adv. Comput. Math.*, 45:213–241, 2019.

[50] J. Li and Y. Huang. *Time-Domain Finite Element Methods for Maxwell's Equations in Metamaterials.* Springer Series in Computational Mathematics, vol.43, Springer-Verlag, Berlin, 2013.

[51] J. Li and S. Shields. Superconvergence analysis of Yee scheme for metamaterial Maxwell's equations on non-uniform rectangular meshes. *Numer. Math.*, 134:741–781, 2016.

[52] J. Li and A. Wood. Finite element analysis for wave propagation in double negative metamaterials. *J. Sci. Comput.*, 32:263–286, 2007.

[53] W. Li, D. Liang and Y. Lin. A new energy-conserved s-fdtd scheme for maxwell's equations in metamaterials. *Int. J. Numer. Anal. Mod.*, 10:775–794, 2013.

[54] Q. Lin and J. Li. Superconvergence analysis for Maxwell's equations in dispersive media. *Math. Comp.*, 77:757–771, 2008.

[55] T. Lu, P. Zhang and W. Cai. Discontinuous Galerkin methods for dispersive and lossy Maxwell's equations and PML boundary conditions. *J. Comput. Phys.*, 200(2):549–580, 2004.

[56] P. Monk. A mixed method for approximating Maxwell's equations. *SIAM J. Numer. Anal.*, 28:1610–1634, 1991.

[57] P. Monk. *Finite Element Methods for Maxwell's Equations.* Oxford University Press, Oxford, UK, 2003.

[58] P. Monk. Analysis of a finite element method for Maxwell's equations. *SIAM J. Numer. Anal.*, 29:714–729, 1992.

[59] P. Monk. A finite element method for approximating the time-harmonic Maxwell equations. *Numer. Math.*, 63(2):243–261, 1992.

[60] P. Monk. A posteriori error indicators for Maxwell's equations. *J. Comput. Appl. Math.*, 100(2):173–190, 1998.

[61] P. Monk and E. Süli. A convergence analysis of Yee's scheme on nonuniform grid. *SIAM J. Numer. Anal.*, 31:393–412, 1994.

[62] J.-C. Nédélec. Mixed finite elements in R^3. *Numer. Math.*, 35:315–341, 1980.

[63] J.-C. Nédélec. A new family of mixed finite elements in R^3. *Numer. Math.*, 50:57–81, 1986.

[64] S. Nicaise and E. Creusé. A posteriori error estimation for the heterogeneous Maxwell equations on isotropic and anisotropic meshes. *Calcolo*, 40(4):249–271, 2003.

[65] I. Perugia, D. Schötzau and P. Monk. Stabilized interior penalty methods for the time-harmonic Maxwell equations. *Comput. Methods Appl. Mech. Engrg.*, 191(41–42):4675–4697, 2002.

[66] A. Quarteroni and A. Valli. *Numerical Approximation of Partial Differential Equations*. Springer Series in Computational Mathematics, vol.23, Springer-Verlag, Berlin, 1994.

[67] A.A. Rodriguez and A. Valli. *Eddy Current Approximation of Maxwell Equations: Theory, Algorithms and Applications*. Springer-Verlag, Italia, 2010.

[68] A. Shelby, D.R. Smith and S. Schultz. Experimental verification of a negative index of refraction. *Science*, 292:489–491, 2001.

[69] P.P. Silvester and R.L. Ferrari. *Finite Elements for Electrical Engineers*. Cambridge University Press, 3rd Edition, Cambridge, UK, 1996.

[70] D.R. Smith, W.J. Padilla, D.C. Vier, S.C. Nemat-Nasser and S. Schultz. Composite medium with simultaneously negative permeability and permittivity. *Phys. Rev. Lett.*, 84:4184–4187, 2000.

[71] A. Taflove and C. Hagness. *Computational Electrodynamics: the Finite-Difference Time-Domain Method*. Artech House, Norwood, MA, 2nd Edition, 2000.

[72] A. Toselli, O.B. Widlund and B.I. Wohlmuth. An iterative substructuring method for Maxwell's equations in two dimensions. *Math. Comp.*, 70(235):935–949, 2001.

[73] K.S. Yee. Numerical solution of initial boundary value problems involving Maxwell's equations in isotropic media. *IEEE Trans. Antennas and Propagation*, 14:302–307, 1966.

10

Meshless Methods with Radial Basis Functions

Since Kansa [34] introduced the radial basis functions (RBFs) to solve PDEs in 1990, there has been a growing interest in this subject. In this chapter, we will introduce this fascinating area to the reader. In Sec. 10.1, we provide a brief overview of the meshless methods related to RBFs. Then in Sec. 10.2, we introduce some widely used RBFs and their mathematical properties. In Sec. 10.3, we present the method of fundamental solutions (MFS) and several techniques by coupling MFS with RBFs. After that, we discuss Kansa's method and its variants in Sec. 10.4. In order to help readers understand the algorithm better, we present some MATLAB codes in Sec. 10.5 to solve both elliptic and biharmonic problems in 2-D. Finally, in Sec. 10.6, we extend the discussion to domain decomposition methods coupled with Kansa's method. One numerical example is also presented.

10.1 Introduction

During the past decade, there has been increasing attention to the development of meshless methods using radial basis functions for the numerical solution of partial differential equations (PDEs). In general, there are two major developments in this direction. The first one is the so-called MFS-DRM which has evolved from the dual reciprocity boundary element method (DRBEM) [52] in the boundary element literature. In the MFS-DRM approach, the method of fundamental solutions is implemented instead of BEM in the solution process. The MFS is attributed to Kupradze and Aaleksidze in 1964 [39] and is often referred to as an indirect boundary method or regular BEM in the engineering literature. More details about MFS can be found in the excellent review papers [18, 27]. In the MFS, the singularity has been avoided by the use of a fictitious boundary outside the domain. As a result, the MFS has the following advantages over its counterpart BEM:
(i) It requires neither domain nor boundary discretization; i.e., it is truly meshless.
(ii) No domain integration is required.

(iii) It converges exponentially for smooth boundary shape and boundary data.

(iv) It is insensitive to the curse of dimension and thus attractive to high dimensional problems.

(iv) It is easy for implementation and coding.

Despite these huge advantages, the MFS has never been seriously considered as a major numerical technique possibly due to its limitation in solving only homogeneous problems. A key factor that the MFS has revived after three decades of dormancy is that it has been successfully extended to solving nonhomogeneous problems and various types of time-dependent problems [27, 54] by coupling the dual reciprocity method (DRM). The DRM was first introduced by Nardini and Brebbia in 1982 [52]. In the BEM literature, the DRM has been employed to transfer the domain integration to the boundary. In the development of the DRM, RBFs play a key role in the theoretical establishment and applications. With the combined features of the MFS and the DRM, a meshless numerical scheme for solving PDEs has been achieved. The second meshless method using RBFs is the so-called Kansa's method, where the RBFs are directly implemented for the approximation of the solution of PDEs. Kansa's method was developed in 1990 [34], in which the concept of solving PDEs by using RBFs, especially MQ, was initiated. Due to its simplicity in implementation and generality, Kansa's method was extended and used to solve different PDEs such as the biphasic mixture model for tissue engineering, the shallow water equation for the tide and currents simulation, the Navier-Stokes equations, and free boundary value problems such as American option pricing.

10.2   The radial basis functions

By definition, an RBF function $\varphi : R^d \to R, d \geq 1$, is a univariate function and can be expressed as $\varphi(r)$, where $r = \|x\|$ is the Euclidean norm. The most widely used radial basis functions are:

(i) The multiquadric (MQ) $\varphi(r) = (r^2 + c^2)^{\beta/2}$ (β is an odd integer);

(ii) The inverse multiquadric $\varphi(r) = (r^2 + c^2)^{-\beta}$ ($\beta > \frac{d}{2}$);

(iii) The Gaussian (GS) $\varphi(r) = e^{-cr^2}$;

(iv) The polyharmonic splines $\varphi(r) = r^{2n} \ln r$ and $\varphi(r) = r^{2n-1}$ for $n \geq 1$.

Note that the above RBFs have infinite support. In recent years, RBFs with compact support were developed by Wu [67], Wendland [62] and Buhmann [3]. Among the compacted supported RBFs, the Wendland's RBFs are well studied and used. Some examples of Wendland's compacted supported RBFs are listed in Table 10.1, where we denote the cutoff function

$$(1-r)^l_+ = \begin{cases} (1-r)^l & \text{if } 1-r \geq 0, \\ 0 & \text{if } 1-r < 0. \end{cases}$$

TABLE 10.1
Wendland's RBFs $\phi_{d,k}$ in $R^d, d \leq 3$

Function	Smoothness
$\phi_{3,0}(r) = (1-r)_+^2$	C^0
$\phi_{3,1}(r) = (1-r)_+^4(4r+1)$	C^2
$\phi_{3,2}(r) = (1-r)_+^6(35r^2+18r+3)$	C^4
$\phi_{3,3}(r) = (1-r)_+^8(32r^3+25r^2+8r+1)$	C^6

To distinguish different RBFs, we have to introduce the concept of (conditionally) positive definite functions. A continuous function $\phi : \Omega \times \Omega \to R$ is said to be *conditionally positive definite of order m* on $\Omega \subset R^d$ if for all $N > 1$, all distinct points $x_1, \cdots, x_N \in \Omega$ and all $\alpha_j \in R \setminus \{0\}$ satisfying

$$\sum_{j=1}^{N} \alpha_j p(x_j) = 0$$

for all polynomials p of degree less than m, the quadratic form

$$\sum_{j=1}^{N} \sum_{i=1}^{N} \alpha_j \alpha_i \phi(\|x_i - x_i\|)$$

is positive. If $m = 0$, then the function ϕ is *positive definite*.

Let $\Pi_m(R^d)$ denote the set of d-variable polynomials of degree at most m. The theorem below shows the connection between the degree of the polynomial p in the interpolation function

$$I_{f,X}(x) = \sum_{j=1}^{N} \alpha_j \phi(\|x - x_j\|) + p(x) \tag{10.1}$$

and the order m of conditional positive definiteness of the radial basis function $\phi(r)$.

THEOREM 10.1
Assume that ϕ is conditionally positive definite (CPD) of order m on $\Omega \subset R^d$, and that the set of points $X = \{x_1, \cdots, x_N\} \subseteq \Omega$ is $\Pi_{m-1}(R^d)$ unisolvent, i.e., the zero polynomial is the only polynomial from $\Pi_{m-1}(R^d)$ that vanishes on X. Then for any $f \in C(\Omega)$, there is exactly one function $I_{f,X}$ of (10.1) with a polynomial $p \in \Pi_{m-1}(R^d)$ such that

$$I_{f,X}(x_i) = f(x_i) \tag{10.2}$$

and

$$\sum_{j=1}^{N} \alpha_j q(x_j) = 0, \quad 1 \leq i \leq N, \ \forall\, q \in \Pi_{m-1}(R^d). \tag{10.3}$$

Denote the smallest integer greater than or equal to β by $[\beta]$. The following results are well known (see e.g., [5, 59, 64]).

THEOREM 10.2
(i) The multiquadrics $\phi(r) = (r^2 + c^2)^{\beta/2}, \beta > 0, \beta \notin 2\mathcal{N}$, are CPD of order $m \geq [\frac{\beta}{2}]$ on R^d for any $d \geq 1$.
(ii) The function $\phi(r) = r^{2k-1}, k \geq 1$, are CPD of order $k + 1$ on R^d for all $d \geq 1$.
(iii) The thin-plate splines $\phi(r) = r^{2k} \ln r, k \geq 1$, are CPD of order $k + 1$ on R^d for all $d \geq 1$.
(iv) The Wendland's RBFs $\phi_{3,k}(r)$ are positive definite on R^d for any $d \leq 3$.
(v) The Gaussian $\phi(r) = e^{-cr^2}, c > 0$, is positive definite on R^d for all $d \geq 1$.
(vi) The inverse multiquadrics $\phi(r) = (r^2+c^2)^{-\beta}, \beta > 0$, is positive definite on R^d for all $d \geq 1$.

To consider the interpolation error estimate, we define the fill distance (also called the minimum separation distance)

$$h \equiv h_{X,\Omega} = \sup_{y \in \Omega} \min_{x \in X} ||y - x||_2, \tag{10.4}$$

for an arbitrary set $X = \{x_1, \cdots, x_N\} \subseteq \Omega$ of distinct points. Then we have [58].

THEOREM 10.3
(i) For Gaussians $\phi(r) = e^{-cr^2}$, we have

$$||I_{f,X} - f||_{L^2(\Omega)} \leq C \cdot e^{-\delta/h^2}, \quad \delta > 0.$$

(ii) For multiquadrics $\phi(r) = (r^2 + c^2)^{\beta/2}, \beta \notin 2\mathcal{N}$, we have

$$||I_{f,X} - f||_{L^2(\Omega)} \leq C \cdot e^{-\delta/h}, \quad \delta > 0.$$

(iii) For thin-plate splines r^{2k-1} and $r^{2k} \ln r, k \geq 1$, we have

$$||I_{f,X} - f||_{L^2(\Omega)} \leq C \cdot h^k.$$

To end this section, below we state some interpolation error estimates in L^∞ norm.
For $\phi(r) = r^2 \ln r$ in R^2, we have [55]

$$||I_{f,X} - f||_{L^\infty(\Omega)} \leq Ch|\ln h|.$$

For $\phi(r) = r$ in R^3, we have [43]

$$||I_{f,X} - f||_{L^\infty(\Omega)} \leq Ch|\ln h|.$$

For $\phi(r) = (r^2 + c^2)^{\beta/2}, \beta \notin 2\mathcal{N}$ and Gaussians $\phi(r) = e^{-cr^2}$, we have the superalgebraic convergence rate [43]

$$\|I_{f,X} - f\|_{L^\infty(\Omega)} \leq Ch^p, \quad p \geq 1.$$

Under certain conditions, Madych [48] and Beatson and Powell [2] showed that $\|I_{f,X} - f\|_{L^\infty(\Omega)}$ is exponentially decreasing in both c and h for Gaussians $\phi(r) = e^{-cr^2}$.

For Wendland's compact RBFs $\phi_{3,k}(r)$, we have [63]

$$\|I_{f,X} - f\|_{L^\infty(\Omega)} \leq Ch^{k+\frac{1}{2}} \quad \text{in } R^2, \ k \geq 1.$$

For example, for $\phi_{3,1}(r) = (1 - r)_+^4 (4r + 1)$, its convergence rate is $h^{\frac{3}{2}}$.

For thin-plate splines r^{2k-1} and $r^{2k} \ln r, k \geq 2$, we have [4, p. 8]

$$\|I_{f,X} - f\|_{L^\infty(\Omega)} \leq Ch^{k-\frac{d}{2}}.$$

10.3 The MFS-DRM

In this section, we present various MFS-DRM methods for solving the elliptic and parabolic equations.

10.3.1 The fundamental solution of PDEs

The fundamental solution (or free-space Green's function) for a differential operator L at some point $\boldsymbol{x}_0 \in R^d$ is defined as $\phi(\boldsymbol{x}, \boldsymbol{x}_0)$, which satisfies

$$L\phi(\boldsymbol{x}, \boldsymbol{x}_0) = \delta(\boldsymbol{x} - \boldsymbol{x}_0),$$

where the high-dimensional delta function is defined as the product of one-dimensional delta functions.

Now let us calculate the well-known fundamental solution of the Laplacian operator in two dimensions (2-D), i.e., we want to solve

$$\triangle G \equiv \left(\frac{\partial^2}{\partial x^2} + \frac{\partial^2}{\partial y^2} \right)G = \delta(x - x_0, y - y_0), \quad -\infty < x, y < \infty. \tag{10.5}$$

By the definition of $\delta(x)$, we have

$$\int_\Omega \delta(x - x_0, y - y_0)dxdy = 1 \tag{10.6}$$

for any domain Ω containing point (x_0, y_0). Integrating (10.5) over Ω and using Green's Theorem, we obtain

$$\int_{\partial\Omega} \frac{\partial G}{\partial n} ds = 1. \tag{10.7}$$

For simplicity, let Ω be a circular domain centered at (x_0, y_0) with radius r_0. Hence the problem (10.5) becomes: Find a function G such that

$$\triangle G = 0, \quad (x, y) \neq (x_0, y_0), \tag{10.8}$$

and

$$\int_{\partial\Omega} \frac{\partial G}{\partial n} ds = 1. \tag{10.9}$$

Changing the 2-D Laplacian operator into radial form, we can rewrite the problem (10.8)-(10.9) as

$$\frac{1}{r}\frac{d}{dr}(r\frac{dG}{dr}) = 0, \quad r \neq 0, \tag{10.10}$$

$$\int_0^{2\pi} \frac{dG}{dr} r d\theta = 1. \tag{10.11}$$

The equation (10.10) has a general solution

$$G(r) = a + b\ln r, \quad r = \sqrt{(x - x_0)^2 + (y - y_0)^2}, \tag{10.12}$$

where a and b are general constants.

Substituting (10.12) into (10.11) leads to $b = \frac{1}{2\pi}$. Letting the constant $a = 0$, we obtain the fundamental solution for the 2-D Laplacian operator as $G = \frac{1}{2\pi}\ln r$.

Similarly, the fundamental solution G for the 3-D Laplacian operator is determined by

$$\triangle G = 0, \quad (x, y, z) \neq (x_0, y_0, z_0), \tag{10.13}$$

$$\int_{\partial\Omega} \frac{\partial G}{\partial n} ds = 1, \tag{10.14}$$

where $\partial\Omega$ is the sphere of radius r_0 centered at (x_0, y_0, z_0).

Changing the Laplacian into spherical coordinates and looking for the radial solution, we can transform (10.13)-(10.14) into

$$\frac{1}{r^2}\frac{d}{dr}(r^2\frac{dG}{dr}) = 0, \quad r \neq 0, \tag{10.15}$$

$$\int_0^{2\pi}(\int_0^{\pi} \frac{dG}{dr} r^2 \sin\phi d\phi)d\theta = 1. \tag{10.16}$$

It is easy to see that (10.15) has the general solution $G = a + \frac{b}{r}$, which along with (10.16) yields $b = -\frac{1}{4\pi}$. Choosing the constant a to be zero, we obtain the fundamental solution for the 3-D Laplacian operator as $G = -\frac{1}{4\pi r}$.

In summary, the fundamental solution for the Laplacian operator $\triangle$ at point Q is given by

$$G(P, Q) = \begin{cases} \frac{1}{2\pi}\ln r & \text{in } R^2, \\ -\frac{1}{4\pi r} & \text{in } R^3, \end{cases}$$

where we denote for $r = ||P - Q||$, i.e., the distance between points P and Q.

Before we discuss the fundamental solution for other operators, let us first review the Bessel functions. For the Bessel's equation of order α:

$$x^2 y'' + xy' + (x^2 - \alpha^2)y = 0, \tag{10.17}$$

it is known that

$$J_\alpha = \sum_{k=0}^{\infty} \frac{(-1)^k}{k!\Gamma(k + \alpha + 1)} \left(\frac{x}{2}\right)^{2k+\alpha} \tag{10.18}$$

is a valid solution of Bessel's equation (10.18) for any $\alpha \geq 0$. Note that $J_\alpha(x)$ is often called the Bessel function of the first kind of order α. Another linearly independent solution of (10.18) is the Bessel function of the first kind of order $-\alpha$:

$$J_{-\alpha} = \sum_{k=0}^{\infty} \frac{(-1)^k}{k!\Gamma(k - \alpha + 1)} \left(\frac{x}{2}\right)^{2k-\alpha}. \tag{10.19}$$

When α is not an integer, we define the function

$$Y_\alpha = [(\cos \pi\alpha)J_\alpha(x) - J_{-\alpha}(x)]/\sin \pi\alpha,$$

which is called the Bessel function of the second kind of order α. In general, $Y_n(x)$ is the form

$$J_n(x)(A \ln x + B) + x^{-n} \sum_{k=0}^{\infty} a_k x^k, \quad n = 1, 2, \cdots,$$

where constants A and a_0 are nonzero.

Similarly, for the modified Bessel's equation of order α:

$$x^2 y'' + xy' - (x^2 + \alpha^2)y = 0, \tag{10.20}$$

we have two independent solutions

$$I_\alpha = \sum_{k=0}^{\infty} \frac{1}{k!\Gamma(k + \alpha + 1)} \left(\frac{x}{2}\right)^{2k+\alpha}, \quad \alpha \geq 0, \tag{10.21}$$

and

$$I_{-\alpha} = \sum_{k=0}^{\infty} \frac{1}{k!\Gamma(k - \alpha + 1)} \left(\frac{x}{2}\right)^{2k-\alpha}, \quad \alpha > 0, \alpha \text{ not an integer.} \tag{10.22}$$

$I_\alpha(x)$ and $I_{-\alpha}(x)$ are often called the modified Bessel functions of the first kind of α and $-\alpha$, respectively.

If $\alpha > 0$ is not an integer, the function

$$K_\alpha = \frac{\pi}{2} \cdot \frac{I_{-\alpha}(x) - I_\alpha(x)}{\sin \pi\alpha} \tag{10.23}$$

is the modified Bessel function of the second kind of order α. For any integer $n = 0, 1, \cdots$, we define

$$K_n(x) = \lim_{\alpha \to n} K_\alpha(x). \tag{10.24}$$

The Hankel functions of order α of the first and second kinds are defined by

$$H_\alpha^{(1)}(x) = J_\alpha(x) + iY_\alpha(x)$$

and

$$H_\alpha^{(2)}(x) = J_\alpha(x) - iY_\alpha(x),$$

respectively.

By similar technique used for the Laplacian operator, it is proved that the fundamental solution for the modified Helmholtz operator $\triangle - \lambda^2$ is

$$G(P, Q) = \begin{cases} \frac{1}{2\pi} K_0(\lambda r) & \text{in } R^2, \\ \frac{1}{4\pi r} \exp(-\lambda r) & \text{in } R^3, \end{cases}$$

where $K_0(\cdot)$ denotes the modified Bessel function of the second kind with order zero.

Similarly, the fundamental solution for the Helmholtz operator $\triangle + \lambda^2$ is given by

$$G(P, Q) = \begin{cases} \frac{i}{4} H_0^{(1)}(\lambda r) & \text{in } R^2, \\ \frac{1}{4\pi r} \exp(-i\lambda r) & \text{in } R^3, \end{cases}$$

where $H_0^{(1)}(\cdot)$ denotes the Hankel function of the first kind of order zero.

Finally, the fundamental solution for the iterated Laplacian operator $\triangle^m$ ($m \geq 2$) is given by

$$G(P, Q) = \begin{cases} r^{2m-d}, & \text{in } R^d, \ d \text{ is odd, and } 2m > d, \\ r^{2m-d} \ln r, & \text{in } R^d, \ d \text{ is even, and } 2m > d. \end{cases}$$

10.3.2   The MFS for Laplace's equation

The basic idea of the MFS is to approximate the solution u to a homogeneous equation

$$\mathcal{L}u(P) = 0 \quad \text{in } \Omega,$$
$$u(P) = g(P) \quad \text{on } \partial\Omega,$$

with a linear combination of the fundamental solution of the $\mathcal{L}$ operator.

For illustration, we start with Laplace's equation

$$\triangle u(P) = 0, \quad P \in \Omega, \tag{10.25}$$
$$u(P) = g(P), \quad P \in \partial\Omega, \tag{10.26}$$

where Ω is a bounded, simply connected domain in R^d, $d = 2, 3$, with boundary $\partial\Omega$.

We can approximate the solution to (10.25) by

$$u_h(P) = \sum_{j=1}^{n} a_j G(P, Q_j), \quad \forall\, P \in \Omega, \tag{10.27}$$

where $G(P, Q)$ is the fundamental solution of (10.25), and Q_j are collocation points located on the artificial boundary $\partial\hat{\Omega}$, which encloses the domain Ω. In practical application, the artificial boundary is often chosen as a circle in R^2 or a sphere in R^3.

The unknown coefficients a_j can be obtained by satisfying the equation

$$\sum_{j=1}^{n} a_j G(P_i, Q_j) = g(P_i), \quad 1 \le i \le n, \tag{10.28}$$

where P_i are collocation points on $\partial\Omega$.

While in R^2, for completeness purposes, we usually add a constant to the solution, i.e., we seek approximate solution as

$$u_h(P) = \sum_{j=1}^{n} a_j G(P, Q_j) + c, \quad \forall\, P \in \Omega \subset R^2, \tag{10.29}$$

in which case we need to collocate at $n+1$ points on $\partial\Omega$ giving

$$\sum_{j=1}^{n} a_j G(P_k, Q_j) = g(P_k), \quad 1 \le k \le n+1. \tag{10.30}$$

But when Ω is a circle, we usually take $c = 0$.

Note that the MFS is equivalent to the so-called regular boundary element method (BEM) in the engineering community. Furthermore, the MFS does not require either boundary mesh or numerical integration.

Though MFS has been widely used, very few theoretical results exist about its convergence. Cheng [13] proved the convergence of the MFS for (10.25)-(10.26) in R^2 when both $\partial\Omega$ and $\partial\hat{\Omega}$ are circles and the collocation points P_k and Q_j in (10.30) are uniformly distributed on $\partial\Omega$ and $\partial\hat{\Omega}$, respectively. Later, Cheng's results were extended to the case where Ω is a simply connected bounded domain with analytic boundary $\partial\Omega$ by Katsurada et al. [37, 38]. Since the proofs are quite technical, we only state the results here. Detailed proofs can be found in the original papers.

THEOREM 10.4
[27, p. 117] Let Ω be a disk of radius r with boundary $\partial\Omega$, i.e.,

$$\Omega = \{(x,y) \in R^2 : \sqrt{x^2 + y^2} < r\}, \quad \partial\Omega = \{(x,y) \in R^2 : \sqrt{x^2 + y^2} = r\}.$$

Furthermore, let $\partial\hat{\Omega} = \{(x, y) \in R^2 : \sqrt{x^2 + y^2} = R\}, R > r,$ *and*

$$P_k = (r\cos\theta_k, r\sin\theta_k), \quad Q_j = (R\cos\theta_j, R\sin\theta_j), \quad 1 \le j, k \le n,$$

where

$$\theta_j = \frac{2\pi}{n} \cdot (j - 1).$$

Then for sufficiently large n, *the MFS solution (10.27) exists and converges uniformly to the analytic solution* u *of (10.25)-(10.26) in* $\Omega \cup \partial\Omega$. *Moveover, if* u *can be extended to a harmonic function, then we have the error estimate*

$$||u - u_h||_{L^\infty(\Omega)} \le C \cdot (\frac{r}{R})^n,$$

where the constant $C > 0$ *is independent of* n.

Theorem 10.4 implies the spectral convergence of the MFS for the special circle domain case, and it also implies that theoretically $R = \infty$ is the best choice. However, the MFS coefficient matrix becomes highly ill-conditioned as R increases. In practical implementation, R is generally limited to above five times the diameter of Ω and the collocation points are often chosen equally distributed on $\partial\Omega$ and $\partial\hat{\Omega}$.

Finally, we would like to remark that quite arbitrary boundary conditions can be incorporated in the MFS. For example, if the Laplace problem (10.25) is imposed with mixed boundary conditions

$$u(P) = g_1(P) \quad \forall\, P \in \partial\Omega_1,$$
$$\frac{\partial u}{\partial n}(P) = g_2(P) \quad \forall\, P \in \partial\Omega_2,$$

where $\partial\Omega = \partial\Omega_1 \cap \partial\Omega_2$ and $\partial\Omega_1 \cup \partial\Omega_2 = \emptyset$. When $\Omega \subset R^2$, we can choose n_1 points on $\partial\Omega_1$ and n_2 points on $\partial\Omega_2$, where $n_1 + n_2 = n + 1$, to satisfy the following equations

$$\sum_{j=1}^{n_1} a_j G(P_i, Q_j) + c = g_1(P_i), \quad 1 \le i \le n_1, \tag{10.31}$$

$$\sum_{j=n_1+1}^{n_1+n_2} a_j \frac{\partial}{\partial n} G(P_i, Q_j) = g_2(P_i), \quad n_1 + 1 \le i \le n_1 + n_2. \tag{10.32}$$

When $\Omega \subset R^3$, we do not need c in (10.31) and we only need $n_1 + n_2 = n$ total points. Experiences show that $n \le 30$ in R^2 and $n \le 100$ in R^3 often produce good enough accuracy [27, p. 109].

Note that in (10.28) and (10.29), the points Q_j are chosen as fixed. Another approach is to choose Q_j along with the coefficients a_j by minimizing the error

$$E = \sum_{i=1}^{n} |\sum_{j=1}^{n} a_j G(P_i, Q_j) - g(P_i)|^2. \tag{10.33}$$

This approach was considered by Fairweather and Karageorghis [18, 33]. Note that this method is quite time-consuming, since (10.33) is a nonlinear minimization problem. As a compromise, we may choose the points Q_j to be fixed on a circle or a sphere with radius R as an unknown parameter in (10.33).

10.3.3 The MFS-DRM for elliptic equations

The basic idea of MFS-DRM is to split the given PDE into two parts: an inhomogeneous equation and a homogeneous equation. To be more specific, consider

$$\mathcal{L}u(P) = f(P), \quad P \in \Omega, \tag{10.34}$$
$$u(P) = g(P), \quad P \in \partial\Omega \tag{10.35}$$

where $\mathcal{L}$ is an elliptic differential operator. Then we look for a particular solution u_p, which satisfies the equation

$$\mathcal{L}u_p(P) = f(P), \tag{10.36}$$

but does not have to satisfy the boundary condition (10.35). Then consider the difference between the real solution u of (10.34)-(10.35) and the particular solution u_p of (10.36), i.e., $v_h = u - u_p$, which satisfies

$$\mathcal{L}v_h(P) = 0, \quad P \in \Omega, \tag{10.37}$$
$$v_h(P) = g(P) - u_p, \quad P \in \partial\Omega. \tag{10.38}$$

Now the homogeneous problem (10.37)-(10.38) can be solved by the MFS. The rest of the issue is to determine the particular solution u_p. The most popular way for obtaining an approximate particular solution is by the dual reciprocity method (DRM), which was introduced by Nardini and Brebbia [52] and extended by many later researchers [53].

The success of the DRM depends on how the approximate particular solutions are evaluated. First, we approximate f by a linear combination of basis functions $\{\varphi_j\}$

$$f(P) \simeq \hat{f} = \sum_{j=1}^{N} a_j \varphi_j(P) \tag{10.39}$$

where the coefficients $\{a_j\}_1^N$ are usually obtained by interpolation; i.e., by solving

$$\sum_{j=1}^{N} a_j \varphi_j(P_k) = f(P_k), \quad 1 \le k \le N, \tag{10.40}$$

where $P_k \in \mathcal{R}^d, d = 2$ or 3. An approximate particular solution $\hat{u}_p$ to (10.36) is given by

$$\hat{u}_p(P) = \sum_{j=1}^{N} a_j \Phi_j(P) \tag{10.41}$$

where $\{\Phi_j\}_1^N$ are obtained by analytically solving

$$\mathcal{L}\Phi_j = \varphi_j, \quad 1 \le j \le N. \tag{10.42}$$

The expectation is that if $\hat{f}$ is a good approximation of f, then $\hat{u}_p$ will be an accurate approximation to a particular solution u_p of (10.36). The simple choice of $\varphi_j = 1 + r_j$, where $r_j = \|P - P_j\|$, is the most popular in the early DRM literature.

Later, theory of RBFs provides a firm theoretical basis to approximate f by $\hat{f}$ in (10.39). Let $\{P_i\}_{i=1}^n$ be a set of distinct points. If the exactness condition $f(P_i) = \hat{f}(P_i), 1 \le i \le n$, is imposed, then the system in (10.40) is uniquely solvable if the matrix

$$A_\varphi = (\varphi_j(P_k - P_j))_{1 \le j, k \le n}$$

is nonsingular. It is well known that positive definiteness is sufficient to guarantee the invertibility of the coefficient matrix A_φ. However, most of the globally defined RBFs are only conditionally positive definite. In order to guarantee the unique solvability of the interpolation problem, one needs to add a polynomial term in (10.40), i.e.,

$$\hat{f}(P) = \sum_{j=1}^N a_j \varphi(\|P - P_j\|) + \sum_{k=1}^l b_k p_k(P), \tag{10.43}$$

where $\{p_k\}_{k=1}^l$ is the complete basis for d-variate polynomials of degree $\le m - 1$. The coefficients a_j, b_k can be found by solving the system

$$\sum_{j=1}^N a_j \varphi(\|P_i - P_j\|) + \sum_{k=1}^l b_k p_k(P_i) = f(P_i), \quad 1 \le i \le N, \tag{10.44}$$

$$\sum_{j=1}^N a_j p_k(P_j) = 0, \quad 1 \le k \le l, \tag{10.45}$$

where $\{x_i\}_{i=1}^N$ are the collocation points on $\bar{\Omega}$.

By linearity, once the coefficients a_j, b_k are obtained, the approximate particular solution $\hat{u}_p$ can be found as

$$\hat{u}_p = \sum_{j=1}^N a_j \Phi_j + \sum_{k=1}^l b_k \Psi_k,$$

where Φ_j is shown in (10.42) and Ψ_k satisfies

$$\mathcal{L}\Psi_k = p_k, \quad \text{for } k = 1, \cdots, l.$$

For example, if we use the thin-plate splines

$$\phi(r) = r^2 \ln r \quad \text{in } R^2; \quad \phi(r) = r \quad \text{in } R^3,$$

in (10.43), we need to add a linear polynomial in (10.43) to guarantee the existence of a solution for the interpolation, i.e., (10.45) becomes

$$\sum_{j=1}^{N} a_j = \sum_{j=1}^{N} a_j x_j = \sum_{j=1}^{N} a_j y_j = 0 \quad \text{in } R^2, \tag{10.46}$$

and

$$\sum_{j=1}^{N} a_j = \sum_{j=1}^{N} a_j x_j = \sum_{j=1}^{N} a_j y_j = \sum_{j=1}^{N} a_j z_j = 0 \quad \text{in } R^3. \tag{10.47}$$

More specifically, the systems (10.44) and (10.46), and (10.44) and (10.47) have a unique solution [17] if P_i are not colinear in R^2 and not coplanar in R^3.

Note that the derivation of Ψ_k can be obtained by the method of indeterminate coefficients. A list of Ψ_k for $\mathcal{L} = \Delta$ can be found in [12]. For the 2-D Helmholtz-type operators, Muleshkov et al. [49] have shown that the particular solution to

$$\Delta u_p + \varepsilon \lambda^2 u_p = x^m y^n \qquad (\varepsilon = \pm 1)$$

is given by

$$u_p(x,y) = \sum_{k=0}^{\left[\frac{n}{2}\right]} \sum_{\ell=0}^{\left[\frac{m}{2}\right]} \varepsilon(-\varepsilon)^{k+\ell} \frac{(k+\ell)!m!n!x^{n-2k}y^{n-2\ell}}{k!\ell!(m-2k)!(n-2\ell)!\lambda^{2k+2\ell+2}}. \tag{10.48}$$

The explicit form of u_p in (10.48) can be easily obtained by symbolic software such as MATHEMATICA or MAPLE.

The determination of Φ_j has proved to be difficult, which depends on the underlining differential operator and the chosen RBF φ_j. For $\mathcal{L} = \Delta$, a list of commonly used Φ_j have been derived as shown in Table 10.2. A list of Φ_j using compactly supported RBFs is also available [7]. For example, for Wendland's RBF

$$\phi_{3,0}(r) = \begin{cases} (1 - \frac{r}{a})^2, & \text{for } 0 \leq r \leq a, \\ 0, & \text{for } r > a, \end{cases}$$

we have the corresponding particular solution

$$\Phi(r) = \begin{cases} \frac{r^2}{4} - \frac{2r^3}{9a} + \frac{r^4}{16a^2}, & \text{for } 0 \leq r \leq a, \\ \frac{13a^2}{144} + \frac{a^2}{12} \ln \frac{r}{a}, & \text{for } r > a. \end{cases}$$

While for the other Wendland's RBF

$$\phi_{3,1}(r) = \begin{cases} (1 - \frac{r}{a})^4 (1 + \frac{4r}{a}, & \text{for } 0 \leq r \leq a, \\ 0, & \text{for } r > a, \end{cases}$$

TABLE 10.2
Φ versus RBFs φ for $\mathcal{L} = \Delta$ in 2-D

φ	Φ
$1+r$	$\frac{r^3}{9} + \frac{r^2}{4}$
$r^2 \ln r$	$\frac{r^4 \ln r}{16} - \frac{r^4}{32}$
$\sqrt{r^2 + c^2}$	$\frac{-c^3}{3}\ln(c\sqrt{r^2+c^2}+c^2) + \frac{1}{9}(r^2 + 4c^2\sqrt{r^2+c^2})$

the corresponding particular solution is

$$\Phi(r) = \begin{cases} \frac{r^2}{4} - \frac{5r^4}{8a^2} + \frac{4r^5}{5a^3} - \frac{5r^6}{12a^4} + \frac{4r^7}{49a^5}, & \text{for } 0 \le r \le a, \\ \frac{529a^2}{5880} + \frac{a^2}{14}\ln\frac{r}{a}, & \text{for } r > a. \end{cases}$$

It is noted that even for a simple differential operator such as Laplacian Δ, the closed form of Φ could be difficult to obtain. For example, for $\varphi = e^{-r^2}$ (Gaussian), we have

$$\Phi(r) = \frac{1}{4}\left(\ln r^2 + \int_{r^2}^\infty \frac{\exp(-t)}{t}dt\right),$$

which is very time-consuming to evaluate. Hence, the success of the implementation of the DRM relies on the derivation of the closed form particular solution Φ.

Recent advances in deriving the closed form Φ for Helmholtz-type operators using polyharmonic splines in both 2-D and 3-D [8, 50] and compactly supported RBFs in 3-D [28] have made it possible for solving time-dependent problems more effectively. A list of Φ using polyharmonic splines for the 2-D modified Helmholtz operator is given in Table 10.3, where $\gamma \simeq 0.5772156649015328$ is the Euler's constant, $K_0(\cdot)$ is the modified Bessel functions of the second kind with order zero.

In the next subsection, we present an example of how to derive the particular solution Φ.

10.3.4 Computing particular solutions using RBFs

Considering the efficiency of using the analytic particular solutions in the MFS-DRM, here we show how to find a particular solution [8, 50] Φ to the equation

$$(\Delta - \lambda^2)\Phi = r^2 \ln r. \tag{10.49}$$

Applying Δ^2 to both sides of (10.49) and using the fact that

$$\Delta^2(r^2 \ln r) = 0, \quad \forall r > 0,$$

we obtain

$$\Delta^2(\Delta - \lambda^2)\Phi = 0. \tag{10.50}$$

TABLE 10.3
Φ versus RBFs φ for $\mathcal{L} = \Delta - \lambda^2$ in 2-D

φ	Φ
$r^2 \ln r$	$\begin{cases} -\frac{4}{\lambda^4}(K_0(\lambda r) + \ln r) - \frac{r^2 \ln r}{\lambda^2} - \frac{4}{\lambda^4}, & r > 0 \\ \frac{4}{\lambda^4}(\gamma + \log(\frac{\lambda}{2})) - \frac{4}{\lambda^4}, & r = 0 \end{cases}$
$r^4 \ln r$	$\begin{cases} -\frac{64}{\lambda^6}(K_0(\lambda r) + \ln r) - \frac{r^2 \ln r}{\lambda^2}\left(\frac{16}{\lambda^2} + r^2\right) - \frac{8r^2}{\lambda^4} - \frac{96}{\lambda^6}, & r > 0 \\ \frac{64}{\lambda^6}(\gamma + \log(\frac{\lambda}{2})) - \frac{96}{\lambda^6}, & r = 0 \end{cases}$
$r^6 \ln r$	$\begin{cases} -\frac{2304}{\lambda^8}(K_0(\lambda r) + \ln r) - \frac{r^2 \ln r}{\lambda^2}\left(\frac{576}{\lambda^4} + \frac{36r^2}{\lambda^2} + r^4\right) \\ \quad -\frac{12r^2}{\lambda^4}\left(\frac{40}{\lambda^2} + r^2\right) - \frac{4224}{\lambda^8}, & r > 0 \\ \frac{2304}{\lambda^8}(\gamma + \log(\frac{\lambda}{2})) - \frac{4224}{\lambda^8}, & r = 0 \end{cases}$

Denote

$$\Delta_r = \frac{1}{r}\frac{d}{dr}\left(r\frac{d}{dr}\right) = \frac{d^2}{dr^2} + \frac{1}{r}\frac{d}{dr}, \tag{10.51}$$

which is the radial part of the Laplacian Δ in two dimensions. Hence the radial solutions of (10.50) satisfying

$$\Delta_r^2(\Delta_r - \lambda^2)\Phi = 0. \tag{10.52}$$

Assume that Φ_1 and Φ_2 are the solutions to equations

$$(\Delta_r - \lambda^2)\Phi_1 = 0, \quad \text{or} \quad (r^2\frac{d^2}{dr^2} + r\frac{d}{dr} - \lambda^2 r^2)\Phi_1 = 0, \tag{10.53}$$

and

$$\Delta_r^2\Phi_2 = 0, \tag{10.54}$$

respectively. Then we can see that

$$\Phi = \Phi_1 + \Phi_2, \tag{10.55}$$

which is a solution to (10.52).

It is known that (10.53) is a modified Bessel's equation of order n, whose general solution can be written as

$$\Phi_1 = AI_0(\lambda r) + BK_0(\lambda r), \tag{10.56}$$

where I_0 and K_0 are the modified Bessel functions of the first and second kind with order zero, respectively.

It is easy to check that (10.54) is equivalent to the Euler differential equation

$$r^4\frac{d^4\Phi_2}{dr^4} + 2r^3\frac{d^3\Phi_2}{dr^3} - r^2\frac{d^2\Phi_2}{dr^2} + r\frac{d\Phi_2}{dr} = 0, \tag{10.57}$$

whose general solution is given by:

$$\Phi_2(r) = a + b\ln r + cr^2 + dr^2\ln r. \tag{10.58}$$

Therefore, we obtain a general particular solution

$$\Phi(r) = AI_0(\lambda r) + BK_0(\lambda r) + a + b\ln r + cr^2 + dr^2 \ln r, \qquad (10.59)$$

whose coefficients can be found by requiring that Φ be continuous at $r = 0$ and satisfying (10.49). Through some tedious algebra, we can obtain one particular solution [49]:

$$\Phi = \begin{cases} -\frac{4}{\lambda^2} - \frac{4\ln r}{\lambda^4} - \frac{r^2\ln r}{\lambda^2} - \frac{4K_0(\lambda r)}{\lambda^4}, & r > 0, \\ -\frac{4}{\lambda^2} + \frac{4\gamma}{\lambda^4} + \frac{4}{\lambda^4}\log(\frac{\lambda}{2}), & r = 0, \end{cases}$$

where γ is the Euler's constant.

10.3.5   The RBF-MFS

Instead of first approximating the right-hand side function of the equation and then finding the approximate particular solution by analytical method, we can find the approximate particular solution directly by the collocation method, i.e., by requiring the approximate particular solution

$$u_p(x,y) = \sum_{j=1}^{N} u_j\varphi_j(x,y), \qquad (10.60)$$

to satisfy the underlying equation $Lu = f$ at the collocation points

$$\sum_{j=1}^{N}(L\varphi_j)(x_i,y_i)u_j = f(x_i,y_i), \quad i = 1,2,\cdots,N. \qquad (10.61)$$

Hence, we end up with solving an $N \times N$ linear system with coefficient matrix $A = [(L\varphi_j)(x_i,y_i)]$ for the unknowns $\{u_j\}_{j=1}^{N}$. Here N is the total number of collocation points.

Note that this technique was first proposed by Golberg in 1995 [24, p. 102] and was restated in the book of Golberg and Chen [26, p. 323]. Among many practitioners in the BEM community, Chen and Tanaka [10] used multiquadric RBF $\varphi(r) = (r^2 + c^2)^{3/2}$ and Kögl and Gaul [36] used RBF $\varphi(r) = r^2 + r^3$ in their applications.

10.3.6   The MFS-DRM for the parabolic equations

For time-dependent diffusion problems, the basic approach is to use the Laplace transform or time-differencing methods to reduce a time-dependent boundary value problem to a sequence of Helmholtz-type equations of the form

$$\Delta u \pm \lambda^2 u = f. \qquad (10.62)$$

In early BEM literature, the DRM was only applied to the case when the major operator $\mathcal{L}$ in $\mathcal{L}u = f$ is kept as the Laplace or harmonic operators.

For instance, $\mathcal{L}u = f$ will be treated as $\Delta u = f \mp \lambda^2 u$. This is primarily due to the difficulty in obtaining particular solutions in closed form for other differential operators. As a result, the DRM is less effective when the right-hand side becomes too complicated. In general, it is preferred to keep the right-hand side as simple as possible so that it can be better approximated by RBFs. Meanwhile, the simpler the right-hand side, the more preserved the differential operator in the left-hand side and the fundamental solution becomes more involved and the better accuracy is expected.

First, let us consider a method that couples the MFS with the Laplace transform [69] for solving the parabolic problem

$$\frac{\partial u}{\partial t}(P,t) = \Delta u(P,t) \quad P \in \Omega \subset R^d, d = 2, 3, t > 0, \tag{10.63}$$

$$u(P,t) = g_1(P) \quad P \in \partial\Omega_1, t > 0, \tag{10.64}$$

$$\frac{\partial u}{\partial n}(P,t) = g_2(P) \quad P \in \partial\Omega_2, t > 0, \tag{10.65}$$

$$u(P,0) = u_0(P) \quad P \in \overline{\Omega}. \tag{10.66}$$

Denote the Laplace transform

$$LT[u(P,t)] \equiv \tilde{u}(P,s) = \int_0^\infty u(P,t)e^{-st}dt, \tag{10.67}$$

for $s \in R$. Using the Laplace transform to (10.63)-(10.66) and the property

$$LT[\frac{\partial u}{\partial t}(P,t)] = s\tilde{u}(P,s) - u_0(P),$$

we obtain the transformed problem

$$(\Delta - s)\tilde{u}(P,s) = -u_0(P) \quad P \in \Omega, \tag{10.68}$$

$$\tilde{u}(P,s) = \tilde{g}_1(P,s) \quad P \in \partial\Omega_1, \tag{10.69}$$

$$\frac{\partial \tilde{u}}{\partial n}(P,s) = \tilde{g}_2(P,s) \quad P \in \partial\Omega_2, \tag{10.70}$$

which becomes a modified Helmholtz equation and can be solved by the MFS-DRM.

Let $\tilde{u}_p$ be a particular solution to

$$(\Delta - \lambda^2)\tilde{u} = -u_0 \equiv f, \quad \lambda = \sqrt{s}. \tag{10.71}$$

Then $v = \tilde{u} - \tilde{u}_p$ satisfies

$$(\Delta - \lambda^2)v(P,s) = 0, \quad P \in \Omega, \tag{10.72}$$

$$v(P,s) = \tilde{g}_1(P,s) - \tilde{u}_p(P), \quad P \in \partial\Omega_1, \tag{10.73}$$

$$\frac{\partial v}{\partial n}(P,s) = \tilde{g}_2(P,s) - \frac{\partial \tilde{u}_p}{\partial n}(P) \quad P \in \partial\Omega_2, \tag{10.74}$$

whose solution can be approximated by

$$v_h = \sum_{j=1}^{n} a_j G(P, Q_j; \lambda), \tag{10.75}$$

where G is the fundamental solution for the operator $\triangle - \lambda^2$.

Hence,

$$\tilde{u}_h = v_h + \tilde{u}_p, \tag{10.76}$$

is an approximation to the solution $\tilde{u}$ of (10.68)-(10.70). Having done this, we can obtain the approximate solution u_h to the original problem (10.63)-(10.66) by the inverse Laplace transform

$$u_h(P, t) = LT^{-1}(\tilde{u}_h), \tag{10.77}$$

which can be evaluated by using Stehfest's method [69].

Another way to solve the parabolic problem (10.63)-(10.66) is to be discrete (10.63) in time first, then use the MFS to solve the modified Helmholtz equation at each time step. For example, using the backward Euler scheme for (10.63), we obtain

$$(\triangle - \frac{1}{\triangle t}) u^{n+1}(P) = -\frac{1}{\triangle t} u^n(P), \tag{10.78}$$

which can be solved using the MFS for the modified Helmholtz equation.

10.4   Kansa's method

In this section, we first introduce how to construct Kansa's method to solve both elliptic and parabolic equations. Then we introduce the Hermite-Birkhoff collocation method to avoid the unsymmetric coefficient matrices resulting from Kansa's method.

10.4.1   Kansa's method for elliptic problems

Without loss of generality, let us consider Poisson's equation

$$\frac{\partial^2 u}{\partial x^2} + \frac{\partial^2 u}{\partial y^2} = f(x, y), \quad (x, y) \in \Omega \tag{10.79}$$

$$u|_{\partial \Omega} = g(x, y), \tag{10.80}$$

Other problems and boundary conditions can be pursued in a similar way.

Let $\{(x_j, y_j)\}_{j=1}^N$ be N collocation points in $\bar{\Omega}$ of which $\{(x_j, y_j)\}_{i=1}^{N_I}$ are interior points and $\{(x_j, y_j)\}_{i=N_I+1}^N$ are boundary points. For simplicity, we consider only Hardy's MQ. For each point (x_j, y_j), let us denote

$$\varphi_j(x, y) = \sqrt{(x - x_j)^2 + (y - y_j)^2 + c^2} = \sqrt{r_j^2 + c^2},$$

where $r_j = \sqrt{(x - x_j)^2 + (y - y_j)^2}$.

It is easy to check that

$$\frac{\partial \varphi_j}{\partial x} = \frac{x - x_j}{\sqrt{r_j^2 + c^2}}, \quad \frac{\partial \varphi_j}{\partial y} = \frac{y - y_j}{\sqrt{r_j^2 + c^2}},$$

$$\frac{\partial^2 \varphi_j}{\partial x^2} = \frac{(y - y_j)^2 + c^2}{(r_j^2 + c^2)^{3/2}}, \quad \frac{\partial^2 \varphi_j}{\partial y^2} = \frac{(x - x_j)^2 + c^2}{(r_j^2 + c^2)^{3/2}}.$$

For the elliptic problem (10.79)-(10.80), we approximate u by $\hat{u}$ by assuming

$$\hat{u}(x, y) = \sum_{j=1}^N c_j \varphi_j(x, y), \tag{10.81}$$

where $\{c_j\}_{j=1}^N$ are the unknown coefficients to be determined.

By substituting (10.81) into (10.79)-(10.80), we have

$$\sum_{j=1}^N \left(\frac{\partial^2 \varphi_j}{\partial x^2} + \frac{\partial^2 \varphi_j}{\partial y^2} \right)(x_i, y_i) c_j = f(x_i, y_i), \quad i = 1, 2, \cdots, N_I, \tag{10.82}$$

$$\sum_{j=1}^N \varphi_j(x_i, y_i) c_j = g(x_i, y_i), \quad i = N_I + 1, N_I + 2, \cdots, N, \tag{10.83}$$

from which we can solve the $N \times N$ linear system of (10.82)-(10.83) for the unknowns $\{c_j\}_{j=1}^N$. Then (10.81) can give us the approximate solution at any point in the domain Ω.

10.4.2 Kansa's method for parabolic equations

For the parabolic problem

$$\frac{\partial u}{\partial t} - \alpha \left(\frac{\partial^2 u}{\partial x^2} + \frac{\partial^2 u}{\partial y^2} \right) = f(x, y, t, u, u_x, u_y),$$

$$(x, y) \in \Omega, 0 \le t \le T, \tag{10.84}$$

$$u|_{\partial \Omega} = g(x, y, t), \quad (x, y) \in \partial \Omega, \tag{10.85}$$

$$u|_{t=0} = h(x, y), \quad (x, y) \in \Omega, \tag{10.86}$$

we can consider the implicit scheme

$$\frac{u^{n+1} - u^n}{\delta t} - \alpha \left(\frac{\partial^2 u^{n+1}}{\partial x^2} + \frac{\partial^2 u^{n+1}}{\partial y^2} \right) = f(x, y, t_n, u^n, u_x^n, u_y^n), \tag{10.87}$$

where δt is the time step, and u^n and u^{n+1} are the solutions at time $t_n = n\delta t$ and $t_{n+1} = (n+1)\delta t$, respectively.

Similar to the elliptic problem, we can assume that the approximate solution for the parabolic problem (10.84)-(10.86) is expressed as

$$u(x, y, t_{n+1}) = \sum_{j=1}^{N} c_j^{n+1} \varphi_j(x, y), \qquad (10.88)$$

where $\{u_j^{n+1}\}_{j=1}^{N}$ are the unknown coefficients to be determined.

By substituting (10.88) into (10.84)-(10.85), we have

$$\sum_{j=1}^{N} (\frac{\varphi_j}{\delta t} - \alpha \frac{\partial^2 \varphi_j}{\partial x^2} - \alpha \frac{\partial^2 \varphi_j}{\partial y^2})(x_i, y_i) c_j^{n+1}$$

$$= \frac{u^n}{\delta t}(x_i, y_i) + f(x_i, y_i, t_n, u^n(x_i, y_i), u_x^n(x_i, y_i), u_y^n(x_i, y_i)),$$

$$i = 1, 2, \cdots, N_I, \qquad (10.89)$$

$$\sum_{j=1}^{N} \varphi_j(x_i, y_i) c_j^{n+1} = g(x_i, y_i, t_{n+1}), \quad i = N_I + 1, N_I + 2, \cdots, N, \quad (10.90)$$

from which we can solve the $N \times N$ linear system of (10.89)-(10.90) for the unknowns $\{c_j^{n+1}\}_{j=1}^{N}$, and obtain the approximate solution (10.88) at any point in the domain Ω. Here we denote

$$u^n(x_i, y_i) = \sum_{j=1}^{N} c_j^n \varphi_j(x_i, y_i),$$

$$u_x^n(x_i, y_i) = \sum_{j=1}^{N} c_j^n \frac{\partial \varphi_j(x_i, y_i)}{\partial x}, \quad u_y^n(x_i, y_i) = \sum_{j=1}^{N} c_j^n \frac{\partial \varphi_j(x_i, y_i)}{\partial y}.$$

10.4.3   The Hermite-Birkhoff collocation method

Kansa's method leads to an unsymmetric coefficient matrix, and no one has been able to prove its nonsingularity so far. Inspired by the scattered Hermite-Birkhoff interpolation method [51, 66], Fasshauer [19] extended it to solving PDEs. In general, let us consider a general linear elliptic problem:

$$Lu(\boldsymbol{x}) = f(\boldsymbol{x}), \quad \boldsymbol{x} \text{ in } \Omega \subset R^d, d = 2, 3, \qquad (10.91)$$
$$u(\boldsymbol{x}) = g(\boldsymbol{x}), \quad \boldsymbol{x} \text{ on } \partial\Omega. \qquad (10.92)$$

The approximate solution u to the problem (10.91)-(10.92) is chosen as

$$u(\boldsymbol{x}) \approx \sum_{j=1}^{N_I} c_j L^\xi \phi(||\boldsymbol{x} - \boldsymbol{\xi}_j||) + \sum_{j=N_I+1}^{N} c_j \phi(||\boldsymbol{x} - \boldsymbol{\xi}_j||), \qquad (10.93)$$

where we denote N_I for the total number of interior collocation points, $N - N_I$ the total number of boundary collocation points, and ϕ some kind of radial basis functions. Here L^ξ indicates that the differential operator L acts on ϕ with respect to the variable ξ.

Substituting (10.93) into (10.91)-(10.92) and satisfying them at the corresponding collocation points, we obtain

$$Lu(\boldsymbol{x}_i) = f(\boldsymbol{x}_i), \quad i = 1, \cdots, N_I, \tag{10.94}$$
$$u(\boldsymbol{x}) = g(\boldsymbol{x}), \quad i = N_I + 1, \cdots, N. \tag{10.95}$$

We can rewrite (10.94)-(10.95) as a linear system

$$A\mathbf{C} = \mathbf{F}, \tag{10.96}$$

where the vectors $\mathbf{C}$ and $\mathbf{F}$ are

$$\mathbf{C} = (c_1, \cdots, c_{N_I}; c_{N_I+1}, \cdots, c_N)',$$
$$\mathbf{F} = (f(\boldsymbol{x}_1), \cdots, f(\boldsymbol{x}_{N_I}); f(\boldsymbol{x}_{N_I+1}), \cdots, f(\boldsymbol{x}_N))',$$

and the coefficient matrix

$$A = \begin{bmatrix} A_{11} & A_{12} \\ A_{21} & A_{22} \end{bmatrix} \tag{10.97}$$

where

$$(A_{11})_{ij} = LL^\xi \phi(||\boldsymbol{x}_i - \boldsymbol{\xi}_j||), \quad 1 \le i, j \le N_I,$$
$$(A_{12})_{ij} = L\phi(||\boldsymbol{x}_i - \boldsymbol{\xi}_j||), \quad 1 \le i \le N_I, N_I + 1 \le j \le N,$$
$$(A_{21})_{ij} = L^\xi \phi(||\boldsymbol{x}_i - \boldsymbol{\xi}_j||), \quad N_I + 1 \le i \le N, 1 \le j \le N_I,$$
$$(A_{22})_{ij} = \phi(||\boldsymbol{x}_i - \boldsymbol{\xi}_j||), \quad N_I + 1 \le i, j \le N.$$

To illustrate the algorithm clearly, let us assume that the differential operator

$$L = \triangle \equiv \frac{\partial^2}{\partial x^2} + \frac{\partial^2}{\partial y^2}$$

and the RBF $\phi = \sqrt{(x - \xi)^2 + (y - \eta)^2 + c^2}$.

By simple calculation, we have

$$\frac{\partial \phi}{\partial \xi} = \frac{-(x - \xi)}{\phi^{1/2}}, \quad \frac{\partial^2 \phi}{\partial \xi^2} = \frac{(y - \eta)^2 + c^2}{\phi^{3/2}},$$

which gives

$$L^\xi \phi = \frac{(x - \xi)^2 + (y - \eta)^2 + 2c^2}{\phi^{3/2}}.$$

Similarly, we can obtain

$$\frac{\partial}{\partial x}(L^\xi \phi) = \frac{-(x - \xi)[(x - \xi)^2 + (y - \eta)^2 + 4c^2]}{\phi^{5/2}},$$

and

$$\frac{\partial^2}{\partial x^2}(L^\xi \phi)$$
$$= \frac{[(x-\xi)^2 + (y-\eta)^2 + c^2][2(x-\xi)^2 - (y-\eta)^2 - 4c^2] + 15(x-\xi)^2 c^2}{\phi^{7/2}},$$

which gives

$$LL^\xi \phi = \{[(x-\xi)^2 + (y-\eta)^2 + c^2][(x-\xi)^2 + (y-\eta)^2 - 8c^2]$$
$$+ 15(x-\xi)^2 c^2 + 15(y-\eta)^2 c^2\}/\phi^{7/2}.$$

From the example, we can see that implementation of the Hermite-Birkhoff collocation method is much more complicated than Kansa's method, and it is not feasible for nonlinear problems. Furthermore, experiences [19, 40, 42] show that both methods achieve similar accuracy. A nice feature about the Hermite-Birkhoff collocation method is that the method is mathematically sound and error analysis is available [23].

10.5   Numerical examples with MATLAB codes

In this section, we present several examples solved by RBF meshless methods. MATLAB source codes are provided for all examples so that readers can grasp such methods easily and use them for other application problems.

10.5.1   Elliptic problems

Here we consider the problem (10.79)-(10.80) with $f(x, y) = 13 \exp(-2x+3y)$ and Dirichlet boundary function $g(x, y)$ chosen such that the exact solution is

$$u(x, y) = \exp(-2x + 3y).$$

This problem was considered previously by Kansa [34, p. 159], where he obtained the maximum norm error 0.4480 with only 10 boundary points and 20 interior points. Here we solved this problem by both the RBF-MFS method and Kansa's method. Both methods give almost the same results. For example, just run the listed codes *Kansa_All.m* and *LiDRM_All.m* without any modification. They all give relative maximum error 0.0066 for the 21 × 21 grid.

The MATLAB code *LiDRM_All.m* is shown below:

```
%-----------------------------------------------------------------
% LiDRM_all.m: use the RBF-DRM method to solve elliptic problem.
%
```

```
% This code implements most RBF functions.
% Running directly the code gives the relative max error 0.0066.
%---------------------------------------------------------------

tt=cputime;

%total intervals in each direction
mn=20;
h=1.0/mn;
%total collocation points
mn1=mn+1;
nn=mn1*mn1;
cs=0.2;  %shape parameter for MQ

% NK: order of r^{2k+1}, k=1,2,3,...
%      or order of TPS r^{2k}*ln(r), k=1,2, ...
%ID=1 for MQ; ID=2 for r^3;
%ID=3 for r^4ln(r); ID=4 for e^(-beta*r^2)
ID=2;
NK=3;
beta=5;

p1=zeros(1,nn);
p2=zeros(1,nn);
for j=1:mn1
  for i=1:mn1
    p1(i+(j-1)*mn1)=h*(i-1);
    p2(i+(j-1)*mn1)=h*(j-1);
  end
end

aa=zeros(nn,nn);
%t=cputime;

% Form matrix 'aa'. Here 'nn' is the # of collocation pts
% aa=[L\phi_j(P_i)]
for i=1:nn
  for j=1:nn
    if ID == 1     % use MQ
      r2=(p1(i)-p1(j))^2+(p2(i)-p2(j))^2+cs*cs;
      aa(i,j)=(r2+cs*cs)/(r2^1.5);
    elseif ID == 2 % use r^{2k+1}, k=1
      tmp=sqrt((p1(i)-p1(j))^2+(p2(i)-p2(j))^2);
      aa(i,j)=(2*NK+1)^2*tmp^(2*NK-1);
    elseif ID == 3
```

```
            tmp= sqrt((p1(i)-p1(j))^2+(p2(i)-p2(j))^2);
            if tmp > 0
                aa(i,j)=16*tmp^2*log(tmp)+8*tmp^2;
            end
        elseif ID == 4
            tmp= sqrt((p1(i)-p1(j))^2+(p2(i)-p2(j))^2);
            aa(i,j)=(-4*beta+4*beta^2*tmp^2)*exp(-beta*tmp^2);
        end
    end
end

% RHS of the original equation
g=zeros(nn,1);
%g=2*exp(p1'-p2');
g=13*exp(-2*p1'+3*p2');
[L,U]=lu(aa);
c=U\(L\g);

% boundary nodes
q1=[0 0 0 0 0 .2 .4 .6 .8 1 1 1 1 1 1 .8 .6 .4 .2 ];
q2=[0 .2 .4 .6 .8 1 1 1 1 1 1 .8 .6 .4 .2 0 0 0 0 0 ];

% here we get particular solutions at all points
n=20;
for i=1:nn+n
    sum=0;
    if i <= nn
        x=p1(i); y=p2(i);
    else
        x=q1(i-nn); y=q2(i-nn);
    end
    for j=1:nn
        if ID == 1 % use MQ
            phi=sqrt((x-p1(j))^2+(y-p2(j))^2+cs*cs);
        elseif ID == 2 % use r^{2k+1}, k=1,2, ...
            phi=sqrt((x-p1(j))^2+(y-p2(j))^2)^(2*NK+1);
        elseif ID == 3
            r2=(x-p1(j))^2+(y-p2(j))^2;
            if r2 > 0
                phi=r2^2*log(r2)/2;
            end
        elseif ID == 4
            r2=(x-p1(j))^2+(y-p2(j))^2;
            phi=exp(-beta*r2);
        end
```

```
       sum=sum+c(j)*phi;
   end
   v2(i)=sum;
end

% BC values for MFS = original BC - particular solution
v3=v2(nn+1:nn+n);
g2=zeros(1,n);
%g2=exp(q1-q2)+exp(q1).*cos(q2)-v3;
g2=exp(-2*q1+3*q2)-v3;

% form the source points for MFS
 m=n-1;
rad=0:2.*pi/m:2.*pi*(1-1/m);
x1=5*cos(rad)+.5;
y1=5*sin(rad)+.5;

% form matrix 'A' for MFS, 'n' is the # of boundary points
A=zeros(n,n);
 for i=1:n
    A(i,n)=1;
    for j=1:n-1
       A(i,j)=log((x1(j)-q1(i))^2+(y1(j)-q2(i))^2);
    end
 end

% solution returned in vector 'z'
[L,U]=lu(A);
z=U\(L\g2');

%   z=A\g2';
   err_max = 0;
   fid2=fopen('out_LiDRM.txt','w');
% evaluate the error at those interesting points only!
    for j=1:nn
% get the points (x,y) we are interested in!
        x=p1(j);
        y=p2(j);
% 'n' is the # of BC points
        v=0;
        for i=1:n-1
           d1=x1(i)-x;
           d2=y1(i)-y;
           v=v+z(i)*log(d1*d1+d2*d2);
        end
```

```
% final solution = particular solution + MFS' solution
        total=v2(j)+v+z(n);
        exact=exp(-2*x+3*y);
        diff=abs(exact-total)/exact;
        fprintf(fid2,'%9.4f %9.4f %10.6e %10.6e %10.6e\n',...
                    x,    y,    total, exact, diff);
        if (err_max < diff)
            err_max = diff;     % find the max relative error
        end
    end
status=fclose(fid2);

disp('max relative err='), err_max,
```

The MATLAB code *Kansa_All.m* is shown below:

```
%------------------------------------------------------------
% KansaAll.m: use the so-called Kansa method to solve
%             an elliptic problem.
% This implementation includes most popular RBFs.
%------------------------------------------------------------

mn=20;
h=1.0/(1.0+mn);
NI=mn*mn;      %total # of interior nodes

NBC=16;        %total BC points
c=0.9;
%c=1.8;        %shape parameter
N=NI+NBC;      %total # of collocation points
Tol=1.0e-15;

% ID=1 for MQ, ID=2 for r^{2k+1},
% ID=3 for r^4ln(r), ID=4 for e^{-beta*r^2}
ID=2;
NK=3;          % the parameter k in r^{2k+1}
beta=10;

XBC=zeros(1,NBC);    %x-coordinate of BC nodes
YBC=zeros(1,NBC);    %y-coordinate of BC nodes
XBC=[0.00 0.25 0.50 0.75 1.00 1.00 1.00 1.00 ...
     1.00 0.75 0.50 0.25 0.00 0.00 0.00 0.00];
YBC=[0.00 0.00 0.00 0.00 0.00 0.25 0.50 0.75 ...
     1.00 1.00 1.00 1.00 1.00 0.75 0.50 0.25];
```

```
X=zeros(1,N);
Y=zeros(1,N);
for j=1:mn
  for i=1:mn
    X(i+(j-1)*mn)=h*i;
    Y(i+(j-1)*mn)=h*j;
  end
end

%after this, 'h' is used for scaling, so h=1 means no scaling!
h=1;
for i=1:NBC
   X(i+NI)=XBC(i);
   Y(i+NI)=YBC(i);
end

A=zeros(N,N);   %the Kansa's matrix
A11=zeros(NI,NI);   %A1 submatrix of A
A12=zeros(NI,NBC);   %submatrix of A
A21=zeros(NBC,NI);    %submatrix of A
A22=zeros(NBC,NBC);    %submatrix of A
GF=zeros(N,1);          %global RHS

%form global matrix A & rescale it
for i=1:NI
   GF(i)=13*exp(-2*X(i)+3*Y(i));
   GF(i)=GF(i)*(h*h);
   for j=1:N
     if ID == 1
       tmp=((X(i)-X(j))^2+(Y(i)-Y(j))^2)+c*c;
       A(i,j)=(tmp+c*c)/(tmp^1.5);
       elseif ID == 2
       tmp=sqrt((X(i)-X(j))^2+(Y(i)-Y(j))^2);
       A(i,j)=(2*NK+1)^2*tmp^(2*NK-1);
       elseif ID == 3
          tmp= sqrt((X(i)-X(j))^2+(Y(i)-Y(j))^2);
          if tmp > Tol
             A(i,j)=16*tmp^2*log(tmp)+8*tmp^2;
          end
       elseif ID == 4
          tmp= sqrt((X(i)-X(j))^2+(Y(i)-Y(j))^2);
          A(i,j)=(-4*beta+4*beta^2*tmp^2)*exp(-beta*tmp^2);
       end
       A(i,j)=A(i,j)*(h*h);
```

```
      end
end

%here we rescale the BC nodes
for i=NI+1:NI+NBC
    GF(i)=exp(-2*X(i)+3*Y(i));
%    GF(i)=GF(i)/(h*h);
    for j=1:N
      if ID == 1
        tmp=(X(i)-X(j))^2+(Y(i)-Y(j))^2+c*c;
        A(i,j)=sqrt(tmp);
      elseif ID == 2
            tmp=sqrt((X(i)-X(j))^2+(Y(i)-Y(j))^2);
            A(i,j)=tmp^(2*NK+1);
      elseif ID == 3
          tmp=sqrt((X(i)-X(j))^2+(Y(i)-Y(j))^2);
          if tmp > Tol
              A(i,j)=tmp^4*log(tmp);
          end
      elseif ID == 4
          tmp=sqrt((X(i)-X(j))^2+(Y(i)-Y(j))^2);
          A(i,j)=exp(-beta*tmp^2);
      end
%      A(i,j)=A(i,j)/(h*h);
    end
end

%Solve the equation: u_xx+u_yy=f(x,y), u=g(x,y) on BC
u=zeros(N,1);      %approx solution at all points
uex=zeros(N,1);    % exact solution

for i=1:N
      uex(i)=exp(-2*X(i)+3*Y(i));
end

%Solving the global system directly gives worse
%accuracy than solved by LU decomposition!
%u=A\GF

%solve by LU decomposition
[L,U]=lu(A);
u=U\(L\GF);

fid2=fopen('out_KansaAll.txt','w');
for i=1:N
```

```
%find the approx solution at each point
uappr=0.0;
for j=1:N
      r2=(X(i)-X(j))^2+(Y(i)-Y(j))^2;
  if ID == 1
   tmp=sqrt(r2+c*c);
  elseif ID == 2
      tmp=sqrt(r2)^(2*NK+1);
  elseif ID == 3
     if r2 > Tol
         tmp=r2^2*log(r2)/2;
     end
  elseif ID == 4
      tmp=exp(-beta*r2);
  end
   uappr=uappr+u(j)*tmp;
end
err(i)=abs(uappr-uex(i))/uex(i);
fprintf(fid2,'%9.4f %9.4f %10.6e %10.6e %10.6e\n',...
               X(i),Y(i), uappr, uex(i),err(i));
end
status=fclose(fid2);
disp('max relative err='), max(err)
```

10.5.2 Biharmonic problems

Here we consider solving the biharmonic equation [45]

$$L\psi \equiv \nabla^4\psi = 0 \quad \text{in } \Omega \tag{10.98}$$

in a closed, bounded domain Ω in the plane, subject to either

$$B_1\psi \equiv \psi = f(x,y), \quad B_2\psi \equiv \frac{\partial\psi}{\partial n} = g(x,y), \quad \text{for } (x,y) \in \partial\Omega, \tag{10.99}$$

or

$$B_1\psi \equiv \psi = f(x,y), \quad B_2\psi \equiv \nabla^2\psi = g(x,y), \quad \text{for } (x,y) \in \partial\Omega. \tag{10.100}$$

Here (10.99) is the so-called clamped boundary condition, while (10.100) is the simply supported boundary condition. $\frac{\partial\psi}{\partial n}$ denotes the normal derivative on the boundary $\partial\Omega$.

To solve the direct problem (10.98)-(10.100) by the RBF meshless method, we assume that the approximate solution can be expressed as

$$\psi_N(x,y) = \sum_{j=1}^{N_I+N_d+N_n} c_j\varphi(\|(x,y)-(x_j,y_j)\|), \tag{10.101}$$

where c_j are the unknown coefficients to be determined, and $\varphi(r_j)$ is some kind of RBF [56, 5]. Here $r_j = ||(x,y)-(x_j,y_j)||$ is the Euclidean distance between points (x,y) and (x_j,y_j). We denote $\{(x_j,y_j)\}_1^{N_I}$ the collocation points inside the domain Ω, while $\{(x_j,y_j)\}_{N_I+1}^{N_I+N_d}$, $\{(x_j,y_j)\}_{N_I+N_d+1}^{N_I+N_d+N_n}$ are the collocation points on the boundary $\partial\Omega$ for boundary conditions B_1 and B_2, respectively. Note that we use different nodes for imposing different types of boundary conditions, even though they are all imposed on $\partial\Omega$.

Substituting (10.101) into (10.98), (10.99), or (10.100), and making the equations hold true at the corresponding collocation points, we have

$$\sum_{j=1}^{N}(L\varphi)(||(x_i,y_i)-(x_j,y_j)||)c_j = 0, \quad i = 1,2,\cdots,N_I, \qquad (10.102)$$

$$\sum_{j=1}^{N}(B_1\varphi)(||(x_i,y_i)-(x_j,y_j)||)c_j = f(x_i,y_i),$$

$$i = N_I+1,\cdots,N_I+N_d, \qquad (10.103)$$

$$\sum_{j=1}^{N}(B_2\varphi)(||(x_i,y_i)-(x_j,y_j)||)c_j = g(x_i,y_i),$$

$$i = N_I+N_d+1,\cdots,N. \qquad (10.104)$$

The resultant linear system (10.102)-(10.104) is solved for the unknowns $\{c_j\}_{j=1}^{N}$, where $N = N_I + N_d + N_n$ denotes the total number of collocation points.

Example 10.1
Our first example is taken from [35, p. 441]: solve the equation $\nabla^4\psi = 0$ in the unit square with the exact solution $\psi = x^2 + y^2$, and boundary conditions ψ and $\nabla^2\psi$ imposed on the boundary $\partial\Omega$. In [35] this problem was solved by the method of fundamental solutions (MFS) and achieved accuracy of order 10^{-4} in the maximum error for ψ.

Here we solved this problem by using 39×39 uniformly distributed interior nodes, 320 uniformly distributed nodes located on $\partial\Omega$ for imposing boundary condition B_1, and another 320 uniformly distributed nodes between B_1 nodes for satisfying boundary condition B_2. Due to the density of this node distribution, a corresponding coarse 19×19 node case is presented in Fig. 10.1 to illustrate how the collocation nodes are distributed. The obtained numerical solution and corresponding maximum error for ψ are given in Fig. 10.2, which shows clearly that our meshless method also produces accuracy comparable to [35]. ▯

Example 10.2
Our second example is also taken from [35, p. 445] and solved there by MFS. This example considers the thin plate bending with clamped boundary condi-

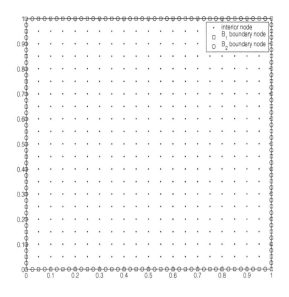

FIGURE 10.1
Example 1. The distribution of the collocation nodes for the case 19×19. [From Fig. 1 of [45]. Copyright 2004. John Wiley & Sons Limited. Reproduced with permission.]

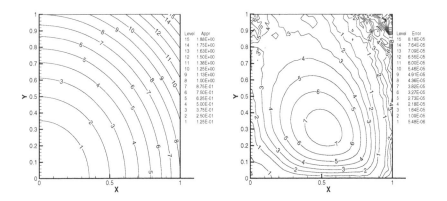

FIGURE 10.2
Example 1. (Left) numerical solution for ψ; (Right) the corresponding maximum error. [From Fig. 2 of [45]. Copyright 2004. John Wiley & Sons Limited. Reproduced with permission.]

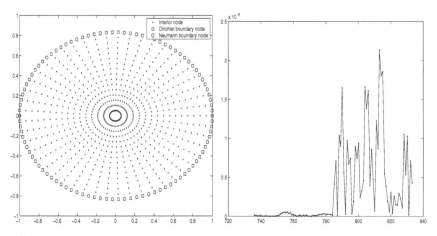

FIGURE 10.3
Example 2. (Left) the distribution of the collocation nodes;   (Right) the
maximum error on the domain boundary. [From Fig. 3 of [45]. Copyright
2004. John Wiley & Sons Limited. Reproduced with permission.]

tion: more specifically we solve the equation $\nabla^4 \psi = 0$ subject to the bound-
ary conditions $\psi = -\frac{(x^2+y^2)^2}{64}$ and $\frac{\partial \psi}{\partial n} = -\frac{\partial}{\partial n}\left(\frac{(x^2+y^2)^2}{64}\right)$. Here domain Ω is an
ellipse, whose semi-major axis and semi-minor axis are 1 and 0.8333, respec-
tively. Note that the exact solution for this problem is unavailable.

We solved this problem by the RBF meshless method using 833 collocation nodes (the first 735 are interior nodes, the remaining 98 are boundary
nodes). The detailed nodal distribution is shown in Fig. 10.3 (Left). The
corresponding maximum error for ψ along the domain boundary is shown in
Fig. 10.3 (Right), which shows clearly that our meshless method produces a
very accurate solution.   ▯

The MATLAB code *ellipse_rbf.m* is listed below.

```
%------------------------------------------------------------------
% ellipse_rbf.m: solve the biharmonic problem over an ellipse.
%
% mod 1: uex=x^3+y^3;
% mod 2: uex=0.5*(x*sin(x)*cosh(y)-x*cos(x)*sinh(y))
% mod 3: uex=(x^2+y^2)/64;
%------------------------------------------------------------------
clear all;

Nc=49;    % # of nodes in theta direction
Nr=16;    % # of nodes in the radial direction
```

```
NI=(Nr-1)*Nc;    % # of interior nodes: one in the center
Nd=Nc;      % # of Diri BC nodes on the circle
Nn=Nc;      % # of Neum Bc nodes on the  circle
Nall=NI+Nd+Nn;   % total # of collocation points
NK=9;      % the degree of the basis function r^NK
MODEL=3;    % indicator which model

dc=2*pi/Nc;
Ra=1; Rb=0.8333;    %semi-major, semi-minor of the ellipse
dra=Ra/Nr; drb=Rb/Nr;

% generate the nodes: 1) inner; 2) Diri BC; 3) Neum BC
for j=1:Nc
    for i=1:Nr-1
        % uniform grid
        ra=i*dra;   rb=i*drb;
        theta=(j-1)*dc;
        ij=(j-1)*(Nr-1)+i;
        XX(ij)=ra*cos(theta);   YY(ij)=rb*sin(theta);
        XC(ij)=XX(ij);   YC(ij)=YY(ij);
      end
end

% Diri BC
for j=1:Nd
    theta=(j-1)*dc;
    ij=NI+j;
    XX(ij)=Ra*cos(theta); YY(ij)=Rb*sin(theta);
    XC(ij)=XX(ij);    YC(ij)=YY(ij);
  end


% Neum BC
for j=1:Nn
    theta=(j-1)*dc+0.5*dc;   % nodes between Diri BC nodes
    ij=NI+Nd+j;
    XX(ij)=Ra*cos(theta); YY(ij)=Rb*sin(theta);
    XC(ij)=XX(ij); YC(ij)=YY(ij);
end

% form the coeff. matrix and HRS
for i=1:NI       % inner nodes
    rhs(i)=0;
  for j=1:Nall
     dist=sqrt((XX(i)-XC(j))^2+(YY(i)-YC(j))^2);
```

```
      matA(i,j)=(NK*(NK-2))^2*dist^(NK-4);
  end
end

for i=1:Nd       % Diri BC nodes
    ii=NI+i;
    if MODEL==1
      % mod 1
      rhs(ii)=XX(ii)^3+YY(ii)^3;
    elseif MODEL==3
      rhs(ii)=-(XX(ii)^2+YY(ii)^2)^2/64;
    elseif MODEL==2
      % mod 2
      rhs(ii)=0.5*XX(ii)*(sin(XX(ii))*cosh(YY(ii)) ...
                        -cos(XX(ii))*sinh(YY(ii)));
    end
    for j=1:Nall
       dist=sqrt((XX(ii)-XC(j))^2+(YY(ii)-YC(j))^2);
      matA(ii,j)=dist^NK;
    end
end


for i=1:Nn       % Neum BC nodes
    ii=NI+Nd+i;
    theta=(i-1)*dc+0.5*dc;
    dxdt=-Ra*sin(theta); dydt=Rb*cos(theta);
    % find the normal at this boundary node
    if abs(theta) < eps
        cosc=1; sinc=0;
    elseif abs(theta-pi) < eps
        cosc=-1; sinc=0;
    else
        psi=atan(dydt/dxdt);
        cosc=sin(psi); sinc=cos(psi);
    end

    if MODEL==1
      % mod 1
      rhs(ii)=3*XX(ii)^2*cosc+3*YY(ii)^2*sinc;   % du/dn
    elseif MODEL==3
      xi=XX(ii); yi=YY(ii);
      t1=2*(xi^2+yi^2)/64*2*xi;
      t2=2*(xi^2+yi^2)/64*2*yi;
      rhs(ii)=-(t1*cosc+t2*sinc);
```

```
      elseif MODEL==2
        % mod 2
        xi=XX(ii); yi=YY(ii);
        c1=cosh(yi); s1=sinh(yi);
        t1=(xi*cos(xi)+sin(xi))*c1-(cos(xi)-xi*sin(xi))*s1;
        t2=xi*sin(xi)*s1-xi*cos(xi)*c1;
        rhs(ii)=0.5*(t1*cosc+t2*sinc);
      end
   for j=1:Nall
        dist=sqrt((XX(ii)-XC(j))^2+(YY(ii)-YC(j))^2);
        phix=NK*dist^(NK-2)*(XX(ii)-XC(j));
        phiy=NK*dist^(NK-2)*(YY(ii)-YC(j));
      matA(ii,j)=phix*cosc+phiy*sinc;
   end
end

% solve the linear system
u=matA\rhs';

plot(XX(1:NI),YY(1:NI),'.',...      % interior node
   XX(NI+1:NI+Nd),YY(NI+1:NI+Nd),'s',...   % Diri node
   XX(NI+Nd+1:NI+Nd+Nn),YY(NI+Nd+1:NI+Nd+Nn),'o'); %Neum node
legend('interior node','Dirichlet boundary node',...
       'Neumann boundary node');

fid=fopen('V1.doc','w');   % save the results
% get the numerical solution and compare with the exact one
%for i=NI+1:Nall   % model 3: only compare boundary errors
for i=1:Nall
    if MODEL==1
      % mod 1
      uex(i)=XX(i)^3+YY(i)^3;      % exact solution
    elseif MODEL==3
       uex(i)=-(XX(i)^2+YY(i)^2)^2/64;
    elseif MODEL==2
      % mod 2
      uex(i)=0.5*XX(i)*(sin(XX(i))*cosh(YY(i))...
                     -cos(XX(i))*sinh(YY(i)));
    end
    uapp(i)=0;
    for j=1:Nall
        dist=sqrt((XX(i)-XC(j))^2+(YY(i)-YC(j))^2);
        uapp(i)=uapp(i)+u(j)*dist^NK;
    end
    err(i)=abs(uex(i)-uapp(i));
```

```
    if abs(uex(i)) > 1e-6
        Rerr(i)=err(i)/abs(uex(i));    % relative error
    else
        Rerr(i)=0;
    end
    fprintf(fid, '%9.4f %9.4f %10.6e %10.6e %10.6e %10.6e\n', ...
        XX(i),YY(i),uapp(i),uex(i),err(i),Rerr(i));
end
disp('Nall=, max err, Relative err='), ...
    Nall, max(err), max(Rerr),
% plot only the errors on the boundaries
figure(2),
plot(NI+1:Nall,err(NI+1:Nall),'b-', ...
    NI+1:Nall,err(NI+1:Nall),'.');
status=fclose(fid);
```

10.6　Coupling RBF meshless methods with DDM

The resultant coefficient matrices from RBF meshless methods are highly ill-conditioned, which hinders the applicability of the RBF method to solve large-scale problems since the ill condition gets worse with increasing collocation points. Hence the domain decomposition method (DDM) can be used to reduce the condition number for the RBF meshless methods by decomposing a global large-domain problem into many small subdomain problems.

The earliest concept of the domain decomposition method was introduced as a classical Schwarz alternating algorithm by Schwarz in 1870, which provided a fast and robust algorithm for solving the linear systems resulting from discretizations of PDEs. With the development of modern supercomputers, DDM has become a common tool for solving linear systems of equations. For a thorough theory on DDM and its applications, readers can consult some review papers [6, 41], books [57, 60, 61], and the proceedings of the international conferences on DDM (see *www.ddm.org*).

However, coupling DDM with RBF meshless methods are very recent work. It seems that the earliest work in this direction is due to Dubal, who investigated DDM coupled with multiquadric function for one-dimensional problems in [16]. Later, Hon and his collaborators applied DDM to higher-dimensional problems [31, 46, 65, 68]. Now, it has become a popular way to use coupled DDM and RBF to solve different PDEs [1, 14, 32, 47].

Below we describe the basic DDM for BRF meshless methods for a general steady-state problem in d-dimension ($d = 1, 2, 3$):

$$L\phi = f(\vec{x}) \quad \text{in } \Omega, \qquad Bu = g(\vec{x}) \quad \text{on } \partial\Omega, \qquad (10.105)$$

where L is an arbitrary differential operator, B is an operator imposed as boundary conditions, such as Dirichlet, Neumann, Robin, and mixing of them.

Let $\{P_i = (\vec{x}_i)\}_{i=1}^{N}$ be N collocation points in Ω, of which $\{(\vec{x}_i)\}_{i=1}^{N_I}$ are interior points; $\{(\vec{x}_i)\}_{i=N_I+1}^{N}$ are boundary points. The approximate solution for the problem (10.105) can be expressed as

$$\phi(\vec{x}) = \sum_{j=1}^{N} \phi_j \varphi_j(\vec{x}), \qquad (10.106)$$

where $\{\phi_j\}_{j=1}^{N}$ are the unknown coefficients to be determined, and $\varphi_j(\vec{x}) = \varphi(\|P - P_j\|)$ can be any radial basis functions we discussed in previous sections.

By substituting (10.106) into (10.105), we have

$$\sum_{j=1}^{N} (L\varphi_j)(\vec{x}_i)\phi_j = f(\vec{x}_i), \quad i = 1, 2, \cdots, N_I, \qquad (10.107)$$

$$\sum_{j=1}^{N} (B\varphi_j)(\vec{x}_i)\phi_j = g(\vec{x}_i), \quad i = N_I + 1, N_I + 2, \cdots, N, \qquad (10.108)$$

which can be solved for the unknown coefficients $\{\phi_j\}_{j=1}^{N}$.

10.6.1 Overlapping DDM

Let Ω be partitioned into two subdomains Ω_1 and Ω_2, where $\Omega_1 \bigcap \Omega_2 \neq \emptyset$. The artificial boundaries Γ_i are the part of the boundary of Ω_i that is interior of Ω, the rest of the boundaries are denoted by $\partial\Omega_i \backslash \Gamma_i$. We can solve the problem on Ω by the classic additive Schwarz algorithm which can be written as

$$L\phi_1^n = f \quad \text{in} \ \ \Omega_1$$
$$B\phi_1^n = g \quad \text{on} \ \ \partial\Omega_1 \backslash \Gamma_1$$
$$\phi_1^n = \phi_2^{n-1} \quad \text{on} \ \ \Gamma_1$$

and

$$L\phi_2^n = f \quad \text{in} \ \ \Omega_2$$
$$B\phi_2^n = g \quad \text{on} \ \ \partial\Omega_2 \backslash \Gamma_2$$
$$\phi_2^n = \phi_1^{n-1} \quad \text{on} \ \ \Gamma_2$$

where on each subdomain, the numerical scheme is carried out as outlined in the last section.

The multiplicative Schwarz algorithm can be written as

$$L\phi_1^n = f \quad \text{in} \ \ \Omega_1$$

$$B\phi_1^n = g \quad \text{on} \quad \partial\Omega_1 \backslash \Gamma_1$$
$$\phi_1^n = \phi_2^{n-1} \quad \text{on} \quad \Gamma_1$$

and

$$L\phi_2^n = f \quad \text{in} \quad \Omega_2$$
$$B\phi_2^n = g \quad \text{on} \quad \partial\Omega_2 \backslash \Gamma_2$$
$$\phi_2^n = \phi_1^n \quad \text{on} \quad \Gamma_2$$

where subdomain 2 uses the newly obtained solution from subdomain 1 as the artificial boundary condition, from which we can imagine that the multiplicative version should converge faster than the additive version.

The generalization of the Schwarz method to many subdomains is straightforward by coloring or grouping [57, 60].

10.6.2   Non-overlapping DDM

Let Ω be partitioned into two non-overlapping subdomains Ω_1 and Ω_2, where $\Omega_1 \bigcap \Omega_2 = \emptyset$. The subdomain interface $\Gamma = \partial\Omega_1 \bigcap \partial\Omega_2$.

Then the non-overlapping iteration algorithm (i.e., the so-called iterative substructuring method) can be written as

$$L\phi_1^n = f \quad \text{in} \quad \Omega_1$$
$$B\phi_1^n = g \quad \text{on} \quad \partial\Omega_1 \backslash \Gamma$$
$$\phi_1^n = \lambda^{n-1} \quad \text{on} \quad \Gamma$$

and

$$L\phi_2^n = f \quad \text{in} \quad \Omega_2$$
$$B\phi_2^n = g \quad \text{on} \quad \partial\Omega_2 \backslash \Gamma$$
$$\frac{\partial\phi_2^n}{\partial n_L} = \frac{\partial\phi_1^{n-1}}{\partial n_L} \quad \text{on} \quad \Gamma$$

where

$$\lambda^n|_\Gamma = \theta\lambda^{n-1}|_\Gamma + (1-\theta)\phi_2^n|_\Gamma, \quad 0 < \theta < 1,$$

and the conormal derivative $\frac{\partial w}{\partial n_L}$ is defined as $\frac{\partial w}{\partial n_L} = aw_x n_x + bw_y n_y$.

Here on the interface we use the Dirichlet-Neumann method. Other methods such as the Neumann-Neumann method and the Robin method can be considered similarly [57, Ch. 1].

Extensions to many subdomains of the Dirichlet-Neumann method can be achieved by using a black- and white-coloring, say Ω is decomposed into many subdomains $\Omega_i, i = 1, \cdots, M$. Denote $\Gamma_{ij} = \partial\Omega_i \bigcap \partial\Omega_j$ the interface between two adjacent subdomains Ω_i and Ω_j, the two groups of subdomains are

$I_B = \{1 \le i \le M : \Omega_i \text{ is black}\}$ and $I_w = \{1 \le i \le M : \Omega_i \text{ is white}\}$. The above algorithm becomes: solve

$$L\phi_i^n = f \quad \text{in} \ \Omega_i, \quad \forall \, i \in I_B$$
$$\phi_i^n = \theta\phi_j^{n-1} + (1-\theta)\phi_i^{n-1} \quad \text{on} \ \Gamma_{ij}, \forall \, j \in I_W : \Gamma_{ij} \ne 0$$

Then solve

$$L\phi_j^n = f \quad \text{in} \ \Omega_j, \quad \forall \, j \in I_W$$
$$\frac{\partial \phi_j^n}{\partial n_L} = \frac{\partial \phi_j^{n-1}}{\partial n_L} \quad \text{on} \ \Gamma_{ij}, \forall \, i \in I_B : \Gamma_{ij} \ne 0$$

This is the so-called block-parallel version [57, p. 25]. We can consider the block-sequential version by using the newly obtained solution from I_B as the interface condition on the white subdomains.

10.6.3 One numerical example

Here we show one numerical example for DDM with RBF meshless methods. The example is from our published work [46]. Consider the Poisson problem

$$-\triangle\phi = xe^y \quad \text{in} \ \Omega,$$
$$\phi = -xe^y \quad \text{on} \ \partial\Omega.$$

This problem was used by Smith et al. [60, p. 8] to demonstrate the numerical convergence of the Schwarz method on a nontrivial domain, which is a union of a square and a circle. They used centered finite difference on the square and piecewise linear finite elements on the circle. The number of iterations and maximum error obtained on several meshes are presented in Table 10.4.

To compare with Smith et al.'s results, we used non-overlapping DDM with a uniform $(Nx+1) \times (Ny+1)$ grid on the square $[0,2]^2$, and a uniform radial grid $Nr \times Nc$ on the circle which centered at $(2,2)$ and radius 1, where $Nx+1, Ny+1$ are the number of points in x and y directions, Nr, Nc are the number of points in the radial and counterclockwise angle directions, respectively. Furthermore, we used a fixed error tolerance of 1.0×10^{-6}, zero initial guess on the artificial boundaries or the subdomain interfaces, and LU decomposition solver. The number of iterations and maximum error obtained by our method in both multiplicative and additive versions on several grid cases are presented in Tables 10.5 and 10.6*.

Tables 10.5 and 10.6 show that the number of iterations for the multiplicative Schwarz method is roughly half of that needed for the additive version.

---

*Tables 10.4-10.6 are reproduced from Table I-III of *Numerical Methods for Partial Differential Equations*, Jichun Li and Y.C. Hon, Domain decomposition for radial basis meshless methods, Vol. 20, p. 450–462. Copyright 2004. John Wiley & Sons Limited. Reproduced with permission.

These phenomena have also been observed in our earlier work [68], and more examples were provided there. Note that such features are also observed for the FEM domain decomposition method [60, p. 25] due to the fact that Gauss-Seidel (resp. Jacobi) algorithm is formally a multiplicative (resp. additive) Schwarz method [41, p. 134].

Comparing Tables 10.5-10.6 with Table 10.4, we can see that such an RBF DDM method is competitive with the classical DDM results obtained by Smith et al. Especially when we use 961 nodes on the square and 1081 nodes on the circle, the accuracy for r^5 is already better than Smith et al.'s accuracy achieved with 5329 nodes on the square and 2145 nodes on the circle. The accuracy for r^7 is much higher than Smith et al.'s with much fewer nodal points.

To see more clearly, we plot a sample collocation grid of 11×11 on the square, and 10×12 on the circle in Fig. 10.4 (Left). The convergence history is plotted in Fig. 10.4 (Right) for both additive and multiplicative Schwarz methods in the grid case of 31×31 on the square and 30×36 on the circle. Also the solution and pointwise error for this case are plotted in Fig. 10.5.

TABLE 10.4

Maximum error obtained by Smith et al.

No. of	nodes	Max.	error	
Square	Circle	Square	Circle	Iter. steps
100	45	1.2×10^{-2}	6.7×10^{-3}	6
361	153	3.1×10^{-3}	2.2×10^{-3}	6
1369	561	1.0×10^{-3}	6.9×10^{-4}	7
5329	2145	2.7×10^{-4}	2.1×10^{-4}	8

TABLE 10.5

Maximum error obtained by our method with basis function r^5: No. of nodes on square is $(Nx + 1) \times (Ny + 1)$, No. of nodes on circle is $Nr \times Nc + 1$

No. of	nodes	Max.	error	Iter. steps	Iter. steps
Square	Circle	Square	Circle	additive	multiplicative
$11^2 = 121$	121	1.22×10^{-2}	1.79×10^{-2}	15	10
$21^2 = 441$	481	6.10×10^{-4}	1.06×10^{-3}	17	10
$21^2 = 441$	721	5.37×10^{-4}	8.80×10^{-4}	17	10
$31^2 = 961$	1081	1.14×10^{-4}	2.09×10^{-4}	17	10

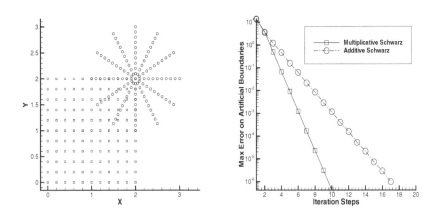

FIGURE 10.4

(Left) sample collocation grid; (Right) convergence history. [From Fig. 1 of [46]. Copyright 2004. John Wiley & Sons Limited. Reproduced with permission.]

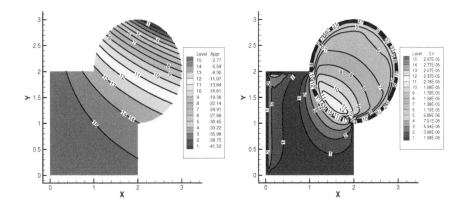

FIGURE 10.5

(Left) numerical solution; (Right) pointwise error. [From Fig. 2 of [46]. Copyright 2004. John Wiley & Sons Limited. Reproduced with permission.]

TABLE 10.6

Maximum error obtained by our method with basis function r^7: No. of nodes on square is $(Nx + 1) \times (Ny + 1)$, No. of nodes on circle is $Nr \times Nc + 1$

No. of nodes		Max. error		Iter. steps	Iter. steps
Square	Circle	Square	Circle	additive	multiplicative
$11^2 = 121$	121	5.17×10^{-3}	8.19×10^{-3}	15	10
$21^2 = 441$	481	1.36×10^{-4}	2.21×10^{-4}	17	10
$21^2 441$	721	1.69×10^{-4}	2.43×10^{-4}	17	10
$31^2 = 961$	1081	2.17×10^{-5}	3.11×10^{-5}	17	10

10.7   Bibliographical remarks

The book by Fasshauer [20] provides an elementary introduction to RBFs and many MATLAB codes are provided for RBF interpolations and solving PDEs. Due to the nice meshless property, recently RBFs have been used in solving inverse problems (cf. papers [11, 15, 29, 30, 44, 45] and references therein). For more recent progress on RBFs for solving various PDEs, readers can consult the review paper by Fornberg and Flyer [21] and recent books [9, 22].

10.8   Exercises

1. Given $\phi = 1 + r$, where $r = \sqrt{x^2 + y^2}$. Prove that $\Phi = \frac{r^3}{9} + \frac{r^2}{4}$ satisfies the equation $\triangle \Phi = \phi$.

2. Let the thin-plate spline $\phi = r^2 \ln r$, where $r = \sqrt{x^2 + y^2}$. Prove that $\Phi = \frac{r^4 \ln r}{16} - \frac{r^4}{32}$ satisfies the equation $\triangle \Phi = \phi$.

3. Implement the RBF-MFS method to solve Poisson's equation

$$\frac{\partial^2 u}{\partial x^2} + \frac{\partial^2 u}{\partial y^2} = 2e^{x-y}, \quad (x, y) \in \Omega,$$

$$u(x, y) = e^{x-y} + e^x \cos y, \quad (x, y) \in \partial\Omega,$$

where Ω is the oval of Golberg and Chen [25] described by the parametric equation equations

$$x = r(\theta) \cos \theta, y = r(\theta) \sin \theta, r(\theta) = \sqrt{\cos 2\theta + \sqrt{1.1 - \sin^2 2\theta}}, 0 \le \theta \le 2\pi.$$

You can use MQ or other RBFs. Note that the exact solution is given by $u(x, y) = e^{x-y} + e^x \cos y$.

4. Prove that in polar coordinates the 2-D Laplacian operator $\triangle \equiv \frac{\partial^2}{\partial x^2} + \frac{\partial^2}{\partial y^2}$ becomes

$$\frac{d^2}{dr^2} + \frac{1}{r}\frac{d}{dr} + \frac{1}{r^2}\frac{d^2}{d\theta^2},$$

i.e., $u_{xx} + u_{yy} = u_{rr} + \frac{1}{r}u_r + \frac{1}{r^2}u_{\theta\theta}$.

5. Recall that the gamma function $\Gamma(x) = \int_0^\infty t^{x-1}e^{-t}dt$. Prove that $\Gamma(x+1) = x\Gamma(x)$ and $\Gamma(\frac{1}{2}) = \sqrt{\pi}$.

6. Let $y = x^\alpha \sum_{n=0}^\infty a_n x^n$ be the solution of Bessel's equation (10.18). Show that

$$a_1 = a_3 = a_5 = \cdots = 0$$

and

$$a_{2k}(\alpha) = \frac{(-1)^k}{2^{2k}k!(k+\alpha)(k-1+\alpha)\cdots(1+\alpha)}a_0 = \frac{(-1)^k\Gamma(1+\alpha)}{2^{2k}k!\Gamma(k+\alpha+1)}a_0, \quad k = 1, 2, \cdots.$$

7. Prove that

$$\frac{d}{dx}[x^\alpha J_\alpha(x)] = x^\alpha J_{\alpha-1}(x)$$

and

$$\frac{d}{dx}[x^{-\alpha}J_\alpha(x)] = -x^{-\alpha}J_{\alpha+1}(x).$$

8. Show that the 3-D Laplace equation can be transformed to

$$\frac{1}{r^2}[(r^2 u_r)_r + \frac{1}{\sin\phi}(u_\phi \sin\phi)_\phi + \frac{1}{\sin^2\phi}u_{\theta\theta}] = 0$$

under spherical coordinates.

9. Show that the general radial solution to the 3-D modified Helmholtz equation $(\triangle - \lambda^2)u = 0$ is given by

$$u = \frac{c}{r}\exp(-\lambda r).$$

References

[1] H. Adibi and J. Es'haghi. Numerical solution for biharmonic equation using multilevel radial basis functions and domain decomposition methods. *Appl. Math. Comput.*, 186(1):246–255, 2007.

[2] R.K. Beatson and M.J.D. Powell. Univariate interpolation on a regular finite grid by a multiquadric plus a linear polynomial. *IMA J. Numer. Anal.*, 12:107–133, 1992.

[3] M.D. Buhmann. Radial functions on compact support. *Proc. Edinburgh Math. Soc. (2)*, 41(1):33–46, 1998.

[4] M.D. Buhmann. Radial basis functions. *Acta Numerica*, 9:1–38, 2000.

[5] M.D. Buhmann. *Radial Basis Functions: Theory and Implementations.* Cambridge University Press, Cambridge, UK, 2003.

[6] T.F. Chan and T.P. Mathew. Domain decomposition algorithms. *Acta Numerica*, 3:61–143, 1994.

[7] C.S. Chen, C.A. Brebbia and H. Power. Dual reciprocity method using compactly supported radial basis functions. *Commun. Numer. Mech. Engrg.*, 15:137–150, 1999.

[8] C.S. Chen and Y.F. Rashed. Evaluation of thin plate spline based particular solutions for Helmholtz-type operators for the DRM. *Mech. Res. Comm.*, 25:195–201, 1998.

[9] W. Chen, Z.-J. Fu and C.S. Chen. *Recent Advances in Radial Basis Function Collocation Methods.* Springer-Verlag, 2014.

[10] W. Chen and M. Tanaka. A meshless, integration-free, and boundary-only RBF technique. *Comp. Math. Appl.*, 43:379–391, 2002.

[11] A.H.-D. Cheng and J. Cabral. Direct solution of ill-posed boundary value problems by radial basis function collocation method. *Internat. J. Numer. Methods Engrg.*, 64(1):45–64, 2005.

[12] A.H.-D. Cheng, O. Lafe and S. Grilli. Dual reciprocity BEM based on global interpolation functions. *Eng. Anal. Boundary Elements*, 13:303–311, 1994.

[13] R.S.C. Cheng. *Delta-Trigonometric and Spline Methods Using the Single-Layer Potential Representation.* Ph.D. dissertation, University of Maryland, 1987.

[14] P.P. Chinchapatnam, K. Djidjeli and P.B. Nair. Domain decomposition for time-dependent problems using radial based meshless methods. *Numer. Methods Partial Differential Equations*, 23(1):38–59, 2007.

[15] M. Dehghan and M. Tatari. The radial basis functions method for identifying an unknown parameter in a parabolic equation with overspecified data. *Numer. Methods for Partial Differential Equations*, 23(5):984–997, 2007.

[16] M.R. Dubal. Domain decomposition and local refinement for multiquadric approximations. *J. Appl. Sci. Comput.*, 1(1):146–171, 1994.

[17] J. Duchon. Splines minimizing rotation-invariant semi-norms in Sobolev spaces. *Constructive Theory of Functions of Several Variables.* (Proc. Conf., Math. Res. Inst., Oberwolfach, 1976), pp. 85–100. Lecture Notes in Math., Vol. 571, Springer, Berlin, 1977.

[18] G. Fairweather and A. Karageorghis. The method of fundamental solutions for elliptic boundary value problems. *Adv. Comput. Math.*, 9(1-2):69–95, 1998.

[19] G.E. Fasshauer. Solving partial differential equations by collocation with radial basis functions, in *Proceedings of Chamonix 1996* (eds. by A. Le Mehaute, C. Rabut and L.L. Schumaker), p. 131–138, Vanderbilt University Press, Nashville, TN, 1997.

[20] G.E. Fasshauer. *Meshfree Approximation Methods with MATLAB*. World Scientific Publishers, Singapore, 2007.

[21] B. Fornberg and N. Flyer. Solving PDEs with radial basis functions. *Acta Numer.*, 24:215–258, 2015.

[22] B. Fornberg and N. Flyer. *A Primer on Radial Basis Functions with Applications to the Geosciences*. SIAM, Philadelphia, PA, 2015.

[23] C. Franke and R. Schback. Convergence order estimates of meshless collocation methods using radial basis functions. *Adv. Comput. Math.*, 8(4):381–399, 1998.

[24] M.A. Golberg. The numerical evaluation of particular solutions in the BEM – a review. *Boundary Elems. Comm.*, 6:99–106, 1995.

[25] M.A. Golberg and C.S. Chen. A bibliography on radial basis function approximation. *Boundary Elems. Comm.*, 7:155–163, 1996.

[26] M.A. Golberg and C.S. Chen. *Discrete Projection Methods for Integral Equations*. Computational Mechanics Publications, Southampton, UK, 1997.

[27] M.A. Golberg and C.S. Chen. The method of fundamental solutions for potential, Helmholtz and diffusion problems, in *Boundary Inetgral Methods – Numerical and Mathematical Aspects*, M.A. Golberg (Ed.), Computational Mechanics Publications, Southampton, UK, pp. 103–176, 1998.

[28] M.A. Golberg, C.S. Chen and M. Ganesh. Particular solutions of 3D Helmholtz-type equations using compactly supported radial basis functions. *Eng. Anal. with Boundary Elements*, 24(7-8):539–548, 2000.

[29] Y.C. Hon and T. Wei. A fundamental solution method for inverse heat conduction problems. *Eng. Anal. with Boundary Elements*, 28(5):489–495, 2004.

[30] Y.C. Hon and Z. Wu. A numerical computation for inverse determination problem. *Eng. Anal. with Boundary Elements*, 24(7-8):599–606, 2000.

[31] Y.C. Hon and Z. Wu. Additive Schwarz domain decomposition with a radial basis approximation. *Int. J. Appl. Math.*, 4(1):81–98, 2000.

[32] M.S. Ingber, C.S. Chen and J.A. Tanski. A mesh free approach using radial basis functions and parallel domain decomposition for solving three-dimensional diffusion equations. *Int. J. Numer. Methods Engrg.*, 60(13):2183–2201, 2004.

[33] R.L. Johnston and G. Fairweather. The method of fundamental solutions for problems in potential flow. *Applied Mathematical Modelling*, 8:265–270, 1984.

[34] E.J. Kansa. Multiquatric – A scattered data approximation scheme with applications to computational fluid dynamics II. *Comp. Math. Appl.*, 19(8-9):147–161, 1990.

[35] A. Karageorghis and G. Fairweather. The method of fundamental solutions for the numerical solution of the biharmonic equation. *J. Comp. Phys.*, 69:434–459, 1987.

[36] M. Kögl and L. Gaul. Dual reciprocity boundary element method for three-dimensional problems of dynamic piezoelectricity. *Boundary Elements XXI*, 537-548, Southampton, UK, 1999.

[37] M. Katsurada and H. Okamoto. A mathematical study of the charge simulation method. *Journal of the Faculty of Science, University of Tokyo, Section 1A*, 35:507–518, 1988.

[38] M. Katsurada and H. Okamoto. The collocation points of the fundamental solution method for the potential problem. *Comp. Math. Appl.*, 31:123–137, 1996.

[39] V.D. Kupradze and M.A. Aaleksidze. The method of functional equations for the approximate solution of certain boundary value problems. *U.S.S.R. Computational Mathematics and Mathematical Physics*, 4:82–126, 1964.

[40] A. La Rocca, A. Hernandez Rosales and H. Power. Radial basis function Hermite collocation approach for the solution of time dependent convection diffusion problems. *Eng. Anal. with Boundary Elements*, 29(4):359–370, 2005.

[41] P. Le Tallec. Domain decomposition methods in computational mechanics. *Comput. Mech. Adv.*, 1:121–220, 1994.

[42] V.M.A. Leitao. A meshless method for Kirchhoff plate bending problems. *Int. J. Numer. Meth. Engng.*, 52:1107–1130, 2001.

[43] J. Levesley. Pointwise estimates for multivariate interpolation using conditionally positive definite functions. *Approximation Theory, Wavelets and Applications* (Maratea, 1994), 381–401, NATO Adv. Sci. Inst. Ser. C Math. Phys. Sci., 454, Kluwer Acad. Publ., Dordrecht, 1995.

[44] J. Li. A radial basis meshless method for solving inverse boundary value problems. *Comm. Numer. Methods Engrg.*, 20(1):51–61, 2004.

[45] J. Li. Application of radial basis meshless methods to direct and inverse biharmonic boundary value problems. *Comm. Numer. Methods Engrg.*, 21(4):169–182, 2005.

[46] J. Li and Y.C. Hon. Domain decomposition for radial basis meshless methods. *Numer. Methods Partial Differential Equations*, 20(3):450–462, 2004.

[47] L. Ling and E.J. Kansa. Preconditioning for radial basis functions with domain decomposition methods. *Math. Comput. Modelling*, 40(13):1413–1427, 2004.

[48] W.R. Madych. Miscellaneous error bounds for multiquadric and related interpolators. *Comput. Math. Appl.*, 24:121–138, 1992.

[49] A.S. Muleshkov, C.S. Chen, M.A. Golberg and A.H-D. Cheng. Analytic particular solutions for inhomogeneous Helmholtz-type equations. *Advances in Computational Engineering & Sciences*, S.N. Atluri, F.W. Brust (eds), Tech Science Press, CA, Vol. 1, pp. 27–32, 2000.

[50] A.S. Muleshkov, M.A. Golberg and C.S. Chen. Particular solutions of Helmholtz-type operators using higher order polyharmonic splines. *Computational Mechanics*, 23:411–419, 1999.

[51] F.J. Narcowich and J.D. Ward. Generalized Hermite interpolation via matrix-valued conditionally positive definite functions. *Math. Comp.*, 63:661–687, 1994.

[52] D. Nardini and C.A. Brebbia. A new approach to free vibration analysis using boundary elements, in *Boundary Element Methods in Engineering*, C.A. Brebbia (Ed.), Springer-Verlag, Berlin, 1982.

[53] P.W. Partridge, C.A. Brebbia and L.C. Wrobel. *The Dual Reciprocity Boundary Element Method*. Computational Mechanics Publications, London, 1992.

[54] P.W. Partridge and B. Sensale. The method of fundamental solutions with dual reciprocity for diffusion and diffusion-convection using subdomains. *Eng. Anal. with Boundary Elements*, 24:633–641, 2000.

[55] M.J.D. Powell. The uniform convergence of thin plate spline interpolation in two dimensions. *Numer. Math.*, 68:107–128, 1994.

[56] M.J.D. Powell. The theory of radial basis function approximation in 1990. *Advances in Numerical Analysis*, Vol. II (Lancaster, 1990), 105–210, Oxford Univ. Press, New York, 1992.

[57] A. Quarteroni and A. Valli. *Domain Decomposition Methods for Partial Differential Equations*. Oxford Science Publications, Oxford, UK, 1999.

[58] R. Schaback. Improved error bounds for scattered data interpolation by radial basis functions. *Math. Comp.*, 68:201–216, 1999.

[59] R. Schaback and H. Wendland. Kernel techniques: from machine learning to meshless methods. *Acta Numerica*, 15:543–639, 2006.

[60] B. Smith and P. Bjorstad and W. Gropp. *Domain Decomposition, Parallel Multilevel Methods for Elliptic Partial Differential Equations.* Cambridge University Press, Cambridge, UK, 1996.

[61] A. Toselli and O. Widlund. *Domain Decomposition Methods – Algorithms and Theory.* Springer, New York, 2004.

[62] H. Wendland. Piecewise polynomial, positive definite and compactly supported radial functions of minimal degree. *Adv. Comput. Math.*, 4:389–396, 1995.

[63] H. Wendland. Error estimates for interpolation by compactly supported radial basis functions of minimal degree. *J. Approx. Theory*, 93:258–272, 1998.

[64] H. Wendland. *Scattered Data Approximation.* Cambridge University Press, Cambridge, UK, 2005.

[65] A.S.M. Wong, Y.C. Hon, T.S. Li, S.L. Chung and E.J. Kansa. Multizone decomposition for simulation of time-dependent problems using the multiquadric scheme. *Comput. Math. Appl.*, 37(8):23–43, 1999.

[66] Z. Wu. Hermite-Birkhoff interpolation of scattered data by radial basis functions. *Approx. Theory Appl.*, 8:1–10, 1992.

[67] Z. Wu. Multivariate compactly supported positive definite radial functions. *Adv. Comput. Math.*, 4:283–292, 1995.

[68] X. Zhou, Y.C. Hon and J. Li. Overlapping domain decomposition method by radial basis functions. *Appl. Numer. Math.*, 44(1-2):241–255, 2003.

[69] S. Zhu and P. Satravaha and X. Lu. Solving linear diffusion equations with the dual reciprocity method in Laplace space. *Eng. Anal. with Boundary Elements*, 13:1–10, 1994.

11

Other Meshless Methods

In this chapter, we introduce some so-called Galerkin-based meshless methods. First, in Sec. 11.1, we focus on how to construct meshless shape functions. Then we introduce the element-free Galerkin method in Sec. 11.2, which is followed by the meshless local Petrov-Galerkin methods in Sec. 11.3.

11.1 Construction of meshless shape functions

In meshless methods, the shape functions are constructed for each node based on its selected local neighboring nodes. The construction does not depend on any fixed relation between nodes, which can be randomly distributed. In this section we discuss several popular techniques for constructing meshless shape functions.

11.1.1 The smooth particle hydrodynamics method

The smooth particle hydrodynamics (SPH) method [11, 20, 21] is claimed to be the oldest meshless method [7]. The basic idea of this method is approximating a function $u(\boldsymbol{x})$ on a domain Ω by

$$u_h(\boldsymbol{x}) \approx \int_\Omega w_h(|\boldsymbol{x} - \boldsymbol{y}|)u(\boldsymbol{y})d\Omega_y \tag{11.1}$$

$$\approx \sum_I w_h(|\boldsymbol{x} - \boldsymbol{x}_I|)u_I \triangle V_I, \tag{11.2}$$

where $w_h(\cdot)$ is a weight function, h is a measure of the support size of the weight function, $u_I \equiv u(\boldsymbol{x}_I)$, $\triangle V_I$ is the domain volume surrounding node I, $|\boldsymbol{x} - \boldsymbol{x}_I|$ denotes the Euclidean distance between node $\boldsymbol{x}$ and node $\boldsymbol{x}_I$. The weight function usually satisfies the following properties:
(a) $w_h(|\boldsymbol{x} - \boldsymbol{y}|) \geq 0$ has a compact support, i.e., zero outside of $|\boldsymbol{x} - \boldsymbol{y}| > \gamma h$, where γ is a scaling factor;
(b) the normality condition: $\int_\Omega w_h(|\boldsymbol{x}|)d\boldsymbol{x} = 1$;
(c) $w_h(|\boldsymbol{x}|) \to \delta(|\boldsymbol{x}|)$ as $h \to 0$, where δ is the Dirac delta function.
 Three commonly used weight functions are the Gaussian function, the cubic spline, and the quartic spline. Let $r = |\boldsymbol{x}|, s = \frac{r}{h}$, we have:

Gaussian: $w_h(r) = \begin{cases} \frac{C_d}{h^d}\exp(-s^2), & \text{if } 0 \le s \le 1, \\ 0 & \text{if } s > 1, \end{cases}$

where $1 \le d \le 3$ is the spatial dimension, $C_d = \frac{1}{\pi^{d/2}}$.

Cubic spline: $w_h(r) = \frac{C_d}{h^d} \begin{cases} 1 - \frac{3}{2}s^2 + \frac{3}{4}s^3, & \text{if } 0 \le s < 1, \\ \frac{1}{4}(2-s)^3, & \text{if } 1 \le s \le 2, \\ 0 & \text{if } s > 2, \end{cases}$

where the normality constant $C_d = \frac{2}{3}, \frac{10}{7\pi}$ and $\frac{1}{\pi}$ for one, two, and three dimensions, respectively.

Quartic spline: $w_h(r) = \frac{C_d}{h^d} \begin{cases} 1 - 6s^2 + 8s^3 - 3s^4, & \text{if } 0 \le s \le 1, \\ 0 & \text{if } s > 1, \end{cases}$

where $C_d = \frac{5}{4}, \frac{5}{\pi}$ and $\frac{105}{16\pi}$ for one, two, and three dimensions, respectively.

Example 11.1

Prove that for the cubic spline, the normality constant $C_d = \frac{2}{3}, \frac{10}{7\pi}$ and $\frac{1}{\pi}$ for one, two, and three dimensions, respectively. □

Proof. In one-dimensional space, by the normality condition, we have

$$1 = \frac{2C_d}{h}\{\int_0^h [1 - \frac{3}{2}(\frac{r}{h})^2 + \frac{3}{4}(\frac{r}{h})^3]dr + \int_h^{2h}\frac{1}{4}(2-\frac{r}{h})^3 dr\}$$

$$= 2C_d\{\int_0^1 (1 - \frac{3}{2}s^2 + \frac{3}{4}s^3)ds + \int_1^2 \frac{1}{4}(2-s)^3 ds\} = 2C_d \cdot \frac{3}{4},$$

which leads to $C_d = \frac{2}{3}$.

In two-dimensional space, the normality condition becomes

$$1 = \frac{C_d}{h^2}\{\int_0^h [1 - \frac{3}{2}(\frac{r}{h})^2 + \frac{3}{4}(\frac{r}{h})^3]2\pi r dr + \int_h^{2h}\frac{1}{4}(2-\frac{r}{h})^3 2\pi r dr\}$$

$$= C_d\{\int_0^1 (1 - \frac{3}{2}s^2 + \frac{3}{4}s^3)2\pi s ds + \int_1^2 \frac{1}{4}(2-s)^3 2\pi s ds\}$$

$$= \pi C_d\{(s^2 - \frac{3}{4}s^4 + \frac{3}{10}s^5)|_{s=0}^1 + (2s^2 - 2s^3 + \frac{3}{4}s^4 - \frac{1}{10}s^5)|_{s=1}^2\},$$

from which we obtain $C_d = \frac{10}{7\pi}$.

Similarly, in three-dimensional space, using the normality condition

$$1 = \frac{C_d}{h^3}\{\int_0^h [1 - \frac{3}{2}(\frac{r}{h})^2 + \frac{3}{4}(\frac{r}{h})^3]4\pi r^2 dr + \int_h^{2h}\frac{1}{4}(2-\frac{r}{h})^3 4\pi r^2 dr\},$$

we have $C_d = \frac{1}{\pi}$. □

In terms of finite element methods, we can rewrite (11.2) as

$$u_h(\boldsymbol{x}) \approx \sum_I u_I \phi_I(\boldsymbol{x}),$$

where $\phi_I(\boldsymbol{x}) = w_h(|\boldsymbol{x} - \boldsymbol{x}_I|)\Delta V_I$ are the SPH shape functions.

11.1.2 The moving least-square approximation

Another approach for developing a meshless shape function is to use a moving least-square (MLS) approximation [15], which originates from data fitting.

In the MLS approximation, we define a local approximation to $u(\boldsymbol{x})$ at any point $\boldsymbol{x}_I$ by

$$u_h(\boldsymbol{x}, \overline{\boldsymbol{x}}) = \sum_{i=1}^{m} p_i(\boldsymbol{x}) a_i(\overline{\boldsymbol{x}}), \tag{11.3}$$

where m is the number of terms in the basis functions $p_i(\boldsymbol{x})$. The unknown coefficients $a_i(\overline{\boldsymbol{x}})$ are obtained by minimizing the difference between the local approximation (11.3) and the function $u(\boldsymbol{x})$, i.e.,

$$J = \sum_{I=1}^{n} w(|\boldsymbol{x} - \boldsymbol{x}_I|) |u_h(\boldsymbol{x}_I, \boldsymbol{x}) - u(\boldsymbol{x}_I)|^2$$

$$= \sum_{I=1}^{n} w(|\boldsymbol{x} - \boldsymbol{x}_I|) [\sum_i p_i(\boldsymbol{x}_I) a_i(\boldsymbol{x}) - u_I]^2.$$

Here $w(|\boldsymbol{x} - \boldsymbol{x}_I|)$ is a weight function with compact support such as that used in SPH, and n is the number of nodes in the neighborhood of $\boldsymbol{x}$, for which the weight function $w(|\boldsymbol{x} - \boldsymbol{x}_I|) \neq 0$.

The coefficients $a(\boldsymbol{x})$ can be found by

$$0 = \frac{\partial J}{\partial a_j} = 2 \sum_{I=1}^{n} w(|\boldsymbol{x} - \boldsymbol{x}_I|)(\sum_i p_i(\boldsymbol{x}_I) a_i(\boldsymbol{x}) - u_I) p_j(\boldsymbol{x}_I), \quad j = 1, \cdots, m,$$

i.e.,

$$A(\boldsymbol{x}) a(\boldsymbol{x}) = B(\boldsymbol{x}) u, \tag{11.4}$$

where we denote

$$A(\boldsymbol{x}) = \sum_{I=1}^{n} w(|\boldsymbol{x} - \boldsymbol{x}_I|) \boldsymbol{p}(\boldsymbol{x}_I) \boldsymbol{p}^T(\boldsymbol{x}_I),$$

$$B(\boldsymbol{x}) = [w(|\boldsymbol{x} - \boldsymbol{x}_1|) \boldsymbol{p}(\boldsymbol{x}_1), \cdots, w(|\boldsymbol{x} - \boldsymbol{x}_n|) \boldsymbol{p}(\boldsymbol{x}_n)],$$

where $\boldsymbol{p}^T(\boldsymbol{x}) = (p_1(\boldsymbol{x}), \cdots, p_m(\boldsymbol{x}))$.

Substituting (11.4) into (11.3), we obtain the MLS approximation

$$u_h(\boldsymbol{x}) = \sum_{i=1}^{m} p_i(\boldsymbol{x})(A^{-1}(\boldsymbol{x}) B(\boldsymbol{x}) u)_i = \sum_{I=1}^{n} \Phi_I(\boldsymbol{x}) u_I, \tag{11.5}$$

where $\Phi_I(\boldsymbol{x}) = \boldsymbol{p}^T(\boldsymbol{x}) A^{-1}(\boldsymbol{x}) w(|\boldsymbol{x} - \boldsymbol{x}_I|) \boldsymbol{p}(\boldsymbol{x}_I)$. The $m \times m$ matrix $A(\boldsymbol{x})$ is often called the moment matrix, which has to be inverted whenever the MLS shape functions are to be evaluated. It should be noted that shape functions Φ_I do not satisfy the Kronecker delta criterion: $\Phi_I(\boldsymbol{x}_J) \neq \delta_{IJ}$.

Hence $u_h(\boldsymbol{x}_J) \neq u_J$. This property makes the imposition of essential boundary conditions more complicated than the traditional finite element method.

Examples of commonly used basis $p(\boldsymbol{x})$ in two dimensions are:

Linear basis: $p^T(\boldsymbol{x}) = (1, x, y)$, in which case $m = 3$;

Quadratic basis: $p^T(\boldsymbol{x}) = (1, x, y, x^2, y^2, xy)$, in which case $m = 6$.

If a zero-order basis $p(\boldsymbol{x}) = 1$ is used, then the MLS shape function becomes

$$\Phi_I^0(\boldsymbol{x}) = w(|\boldsymbol{x} - \boldsymbol{x}_I|)/\sum_{I=1}^n w(|\boldsymbol{x} - \boldsymbol{x}_I|), \qquad (11.6)$$

which is the Shepard function.

11.1.3   The partition of unity method

The essense of the partition of unity method (PUM) [5, 6] is to take a partition of unity and multiply it by some local basis approximating the exact solution of the underlying partial differential equations. Hence this method can be more efficient than the usual finite element methods, when the users have some prior knowledge about the underlying problem.

DEFINITION 11.1   *Let $\Omega \subset R^d$ be an open set, $\{\Omega_i\}$ be an open cover of Ω satisfying a pointwise overlap condition: there exists a positive integer M such that for any $\boldsymbol{x} \in \Omega$,*

$$card\{ i : \boldsymbol{x} \in \Omega_i\} \leq M.$$

Furthermore, let $\{\phi_i\}$ be a Lipschitz partition of unity subordinate to the cover $\{\Omega_i\}$ satisfying

$$supp(\phi_i) \subset closure(\Omega_i), \quad \forall i \in \Lambda,$$
$$\sum_{i \in \Lambda} \phi_i = 1, \quad \forall \boldsymbol{x} \in \Omega,$$
$$\|\phi_i\|_{L^\infty(R^d)} \leq C_\infty,$$
$$\|\nabla\phi_i\|_{L^\infty(R^d)} \leq C_g/diam(\Omega_i),$$

where C_∞ and C_g are two constants. Then $\{\phi_i\}$ is called a (M, C_∞, C_g) partition of unity [6] subordinate to the cover $\{\Omega_i\}$. If $\{\phi_i\} \subset C^m(R^d)$, then the partition of unity $\{\phi_i\}$ is said to be of degree m. The covering sets $\{\Omega_i\}$ are called patches.

Below we present some one-dimensional partition of unity on the interval $(0, 1)$. Considering patches $\Omega_i = (x_{i-1}, x_{i+1}), i = 0, 1, \cdots, n$, where the mesh size $h = \frac{1}{n}, x_i = ih, i = 0, \cdots, n$. Furthermore, we denote $x_{-1} = -h, x_{n+1} = 1 + h$.

Example 11.2
Let the usual piecewise linear hat function

$$\phi(x) = \begin{cases} 1 + \frac{x}{h}, & \text{if } x \in (-h, 0], \\ 1 - \frac{x}{h}, & \text{if } x \in (0, h), \\ 0 & \text{elsewhere.} \end{cases}$$

Then $\phi_i(x) = \phi(x - x_i), i = 0, \cdots, n$, form a partition of unity subordinate to the cover $\{\Omega_i\}$. ☐

Example 11.3
A partition of unity of higher regularity can be constructed similarly. Consider a C^1 function

$$\phi(x) = \frac{1}{h^3} \begin{cases} (x + h)^2(h - 2x), & \text{if } x \in (-h, 0], \\ (h - x)^2(h + 2x), & \text{if } x \in (0, h), \\ 0 & \text{elsewhere.} \end{cases}$$

Then $\phi_i(x) = \phi(x - x_i), i = 0, \cdots, n$, form a partition of unity subordinate to the cover $\{\Omega_i\}$. ☐

Example 11.4
Let Ω_i be any cover of Ω satisfying an overlap condition (i.e., no more than M patches overlap in any given point $x \in \Omega$). Let w_j be Lipschitz continuous functions with support Ω_j. Then the Shepard function

$$\phi_j(x) = w_j(x) / \sum_i w_i(x)$$

forms a partition of unity subordinate to the cover $\{\Omega_i\}$. ☐

As we mentioned earlier, in the partition of unity method, some prior information about the underlying problem is used for the approximation. For example, for the one-dimensional Helmholtz equation

$$-u'' + k^2 u = f, \quad \text{on } (0, 1), \tag{11.7}$$

Babuska and Melenk [6] introduced the approximation

$$u_h(x) = \sum_{I=1}^{N} \Phi_I^0(x)(a_{0I} + a_{1I}x + \cdots + a_{mI}x^m + b_{1I}\sinh kx + b_{2I}\cosh kx)$$

$$= \sum_{I=1}^{N} \Phi_I^0(x) \sum_i \beta_{iI} p_i(x),$$

where $\Phi_I^0(x)$ is the Shepard function (11.6), and

$$\beta_{iI} = [a_{0I}, \cdots, a_{mI}, b_{1I}, b_{2I}],$$

$$\boldsymbol{p}^T(x) = [1, x, \cdots, x^m, \sinh kx, \cosh kx].$$

Note that $\sinh kx$ and $\cosh kx$ are the fundamental solution of (11.7). The unknown coefficients can be determined by a Galerkin procedure.

Babuska and Melenk [6] also introduced a more general approximation

$$u_h(x) = \sum_{I=1}^{N} \Phi_I^0(x)(\sum_J a_J L_{JI}(x))$$

$$= \sum_J \sum_{I:x_J \in \Omega_I} \Phi_I^0(x)L_{JI}(x)a_J,$$

where $L_{JI}(x)$ are the Lagrange interpolants (i.e., $L_{JI}(x_k) = \delta_{IK}$ for any J), and a_J is the approximate value associated with node x_J.

Duarte and Oden [10] used the partition of unity concept in a more general manner and constructed an approximation as

$$u_h(x) = \sum_{I=1}^{N} \Phi_I^k(x)(u_I + \sum_{j=1}^{m} b_{jI}q_j(x)),$$

where the extrinsic basis $q_j(x)$ is a monomial basis of any order greater than k, and $\Phi_I^k(x)$ is the MLS shape function of order k.

11.2 The element-free Galerkin method

Let us consider a two-dimensional elastostatic problem on the domain Ω bounded by Γ:

$$\nabla \cdot \sigma + \boldsymbol{b} = 0 \quad \text{in } \Omega, \tag{11.8}$$

subject to the boundary conditions

$$\sigma \cdot \boldsymbol{n} = \bar{t} \quad \text{on } \Gamma_t, \tag{11.9}$$

$$\boldsymbol{u} = \bar{\boldsymbol{u}} \quad \text{on } \Gamma_u, \tag{11.10}$$

where σ is the stress tensor, $\boldsymbol{b}$ is the body force, $\boldsymbol{u}$ is the displacement field, $\boldsymbol{n}$ is the unit outward normal to $\Gamma \equiv \Gamma_t \cup \Gamma_u, \Gamma_t \cap \Gamma_u = 0$, $\bar{t}$ and $\bar{\boldsymbol{u}}$ are given boundary values. More specifically, the governing equations for elastostatics can be rewritten using components as:

$$\sigma_{ij,j} + b_i = 0 \quad \text{in } \Omega, \tag{11.11}$$

$$t_i \equiv \sigma_{ij}n_j = \bar{t}_i \quad \text{on } \Gamma_t, \tag{11.12}$$

$$u_i = \bar{u}_i \quad \text{on } \Gamma_u, \tag{11.13}$$

where a subscript comma denotes the spatial derivative. Furthermore, we denote

$$\sigma_{ij}(u) = C_{ijkl}\epsilon_{kl}(u), \quad \epsilon_{kl}(u) = \frac{1}{2}(u_{k,l} + u_{l,k}) = \frac{1}{2}(\frac{\partial u_k}{\partial x_l} + \frac{\partial u_l}{\partial x_k}). \quad (11.14)$$

Multiplying (11.8) by test function v and using integration by parts, we obtain

$$\int_\Gamma v_i\sigma_{ij}n_j d\Gamma - \int_\Omega \sigma_{ij}v_{i,j}d\Omega + \int_\Omega v_ib_i d\Omega = 0. \quad (11.15)$$

Noting that $\sigma_{ij} = \sigma_{ji}$, we have

$$\sigma_{ij}v_{i,j} = \sigma_{ij}\frac{1}{2}(v_{i,j} + v_{j,i}) = \sigma_{ij}(u)\epsilon_{ij}(u) = C_{ijkl}\epsilon_{kl}\epsilon_{ij}, \quad (11.16)$$

where in the last step we used the constitutive law (11.14).

Substituting (11.16) and the boundary condition (11.12) into (11.15), we have

$$\int_\Omega \epsilon^T(u)C\epsilon(v)d\Omega = \int_\Gamma \bar{t} \cdot v d\Gamma + \int_\Omega b \cdot v d\Omega. \quad (11.17)$$

The essential boundary condition (11.13) can be imposed using the Lagrange multiplier method, which leads to

$$\int_\Omega \epsilon^T(u)C\epsilon(v)d\Omega = \int_\Gamma \bar{t} \cdot v d\Gamma + \int_\Omega b \cdot v d\Omega + \int_{\Gamma_u} \delta\lambda \cdot (u - \bar{u})d\Gamma + \int_{\Gamma_u} v \cdot \lambda, \quad (11.18)$$

where the Lagrange multiplier λ is expressed by

$$\lambda(x) = N_i(s)\lambda_i, \quad \delta\lambda(x) = N_i(s)\delta\lambda_i, \quad x \in \Gamma_u, \quad (11.19)$$

where $N_i(s)$ denotes a Lagrange interpolant and s is the arc length along the boundary Γ_u.

Substituting the approximate solution

$$u_h(x) = \sum_J u_J\phi_J(x)$$

and the test function $\phi_I(x)$ into (11.18), we obtain the final discrete equation of (11.18):

$$\begin{bmatrix} K & G \\ G^T & 0 \end{bmatrix} \begin{bmatrix} u \\ \lambda \end{bmatrix} = \begin{bmatrix} f \\ q \end{bmatrix},$$

where

$$K_{IJ} = \int_\Omega B_I^T CB_J d\Omega, \quad G_{IJ} = -\int_{\Gamma_u} \phi_I N_J d\Gamma,$$

$$f_I = \int_{\Gamma_t} \phi_I \bar{t} d\Gamma + \int_\Omega \phi_I b d\Omega, \quad q_J = -\int_{\Gamma_u} N_J \bar{u} d\Gamma,$$

$$B_I = \begin{bmatrix} \phi_{I,x} & 0 \\ 0 & \phi_{I,y} \\ \phi_{I,y} & \phi_{I,x} \end{bmatrix}, \quad N_J = \begin{bmatrix} N_I & 0 \\ 0 & N_J \end{bmatrix}, \quad C = \frac{E}{1-\nu^2} \begin{bmatrix} 1 & \nu & 0 \\ \nu & 1 & 0 \\ 0 & 0 & \frac{1-\nu}{2} \end{bmatrix}.$$

Here our C is for plane stress, E and ν are Young's modulus and Poisson's ratio.

The implementation of the element-free Galerkin (EFG) method is basically similar to the classic finite element method. The basic procedure using EFG to solve the model problem (11.8)-(11.10) is as follows [9]:

1. Define the given physical parameters.
2. Create the nodal coordinates of the physical domain.
3. Create the domain of influence for each node.
4. Build a background mesh for performing numerical integration.
5. Set up Gauss points, weights, and Jacobian for each mesh cell.
6. Loop over all Gauss points:
   • Locate all nodes in the neighborhood of the Gauss point.
   • Calculate the weight functions, shape functions, and derivatives of the shape functions.
   • Calculate matrices B and K.
7. Locate both truncation and essential boundary nodes.
8. Create Gauss points along the traction and essential boundaries.
9. Integrate over the traction boundary to form vector f.
10. Integrate over the essential boundary to form matrix G and vector q.
11. Solve for u_I.
12. Postprocess to compare exact and analytical stresses at quadrature points.

A very clearly explained MATLAB code for solving the model problem (11.8)-(11.9) is provided in the paper by Dolbow and Belytschko [9].

11.3   The meshless local Petrov-Galerkin method

In 1998, Atluri and Zhu [3, 4] developed the meshless local Petrov-Galerkin (MLPG) method, which does not require a background mesh for integration. To illustrate the method, let us first consider the Poisson's equation

$$\nabla^2 u(\boldsymbol{x}) = f(\boldsymbol{x}) \quad \boldsymbol{x} \in \Omega \subset R^2, \tag{11.20}$$

where f is a given source function, the domain Ω is enclosed by boundary $\Gamma = \Gamma_D \cup \Gamma_N$ with boundary conditions

$$u(\boldsymbol{x}) = \bar{u}(\boldsymbol{x}) \quad \text{on } \Gamma_D, \tag{11.21}$$

$$q \equiv \frac{\partial u}{\partial n} = \bar{q} \quad \text{on } \Gamma_N, \tag{11.22}$$

where $\bar{u}$ and $\bar{q}$ are the prescribed potential and normal flux, respectively, and $\boldsymbol{n}$ is the unit outward normal to Γ.

Multiplying (11.20) by a test function v and integrating by parts over a local subdomain Ω_s, we obtain

$$\int_{\Omega_s} (u_{,i}v_{,i} + fv)d\Omega - \int_{\partial\Omega_s} \frac{\partial u}{\partial n}vd\Gamma + \alpha \int_{\Gamma_D} (u - \bar{u})vd\Gamma = 0, \qquad (11.23)$$

where we impose the Dirichlet boundary condition (11.21) by the penalty method, and α is a penalty parameter.

Splitting the boundary $\partial\Omega_s = \Gamma_{sD} \cup \Gamma_{sN}$, we can obtain the local weak formulation

$$\int_{\Omega_s} u_{,i}v_{,i}d\Omega - \int_{\Gamma_{sD}} \frac{\partial u}{\partial n}vd\Gamma + \alpha \int_{\Gamma_D} uvd\Gamma$$
$$= -\int_{\Omega_s} fvd\Omega + \int_{\Gamma_{sN}} \bar{q}vd\Gamma + \alpha \int_{\Gamma_D} \bar{u}vd\Gamma. \qquad (11.24)$$

Note that the subdomains Ω_s in a 2-D problem is usually chosen as a circle, an ellipse, or a rectangle, while for a 3-D problem, Ω_s is chosen as a sphere, an ellipsoid, or a cube. In the MLPG method, the trial function and the test function are chosen differently and they may be MLS, PUM, the Shepard function, or even radial basis functions.

Substituting the approximate solution

$$u_h(\boldsymbol{x}) = \sum_{J=1}^{M} \phi_J(\boldsymbol{x})\hat{u}_J, \qquad (11.25)$$

and test function $v = v(\boldsymbol{x}, \boldsymbol{x}_I)$ into (11.24), we obtain the system of linear equations

$$K\hat{u} = f, \qquad (11.26)$$

where the matrix element

$$K_{IJ} = \int_{\Omega_s} \phi_{J,k}(\boldsymbol{x})v_{,k}(\boldsymbol{x}, \boldsymbol{x}_I)d\Omega - \int_{\Gamma_{sD}} \frac{\partial\phi_J(\boldsymbol{x})}{\partial n}v(\boldsymbol{x}, \boldsymbol{x}_I)d\Gamma$$
$$+\alpha \int_{\Gamma_D} \phi_J(\boldsymbol{x})v(\boldsymbol{x}, \boldsymbol{x}_I)d\Gamma, \qquad (11.27)$$

and the vector component

$$f_I = -\int_{\Omega_s} f(\boldsymbol{x})v_{,k}(\boldsymbol{x}, \boldsymbol{x}_I)d\Omega + \int_{\Gamma_{sN}} \bar{q}v(\boldsymbol{x}, \boldsymbol{x}_I)d\Gamma + \alpha \int_{\Gamma_D} \bar{u}(\boldsymbol{x})v(\boldsymbol{x}, \boldsymbol{x}_I)d\Gamma. \qquad (11.28)$$

In the above M denotes the total number of nodes in Ω, and $\hat{u} = [\hat{u}_1, \cdots, \hat{u}_M]$ is the vector for the unknown fictitious nodal values.

The implementation of the MLPG method can be carried out as follows:

---

1. Generate all nodes M in the domain Ω and on its boundary Γ. Then choose the weight functions and shape functions.

2. For each node, choose the proper shape and the size of the local subdomain Ω_s and its corresponding local boundary $\partial\Omega_s$.

3. Loop over all nodes inside Ω and at the boundary Γ:
   - Choose Gauss quadrature points x_Q in Ω_s and on $\partial\Omega_s$.
   - Loop over all quadrature points x_Q in Ω_s and on $\partial\Omega_s$:
   (a) Locate all nodes $x_J (J = 1, \cdots, N)$ in the neighborhood of x_Q such that $w_J(x_Q) > 0$;
   (b) Calculate $\phi_J(x_Q)$ and its derivatives $\phi_{J,k}(x_Q)$ for all $x_J (J = 1, \cdots, N)$;
   (c) Evaluate the numerical integrals of (11.27) and (11.28).

4. Solve the linear system (11.26) for $\hat{u}_J$. Then postprocess for approximate solution and compare errors.

Similarly, we can develop MLPG for the two-dimensional elastostatic problem (11.11)-(11.13). Multiplying (11.11) by a test function v_i and integrating over a subdomain Ω_s, and then imposing the essential boundary condition (11.13) by the penalty method, we obtain

$$\int_{\Omega_s} (\sigma_{ij,j} + b_i)v_i d\Omega - \alpha \int_{\Gamma_{su}} (u_i - \bar{u}_i)v_i d\Gamma = 0, \tag{11.29}$$

where $\Gamma_{su} = \Gamma_u \cap \partial\Omega_s$.

Using integration by parts as we did in the EFG method, we can rewrite (11.29) as follows:

$$\int_{\Omega_s} \sigma_{ij}v_{i,j} d\Omega - \int_{\partial\Omega_s} \sigma_{ij}n_j v_i d\Gamma + \alpha \int_{\Gamma_{su}} u_i v_i d\Gamma$$
$$= \int_{\Omega_s} b_i v_i d\Omega + \alpha \int_{\Gamma_{su}} \bar{u}_i v_i d\Gamma. \tag{11.30}$$

We can split $\partial\Omega_s = \Gamma_{su} \cup \Gamma_{st} \cup L_s$, where $\Gamma_{st} = \partial\Omega_s \cap \Gamma_t$, and L_s is the other part of the local boundary $\partial\Omega_s$ over which no boundary condition is specified. Note that for a subdomain located totally inside the physical domain Ω, the Γ_{su} and Γ_{st} are empty. Imposing the natural boundary condition (11.12) in (11.30), we obtain the local weak formulation:

$$\int_{\Omega_s} \sigma_{ij}v_{i,j} d\Omega + \alpha \int_{\Gamma_{su}} u_i v_i d\Gamma - \int_{\Gamma_{su}+L_s} t_i v_i d\Gamma$$
$$= \int_{\Omega_s} b_i v_i d\Omega + \alpha \int_{\Gamma_{su}} \bar{u}_i v_i d\Gamma + \int_{\Gamma_{st}} \bar{t}_i v_i d\Gamma. \tag{11.31}$$

Substituting the approximate solution (11.25) and test function $v_i = v(x, x_I)$ into (11.31), we obtain the following system of linear equations in $\hat{u}_J$:

$$\sum_{J=1}^{M} K_{IJ}\hat{u}_J = f_I, \quad I = 1, 2, \cdots, M, \tag{11.32}$$

where

$$K_{IJ} = \int_{\Omega_s} \epsilon(\boldsymbol{v}) C B_J d\Omega + \alpha \int_{\Gamma_{su}} \boldsymbol{v}(\boldsymbol{x}, \boldsymbol{x}_I) \phi_J d\Gamma - \int_{\Gamma_{su}+L_s} \boldsymbol{v}(\boldsymbol{x}, \boldsymbol{x}_I) N C B_J d\Gamma,$$

$$f_I = \int_{\Omega_s} \boldsymbol{b} \cdot \boldsymbol{v}(\boldsymbol{x}, \boldsymbol{x}_I) + \alpha \int_{\Gamma_{su}} \bar{u} \cdot \boldsymbol{v}(\boldsymbol{x}, \boldsymbol{x}_I) d\Gamma + \int_{\Gamma_{st}} \bar{t} \cdot \boldsymbol{v}(\boldsymbol{x}, \boldsymbol{x}_I) d\Gamma,$$

$$B_I = \begin{bmatrix} \phi_{I,x} & 0 \\ 0 & \phi_{I,y} \\ \phi_{I,y} & \phi_{I,x} \end{bmatrix},$$

$$\epsilon(\boldsymbol{v}) = \begin{bmatrix} \epsilon_{11}(\boldsymbol{v}) & \epsilon_{22}(\boldsymbol{v}) & \epsilon_{12}(\boldsymbol{v}) \\ \epsilon_{11}(\boldsymbol{v}) & \epsilon_{22}(\boldsymbol{v}) & \epsilon_{12}(\boldsymbol{v}) \end{bmatrix}, \quad \epsilon_{ij}(\boldsymbol{v}) = \frac{1}{2}\left(\frac{\partial v_i}{\partial x_j} + \frac{\partial v_j}{\partial x_i}\right),$$

and the matrix $C = \frac{E}{1-\nu^2} \begin{bmatrix} 1 & \nu & 0 \\ \nu & 1 & 0 \\ 0 & 0 & \frac{1-\nu}{2} \end{bmatrix}$ (for plane stress).

11.4 Bibliographical remarks

Readers interested in a detailed introduction to general meshless methods can consult the book by Liu [18]. More detailed discussions on the MLPG method can be consulted in the books by Atluri and Atluri and Shen [1, 2]. Details about the Smoothed Particle Hydrodynamics method and its implementation can be found in books [17, 19]. Readers can find some more recent advances in meshless methods in conference proceedings [8, 16, 22], and especially the series proceedings edited by Griebel and Schweitzer since 2002 [12, 13, 14].

11.5 Exercises

1. Prove that for the quartic spline, the normalization constant $C_d = \frac{5}{4}, \frac{5}{\pi}$, and $\frac{105}{16\pi}$ for one, two, and three dimensions, respectively.
2. Assume that a quadratic weight function has the following form

$$w_h(r) = \begin{cases} a + b(\frac{r}{h})^2, & \text{if } 0 \le \frac{r}{h} \le 1, \\ 0 & \text{if } \frac{r}{h} > 1. \end{cases}$$

Prove that $w_h(r) = \frac{C_d}{h^d}[1 - (\frac{r}{h})^2]$, where $C_d = \frac{3}{4}, \frac{2}{\pi}$, and $\frac{15}{8\pi}$ for one, two, and three dimensions, respectively.
Hint: Use the normality condition and the property $w_h(h) = 0$.

3. Functions that are identically 1 on a subset of their support can form a partition of unity. Prove that $\phi(x - x_i), i = 0, \cdots, n$, form a partition of unity subordinate to patches Ω_i, where

$$\phi(x) = \begin{cases} \frac{3}{2} + 2\frac{x}{h}, & \text{if } x \in (-\frac{3}{4}h, -\frac{h}{4}], \\ 1, & \text{if } x \in (-\frac{h}{4}, \frac{h}{4}], \\ \frac{3}{2} - 2\frac{x}{h}, & \text{if } x \in (\frac{h}{4}, \frac{3}{4}h), \\ 0 & \text{elsewhere.} \end{cases}$$

4. Recall that the Lagrange multiplier method is to convert a constrained minimization problem to an unconstrained one. Use the Lagrange multiplier method to find the minimizer of

$$f(\boldsymbol{x}) = x_1^2 - 2x_1x_2 + 2x_2^2 + 10x_1 + 2x_2$$

under the constraint that $g(\boldsymbol{x}) \equiv x_1 - x_2 = 0$.

5. Develop a local weak formulation for the convection-diffusion problem:

$$-\frac{\partial}{\partial x_j}(K\frac{\partial u}{\partial x_j}) + v_j\frac{\partial u}{\partial x_j} = f \quad \text{in } \Omega,$$

$$u = \overline{u} \quad \text{on } \Gamma_D,$$

$$K\frac{\partial u}{\partial x_j}n_j = \overline{q} \quad \text{on } \Gamma_N,$$

where v_j is the flow velocity, K is the diffusivity, f is the given source.

References

[1] S.N. Atluri. *The Meshless Method (MLPG) for Domain and BIE Discretizations.* Tech Science Press, Forsyth, GA, 2004.

[2] S.N. Atluri and S. Shen. *The Meshless Local Petrov-Galerkin (MLPG) Method.* Tech Science Press, Forsyth, GA, 2002.

[3] S.N. Atluri and T. Zhu. A new meshless local Petrov-Galerkin (MLPG) approach in computational mechanics. *Comput. Mech.*, 22:117–127, 1998.

[4] S.N. Atluri and T. Zhu. A new meshless local Petrov-Galerkin (MLPG) approach to nonlinear problems in computational modeling and simulation. *Comput. Modeling Simulation in Engrg.*, 3:187–196, 1998.

[5] I. Babuska, U. Banerjee and J.E. Osborn. Survey of meshless and generalized finite element methods: a unified approach. *Acta Numerica*, 12:1–125, 2003.

[6] I. Babuska and J.M. Melenk. The partition of unity method. *Int. J. Num. Meth. Engrg.*, 40:727–758, 1997.

[7] T. Belytschko, Y. Krongauz, D. Organ, M. Fleming and P. Krysl. Meshless methods: an overview and recent developments. *Comput. Methods Appl. Mech. Engrg.*, 139:3–47, 1996.

[8] C.A. Brebbia, D. Poljak and V. Popov (eds.). *Boundary Elements and Other Mesh Reduction Methods XXIX.* WIT Press, Southampton, UK, 2007.

[9] J. Dolbow and T. Belytschko. An introduction to programming the meshless element free Galerkin method. *Archives of Computational Methods in Engineering*, 5(3):207–241, 1998.

[10] C.A. Duarte and J.T. Oden. H-p clouds – an h-p meshless method. *Numerical Methods for Partial Differential Equations*, 12:673–705, 1996.

[11] R.A. Gingold and J.J. Monaghan. Smoothed particle hydrodynamics: theory and application to non-spherical stars. *Mon. Not. Roy. Astron. Soc.*, 181:375–389, 1977.

[12] M. Griebel and M.A. Schweitzer (eds.). *Meshfree Methods for Partial Differential Equations.* Springer-Verlag, Berlin, 2002.

[13] M. Griebel and M.A. Schweitzer (eds.). *Meshfree Methods for Partial Differential Equations III.* Springer-Verlag, Berlin, 2006.

[14] M. Griebel and M.A. Schweitzer (eds.). *Meshfree Methods for Partial Differential Equations VIII.* Springer-Verlag, Berlin, 2017.

[15] P. Lancaster and K. Salkauskas. Surfaces generated by moving least squares methods. *Math. Comp.*, 37(155):141–158, 1981.

[16] V.M.A. Leitao, C.J.S. Alves and C. Armando Duarte (eds.). *Advances in Meshfree Techniques.* Springer, New York, NY, 2007.

[17] S. Li and W.-K. Liu. *Meshfree Particle Methods.* Springer-Verlag, Berlin, 2004.

[18] G.R. Liu. *Mesh Free Methods: Moving beyond the Finite Element Method.* CRC Press, Boca Raton, FL, 2003.

[19] G.R. Liu and M.B. Liu. *Smoothed Particle Hydrodynamics.* World Scientific, Hackensack, NJ, 2003.

[20] L.B. Lucy. A numerical approach to the testing of the fission hypothesis. *The Astron. J.*, 8(12):1013–1024, 1977.

[21] J.J. Monaghan. Why particle methods work. *SIAM J. Sci. Stat. Comput.*, 3(4):422–433, 1982.

[22] J. Sladek and V. Sladek (eds.). *Advances in Meshless Methods.* Tech Science Press, Forsyth, GA, 2006.

Appendix A

Answers to Selected Problems

Chapter 2

5. Rewrite (2.77) as

$$u_j^{n+1} = (1 - a_{j+\frac{1}{2}}^n \frac{\Delta t}{(\Delta x)^2} - a_{j-\frac{1}{2}}^n \frac{\Delta t}{(\Delta x)^2})u_j^n$$
$$+ a_{j+\frac{1}{2}}^n \frac{\Delta t}{(\Delta x)^2} u_{j+1}^n + a_{j-\frac{1}{2}}^n \frac{\Delta t}{(\Delta x)^2} u_{j-1}^n. \qquad (A.1)$$

Under the condition (2.78), we have

$$(a_{j+\frac{1}{2}}^n \frac{\Delta t}{(\Delta x)^2} + a_{j-\frac{1}{2}}^n \frac{\Delta t}{(\Delta x)^2}) \le 2a^* \frac{\Delta t}{(\Delta x)^2} \le 1,$$

substituting which into (A.1), we obtain

$$u_j^{n+1} \le (1 - a_{j+\frac{1}{2}}^n \frac{\Delta t}{(\Delta x)^2} - a_{j-\frac{1}{2}}^n \frac{\Delta t}{(\Delta x)^2}) \max_{0 \le j \le J} u_j^n$$
$$+ a_{j+\frac{1}{2}}^n \frac{\Delta t}{(\Delta x)^2} \max_{0 \le j \le J} u_j^n + a_{j-\frac{1}{2}}^n \frac{\Delta t}{(\Delta x)^2} \max_{0 \le j \le J} u_j^n$$
$$= \max_{0 \le j \le J} u_j^n.$$

The other part can be proved similarly.

8. From (2.82), it is easy to see that

$$u_t = -(ik\pi a + k^2\pi^2)u,$$
$$u_x = ik\pi u, \quad u_{xx} = -k^2\pi^2 u,$$

which leads to $u_t + au_x - u_{xx} = 0$.

Substituting $u_j^n = \lambda^n e^{ik(j\Delta x)}$ into the upwind scheme, we obtain

$$(\lambda - 1) + a\frac{\Delta t}{\Delta x}(1 - e^{-ik\Delta x})\lambda = \frac{\Delta t}{(\Delta x)^2}(e^{ik\Delta x} - 2 + e^{-ik\Delta x})\lambda$$

or

$$\lambda = 1/[1 + a\frac{\Delta t}{\Delta x}(2\sin^2\frac{k\Delta x}{2} + i\sin k\Delta x) + 4\frac{\Delta t}{(\Delta x)^2}\sin^2\frac{k\Delta x}{2}],$$

which yields $|\lambda| \le 1$ for any Δt and Δx. Hence the scheme is unconditionally stable.

Chapter 3

5.

$$\frac{dE}{dt} = \int_0^1 (2a^2 u_x u_{xt} + 2u_t u_{tt})dx$$

$$= (2a^2 u_x u_t)|_{x=0}^1 - \int_0^1 2a^2 u_{xx} u_t dx + \int_0^1 2u_{tt} u_t dx$$

$$= 0,$$

where we used the boundary condition $u|_{x=0,1} = 0$ and the governing equation $u_{xx} = a^2 u_{tt}$.

9. From the definition of the flux, we can easily check that

$$h_{k+\frac{1}{2}}^n - h_{k-\frac{1}{2}}^n$$

$$= \frac{1}{2}(f(u_{k+1}^n) - f(u_{k-1}^n)) - \frac{\triangle x}{2\triangle t}(u_{k+1}^n - 2u_k^n + u_{k-1}^n),$$

from which we obtain

$$u_k^n - \frac{\triangle t}{\triangle x}(h_{k+\frac{1}{2}}^n - h_{k-\frac{1}{2}}^n)$$

$$= u_k^n - \frac{\triangle t}{2\triangle x}(f(u_{k+1}^n) - f(u_{k-1}^n)) + \frac{1}{2}(u_{k+1}^n - 2u_k^n + u_{k-1}^n)$$

$$= \frac{1}{2}(u_{k+1}^n - u_{k-1}^n) - \frac{\triangle t}{2\triangle x}(f(u_{k+1}^n) - f(u_{k-1}^n)),$$

which equals the right-hand side of the Lax-Friedrichs (LF) scheme, i.e., the LF scheme can be cast in the form of (3.62). Hence the LF scheme is conservative.

Furthermore, it is easy to check that the flux h satisfies the property $h(u, u) = f(u)$, which means that the LF scheme is consistent.

Chapter 4

4. Let $v_\epsilon(x, y) = u(x, y) + \epsilon(x^2 + y^2)$. It is easy to check that

$$\triangle v_\epsilon = 4\epsilon > 0 \quad \forall \ (x, y) \in \Omega. \tag{A.2}$$

Assume that v_ϵ achieves its maximum at an interior point (x_*, y_*) of Ω, then

$$\frac{\partial^2 v_\epsilon}{\partial x^2}(x_*, y_*) < 0, \quad \frac{\partial^2 v_\epsilon}{\partial y^2}(x_*, y_*) < 0,$$

which contradicts (A.2). Hence v_ϵ has no interior maximum point, i.e.,

$$v_\epsilon(x, y) \le \max_{(x,y)\in\partial\Omega} v_\epsilon(x, y) \quad \forall \ (x, y) \in \Omega.$$

Taking the limit $\epsilon \to 0$ on both sides, we obtain

$$u(x, y) \le \max_{(x,y)\in\partial\Omega} u(x, y).$$

For the other part, just consider $v_\epsilon(x, y) = u(x, y) - \epsilon(x^2 + y^2)$.

7. Assume that the grid is uniform and has mesh size $h = \frac{1}{J}$, i.e.,

$$0 = x_0 < x_1 < \cdots < x_J = 1, \quad 0 = y_0 < y_1 < \cdots < y_J = 1.$$

Then we can construct a second-order difference scheme

$$\delta_{x_1}^4 u_{ij} + 2\delta_{x_1}^2 \delta_{x_2}^2 u_{ij} + \delta_{x_2}^4 u_{ij} = 0,$$

or

$$\begin{aligned} &20u_{ij} - 8(u_{i+1,j} + u_{i-1,j} + u_{i,j+1} + u_{i,j-1}) \\ &+2(u_{i+1,j+1} + u_{i+1,j-1} + u_{i-1,j+1} + u_{i-1,j-1}) \\ &+u_{i+2,j} + u_{i-2,j} + u_{i,j+2} + u_{i,j-2} = 0, \quad \forall 2 \le i, j \le J - 2, \quad \text{(A.3)} \end{aligned}$$

which is a 13-point scheme.

For those near boundary points, the scheme (A.3) has to be changed. For example, for those nodes adjacent to the bottom side (i.e., $j = 1, 2 \le i \le J - 2$), we have to replace $u_{i,-1}$ by $u_{i,1} + 2hg_{i,0}$ and $u_{i,0}$ by $f_{i,0}$, in which case, the scheme (A.3) becomes

$$\begin{aligned} &21u_{i1} - 8(u_{i+1,1} + u_{i-1,1} + u_{i,2}) \\ &+2(u_{i+1,2} + u_{i-1,2}) + u_{i+2,1} + u_{i-2,1} + u_{i,3} \\ &= 8f_{i,0} - 2f_{i+1,0} - 2f_{i-1,0} - 2hg_{i,0}. \ \forall 2 \le i \le J - 2. \end{aligned}$$

For the near corner point $i = j = 1$, we have to replace $u_{1,-1}$ by $u_{1,1} + 2hg_{1,0}$ and $u_{-1,1}$ by $u_{1,1} + 2hg_{0,1}$, in which case, the scheme (A.3) becomes

$$\begin{aligned} &22u_{11} - 8(u_{2,1} + u_{1,2}) + 2u_{2,2} + u_{3,1} + u_{1,3} \\ &= 8(f_{0,1} + f_{1,0}) - 2(f_{2,0} + f_{0,2} + f_{0,0}) - 2h(g_{0,1} + g_{1,0}). \end{aligned}$$

Similar procedures apply for other nodes adjacent to the boundary.

Chapter 5

6. By Taylor expansion, we have

$$R(u^n + \Delta t R(u^n)) = R(u^n) + \Delta t R(u^n)R'(u^n) + O(\Delta t^2),$$

from which we see that

$$\frac{u(x, t^{n+1}) - u(x, t^n)}{\Delta t} - \frac{1}{2}[R(u^n) + R(u^n + \Delta t R(u^n))]$$

$$= u'(x, t^n) + \frac{\Delta t}{2} u''(x, t^n) + O(\Delta t^2)$$

$$- \frac{1}{2}[2R(u^n) + \Delta t R(u^n)R'(u^n) + O(\Delta t^2)]$$

$$= O(\Delta t^2),$$

where we used $u'(x, t^n) = R(u^n)$ and $u''(x, t^n) = R'(u^n)R(u^n)$.

9. By Taylor expansion, we have

$$f(x_{i+\frac{1}{2}}) - f(x_{i-\frac{1}{2}})$$

$$= 2[\frac{h}{2}f'(x_i) + \frac{1}{3!}(\frac{h}{2})^3 f^{(3)}(x_i) + \frac{1}{5!}(\frac{h}{2})^5 f^{(5)}(x_i)] + O(h^7), \qquad (A.4)$$

$$f(x_{i+\frac{3}{2}}) - f(x_{i-\frac{3}{2}})$$

$$= 2[\frac{3h}{2}f'(x_i) + \frac{1}{3!}(\frac{3h}{2})^3 f^{(3)}(x_i) + \frac{1}{5!}(\frac{3h}{2})^5 f^{(5)}(x_i)] + O(h^7), \quad (A.5)$$

$$f'(x_{i-1}) + f'(x_{i+1})$$

$$= 2[f'(x_i) + \frac{h^2}{2}f^{(3)}(x_i) + \frac{h^4}{4!}f^{(5)}(x_i)] + O(h^6). \qquad (A.6)$$

Denote the truncation error

$$R = \alpha[f'(x_{i-1}) + f'(x_{i+1})] + f'(x_i) \qquad (A.7)$$

$$- \frac{a}{h}(f(x_{i+\frac{1}{2}}) - f(x_{i-\frac{1}{2}})) - \frac{b}{3h}(f(x_{i+\frac{3}{2}}) - f(x_{i-\frac{3}{2}})). \qquad (A.8)$$

Substituting (A.4)-(A.6) into (A.8), we obtain

$$R = (2\alpha + 1 - a - b)f'(x_i) + (\alpha - \frac{a}{24} - \frac{3b}{8})h^2 f^{(3)}(x_i)$$

$$+ (\frac{\alpha}{12} - \frac{a}{1920} - \frac{81b}{1920})h^4 f^{(5)}(x_i) + O(h^6). \qquad (A.9)$$

Let the first coefficients of (A.9) be zero, i.e.,

$$2\alpha + 1 - a - b = 0$$

$$-24\alpha + a + 9b = 0,$$

which leads to the solution

$$b = \frac{22\alpha - 1}{8}, \qquad a = \frac{-6\alpha + 9}{8}.$$

Hence the leading term in the truncation error (A.9) becomes

$$R = [\frac{\alpha}{12} + \frac{6\alpha - 9}{8 \cdot 1920} - \frac{81(22\alpha - 1)}{8 \cdot 1920}]h^4 f^{(5)}(x_i)$$

$$= (\frac{\alpha}{12} - \frac{-9 + 222\alpha}{1920})h^4 f^{(5)}$$

$$= \frac{9 - 62\alpha}{1920}h^4 f^{(5)}.$$

Chapter 6

1. Using integration by parts and the boundary condition, we obtain

$$\int_\Omega \frac{\partial^2 v}{\partial x_1^2} \frac{\partial^2 v}{\partial x_2^2} dx_1 dx_2 = \int_{\partial\Omega} \frac{\partial^2 v}{\partial x_1^2} \frac{\partial v}{\partial x_2} n ds - \int_\Omega \frac{\partial^3 v}{\partial x_1^2 \partial x_2} \frac{\partial v}{\partial x_2} dx_1 dx_2$$

$$= -\int_{\partial\Omega} \frac{\partial^2 v}{\partial x_1 \partial x_2} n \frac{\partial v}{\partial x_2} ds + \int_\Omega \frac{\partial^2 v}{\partial x_1 \partial x_2} \frac{\partial^2 v}{\partial x_1 \partial x_2} dx_1 dx_2$$

$$= \int_\Omega |\frac{\partial^2 v}{\partial x_1 \partial x_2}|^2 dx_1 dx_2,$$

from which we have

$$|\triangle v|^2_{L^2(\Omega)} = \int_\Omega (|\frac{\partial^2 v}{\partial x_1^2}|^2 + |\frac{\partial^2 v}{\partial x_2^2}|^2 + 2|\frac{\partial^2 v}{\partial x_1^2} \frac{\partial^2 v}{\partial x_2^2}|) dx_1 dx_2$$

$$= \int_\Omega (|\frac{\partial^2 v}{\partial x_1^2}|^2 + |\frac{\partial^2 v}{\partial x_2^2}|^2 + 2|\frac{\partial^2 v}{\partial x_1 \partial x_2}|^2) dx_1 dx_2 = |v|^2_{H^2(\Omega)}.$$

6. We only need to show that if

$$v(a_i) = v(a_{ij}) = v(a_0) = 0,$$

then $v \equiv 0$. Since $v = 0$ at those four nodes on edge $a_2 a_3$, hence $v \equiv 0$ on edge $a_2 a_3$. Therefore, v must contain factor $\lambda_1(x)$. A similar argument concludes that

$$v = c\lambda_1(x)\lambda_2(x)\lambda_3(x),$$

which leads to

$$v(a_0) = c\lambda_1(a_0)\lambda_2(a_0)\lambda_3(a_0) = c \cdot (\frac{1}{3})^3.$$

But by assumption $v(a_0) = 0$, we conclude that $c = 0$, i.e., $v \equiv 0$.

Chapter 7

6. Note that

$$\frac{\partial F_1}{\partial \hat{x}_1} = [-(1 - \hat{x}_2)a_1 + (1 - \hat{x}_2)a_2 + \hat{x}_2 a_3 - \hat{x}_2 a_4]|1\text{st component}$$

$$= 2(1 - \hat{x}_2) + 2\hat{x}_2 = 2.$$

Similarly,

$$\frac{\partial F_1}{\partial \hat{x}_2} = [-(1 - \hat{x}_1)a_1 - \hat{x}_1 a_2 + \hat{x}_1 a_3 + (1 - \hat{x}_1)a_4]|_{\text{1st component}}$$
$$= -2\hat{x}_1 + 2\hat{x}_1 = 0.$$

$$\frac{\partial F_2}{\partial \hat{x}_1} = [-(1 - \hat{x}_2)a_1 + (1 - \hat{x}_2)a_2 + \hat{x}_2 a_3 - \hat{x}_2 a_4]|_{\text{2nd component}}$$
$$= 3\hat{x}_2 - 5\hat{x}_2 = -2\hat{x}_2.$$

$$\frac{\partial F_2}{\partial \hat{x}_2} = [-(1 - \hat{x}_1)a_1 - \hat{x}_1 a_2 + \hat{x}_1 a_3 + (1 - \hat{x}_1)a_4]|_{\text{2nd component}}$$
$$= 3\hat{x}_1 + 5(1 - \hat{x}_1) = 5 - 2\hat{x}_1.$$

Hence we have

$$J = \begin{vmatrix} \frac{\partial F_1}{\partial \hat{x}_1} & \frac{\partial F_1}{\partial \hat{x}_2} \\ \frac{\partial F_2}{\partial \hat{x}_1} & \frac{\partial F_2}{\partial \hat{x}_2} \end{vmatrix} = \begin{vmatrix} 2 & 0 \\ -2\hat{x}_2 & 5 - 2\hat{x}_1 \end{vmatrix} = 2(5 - 2\hat{x}_1),$$

which is always positive.

7.

$$\frac{\partial F_1}{\partial \hat{x}_1} = -2(1 - \hat{x}_2) + 3(1 - \hat{x}_2) + 5\hat{x}_2 - 2\hat{x}_2 = 1 + 2\hat{x}_2.$$

$$\frac{\partial F_1}{\partial \hat{x}_2} = -2(1 - \hat{x}_1) - 3\hat{x}_1 + 5\hat{x}_1 + 2(1 - \hat{x}_1) = 2\hat{x}_1.$$

$$\frac{\partial F_2}{\partial \hat{x}_1} = 2(1 - \hat{x}_2) + 3\hat{x}_2 - 3\hat{x}_2 = 2(1 - \hat{x}_2).$$

$$\frac{\partial F_2}{\partial \hat{x}_2} = -2\hat{x}_1 + 3\hat{x}_1 + 3(1 - \hat{x}_1) = 3 - 2\hat{x}_1.$$

Hence

$$J = \begin{vmatrix} 1 + 2\hat{x}_2 & 2\hat{x}_1 \\ 2(1 - \hat{x}_2) & 3 - 2\hat{x}_1 \end{vmatrix} = 3 - 6\hat{x}_1 + 6\hat{x}_2,$$

which equals zero when $\hat{x}_1 - \hat{x}_2 = \frac{1}{2}$ (e.g., the point $(\frac{1}{2}, 0)$). Hence the mapping is not invertible.

Chapter 8

2. Note that

$$\mathcal{L}(u + \delta u, p) = \frac{1}{2}(u + \delta u, u + \delta u) - (u_x + (\delta u)_x, p) + (f, p)$$

$$= \mathcal{L}(u, p) + \frac{1}{2}(\delta u, \delta u) + (\delta u, u) - ((\delta u)_x, p). \qquad \text{(A.10)}$$

Using integration by parts and the boundary condition $p|_{\partial\Omega} = 0$, we have

$$((\delta u)_x, p) = -(\delta u, p_x),$$

substituting which into (A.10) and using $u = -p_x$ yields

$$\mathcal{L}(u + \delta u, p) = \mathcal{L}(u, p) + \frac{1}{2}(\delta u, \delta u) \geq \mathcal{L}(u, p),$$

i.e.,

$$\mathcal{L}(u, p) \leq \mathcal{L}(v, p) \quad \text{for any } v \in V. \qquad \text{(A.11)}$$

On the other hand,

$$\mathcal{L}(u, p + \delta p) = \frac{1}{2}(u, u) - (u_x, p + \delta p) + (f, p + \delta p)$$

$$= \mathcal{L}(u, p) - (u_x, \delta p) + (f, \delta p) = \mathcal{L}(u, p),$$

which together with (A.11) concludes that (u, p) is a saddle-point of $\mathcal{L}(\cdot, \cdot)$.

4. First note that the total number of degrees of freedom is

$$3(r + 1) + 2 \cdot \frac{r(r + 1)}{2} = (r + 1)(r + 3),$$

which equals $dim(RT_r(K))$. Hence the uniqueness is assured if we can prove that if all degrees of freedom are zero

$$< \boldsymbol{v} \cdot \boldsymbol{n}, w >_e = 0, \quad \forall \, w \in P_r(e), \qquad \text{(A.12)}$$

$$(\boldsymbol{v}, \boldsymbol{w})_K = 0, \quad \forall \, \boldsymbol{w} \in (P_{r-1}(K))^2, \qquad \text{(A.13)}$$

then $\boldsymbol{v} \equiv 0$.

Since $\boldsymbol{v} \cdot \boldsymbol{n} \in P_r(e)$, (A.12) leads to $\boldsymbol{v} \cdot \boldsymbol{n} = 0$ on e.

Integration by parts gives us

$$\int_K (\nabla \cdot \boldsymbol{v})w = \int_{\partial K} (\boldsymbol{v} \cdot \boldsymbol{n})w - \int_K \boldsymbol{v} \cdot \nabla w. \qquad \text{(A.14)}$$

Letting $w \in P_r(K)$ in (A.14) and using (A.12)-(A.13), we have

$$\int_K (\nabla \cdot \boldsymbol{v})w = 0, \quad \forall \, w \in P_r(K),$$

which yields
$$\nabla \cdot \boldsymbol{v} = 0 \qquad\qquad\qquad (A.15)$$
due to the fact that $\nabla \cdot \boldsymbol{v} \in P_r(K)$ for any $\boldsymbol{v} \in RT_r(K)$.

(A.15) guarantees that there exists a stream function $\phi \in H^1(K)$ such that
$$\boldsymbol{v} = \mathrm{curl}\phi \equiv (-\frac{\partial \phi}{\partial y}, \frac{\partial \phi}{\partial x})'.$$

For any $\boldsymbol{v} \in RT_r(K)$, we can decompose $\boldsymbol{v}$ as
$$\boldsymbol{v} = \boldsymbol{p} + \begin{pmatrix} x \\ y \end{pmatrix} q, \quad \boldsymbol{p} \in (P_r(K))^2, \quad q \in P_r(K).$$

Without loss of generality, we can assume that q is a homogeneous polynomial of degree r (since other terms can be absorbed into $\boldsymbol{p}$). Hence
$$\nabla \cdot \boldsymbol{v} = \nabla \cdot \boldsymbol{p} + 2q + x\partial_x q + y\partial_y q = \nabla \cdot \boldsymbol{p} + (2+r)q,$$

which together with (A.15) and the fact $\nabla \cdot \boldsymbol{p} \in P_{r-1}(K)$ yields $q = 0$. Therefore $\boldsymbol{v} = \boldsymbol{p}$, and furthermore $\phi \in P_{r+1}(K)$.

On the other hand, $0 = \boldsymbol{v} \cdot \boldsymbol{n} = -\frac{\partial \phi}{\partial y} \cdot n_x + \frac{\partial \phi}{\partial x} \cdot n_y$ is equivalent to the tangential derivative $\frac{\partial \phi}{\partial t}$ by noting that $\boldsymbol{t} = (-n_y, n_x)$ if $\boldsymbol{n} = (n_x, n_y)$. Hence, $\phi = const$ on all edges of K.

Using (A.13) and the Stokes' formula, we have
$$0 = \int_K \boldsymbol{v} \cdot \boldsymbol{w} = \int_K \mathrm{curl}\phi \cdot \boldsymbol{w}$$
$$= \int_{\partial K} \boldsymbol{n} \times \phi \cdot \boldsymbol{w} + \int_K \phi \cdot \mathrm{curl}\boldsymbol{w}, \quad \forall \boldsymbol{w} \in (P_{r-1}(K))^2, \qquad (A.16)$$

where the scalar curl is defined as $\mathrm{curl}\boldsymbol{w} = \frac{\partial w_1}{\partial y} - \frac{\partial w_2}{\partial x}$.

Letting $\boldsymbol{w} = (1,1)'$ in (A.16), we obtain $\boldsymbol{n} \times \phi = 0$ on ∂K, i.e., $\phi = 0$ on all edges of K. Therefore, we can represent ϕ as
$$\phi(\boldsymbol{x}) = \lambda_1(\boldsymbol{x})\lambda_2(\boldsymbol{x})\lambda_3(\boldsymbol{x})\psi(\boldsymbol{x}), \qquad\qquad (A.17)$$

where λ_i are the barycentric coordinates of the triangle K. Hence, $\psi(\boldsymbol{x}) \in P_{r-2}(K)$.

Substituting (A.17) into (A.16) and choosing $\boldsymbol{w} \in P_{r-1}(K)$ such that $\psi = \mathrm{curl}\boldsymbol{w} \in P_{r-2}(K)$, we obtain
$$\int_K \lambda_1\lambda_2\lambda_3\psi^2 = 0,$$

which leads to $\psi = 0$. Hence $\phi = 0$ and $\boldsymbol{v} \equiv 0$.

Chapter 9

7. Multiplying (9.166) by $\boldsymbol{E}$, and integrating over domain Ω, we have

$$\epsilon_0 \frac{d}{dt}||\boldsymbol{E}||_0^2 = (\nabla \times \boldsymbol{H}, \boldsymbol{E}) - (\boldsymbol{J}, \boldsymbol{E}). \qquad (A.18)$$

Multiplying (9.167) by $\boldsymbol{H}$, integrating over domain Ω, and using the Stokes' formula and the boundary condition $\boldsymbol{n} \times \boldsymbol{E} = 0$, we have

$$\mu_0 \frac{d}{dt}||\boldsymbol{H}||_0^2 = -(\nabla \times \boldsymbol{E}, \boldsymbol{H}) = -(\boldsymbol{E}, \nabla \times \boldsymbol{H}). \qquad (A.19)$$

Multiplying (9.168) by $\frac{1}{\epsilon_0 \omega_p^2}\boldsymbol{J}$, and integrating over domain Ω, we have

$$\frac{1}{\epsilon_0 \omega_p^2}\frac{d}{dt}||\boldsymbol{J}||_0^2 + \frac{\nu}{\epsilon_0 \omega_p^2}||\boldsymbol{J}||_0^2 = (\boldsymbol{E}, \boldsymbol{J}). \qquad (A.20)$$

Adding (A.18)-(A.20) together, we obtain

$$\frac{d}{dt}[\epsilon_0||\boldsymbol{E}||_0^2 + \mu_0||\boldsymbol{H}||_0^2 + \frac{1}{\epsilon_0 \omega_p^2}||\boldsymbol{J}||_0^2] + \frac{\nu}{\epsilon_0 \omega_p^2}||\boldsymbol{J}||_0^2 = 0,$$

integrating which from 0 to t and using the fact that $\nu \geq 0$ concludes the proof.

10. Squaring both sides of the integral identity

$$\frac{1}{2}(\boldsymbol{u}^k + \boldsymbol{u}^{k-1}) - \frac{1}{\tau}\int_{t^{k-1}}^{t^k} \boldsymbol{u}(t)dt = \frac{1}{2\tau}\int_{t^{k-1}}^{t^k} (t - t^{k-1})(t^k - t)\boldsymbol{u}_{tt}(t)dt, \quad (A.21)$$

we can obtain

$$|\overline{\boldsymbol{u}}^k - \frac{1}{\tau}\int_{t^{k-1}}^{t^k} \boldsymbol{u}(t)dt|^2 \leq \frac{1}{4\tau^2}(\int_{t^{k-1}}^{t^k} (t - t^{k-1})^2(t^k - t)^2 dt)(\int_{t^{k-1}}^{t^k} |\boldsymbol{u}_{tt}(t)|^2 dt)$$

$$\leq \frac{1}{4}\tau^3 \int_{t^{k-1}}^{t^k} |\boldsymbol{u}_{tt}(t)|^2 dt.$$

Chapter 10

4. Note that

$$\frac{\partial}{\partial x} = \frac{\partial}{\partial r}\frac{\partial r}{\partial x} + \frac{\partial}{\partial \theta}\frac{\partial \theta}{\partial x}$$

$$= \frac{\partial}{\partial r} \cdot \frac{x}{r} + \frac{\partial}{\partial \theta} \cdot \frac{-y}{x^2 + y^2} = \frac{\partial}{\partial r} \cdot \cos\theta - \frac{\partial}{\partial \theta} \cdot \frac{\sin\theta}{r},$$

from which we obtain

$$\frac{\partial^2}{\partial x^2} = \frac{\partial}{\partial r}[\frac{\partial}{\partial r} \cdot \cos\theta - \frac{\partial}{\partial \theta} \cdot \frac{\sin\theta}{r}] \cdot \cos\theta$$

$$- \frac{\partial}{\partial \theta}[\frac{\partial}{\partial r} \cdot \cos\theta - \frac{\partial}{\partial \theta} \cdot \frac{\sin\theta}{r}] \cdot \frac{\sin\theta}{r}$$

$$= [\frac{\partial^2}{\partial r^2} \cdot \cos\theta - \frac{\partial^2}{\partial r \partial \theta} \cdot \frac{\sin\theta}{r} + \frac{\partial}{\partial \theta} \cdot \frac{\sin\theta}{r^2}] \cdot \cos\theta$$

$$- [\frac{\partial^2}{\partial r \partial \theta} \cdot \cos\theta - \frac{\partial}{\partial r} \cdot \sin\theta - \frac{\partial^2}{\partial \theta^2} \cdot \frac{\sin\theta}{r} - \frac{\partial}{\partial \theta} \cdot \frac{\cos\theta}{r}] \cdot \frac{\sin\theta}{r}$$

$$= \cos^2\theta \cdot \frac{\partial^2}{\partial r^2} - \frac{2\sin\theta\cos\theta}{r} \frac{\partial^2}{\partial r \partial \theta} + \frac{2\sin\theta\cos\theta}{r^2} \frac{\partial}{\partial \theta} + \frac{\sin^2\theta}{r} \frac{\partial}{\partial r} + \frac{\sin^2\theta}{r^2} \frac{\partial^2}{\partial \theta^2}.$$

Similarly, we have

$$\frac{\partial}{\partial y} = \frac{\partial}{\partial r} \frac{\partial r}{\partial y} + \frac{\partial}{\partial \theta} \frac{\partial \theta}{\partial y}$$

$$= \frac{\partial}{\partial r} \cdot \frac{y}{r} + \frac{\partial}{\partial \theta} \cdot \frac{x}{x^2 + y^2} = \frac{\partial}{\partial r} \cdot \sin\theta + \frac{\partial}{\partial \theta} \cdot \frac{\cos\theta}{r},$$

from which we obtain

$$\frac{\partial^2}{\partial y^2} = \frac{\partial}{\partial r}[\frac{\partial}{\partial r} \cdot \sin\theta + \frac{\partial}{\partial \theta} \cdot \frac{\cos\theta}{r}] \cdot \sin\theta$$

$$+ \frac{\partial}{\partial \theta}[\frac{\partial}{\partial r} \cdot \sin\theta + \frac{\partial}{\partial \theta} \cdot \frac{\cos\theta}{r}] \cdot \frac{\cos\theta}{r}$$

$$= [\frac{\partial^2}{\partial r^2} \cdot \sin\theta + \frac{\partial^2}{\partial r \partial \theta} \cdot \frac{\cos\theta}{r} + \frac{\partial}{\partial \theta} \cdot \frac{-\cos\theta}{r^2}] \cdot \sin\theta$$

$$+ [\frac{\partial^2}{\partial r \partial \theta} \cdot \sin\theta + \frac{\partial}{\partial r} \cdot \cos\theta + \frac{\partial^2}{\partial \theta^2} \cdot \frac{\cos\theta}{r} + \frac{\partial}{\partial \theta} \cdot \frac{-\sin\theta}{r}] \cdot \frac{\cos\theta}{r}$$

$$= \sin^2\theta \cdot \frac{\partial^2}{\partial r^2} + \frac{2\sin\theta\cos\theta}{r} \frac{\partial^2}{\partial r \partial \theta} - \frac{2\sin\theta\cos\theta}{r^2} \frac{\partial}{\partial \theta} + \frac{\cos^2\theta}{r} \frac{\partial}{\partial r} + \frac{\cos^2\theta}{r^2} \frac{\partial^2}{\partial \theta^2}.$$

Hence, adding up $\frac{\partial^2}{\partial x^2}$ and $\frac{\partial^2}{\partial y^2}$, we obtain

$$\frac{\partial^2}{\partial x^2} + \frac{\partial^2}{\partial y^2} = \frac{\partial^2}{\partial r^2} + \frac{1}{r} \frac{\partial}{\partial r} + \frac{1}{r^2} \frac{\partial^2}{\partial \theta^2}.$$

7.

$$\frac{d}{dx}[x^\alpha J_\alpha(x)] = \frac{d}{dx} \sum_{k=0}^\infty [\frac{(-1)^k}{k! \Gamma(k + \alpha + 1)} (\frac{x}{2})^{2k+\alpha} \cdot x^\alpha]$$

$$= \sum_{k=0}^\infty \frac{(-1)^k}{k! \Gamma(k + \alpha + 1)} \cdot \frac{1}{2^{2k+\alpha}} \cdot (2k + 2\alpha) x^{2k+2\alpha-1}$$

$$= x^\alpha \sum_{k=0}^{\infty} \frac{(-1)^k \cdot (k+\alpha)}{k! \Gamma(k+\alpha+1)} \cdot (\frac{x}{2})^{2k+\alpha-1}$$

$$= x^\alpha \sum_{k=0}^{\infty} \frac{(-1)^k \cdot (k+\alpha)}{k!(k+\alpha)\Gamma(k+\alpha)} \cdot (\frac{x}{2})^{2k+\alpha-1}$$

$$= x^\alpha J_{\alpha-1}(x).$$

The other identity can be proved similarly.

Chapter 11

2. In one dimension, the normality condition gives

$$2 \int_0^h [a + b(\frac{r}{h})^2] dr = 1,$$

or

$$2h[a + b \cdot \frac{1}{3}] = 0. \tag{A.22}$$

By $w_h(h) = 0$, we have

$$a + b = 0. \tag{A.23}$$

Solving (A.22)-(A.23), we obtain

$$a = \frac{3}{4h} = -b,$$

which leads to $w_h(r) = \frac{3}{4h}[1 - (\frac{r}{h})^2]$.
In two dimensions, we have

$$\int_0^h [a + b(\frac{r}{h})^2] 2\pi r dr = 1,$$

or

$$\pi h^2 [a + \frac{b}{2}] = 1,$$

which coupling with (A.23) yields

$$a = \frac{2}{\pi h^2} = -b.$$

While in three dimensions, we have

$$\int_0^h [a + b(\frac{r}{h})^2] 4\pi r^2 dr = 1,$$

or

$$4\pi h^3 [\frac{a}{3} + \frac{b}{5}] = 1,$$

which coupling with (A.23) yields

$$a = \frac{15}{8\pi h^3} = -b.$$

4. Let

$$F = f + \lambda g$$
$$= (x_1^2 - 2x_1 x_2 + 2x_2^2 + 10x_1 + 2x_2) + \lambda(x_1 - x_2).$$

To achieve a minimum value, it must satisfy

$$0 = \frac{\partial F}{\partial x_1} = 2x_1 - 2x_2 + 10 + \lambda,$$

$$0 = \frac{\partial F}{\partial x_2} = -2x_1 + 4x_2 + 2 - \lambda,$$

$$0 = \frac{\partial F}{\lambda} = x_1 - x_2,$$

which leads to the solution

$$x_1 = x_2 = -6.$$

Hence the minimizer is $f(x_1, x_2) = -36.$

Index